VLADIMIR I. ARNOLD

Collected Works

Springer

Vladimir I. Arnold at the Moscow State University, circa 1985

Photo by Boris Khesin

VLADIMIR I. ARNOLD

Collected Works

VLADIMIR I. ARNOLD

Collected Works

VOLUME IV

Singularities in Symplectic
and Contact Geometry 1980–1985

VLADIMIR I. ARNOLD

Collected Works

VOLUME IV

Singularities in Symplectic
and Contact Geometry 1980–1985

Edited by

Alexander B. Givental
Boris A. Khesin
Mikhail B. Sevryuk
Victor A. Vassiliev
Oleg Ya. Viro

Springer

Vladimir I. Arnold

June 12, 1937 – June 3, 2010

Editors

Alexander B. Givental
Department of Mathematics
University of California
Berkeley, CA, USA

Boris A. Khesin
Department of Mathematics
University of Toronto
Toronto, ON, Canada

Mikhail B. Sevryuk
V.L. Talroze Institute of Energy
Problems of Chemical Physics
Russian Academy of Sciences
Moscow, Russia

Victor A. Vassiliev
Steklov Mathematical Institute
Russian Academy of Sciences
Moscow, Russia

Oleg Ya. Viro
Institute for Mathematical Sciences
Stony Brook University
Stony Brook, NY, USA

ISBN 978-3-662-56188-1

Library of Congress Control Number: 2017962565

Mathematics Subject Classification: 35A18, 70S05, 53D05, 53D12, 57N70

Printed on acid-free paper

This Springer imprint is published by Springer Nature
The registered company is Springer-Verlag GmbH, DE
The registered company address is: Heidelberger Platz 3, 14197 Berlin, Germany

Preface

This volume of the Collected Works includes papers written by V.I. Arnold mostly during the period from 1980 to 1985. Arnold's work of this period is so multifaceted that it is almost impossible to give a single unifying theme for it. It ranges from properties of integral convex polygons to the large-scale structure of the Universe. Also during this period Arnold wrote eight papers related to magnetic dynamo problems, which were included in Volume II, mostly devoted to hydrodynamics. We have chosen the topic of singularities in symplectic and contact geometry only as a "marker" for this Volume IV.

There are many articles specifically translated for this volume. They include problems for the Moscow State University alumni conference, papers on magnetic analogues of Newton's and Ivory's theorems, on attraction of dust-like particles, on singularities in variational calculus, on Poisson structures, and others. We would like to draw the reader's attention to the translations of Arnold's comments to Selected works of H. Weyl and those of A.N. Kolmogorov. The latter were included along with short prefaces by A.N. Kolmogorov himself.

As usual, we reiterate the disclaimer that many reprinted translations of the 60s-80s may contain translation errors, and the interested reader should take this into account.

September 2017

Alexander Givental
Boris Khesin
Mikhail Sevryuk
Victor Vassiliev
Oleg Viro

Acknowledgements

The Editors thank the Göttingen State and University Library, the Mathematics Department Library of Heidelberg University, and the Mathematisches Forschungsinstitut Oberwolfach for providing the original articles for this edition. They are also grateful to Andrey Khesin and Masha Zorich for their help with the translation and typesetting of several papers in this volume. The Editors also thank the Springer office in Heidelberg, and in particular Ruth Allewelt and Martin Peters, for their extensive help and tireless support with this project.

Contents

Books and surveys by V.I. Arnold written in 1980–1985 but not included in the "Collected Works"

1. (with A.N. Varchenko and S.M. Guseĭn-Zade) Osobennosti differentsiruemykh otobrazheniĭ (Singularities of Differentiable Mappings) (in Russian). Vol. I. Klassifikatsiya kriticheskikh tochek, kaustik i volnovykh frontov (Classification of Critical Points, Caustics, and Wave Fronts), Nauka, Moscow, 1982, 304 pp. Vol. II. Monodromiya i asimptotiki integralov (Monodromy and Asymptotics of Integrals), Nauka, Moscow, 1984, 336 pp. The subsequent Russian editions (as a single book, by the MTsNMO Publishing House, essentially unrevised): 2004 (672 pp.), 2009 (672 pp.). The first English translation: Singularities of Differentiable Maps. Vol. I. The Classification of Critical Points, Caustics and Wave Fronts. Translated from the Russian by I. Porteous and M. Reynolds. Monographs in Mathematics, 82. Birkhäuser Boston, Inc., Boston, MA, 1985, xi+382 pp. Vol. II. Monodromy and Asymptotics of Integrals. Translated from the Russian by H. Porteous. Translation revised by the authors and J. Montaldi. Monographs in Mathematics, 83. Birkhäuser Boston, Inc., Boston, MA, 1988, viii+492 pp. The subsequent English edition in the Modern Birkhäuser Classics series: Vol. I (2012), Vol. II (2012).

2. Teoriya katastrof (Catastrophe Theory) (in Russian), Znanie, Moscow, 1981, 63 pp. ("Novelties in Life, Science, Technology", the series "Mathematics, Cybernetics", 1981, no. 9). The subsequent Russian editions: 1983 (an expanded edition, Moscow State University Press, Moscow, 80 pp., the series "New Ideas in Natural Sciences"), 1990 (a further expanded edition, Nauka, Moscow, 128 pp.), then by various publishing houses, almost unrevised: 2004, 2007–2009, 2013–2014, 2015–2016. The first English translation: Catastrophe Theory. Translated from the Russian by R.K. Thomas. Springer, Berlin, 1984, iv+79 pp. The second English edition: Catastrophe Theory. Translated from the Russian by G.S. Wassermann. Based on a translation by R.K. Thomas. Springer, Berlin, 1986, xiv+108 pp. The third English edition: Catastrophe Theory. Translated from the Russian by G.S. Wassermann. Based on a translation by R.K. Thomas. Springer, Berlin, 1992, xiv+150 pp. There are also translations into many other languages.

3. (with Yu.S. Il'yashenko) Obyknovennye differentsial'nye uravneniya (Ordinary differential equations) (in Russian). In: Current Problems in Mathematics. Fundamental Directions, Vol. 1 (Dynamical Systems – 1), Itogi Nauki i Tekhniki, Akad. Nauk SSSR, Vsesoyuz. Inst. Nauchn. i Tekhn. Inform., Moscow, 1985, 7–149, 244. The English translation: Ordinary differential equations. Translated from the Russian by E.R. Dawson and D. O'Shea. Encyclopaedia Math. Sci., Vol. 1, Dynamical systems I, Springer, Berlin, 1988, 1–148. The subsequent printing: 1997.

4. (with V.V. Kozlov and A.I. Neĭshtadt) Matematicheskie aspekty klassicheskoĭ i nebesnoĭ mekhaniki (Mathematical Aspects of Classical and Celestial Mechanics) (in Russian). Current Problems in Mathematics. Fundamental Directions, Vol. 3 (Dynamical Systems – 3), Itogi Nauki i Tekhniki, Akad. Nauk SSSR, Vsesoyuz. Inst. Nauchn. i Tekhn. Inform., Moscow, 1985, 304 pp. The second (significantly expanded and revised) Russian edition: Editorial URSS, Moscow, 2002, 414 pp. The subsequent Russian editions (also by Editorial URSS, almost unrevised): 2009, 2015, 2016, 2017. The first English translation: Mathematical Aspects of Classical and Celestial Mechanics. Translated from the Russian by A. Iacob. Encyclopaedia Math. Sci., Vol. 3, Dynamical systems III, Springer, Berlin, 1988, xiv+291 pp. The subsequent printings: 1993, 1997. The third (significantly expanded and revised) English edition: Mathematical Aspects of Classical and Celestial Mechanics. Translated from the Russian by E.I. Khukhro. Encyclopaedia Math. Sci., Vol. 3, Dynamical systems III, Springer, Berlin, 2006, xiv+518 pp.

5. (with A.B. Givental') Simplekticheskaya geometriya (Symplectic geometry) (in Russian). In: Current Problems in Mathematics. Fundamental Directions, Vol. 4 (Dynamical Systems – 4), Itogi Nauki i Tekhniki, Akad. Nauk SSSR, Vsesoyuz. Inst. Nauchn. i Tekhn. Inform., Moscow, 1985, 5–139, 291. The second Russian edition (as a separate book, essentially unrevised) in the series "Regulyarnaya i Khaoticheskaya Dinamika" (Regular and Chaotic Dynamics), Vol. 13: Udmurt University Publishing House, NITs "Regulyarnaya i Khaoticheskaya Dinamika", Izhevsk, 2000, 168 pp. The English translation: Symplectic geometry. Translated from the Russian by G.S. Wassermann. Encyclopaedia Math. Sci., Vol. 4, Dynamical systems IV, Springer, Berlin, 1990, 1–136. The subsequent printing: 2001.

During this period 1980-1985 Arnold also wrote papers #33-40 in Volume II. Note that paper #39 published in Volume II in Russian, had been translated into English: "On the evolution of a magnetic field under the action of transport and diffusion," Amer. Math. Soc. Transl., II Ser., Amer. Math. Soc., Providence, RI, 137, 119–129, 1987.

STATISTICS OF INTEGRAL CONVEX POLYGONS

V. I. Arnol'd

UDC 513.19

Two convex polygons with vertices at integral points of the plane are equivalent if they can be carried into one another by an affine transformation preserving the lattice of integral points.

We denote by $H(\mu)$ the logarithm of the number of equivalence classes of polygons of area $\mu/2$ (μ is an integer).

THEOREM 1. For μ sufficiently large, $c_1\mu^{1/3} \le H(\mu) \le c_2\mu^{1/3}\ln\mu$.

The proof is based on the following estimate.

THEOREM 2. The number of vertices of a convex polygon of area $\mu/2$ with vertices at integral points of the plane does not exceed $16\mu^{1/3}$.

1. Estimate of the Number of Vertices

We orient the sides of a convex N-gon M as the chain ∂M and we consider the N vectors of the oriented sides. We denote $[(N+1)/2]$ by P. We draw through O a line, strictly on one side of which lie not less than P vectors of the sides. We denote the sequence of oriented sides of the N-gon by $e_1, \ldots, e_N$, taking e_1 so that the first P vectors e_i should lie strictly in the chosen half plane, while $[e_i, e_j] > 0$ for $1 \le i < j \le P$ (here $[a, b] = a_1b_2 - a_2b_1$ is the area). We denote twice the area of the polygon M by μ.

LEMMA 1. $\mu = \Sigma[e_i, e_j]$, where $1 \le i < j \le N-1$.

Proof. We divide M into triangles by diagonals issuing from the start of the side e_1. We get $\mu = [e_1, e_2] + [e_1 + e_2, e_3] + \ldots$, which is what was needed.

LEMMA 2. $\mu \ge \Sigma\,|[e_i, e_j]|$, where $1 \le i < j \le P$.

Proof. Lemma 2 follows from Lemma 1, applied to the part cut out by the diagonal $e_1 + \ldots + e_P$, since $[e_i, e_j] > 0$ for $1 \le i < j \le P$.

LEMMA 3. If $x_i \ge 0$, then $(\max x_i)(x_1 + 2x_2 + \ldots + nx_n) \ge (x_1 + \ldots + x_n)^2/2$.

Proof. $(\max x_i)(x_1 + 2x_2 + \ldots) \ge x_1(x_1 + x_2 + \ldots) + x_2(x_2 + \ldots) + \ldots \ge (x_1 + \ldots + x_n)^2/2$.

LEMMA 4 (Basic). Let $f_1, \ldots, f_m \in \mathbf{Z}^2$ be distinct and not collinear, the vector $f \in \mathbf{Z}^2$, and let $|f_i| \ge |f|$. Then $\Sigma\,|[f_i, f]| \ge m^2/8$.

Proof. Let $f = k\varphi$, where $k > 0$ and φ is indivisible. We denote by x_s the number of vectors f_i, for which $|[f_i, \varphi]| = s$; it is clear that $\Sigma x_s = m$. The number of integral points on the diameter D containing f of the circle of radius $|f|$ with center O is equal to $2k+1$. On a chord parallel to D the number of integral points does not exceed $2k$. Hence $x_s \le 4k$. By Lemma 3, $\Sigma\,|[f_i, f]| = k\Sigma s x_s \ge k(\Sigma x_s)^2/8k = m^2/8$, which is what was required.

Proof of Theorem 2. Let us arrange the vectors $e_1, \ldots, e_P$ in order of increasing length. We denote the sequence obtained by $f_1, \ldots, f_P$. By Lemma 4, for $2 \le j \le P$, we have $|[f_1, f_j]| + \ldots + |[f_{j-1}, f_j]| \ge (j-1)^2/8$. Consequently, for $1 \le i < j \le P$, $\Sigma\,|[e_i, e_j]| = \Sigma|[f_i, f_j]| \ge (1^2 + \ldots + (P-1)^2)/8 \ge (P-1)^3/24$. By Lemma 2 and since $P - 1 \ge N/4$, we have $\mu \ge (P-1)^3/48 \ge N^3/3072$, which is what was required.

Remark. There is a well-known estimate of the number of vertices of a convex integral N-gon in terms of the diameter ($N \le cd^{2/3}$). Our estimate of N in terms of the area does not follow from this (the rectangle with sides 1 and a has area a).

Moscow State University. Translated from Funktsional'nyi Analiz i Ego Prilozheniya, Vol. 14, No. 2, pp. 1-3, April-June, 1980. Original article submitted January 21, 1979.

Originally publ. in: Funkts. Anal. Prilozh. 14(2), 1-3, © 1980 Russ. Acad. Sciences
English transl.: Funct. Anal. Appl. 14(2), 79-81, © 1980 Consultants Bureau, New York.

2. Upper Bound for Entropy

LEMMA 1. A convex integral polygon of area S can be covered by an integral parallelogram of area no more than 6S.

Proof. Let AB be a diameter of the polygon M, and let C and D be vertices fartest from the line AB (on different sides). We consider the parallelogram Π with vertices $2B - A \pm (C - D)$, $2A - B \pm (C - D)$. M is contained in it and contains the quadrilateral ABCD, whose area $|[B - A, C - D]|$ is six times less than the area of Π.

LEMMA 2. An integral parallelogram of area S is equivalent with a parallelogram of diameter no more than 3S.

Proof. Let integral points divide the side PQ into a parts and let PR be the first of these. We denote by X an integral point closest to PQ and not lying on it. The parallelogram with sides (PR, PX) is fundamental. Hence $S = ab$, where the integer b is equal to the ratio of the height of the original parallelogram with the sides PQ, PP' and the constructed fundamental one. Adding to the vector PX an appropriate multiple of PR, we get a fundamental parallelogram (PR, PY), for which Y lies inside the original parallelogram. We estimate the diagonal of the original parallelogram in the metric in which the basis (PR, PY) is orthonormal. Let Z be the point of the side PP' for which the line ZY is parallel to PR. The length of ZY does not exceed a, so $|ZP| \leq a + 1$. Consequently, $|PP'| \leq (a + 1)b$, so the diameter does not exceed $ab + a + b \leq 3S$.

Proof of the Upper Bound. By Lemmas 1 and 2, each convex integral polygon of area S is equivalent with a polygon lying in the square with side 36S. The number of integral points in it does not exceed $T = (36S + 1)^2$. By Theorem 1, the number of vertices of the polygon does not exceed $N = 16^3\sqrt{2S}$. The number of classes of polygons of area S does not exceed T^N, which is what is required.

3. Lower Bound for Entropy

1°. We denote by M_a ($a > \sqrt{2}$) a convex polygon, whose oriented sides are all indivisible integral vectors (x, y) of the semicircle $x < 0$, $x^2 + y^2 \leq a^2$ and taken with sign minus the sum of these vectors equal to $(2\rho, 0)$, $\rho > 0$. A polygon M_a with large a is like a semicircle whose radius ρ has, as $a \to \infty$, order a^3, and the number of vertices has order a^2; the lengths of all sides except the largest do not exceed a. M_a changes only when a^2 passes through integral values. The ratio of the areas of two successive M_a is bounded by a constant independent of a.

2°. We consider a polygon $2M_a$, on each of N_a short sides of which three integral points lie: the beginning, the middle, and the end. On each of $N_a - 1$ short sides (from the first to the penultimate) we choose an integral point from the pair (middle, end). Adding the beginning of the first and the end of the last short side, we get a convex polygon Λ inscribed in $2M_a$. The number of polygons Λ is equal to $2^{N_a - 1}$. The ratio of areas $S(2M_a)/S(\Lambda)$ is bounded by a constant independent of a and of the choice of Λ.

3°. We call a polygon d-admissible if it is situated in the half strip $0 \leq x \leq d$, $y \geq 0$ and has the base $y = 0$ as one of its sides.

LEMMA. For any natural numbers d and $\lambda \geq d$ there exists a d-admissible convex integral polygon of area $\lambda/2$.

Proof. The areas of d-admissible integral polygons form an additive semigroup. Admissible polygons of height 1 have as doubled areas all the integrals (d, ..., 2d). All integers $\lambda \geq d$ can be represented as sums of such summands, which is what was required.

4°. Let μ be a (sufficiently large) integer. We choose as a the largest of the numbers for which the sum of the doubled area and the length of the largest side of the polygon $2M_a$ does not exceed μ. It is clear that $a \geq c\mu^{1/6}$ for sufficiently large μ.

We construct $2^{N_a - 1}$ polygons Λ inscribed in $2M_a$ as in 2°. All the polygons Λ are d-admissible, where $0 \leq x \leq d$, $y = 0$ is a large side of $2M_a$. Let $\lambda = \mu - 2S(\Lambda)$. Then $\lambda \geq d_0$. According to Lemma 3 of Sec. 1, there exists a d-admissible polygon of area $\lambda/2$. The polygon symmetric to it with respect to the large side ($y = 0$) forms together with Λ a convex polygon Σ of area $\mu/2$.

5°. Among the $2^{N_a - 1}$ polygons Σ the number equivalent to the given one does not exceed a constant independent of a.

In fact, one can insert in Σ and Σ can be covered by circles of radii r and R of order a^3 (so that R/r is bounded by a constant independent of a and Σ). Whence follows the boundedness of the norm of the transformation carrying one of the polygons Σ into another, which means also the boundedness of the number of transformations.

Since $N_a \geqslant c_3 a^2 = c_3 \sqrt[3]{a^6} \geqslant c_4 \sqrt[3]{\mu}$ for sufficiently large μ, the bound is proved.

Remark. In Z^n, 1/3 is probably replaced by $(n-1)/(n+1)$. Proof of the lower bound: let $x_1^2 + \dots + x_{n-1}^2 \leq x_n \leq A$.

On some problems in singularity theory

By

V ARNOLD

Much progress in singularity theory of differentiable maps is based on empirical data. Some of these empirical facts later become theorems. We discuss here some facts, that are not explained today and some conjectures, related to them.

1. Semicontinuity of singularity spectrum

Let $f: (C^n, 0) \to (C, 0)$ be a holomorphic function-germ of finite multiplicity μ. One can associate to such a germ a set of μ rationals (not necessarily all different), which we shall call *singularity spectrum*.

Following Steenbrink [14], we denote spectrum points l_k, $k = 1, \ldots, \mu$. The singularity spectrum has the following properties.

(1) Eigenvalues λ_k of the monodromy are related to the spectrum by exponentiation : $\lambda_k = \exp(2\pi i l_k)$.

(2) Let f be quasi-homogeneous, and let $\{x^{m_k}\}$ be a monomial C-basis of the local algebra

$$Q_f = C[[x_1, \ldots, x_n]]/(\partial f/\partial x_1, \ldots, \partial f/\partial x_n).$$

Then $\{l_k\}$ is the set of weights of μ differential forms $x^{m_k} dx_1 \ldots dx_n$ (here $\deg x_k = \deg dx_k = a_k$, $\deg f = 1$).

(3) Let f be a function in two variables, which is generic among functions with a given Newton diagram Γ (figure 1).

Then the spectrum consists of orders (for the Newton filtration) of monomials, whose exponents are detectable from figure 1 (the order of a monomial x^m in the Newton filtration is the coefficient λ, for which $\lambda m \in \Gamma$).

(4) In the general case l_k mod 1 is defined by 1, and the integer part of l_k — by the Steenbrink conventions :

$[l_k] = q$ if l_k corresponds to an eigenvalue of the monodromy on the space $H^{p,q}$ of the mixed Hodge structure on the vanishing cohomology group.

(5) The spectrum is symmetric, with centre $l = n/2$.

Many examples led to the following conjecture. Let the spectra of singularities be ordered : $l_1 \leqslant l_2 \leqslant \cdots \leqslant l_\mu$.

Originally publ. in: Proc. Indian Acad. Sci. (Math. Sci.) 90(1), 1-9, © 1981 Indian Acad. Sciences.

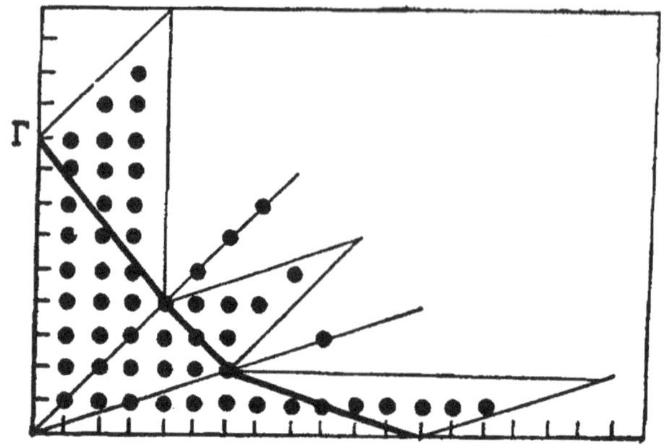

Figure 1. The spectrum of a Newton diagram.

Conjecture. The spectrum is semicontinuous in the following sense: let a singularity S be adjacent to a (simpler) singularity S' (with $\mu' < \mu$), then $l_k \leqslant l'_k$ for $k = 1, \ldots, \mu'$.

Remarks. (1) Even in simple and explicitly calculable cases, like (2) and (3) above, this conjecture is a nontrivial arithmetical assertion on integer points in convex polyhedra.

(2) The conjecture implies the semicontinuity of dimensions of spaces of the Hodge filtration and of the filtration dual to the Hodge filtration in vanishing cohomology, that is the semicontinuity of numbers

$$h^r = \sum_{p \leqslant r} \sum_q h^{p,q} \quad \text{and} \quad h_r = \sum_{p > r} \sum_q h^{p,q}.$$

(3) In particular, for the case of functions in two variables, these semicontinuities are reduced to the semicontinuities of genus g of the Riemann surface (of the "vanishing cycles manifold") and of the "cogenus" $\mu - g$. Semicontinuity of both numbers g and $\mu - g$ is clear (the last, as Varchenko has explained to the present author, reduced to the fact that the inclusion of the vanishing cycles manifold of a simpler singularity, to that of a more complicated one is monomorphical at the homology level).

(4) The semicontinuity of l_1, that is of the first spectrum point, is very important for the theory of integral asymptotics. Probably, our conjecture on the semicontinuity of l_1 implies (or is equivalent to) the conjecture on the semicontinuity of oscillating integral asymptotics, which originated from [1], was disproved by Varchenko in [16] and was finally reformulated by Pham [13].

(5) The relation of the spectrum to the set of zeros of Bernstein polynomial (see [3], [12]) seems less proved that one should like, but our conjecture can be reformulated in this setting too (?).

(6) It follows from the spectrum symmetry, that the conjecture implies the two-sided inequalities

$$l_k \leqslant l'_k \leqslant l_{k+(\mu-\mu')}.$$

For instance, if only one point bifurcates from a complicated singularity (S), so that $\mu = \mu' + 1$, then the deformed singularity (S') has a spectrum which divides the spectrum of (S).

The relation between spectra of (S) and (S') is the same as between semiaxes of an ellipsoid in R^{μ} and of its section by $R^{\mu'}$. Does there exist a quadratic form, associated with the singularity, whose eigenvalues (in some Euclidean space) are values of a monotonic function at spectrum points?

(7) We can deform the loop, defining the monodromy in the base space of the versal deformation of a complicated singularity into a product of the loop, defining the monodromy of the simpler one and of some "positive" loops, along which the discriminant argument increases. One can conjecture that the image of the product in the matrix group "rotates more than for the simpler monodromy". This gives some heuristic explanation for the semicontinuity. While these ideas are not at all clear, they are sometimes useful; for instance it was precisely these ideas that have led the present author to the conjecture on the formula for the quasi-homogeneous singularity signature (see [6]). In this case, the "rotation" was defined in terms of two-dimensional invariant planes of the symplectic mapping

$$\begin{pmatrix} 0 & \mathrm{Var} \\ -(\mathrm{Var}^{t})^{-1} & 0 \end{pmatrix}$$

in $H^* + H_*$. The conjecture was related to the positivity of some eigenvalues of this symplectic mapping in the sense of Krein's parametrical resonance theory. However the proof of the conjecture, given later by Steenbrink [15], is quite different.

2. Bifurcation diagrams of complex singularities

Bifurcation diagrams of real functions at critical points of series A ($A_2 = x^3 + y^2$, $A_3 = x^4 + y^2, \ldots$) are very useful for the calculations (and definitions) of generalised Whitehead groups in algebraic K-theory (Cerf [5], Hatcher [9], Wagoner [18], Volodin [17]).

This led to the question, what is the "complex analogue" of these algebraical objects? Such "complex analogues" are perhaps quite different from K-theory. For instance, complex analogue for Morse theory is Picard-Lefschetz theory, but it would be very difficult to reconstruct the second theory, knowing nothing but the first. We also know that the complex analogue for $K(\pi, 0)$ is $K(\pi, 1)$, and for the symmetric group—the braids group. The complex analogue for "boundary" is "two-fold ramified covering". But we have no general methods or axioms for finding such analogues.

As one of the candidates, arising from the complex bifurcation diagrams, we describe a "quasi-resolvent" of the fundamental group of the complement to the singularity bifurcation diagram.

Let Γ_0 be a group, presented as the quotient F_0/R_0, where F_0 is free and R_0 is an invariant subgroup, generated (as an *invariant* subgroup) by elements of the form $(af)f^{-1}$, where $f \in F_0$ and $a \in \mathrm{Aut}\, F_0$, group of automorphisms of F_0.

Let Γ_1 be a group of automorphisms of F_0, leaving invariant every class $f R_0$, and large enough to generate R_0. Let us suppose, that Γ_1 is represented as $\Gamma_1 = F_1/R_1$, and so on : we have a chain of groups $\Gamma_i = F_i/R_i$. We call such a chain *quasi-resolvent* of Γ_0.

Something of this kind arises from the fundamental group Γ_0 of the complement $C^\mu - \Sigma$ to the bifurcations diagram Σ in the versal deformation base space C^μ of a singularity. Let F_0 be $\pi_1 (C^1 \setminus \Sigma)$, where C^1 is a generic line in C^μ. Let us consider in C^μ/C^1 the set Σ_1 of non-generic (with respect to Σ) lines. The group

$$\Gamma_1 = \pi_1 ((C^\mu/C^1) \setminus \Sigma_1)$$

acts on $\pi_1 (C^1 \setminus \Sigma)$, leaving invariant elements of $\pi_1 (C^\mu \setminus \Sigma)$. Choose in C^μ/C^1 a generic line (that is, a generic C^2 containing C^1 in C^μ). We obtain a set of generators for Γ_1, that is we consider the free group

$$F_1 = \pi_1 ((C^2/C^1) \setminus \Sigma_1),$$

and so on, finishing at $F_{\mu-1}$.

In the case of A_μ singularities, the group Γ_0 is the Artin braids group with $\mu + 1$ strings, the free group F_0 is generated by the μ standard generators of the braids group, the group Γ_1 can be considered as a "quasi-relations" group for Γ_0 (not to be confused with R_0). In the same way Γ_2 corresponds to the "quasi-relations between quasi-relations" and so on. But even in the A_μ case it is not clear whether Γ_1 coincides with the whole group of automorphisms of F_0, which conserves all elements of Γ_0 (and which belongs to the group of automorphisms, generated by orientation-preserving plane with holes homeomorphisms).

Perhaps for the study of $\{\Gamma_i\}$ the classification of all decompositions of simple singularities A, D, E into simpler ones will be useful. Such a classification for functions (not just for levels, which is much easier) is recently found by Ljashko [10].

Very little is known on the topology of the complements to more complicated singularities bifurcations sets. Looijenga [11] has reported that the complements are $K(\pi, 1)$ for parabolical singularities, but his arguments are not clear.

3. Cohomology of complements to bifurcation diagrams

The imbedding of the versal deformation of a simpler singularity (S') base space $C^{\mu'}$ into the base space for a more complicated singularity (S) defines a cohomology homomorphism :

$$H^* (C^\mu \setminus \Sigma) \leftarrow H^* (C^{\mu'} \setminus \Sigma') \quad (\mu > \mu'),$$

between the cohomologies of complements to bifurcation diagrams.

A question naturally arises whether these homomorphisms are canonical and whether one can define a stable cohomology ring (which is, in a sense, the ring of cohomologies of the complement to the bifurcations diagram for $f \equiv 0$ in the infinite-dimensional versal deformation space). Even if this programme cannot be completely realised, one still can associate stable cohomology classes to (at least) some strata of the natural stratification of the set of function-germs set (or hyperface-germs set).

Since the cohomology classes of complements to bifurcation diagrams for versal deformations define corresponding classes in the base spaces of arbitrary deformation complement to bifurcation sets, one can hope that any information on the "stable ring" can be useful to obtain information on global properties of bifurcation sets for arbitrary families of functions (hypersurfaces, mappings, ...).

4. Modality

The modality (moduli number) of a Lie group action at a point of a manifold is the minimal integer m for which orbits in some neighbourhood of the point can be arranged in a finite number of families with $\leqslant m$ parameters.

Problem. Let $f: (R^n, 0) \to (R, 0)$ be a finite multiplicity real function-germ. Is its modulus number in the real jets space equal to its modulus number in the complex jets space? [The corresponding groups are the groups of jets $(R^n, 0) \to (R^n, 0)$ and $(C^n, 0) \to (C^n, 0)$ acting as "right equivalence" $(gf)(x) = f(g(x))$].

I was told by Professor E. B. Vinberg that there exist real representations of real Lie groups, such that the modality of their complexification is larger than the modality of the initial real action. But it is unknown whether such a case is possible for the right equivalence action of the diffeomorphism group on the functions space.

For quasi-homogeneous singularities there exists a notion of "inner modality" which can be calculated as the number of monomials of positive degrees in the monomial basis of the corresponding versal deformations module

$$T_f = (C[x_1, \ldots, x_n])^p \Big/ \Big(\sum_{j=1}^{p} \frac{\partial f_j}{\partial x_i} \frac{\partial}{\partial y_j}, \ f_r \frac{\partial}{\partial y_q} \Big), \quad i = 1, \ldots, n;$$
$$r, q = 1, \ldots, p.$$

Here $f_j \in C[x_1, \ldots, x_n]$ are quasi-homogeneous polynomials of degrees D_j, where $\deg x_i = A_i$; the module $(C[x])^p$ is generated by p free generators $\partial/\partial y_q$, whose degrees are $-D_q$; one supposes that $p \leqslant n$ and that $\dim_C T < \infty$.

One conjecture that the inner modality for $p < n$ is equal to the modality (of the contact group action), but it is not proved (as it is a known problem for the case of right-equivalence of functions). The above conjecture is partially confirmed by the theorem (due to I. G. Scherbak) that inner modality 0 complete intersections (quasi-homogeneous curves) in C^3 coincide with contact modality 0 complete intersections (which are all quasi-homogeneous). The classification of these as a standard exercise on Newton diagrams, was at Moscow a known examination problem (1973). By the way, this classification disproves some of the classification results in the sixth part of J. Mather's celebrated paper on singularities.

The Scherbak theorem (containing also quasi-homogeneous unimodular singularities list) was proved with the help of the following formula for the Poincaré polynomial of the graded module T_f for $p = n - 1$ (that is, for singularities of generic quasi-homogeneous curves) :

$$p(t) = \frac{\Pi(1 - t^{D_j})}{\Pi(1 - t^{A_i})}(\Sigma\, t^{-D_j} - \Sigma\, t^{-A_i} + 1) + t^{\Sigma D_j - \Sigma A_i}.$$

One can rewrite this as

$$(*)\ p(t) = t^{\Sigma D_j - \Sigma A_i} h(1/t)$$

denoting by h the Poincaré polynomial of the relative differential forms graded module H-, calculated by Hamm [7]. The formula (*) was found empirically for $n = 3$, $p = 2$. It seems that Ljashko can prove it, but his proof does not explain the duality between modules T and H for quasi-homogeneous curves singularities.

In the more general case of generic complete intersections of positive dimension the Ljashko's formula for the Poincaré polynomial of T_f is

$$p(t) = (-1)^{n-p}$$

$$t^{\Sigma D_j - \Sigma A_i}\left[-1 + \operatorname*{res}_{s=0} \prod \frac{1 - st^{A_i}}{s - st^{A_i}} \prod \frac{s - st^{D_j}}{1 - st^{D_j}} \frac{ds}{1 - s}\right]$$

$$+ \frac{\Pi(1 - t^{D_j})}{\Pi(1 - t^{A_i})}[\Sigma\, t^{-D_j} - \Sigma\, t^{-A_i} + 1].$$

Recently Hamm and Gruel have proved $p(1) = h(1)$ for $n - p > 0$, but they give no formula for $p(t)$.

5. Real singularities topology

Mixed Hodge structure defines for every singularity of a function a large set of integers $h_\lambda^{p,q}$ (where λ are monodromy eigenvalues). It seems that these numbers are closely related to the real geometry of the function, its level sets and its morsifications and bifurcations diagrams topology.

A simple example of this is the generalised Petrovski inequality

$$|\operatorname{ind}| \leqslant h_1^{n/2,\, n/2},$$

for the local degree of the gradient mapping of a real smooth function-germ $f : (R^n, 0) \to (R, 0)$, n even : see [1] for more details.

Other geometrical invariants of the real singularity, which would be interesting to compare with Hodge numbers, are, for instance :

(i) Betti numbers of real non-singular neighbour level set (or their partial sums, may be alternate);

(ii) Numbers of critical points of different indexes, arising from different real morsifications (or their partial sums, may be alternate);

(iii) Numbers describing the possible complication of one real level set of a morsification, for instance, the maximal number of singular points on the same level set.

For empirical works in this direction it is possible to use the Steenbrink conjecture [14] (while this conjecture is wrong as it is stated). A comparison of Hodge numbers, calculated from the Newton diagram by formulas of Steenbrink, Danilov and Kirillov, with the values of geometrical invariants, like those above will, perhaps, generate new (and best possible) inequalities for the real geometry invariants.

However even the Petrovski inequality is known to be best possible only in the simplest cases. For example, it is not known how large the Poincaré index (= the local degree of the gradient map) can be for a real homogeneous function f of degree m in n variables, $f : R^n \to R$.

We only know that the Petrovski inequality gives the exact maximum for $n = 3$ (that is, for curves in RP^2) or for $m = 3$ [the extremal function $f = (x_1 + \ldots + x_n)^3 - x_1^3 - \ldots - x_n^3$ was constructed by D B Fuks]. For $m = n = 4$ (surfaces of degree 4 in RP^3) the inequality is the best possible too.

The classification of real projective surface of degree 4 in RP^3 was one of the questions in the 16th Hilbert problem, at present this classification is known completely, after the works of V M Harlamov (see [8]) and V V Nikulin; they find not only the topological types, but also all possible isotopical types in RP^3 and even classify components of the complement to the degenerate surfaces set in the space of all real surfaces of degree 4 in RP^3.

The Petrovski inequality is still true for nongradient vector fields (see [1]).

It gives the best possible bound for the Poincaré index of a vector field in R^n, whose components are homogeneous polynomials of degree $m - 1$ (A G Hovanski).

6. Maxima singularities

Let

$$F(y) = \max f(x, y)$$

be the maxima function $F : B \to R$ of a family $f : M \times B \to R$ of real functions on a compact closed manifold M, depending on a parameter y, belonging to a "base space" B, which is an (open) manifold of dimension n.

The maxima function is continuous, but generically, is not smooth. Empirical data lead to a conjecture : the maxima function for a generic family is topologically equivalent (in some neighbourhood of every base space point y) to a Morse function (that is, either to a non-zero linear function or to a sum of a constant and of a non-degenerate quadratic form at point 0).

For $n \leqslant 6$ this is proved by Brisgalova (see [4]).

For the general case, the arguments are :

(i) Suppose for a given $y \in B$ there is only one maximum point x (may be degenerate). In this case the graph of F has at y a tangent plane, and the maxima function is generically topologically linear.

(ii) Suppose for a given $y \in B$ there are $n+1$ maxima points. Then the graph of F is a pyramid with $n+1$ faces, each face having at y a tangent plane. The maxima function is generically either topologically linear or topologically equivalent to a Morse function at a minimum point.

(iii) In a generic family the maximum is obtained at s points for y on a manifold of codimension $s-1$. Along this submanifold S the graph of $F|_S$ has tangent planes and in the transversal direction we can use argument (ii).

(iv) In the particular case of a germ f, which is R^+-equivalence stable (see [4]) one can prove that the set $\{y, z : z \geqslant F(y)\}$ is locally diffeomorphic to a convex body. For instance, stability is generic for families with $n \leqslant 6$ parameters. It is natural to ask whether the set described above is still locally diffeomorphic to a convex body for generical families maxima function singularities, if $n > 6$.

If it is true, this will be one more confirmation for a general principle of fragility of all good things.

To explain this principle let us consider, for example, the set of all polynomials $x^n + a_1x^{n-1} + \cdots + a_n$ (a real), having only real zeros (or only non-real, or only zeros with negative real parts). This set, at singularities of its boundary, fills *less than one half* of the neighbourhood space. Thus under deformation the corresponding property of being good (elliptic, hyperbolic, stable and so on) will rather disappear than persist. The theorems, generated by this principle, (and describing the cones of velocities of curves, moving from the boundary point inside the good set) were recently proved by L. V. Levantovski.

References

[1] Arnold V I 1973 Remarks on stationary phase method and Coxeter number; *Usp. Mat. Nauk* **28**: 5 17–44

[2] Arnold V I 1978 Index of vector field singular point, Petrovski-Oleinik inequalities and mixed Hodge structures; *Funkt. Analysis* **12**: 1 1–14

[3] Bernstein I N 1972 Analytical continuation of generalised functions; *Funkt. Analysis* **6**: 4 26–40

[4] Brisgalova L N 1978 On maxima functions of families of functions; *Funkt. Analysis* **12**: 1 66–67

[5] Cerf J 1970 La stratification naturelle des espaces des fonctions differentiables reelles et le theoreme de la pseudo-isotopies; *Institut des Hautes Etudes Sci. Publ Math.* **39** 5–173

[6] Gousein-Zade S M 1977 Monodromy groups of isolated singularities of hypersurfaces; *Usp. Math. Nauk* **32**: 2 23–65

[7] Hamm H 1977 Genus χ_y for quasi-homogeneous complete intersections; *Funkt. Analysis* **11**: 1 87–88

[8] Harlamov V M 1978 Isotopical classes of nonsingular surfaces of degree 4 in RP^3; *Funkt. Analysis* **12**: 1 86–87

[9] Hatcher A E 1973 Parametrised *h*-cobordism theory; *Ann. Inst. Fourier (Grenoble)* **23**: 2 61–74

[10] Ljashko O V 1976 Decompositions of simple singularities; *Funkt. Analysis* **10**: 2 49–56

[11] Looijenga E 1977 On the semiuniversal deformations of a simple elliptic hypersurface singularity I. *Topology* **16**: 3 257–262

[12] Malgrange B 1974 Integrales asymptotiques et monodromie; *Ann. Sci. Ec. Norm. Suppl. Ser.* **47**: 3 405–430

[13] Pham F 1977 *Caustics and microfunctions* (Kyoto Univ : RIMS) 12 Suppl. pp. 357–366

[14] Steenbrink J H M 1976 Mixed Hodge structure on vanishing cohomology, *Real and complex singularities*, Oslo

[15] Steenbrink J H M 1977 Intersections form for quasi-homogeneous singularities; *Compositio Math.* **34** 211–223

[16] Varchenko A N 1976 Newton polyhedra and estimations of oscillating integrals; *Funkt. Analysis* **10**: 3 13–38

[17] Volodin I A 1972 Generalised Whitehead groups and pseudo-isotopies; *Usp. Math. Nauk* **27**: 5 229–230

[18] Wagoner J B 1971 Algebraic invariants for pseudo-isotopies; *Lecture Notes in Mathematics* (Berlin-Heidelberg—New York: Springer) **209** 164–190

Department of Mathematics,
University of Moscow,
Moscow 117 234, USSR.

On some nonlinear problems

by

V. ARNOLD

Moscow State University, Moscow, U.S.S.R.

A small piece of every curve resembles a straight line segment. One calls linearisation the local approximation of a curve by a straight line. In many instances such an approximation suffices. For example, small oscillations of a pendulum or small deviations of a satellite from its standard orbit can be described with sufficient precision by linear equations.

Nonlinear equations appear where the approximation of the relations by simple proportionality does not suffice. For instance, light is described as linear electromagnetic oscillations, but with growing intensity of light nonlinear phenomena become crucial and thus laser optics is mainly nonlinear.

We may usually consider mechanical oscillations as linear ones, but nonlinear phenomena become important if we are interested in the behavior of nondamped oscillations after many oscillation periods, because the accumulation of small nonlinear effects may completely alter the system's long time behaviour.

One encounters similar difficulties in celestial mechanics. The perturbative effect of each planet on the others is small, but it might in principle alter their orbits completely if it would accumulate during, say, 10000 years. One must take into account the accumulation of nonlinear effects to explain why the planets do not collide, fall onto the sun or escape to infinity.

A classical example of a nonlinear theory is incompressible fluid dynamics, with its still unsolved problem of the mathematical description of turbulence.

The difficulties of nonlinear problems constantly require new mathematical methods. Thus such powerful mathematical devices were invented as symplectic geometry machinery, introduced in Hamilton's work on optics and mechanics, and topology, founded by Poincaré who was led to it by the problems of nonlinear ordinary differential equation theory and of celestial mechanics.

In this talk I shall speak on three nonlinear problems: on evolution of Hamiltonian systems, on turbulence and the long term weather prediction problem, and on the application of the theory of caustics to the explanation of the large scale structure of the Universe.

(1) The problem of the description of the mutual perturbations of the planets, formulated by Newton, was solved "in the first approximation" by Lagrange and Laplace. Laplace has stated his "Solar System Stability Theorem": the planetary orbits close to coplanar circles will remain close to the same circles infinitely long.

This theorem was proved by Laplace for the planet motions governed not by Newton's nonlinear equation but by some linearisation of it. The nonlinear effects

Originally publ. in: Crafoord Prize in Mathematics 1982. Crafoord Lectures, 1-7,

Reprint by Springer-Verlag Berlin Heidelberg 2018 with kind permission

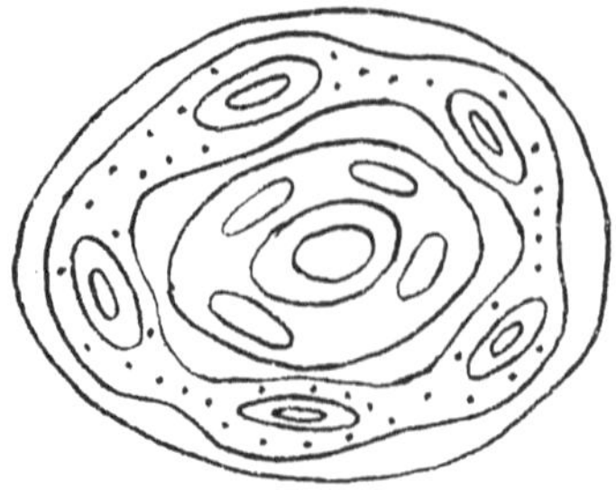

Fig. 1.

not taken into account produce only small perturbations at time intervals of many thousands of years. But they might completely alter the orbits when accumulated during millions of years—if the orbit evolution due to them were a systematic drift rather than oscillations.

The attempts to justify Laplace's conclusions were very fruitful for the theory of differential equations (I shall only quote the works of H. Poincaré, D. Birkhoff, C. L. Siegel and J. Littlewood). Birkhoff formulated a simple model problem with the same difficulty. For a plane rotation all the consecutive images of a point close to the rotation center remain close to it, while we iterate the rotation. Birkhoff asked whether the same is true for the iterations of a generic area preserving plane mapping close to a rotation (the opposite would mean that accumulation of nonlinear perturbations may alter the stability of the rotation center, which is obviously the case for generic mappings which do not preserve area).

Strict conclusions on the behaviour of orbits on infinite time intervals have become accessible only after Kolmogorov (1954) introduced his "Newton method" and proved the persistence of the n-dimensional tori in the $2n$-dimensional phase space of an analytical Hamiltonian system under small perturbations of the Hamiltonian function.

Birkhoff's problem was solved positively by this method in 1961, and the stability of planetary motions for the majority of initial conditions close to coplanar circular orbits was proved for sufficiently small ratios of the planet masses to that of the central body during the spring of 1962.

It is still unknown, whether the actual mass ratios, orbit excentricities and inclinations are sufficiently small for the existence of Kolmogorov tori. (The proof requires unrealistically small values for these quantities.)

The analyticity requirement in Kolmogorov's theorem was relaxed by J. Moser, who has also relaxed the irrationality condition in my solution of Birkhoff's problem and added many other improvements. He has used the smoothing technique invented by J. Nash to solve a nonlinear problem in differential geometry (that of the realisation of a given Riemannian metric by a submanifold of a Euclidean space).

To understand Kolmogorov's theorem let us consider its application to the study of magnetic lines in toroidal systems such as a plasma trap. In this case the theorem asserts the stability of the magnetic surfaces formed by the magnetic lines. Figure 1 represents a plane section of the system. The closed curves are the sections of magnetic surface (or Kolmogorov tori). All the points drawn in the space between them are the consecutive intersections of a single magnetic line with the section plane. This line does not lie on any magnetic surface, but it is trapped between two of them.

The situation is similar for mechanical systems with two degrees of freedom close to integrable ones: the orbits are trapped between tori on the energy level surfaces.

For systems with more than two degrees of freedom the Kolmogorov tori do not divide the energy level manifold, and the orbit may escape from their neighborhoods through the chinks between them. The mechanism for such an escape was studied in 1964, and examples of the unbounded accumulation of perturbations were constructed. The rate of escape in these examples decreases exponentially (as $\exp(-1/\sqrt{\varepsilon})$ when the perturbations decrease as ε). Later (in 1971) N. N. Nechoroshev proved the impossibility of a faster escape. The exponential decrease of the escape velocity explains why the escape has not been detected in any order approximation of perturbation theory: $\exp(-1/\varepsilon^a)$ decreases faster than any power of ε.

It seems that the motions of a generic analytic system with three or more degrees of freedom close to an integrable one are able to drift from any neighborhood of any Kolmogorov torus to any neighbourhood of any other one at the same energy level, as a result of the accumulation of perturbations. For instance, it is probable that the orbits of the planets subject to the Newtonian attractive force are able to change their shapes drastically as a result of an appropriately chosen arbitrarily small perturbation of the initial conditions, but such perturbations are extremely rare and the time required for a considerable orbital change is greater than the age of the solar system. Of course, for such a long time period one should not neglect the effect of nonconservative forces not taken into account in the Newtonian approach.

(2) A short time after the discovery of Kolmogorov tori a new important discovery was made in the theory of nonlinear differential equations, that of the stability of chaotic motions in deterministic dynamical systems (S. Smale, D. Anosov and others, in the early sixties).

Before this discovery most mathematicians believed that a generic (nonhamiltonian) dynamical system would evolve either toward a stable stationary state or to a strictly periodic oscillatory state. This belief was confirmed by many examples and by a theorem, due to Andronov and Pontriagin, saying that there are no other possibilities for a generic system on a two dimensional phase plane.

Now it has been discovered that the evolution of a dynamical system may lead to a type of motion quite different from stationary or periodic. One characteristic feature of these new limiting types of motion is the extreme sensitivity of the detailed motion to the initial conditions. On the other hand, the averaged characteristics of the limiting behaviour are stable under small changes of the initial state or even of the whole dynamical system.

An example of such complicated behaviour is given by the geodesics on a negatively curved surface (an example of a negatively curved surface is the surface of a saddle). It was proved many decades ago (by Hadamard and E. Hopf) that the geodesics on a negatively curved surface are extremely sensitive to the choice of the initial point. Neighbouring geodesics diverge exponentially and may fill up the whole surface densely. It has now become clear that all this complicated behaviour is stable and that a generic dynamical system with more than two dimensional phase space may exhibit such a complicated behaviour as the final state of its evolution.

Two areas of applications of these exponentially unstable final motions appeared in 1963: the verification of the Boltzmann ergodic conjecture and in the modelling of turbulence. I believe that Kolmogorov was thinking of both those problems when in 1958 he started a seminar on the interrelations between dynamical systems theory, ergodic theory and hydrodynamic instability theory.

To prove ergodicity in the Boltzmann model of elastic balls in a box I suggested that this model might be considered as a limit case of a geodesic flow on a negatively curved manifold. This idea became the starting point of a long series of works by Ja. G. Sinai and his students who have proved ergodicity for some nontrivial cases (say, for two balls), but the general problem is still unsolved.

In 1963 I was very optimistic on the hydrodynamical applications of the new theory. I had the naive hope of finding smooth manifolds covered by exponentially divergent orbits and attracting the neighbouring orbits in the phase space of the Navier-Stokes equations and to recognise them with the help of a computer. I performed a computer experiment in 1964, with a 6-mode approximation of the Navier-Stokes equation on a 2-torus. I was not able to distinguish the attractor obtained from a 3-torus.

Trying to understand the appearance of these 3-tori I was led (after some intermediate steps) to geodesic flows: the ideal fluid motion is nothing other than the motion along the geodesics on an infinite-dimensional manifold (this manifold is the group of transformations of the motion domain).

I calculated the curvature of this infinite dimensional manifold. It turned out that it is mostly negative. The negative curvature makes the fluid motion exponentially unstable. It is even possible to estimate from the curvature value the maximum time over which an adequate weather prediction can be made. For an oversimplified model where the atmosphere is an ideal two-dimensional fluid I find that to forecast weather 2 months ahead one must know its state today with a precision 10^5 times better than that of the forecast. In other words long term dynamical weather prediction is impossible and will remain impossible under all computer improvements.

I hoped to be able to use the negative curvature arguments to deduce for a weakly viscous fluid the exponential instability of the set of final motions from that of the ideal, nonviscous fluid. I found no such proof. As far as I know, there is no proof even today, while the exponential instability itself has been observed in computer experiments many times since the 1963 work of Lorenz appeared. (This was unknown to mathematicians till the seventies.)

In 1965 the reaction of most mathematicians to the idea that turbulence is the exponential instability of orbits on finite dimensional sets of limit motions was almost unanimously negative. J. Leray, and later O. A. Ladyzhenskaya, were trying to persuade me that 2-dimensional turbulence is impossible, that turbulence is either the nonexistence or the nonuniqueness of the solutions of the Navier-Stokes equations, and that the correct mathematical description of turbulence requires a modification of these equations. The only two exceptions were S. P. Novikov and R. Thom, at whose seminar I presented a talk on exponential instability of fluid motions in 1965. I remember that Thom was much impressed by the sequence of bifurcations occurring at the loss of stability, and that the works by A. A. Andronov, E. Hopf and their followers on the bifurcation to circles, tori and more complicated attractors were then practically unknown in occidental Europe.

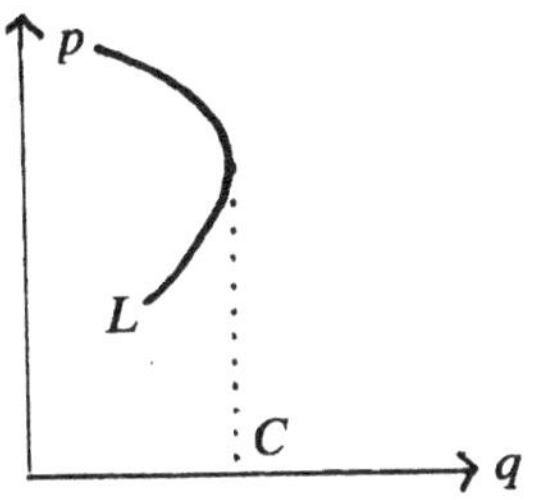

Fig. 2.

The situation changed drastically with the introduction by D. Ruelle and F. Takens of a very successful term "attractor". In the seventies the conjecture of the exponential instability of motions on finite dimensional attractors of the Navier-Stokes equations became widely known. However, as far as I know, it is still as unproved as it was at the time of my 1965 unsuccessful attempts to deduce it from the negativity of curvature. The only recent progress I know of is an upper estimate of the attractor dimensions for the 2-dimensional case ($\dim < CRe^4$, for periodic boundary conditions, Ju. S. Ilyashenko; $\dim < C\exp(Re^{4+\varepsilon})$ for the boundary case, M. I. Vishik and A. V. Babin). [Latest (1983) Vishik–Babin estimations: $CRe^2 \ln^{1/2} Re$, CRe^4.]

For the 3-dimensional case even the finiteness of the attractor dimensions is still unproved. For the case of 2 dimensions the old conjecture that the minimum of the attractor dimensions grows indefinitely with the decrease of viscosity to zero has still not been rejected, and even the obvious growth of the maximum of the minimal attractors dimensions has never been proved for any example of fluid motion (in the mathematical sense of the word "proved").

(3) The theory of focal points and light ray caustics goes back to Archimedes and I. Barrow, Newton's teacher in optics and calculus. Light is concentrated at the caustics. From the point of view of symplectic geometry a ray system (say, the system of all normals to a given surface) defines a Lagrangian submanifold of the phase space, and the caustic is the set of critical values of its projection to the configuration space (figure 2). One calls singularities of such a projection Lagrangian singularities.

The caustics themselves may have singularities. For instance, the generic plane caustics have cusp points, and the generic space caustics—cuspidal edges and

Fig. 3.

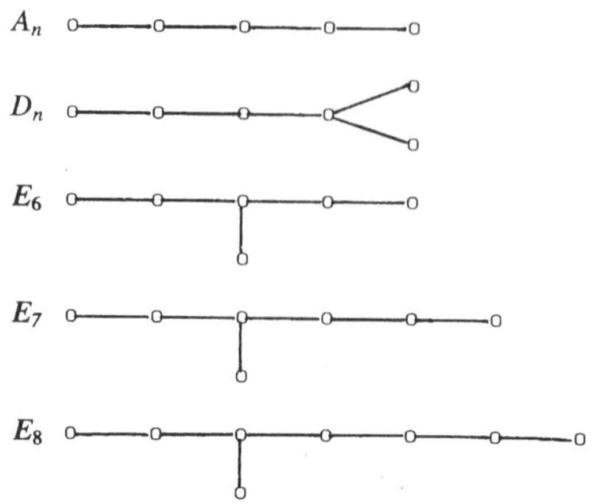

Fig. 4.

singular points of three types, called the swallowtail, the pyramid and the purse by R. Thom (figure 3).

At the beginning of 1972 I found that the classification of simple caustic singularities is closely related to the classification of simple objects in many other problems, such as the classification of regular polytopes, that of simple Lie algebras, of simple quivers (figure 4) and so on. For instance, the swallowtail corresponds to the Lie group $SU(5)$, the pyramid and the purse to $O(8)$, the tetrahedron to E_6 and the icosahedron to E_8.

These relations look like Kepler's mysterious law on the distribution of the planetary distances. Part of them (for instance, the relations of the caustics to the regular polytopes) have even today no satisfactory explanation.

It was impossible for me not to discover these relations because of my previous work on some topics that at that time seemed to be unrelated.

While reading in 1966 the thesis of V. P. Maslov, where he introduced the index which is now known under his name, I started to classify the Lagrangian singularities, inspired by H. Whitney's general singularity theory, far developed just at that time by Thom and J. Mather.

On the other hand, since my early student years I was thinking from time to time of the Hilbert problem on the representation of algebraic functions by superpositions. In 1968 this problem led me to the topology of the swallowtails and to their relation to braids theory.

In 1971 E. Brieskorn, while presenting these works at the Bourbaki seminar, extended them to all finite reflection groups (the exceptional cases were settled by P. Deligne). At that moment it would have been sufficient simply to compare the two lists I had obtained—but I did this only a year later, while calculating the singularities of the oscillatory integrals which represent the light intensity at the caustic singularities. I first appreciated the importance of the relation between the caustics and the finite reflection groups when B. Kostant, who was then in Moscow, helped me to recognize the Coxeter numbers in terms of the asymptotic formulas for the light intensity.

Later it came to my mind that I had encountered the same lists under other disguises many times, for instance in the book by J. Milnor and in a paper by G. N.

Tjurina on the singular points of complex hypersurfaces (1968), in Brieskorn's talk at the Nice congress (1970) and in my own paper on the bifurcations of linear differential equations (1971); the relation to regular polytopes is already discussed in the "Vorlesungen über die Entwicklung der Mathematik im 19. Jahrhundert" by F. Klein.

The development of singularity theory, which resulted from the discovery of the unexpected relations of caustics to reflection groups, braids and so on has been so rapid and has led to so many discoveries by so many mathematicians that it is impossible to describe it here.

Instead I shall briefly present one of the physical applications of Lagrangian singularities theory, its application to the explanation of the large scale structure of the Universe.

The distribution of matter in the Universe is highly nonuniform: there exist stars, galaxies, clusters of galaxies and so on. According to modern astrophysics, such nonuniformity did not exist when the Universe was 1 000 times smaller than now. In 1970 Ja. B. Zeldovitch suggested a mechanism for the clustering of matter due to small nonuniformity of the initial potential velocities field. Zeldovitch's theory, called later "the pancakes theory", coincides mathematically with Lagrangian singularities theory (at least for the case of collision free particles).

The theory of Lagrangian singularities predicts the formation of a cellular large scale structure, with low density inside the cells, high density at the cell walls, higher density at wall cuspidal edges and the highest one—at the most singular points (the swallowtails, the pyramids and the purses in the oversimplified free particles model). Recently S. F. Shandarin has compared this model prediction with the actual distribution of galaxies. To obtain an objective criterion of the existence of cellular structure he has used percolation theory.

Let us consider N particles (each representing a galaxy) in a cube with an edge of length L. The average part of the cube per particle is then a cubicle with an edge of length $l=L/N^{1/3}$. Let us surround every particle by a sphere of small radius r. When r grows, these spheres form chains; their lengths exceed L for some value of r. From this moment on one can percolate from one side of the initial cube to the opposite one through the union of the spheres. The percolation parameter is the ratio of the minimal r sufficient to percolate to the length of our cubicle, that is, r/l.

For a particle distribution of cellular structure the percolation parameter is lower and for a hierarchical structure (where particles form clusters rather than surfaces or lines) it is higher than for a random distribution of independent particles.

For the actual distribution of galaxies Shandarin found a much lower percolation parameter than for the distribution of independent random particles; this proves the tendency of galaxies to form a cellular structure, or at least a net (one can distinguish these cases using a more sophisticated percolation parameter).

THE PROBLEMS PROPOSED AT THE SECOND INTERNATIONAL CONFERENCE OF ALUMNI OF MOSCOW STATE UNIVERSITY (FACULTY OF MECHANICS AND MATHEMATICS) ON THE TOPIC "DIFFERENTIAL EQUATIONS AND THEIR APPLICATIONS"

V.I. ARNOLD

(Translation of Uspekhi Mat. Nauk, **37** (1982), no. 6, 264–269.)

V.I. Arnold's problems

1. Zeroes of integrals. *Let $I(h) = \oint P\,\mathrm{d}x + Q\,\mathrm{d}y$ be an integral over an oval of the curve $H(x,y) = h$, where P and Q are polynomials of degree n while H is a polynomial of degree d. How many isolated zeroes can the integral $I(h)$ have?*

This question is an intermediate one between the two parts of Hilbert's 16th problem since limit cycles are born from the zeroes of $I(h)$ under perturbations of the Hamiltonian system with the Hamilton function H. The question is solved only for the case where H is a polynomial of degree 2 or a quasi-homogeneous polynomial. In the general case, even the existence of a number $N(n,d)$ that bounds the number of zeroes for all H, P, and Q of a given degree has not yet been proven. Some cases with $d = 3$ were studied in [1].

2. The birth of cycles under bifurcations. *How many limit cycles can be born from the origin under small perturbations of a system of the form*

$$\begin{aligned} \dot{x} &= x(\alpha + ax + by + \cdots), \\ \dot{y} &= y(\beta + cx + dy + \cdots), \end{aligned} \qquad \alpha = \beta = 0$$

that is, for small α and β (the remaining coefficients of the right-hand side are generic and fixed)?

The above system appears, firstly, in the Lotka–Volterra theory of the predator and prey interaction, and secondly, as a normal form of the system of equations for the amplitudes in the problem of passage of two pairs of eigenvalues of an oscillating system into the unstable half-plane [2]. The question can be reduced to determining the number of zeroes of integrals over the curves $u^\lambda v^\mu w^\nu = h$, where u, v, and w are linear functions. The conjectural answer is "1 cycle", but this has not yet been proven.

3. An upper bound for the index.[1] *What is the maximal Poincaré index that a singular point of the vector field* grad f *can have, if f is a polynomial in n variables of degree d?*

The question is only solved for $n \leqslant 3$ as well as for $d \leqslant 3$. In these cases the answer is the number $\Pi_n(d)$ of integer points lying strictly inside the cube $(0,d)^n$ and belonging to the hyperplane that passes through the cube center orthogonally to

[1]The "lower bound" in the original title was a typo. (Editors' remark)

Translated by Boris Khesin and Andrey Khesin
Springer-Verlag Berlin Heidelberg 2018 with kind permission

the main diagonal. In the general case the only known upper bound is $|\mathrm{ind}| \leqslant \Pi_n(d)$, after I.G. Petrovskiĭ and O.A. Oleĭnik [3]. This estimate was extended by V.I. Arnold [4] to non-gradient fields, and for those fields it cannot be improved, as was shown by A.G. Khovanskiĭ [5].

4. The topological Morse property of the increment. *Is it true that for generic families of matrices depending on an arbitrary number of parameters, the maximum of the real parts of eigenvalues of the matrix, as a function of the parameters, is topologically equivalent to a Morse function?*

This is so in the case of one and two parameters [6], and this is also supported by the fact that the dependence on parameters of the maximum of a smooth function possesses the above topological Morse property for any number of parameters (see [7], [8]).

5. Stationary magnetic dynamo. This problem, long known to physicists, is formulated below in the simplest case of periodic boundary conditions.

Let v be a fixed divergence-free vector field in the three-dimensional Euclidian space with a period of 2π in each of the Cartesian coordinates (x, y, z), and let $H(x, y, z; t)$ be the desired vector field of the same class. The transport equation for the magnetic field H by means of the velocity field v in the presence of magnetic viscosity ε has the form

$$\dot{H} = \{v, H\} + \varepsilon \Delta H,$$

where $\{\ ,\ \}$ is the Poisson brackets. A solution of the form $\mathrm{e}^{\gamma t} H_0(x, y, z)$ where $\mathrm{Re}\,\gamma > 0$ is called an increasing mode; the value $\mathrm{Re}\,\gamma$ is called the increment of the mode.

The problem is *to figure out whether there exists a strong dynamo, i.e., a field v for which for sufficiently small $\varepsilon > 0$ there is an increasing mode with the increment not tending to 0 as $\varepsilon \to 0$*. It is known that no two-dimensional flow can make dynamo, but in the three-dimensional case the question is open.

In particular, it is unknown *whether the following simplest flow with the exponential stretching of particles*

$$v = (A \cos y + B \sin z,\ B \cos z + C \sin x,\ C \cos x + A \sin y)$$

possesses the strong dynamo property (see [9], [10]).

6. The evolution of differential forms. Consider the evolution equation for differential closed k-forms on an n-dimensional closed Riemannian manifold under the transport by means of the flow of a field v and diffusion, $\dot{\omega} = L_v \omega + \varepsilon \Delta \omega$, where L is the Lie derivative and $\Delta = \mathrm{d}\delta + \delta\mathrm{d}$ is the Laplacian. For large viscosity ε in each cohomology class there exists a unique stationary form to which a solution of this equation with any initial condition from this class tends as $t \to +\infty$.

For any positive viscosity the dimension of the space of stationary solutions is no less than the kth Betti number. However, the question is *whether each cohomology class contains a stationary solution.*

In any case the attraction of any solution to a stationary one can fail for large Reynolds numbers $1/\varepsilon$ if $k \neq 0$, 1, and n (for instance, in the physically interesting case $k = 2$, $n = 3$) [10].

7. Singularities of clusters of dust-like particles. Consider the potential motion of a medium of non-interacting particles

$$x \mapsto x + tv(x), \quad v(x) = \partial S/\partial x.$$

This motion defines a one-parameter family of Lagrangian submanifolds in the phase space. The critical values of its projection to the configuration space $\{x\}$ form a caustic (the place of infinite density of particle clusters).

The singularities of caustics and metamorphoses that they undergo with time are known (see [11]). These are the singularities and bifurcations of generic Lagrangian mappings.

We assume now that there is gravitational interaction between the particles. In this case the motion remains potential but the corresponding Lagrangian submanifold of the phase space loses its smoothness once the caustics are formed.

The problem is *to study the singularities of Lagrangian mappings appearing this way, as well as the singularities of caustics and density functions on the configuration space.*

This problem has not been solved even in the case of a one-dimensional configuration space (the gravity force in $\mathbb{R}^n$ is proportional to r^{1-n}). In the one-dimensional case the conjectural singularities of density at generic moments are of order $x^{-1/2}$, i.e., the same as for projections of smooth Lagrangian manifolds. One may assume that the Lagrangian variety yielding such a density is formed by two smooth curves with a common vertical tangent at the connection point.

8. Non-integrability. Lately, the art of proving "non-integrability" of dynamical systems, which was going back to Poincaré, achieved substantial success (mostly in the works of V.V. Kozlov and S.L. Ziglin, see [12], [13]). However, among various versions of the definition of non-integrability which were discussed so far the non-integrability in the sense of foliation theory is apparently missing. On the other hand, the developed methods can probably *prove not only the non-existence of first integrals but (at least in certain cases) even the non-existence of invariant foliations.* Note here that the theory of integrability in quadratures, going back to Liouville, apparently even today does not possess a precise definition of integrability in quadratures on a manifold, with the exception of the case of an algebraic phase space.

9. Foliations. The next question is an attempt to mathematically formulate the physical understanding of the connection of "local helicity" with the exponential stretching of particles. On a three-dimensional sphere, consider a differential 1-form α satisfying the Frobenius integrability condition $\alpha \wedge d\alpha \equiv 0$. Given a Riemannian metric on the sphere, associate to the form $d\alpha$ a divergence-free vector field. *Is it true that its characteristic Lyapunov exponents vanish almost everywhere?*

10. Stretching direction. Consider a volume-preserving diffeomorphism A on a compact Riemannian manifold, e.g. the "standard" diffeomorphism of the torus $(x, y) \mapsto \big(x + y,\, y + \alpha \sin(x + y)\big)$. At each point of the manifold, construct the direction along which the diffeomorphism A^N is contracting in the strongest way. *Does the direction field thus obtained have a limit as $N \to +\infty$ (in any sense)? How strongly does the direction of this field vary under small perturbations of the initial point? (Of particular interest is the interrelation between N and the correlation* *radius of the field.*) Solutions to these questions can be useful for the dynamo studies (problem 5), since the evolution of a magnetic field can be regarded as consisting of alternating periods of frozenness (when the field is transported by the flow) and the diffusion, whose speed is determined by the serration arising from stretching of the field by the flow.

11. Dimensions of attractors. Consider the three-dimensional Navier–Stokes equation on a compact manifold (with a pumping force stimulating the motion). In spite of the contemporary success in proving conjectures (going back to the beginning of the 60's) on attractors for these equations (in the modern terminology), the following two old questions remain open as of today:

(a) *Does a finite-dimensional attractor exist for any Reynolds number?*

(b) *Does the minimal dimension of an attractor grow unboundedly as the Reynolds number increases?*

The answer to question (b) is unknown even in the two-dimensional case, for which question (a) has been solved affirmatively. The dimension under discussion can be understood in different ways: for instance, as the Hausdorff dimension or as the number of non-negative Lyapunov exponents. In 1965 V.I. Arnold computed the Riemannian curvature of the diffeomorphism group [14] with the hope to use its negativity for the affirmative solution of question (b). While the curvature in many directions is indeed negative (recently A.M. Lukatskiĭ [15], [16] verified this for the flows on an n-torus and on S^2, having also proven the negativity of the Ricci curvature of the diffeomorphism group, the concept he had introduced), the above hopes remain unrealized.

12. Symplectic topology. Since the time of the Poincaré and Birkhoff "geometric theorem" there continue attempts to construct a symplectic version of algebraic topology (which should relate to the conventional topology similarly to how the theory of Hamiltonian dynamical systems relates to the general theory of differential equations). One of the most simply formulated problems in this direction is as follows: *does there exist a compact embedded Lagrangian manifold in $\mathbb{R}^4$ with a univalued generating function $\int p\,dq$?*

In $\mathbb{R}^2$ there is no such manifold since any curve bounding zero area has points of self-intersection.

In the language of wave fronts the same problem is formulated as follows: *consider such a compact surface in the three-dimensional Euclidean space that all its singularities are swallowtails, cuspidal edges, and self-intersections. Assume that the tangent planes to the surface are nowhere vertical. Does there exist a vertical chord such that the tangent planes to the surface at its endpoints are parallel to each other?*

A closed curve on the plane with everywhere non-vertical tangents whose singularities are only cusps and self-intersections always has a vertical chord with parallel tangents at the endpoints, although it is not easy to prove this (until one realizes that one should pass to a Lagrangian curve in $T^*\mathbb{R}$).

13. The Poincaré series. Consider some analytical object (for instance, a differential equation, a smooth mapping, a Riemannian manifold, or a solution to the Einstein equations). Groups of diffeomorphisms ("coordinate changes") usually act on the spaces of such objects.

Let us fix a number k and consider the action of the group of coordinate changes on the manifold of k-jets of the objects. Let m_k be the number of moduli of this action (the dimension of the orbit space); this number depends on the k-jet of the object. Now fix this object itself and define the Poincaré series as

$$p(t) = \sum m_k t^k.$$

Is it true that, for a generic object, p is a rational function?

In particular, *what is the series $p(t)$ for the action of the diffeomorphism group of the (x, y)-plane on the space of germs of differential equations $\mathrm{d}^2y/\mathrm{d}x^2 = F(x, y, \mathrm{d}y/\mathrm{d}x)$?*

The rationality of p is one of the possible formalizations of the idea that "normal forms" of the objects in analytical problems contain "arbitrary functions" and simultaneously this is one of the approaches to problems with infinite-dimensional kernels and co-kernels (similar to problems with oblique derivative in a space of dimension greater than 2).

14. The insolvability of the stability problem. Consider a vector field whose components are polynomials with integer coefficients. *Is the stability problem for the singular point 0 solvable algorithmically?* It could be that a more meaningful question in this direction is the following one. *Consider the decomposition of the space of $(k+s)$-jets of vector fields with the given k-jet into the part corresponding to stability according to the $(k+s)$-jet, the part corresponding to instability according to the $(k+s)$-jet, and the part consisting of susceptible $(k+s)$-jets (i.e. of those which do not guarantee either stability or instability). How complicated can such a decomposition be in the set-theoretical sense?*

In any case, it is known that this decomposition is not semi-analytic [17].

15. Averaging in multi-frequency systems. Here we discuss systems of the form

$$\dot{\varphi} = \omega(I) + \varepsilon f(I, \varphi, \varepsilon),$$
$$\dot{I} = \varepsilon g(I, \varphi, \varepsilon)$$

with k fast angular variables φ mod 2π and n slow variables I, where $\varepsilon \ll 1$ is a small parameter. If the number of fast variables $k \leqslant n+1$, then the change of I during a time interval of order $1/\varepsilon$ differs but slightly from the change of J in the averaged system $\dot{J} = \varepsilon\overline{g}(J, 0)$ for most of the initial conditions (under the assumption that the right-hand sides are generic) (see [18]).

Does there hold a similar conclusion justifying the averaging method when the number of slow variables is smaller (the simplest case is $n = 1$, $k = 3$)?

References

[1] Yu.S. Il′yashenko. The multiplicity of limit cycles that arise in a perturbation of a Hamiltonian equation of the class $w' = P_2/Q_1$ in the real and complex domain. *Trudy Sem. Im. I.G. Petrovskogo*, no. 3 (1978), 49–60. English translation: American Mathematical Society Translations. Series 2. Volume 118. Sixteen Papers on Differential Equations. Amer. Math. Soc., Providence, RI, 1982, 191–202.

[2] E.I. Khorozov. Bifurcations of a vector field in a neighborhood of a singular point in the case of two pairs of purely imaginary eigenvalues. II. *C. R. Acad. Bulgare Sci.*, 35 (1982), no. 2, 149–152.

[3] I.G. Petrovskiĭ, O.A. Oleĭnik. On the topology of real algebraic surfaces. *Izvestiya Akad. Nauk SSSR. Ser. Mat.*, 13 (1949), no. 5, 389–402. English translation: Translations. Series One. Volume 7. Algebraic Topology. Amer. Math. Soc., Providence, RI, 1962, 399–417.

[4] V.I. Arnold. The index of a singular point of a vector field, the Petrovskiĭ–Oleĭnik inequalities, and the mixed Hodge structures. *Funktsional. Anal. i Prilozhen.*, 12 (1978), no. 1, 1–14. English translation: *Funct. Anal. Appl.*, 12 (1978), no. 1, 1–12.

[5] A.G. Khovanskiĭ. The index of a polynomial vector field. *Funktsional. Anal. i Prilozhen.*, 13 (1979), no. 1, 49–58. English translation: *Funct. Anal. Appl.*, 13 (1979), no. 1, 38–45.

[6] V.I. Arnold. Lectures on bifurcations and versal families. *Uspekhi Mat. Nauk*, 27 (1972), no. 5, 119–184. English translation: *Russian Math. Surv.*, 27 (1972), no. 5, 54–123.

[7] V.I. Arnold. On some problems in singularity theory. *Proc. Indian Acad. Sci. Math. Sci.*, 90 (1981), no. 1, 1–9.

[8] V.I. Matov. Topological classification of the germs of functions of maximum and minimax for generic families of functions. *Uspekhi Mat. Nauk*, 37 (1982), no. 4, 167–168. English translation: *Russian Math. Surv.*, 37 (1982), no. 4, 127–128.

[9] V.I. Arnold, Ya.B. Zeldovich, A.A. Ruzmaĭkin, D.D. Sokolov. A magnetic field in a stationary flow with stretching in a Riemannian manifold. *Zh. Èksper. i Teor. Fiz.*, 81 (1981), no. 6, 2052–2058. English translation: *Soviet Phys. JETP*, 54 (1981), no. 6, 1083–1086.

[10] V.I. Arnold. Several remarks on the anti-dynamo theorem. *Vestnik Moskovsk. Univ. Ser. I Mat. Mekh.*, 1982, no. 6, 50–57. English translation: *Moscow Univ. Math. Bull.*, 37 (1982), no. 6, 57–66.

[11] V.I. Arnold. Bifurcations of singularities of potential flows in a collision-free medium and metamorphoses of caustics in a three-dimensional space. *Trudy Sem. Im. I.G. Petrovskogo*, no. 8 (1982), 21–57. English translation: *J. Soviet Math.*, 32 (1986), no. 3, 229–258.

[12] V.V. Kozlov. Integrability and non-integrability in Hamiltonian mechanics. *Uspekhi Mat. Nauk*, 38 (1983), no. 1, 3–67. English translation: *Russian Math. Surv.*, 38 (1983), no. 1, 1–76.

[13] S.L. Ziglin. Ramification of solutions and the non-existence of first integrals in Hamiltonian mechanics. I. *Funktsional. Anal. i Prilozhen.*, 16 (1982), no. 3, 30–41. English translation: *Funct. Anal. Appl.*, 16 (1982), no. 3, 181–189.

[14] V.I. Arnold. Sur la géométrie différentielle des groupes de Lie de dimension infinie et ses applications à l'hydrodynamique des fluides parfaits. *Ann. Inst. Fourier (Grenoble)*, 16 (1966), fasc. 1, 319–361. English translation: V.I. Arnold. Collected Works. Volume II. Hydrodynamics, Bifurcation Theory, and Algebraic Geometry (1965–1972). Springer, Berlin, 2014, 33–69.

[15] A.M. Lukatskiĭ. On the curvature of the group of diffeomorphisms that preserve the measure of a two-dimensional sphere. *Funktsional. Anal. i Prilozhen.*, 13 (1979), no. 3, 23–27. English translation: *Funct. Anal. Appl.*, 13 (1979), no. 3, 174–177.

[16] A.M. Lukatskiĭ. On the curvature of the group of diffeomorphisms that preserve the measure of an n-torus. *Uspekhi Mat. Nauk*, 36 (1981), no. 2, 187–188. English translation: *Russian Math. Surv.*, 36 (1981), no. 2, 179–180.

[17] Yu.S. Il'yashenko. Analytic insolvability of the stability problem and of the problem of the topological classification of singular points of analytic systems of differential equations. *Mat. Sbor. (N. S.)*, 99(141) (1976), no. 2, 162–175. English translation: *Math. USSR, Sbor.*, 28 (1976), no. 2, 140–152.

[18] A.I. Neĭshtadt. On averaging in multi-frequency systems. II. *Dokl. Akad. Nauk SSSR*, 226 (1976), no. 6, 1295–1298. English translation: *Sov. Phys., Dokl.*, 21 (1976), 80–82.

LAGRANGE AND LEGENDRE COBORDISMS. I

V. I. Arnol'd

UDC 513.83

A Lagrange manifold* of a phase space is a description at the level of geometric optics of a wave system (a wave state) in a configuration space. A wave state in a domain defines the wave state at its border. This enables us to define the Lagrange border of a Lagrange manifold as a Lagrange manifold of dimension one less than that of the original manifold.

Lagrange manifolds are said to be Lagrange cobordant if their difference can be represented as a Lagrange border.

The simplest problem in which Legendre cobordisms arise is the following. Consider a wave front propagating in air. The tracks of this front on the earth's surface at different instants may be topologically different, but they are cobordant.

More precisely, a wave front defines a Legendre manifold† in the manifold of contact elements of a space. The front's track on the earth defines a moving Legendre submanifold in the manifold of contact elements on its surface. Legendre submanifolds corresponding to instantaneous fronts at two different instants are Legendre cobordant: their difference comprises a Legendre border of a suitable Legendre manifold of dimension greater by one.

The present paper's main results consist in the classification of one-dimensional immersed Lagrange (Legendre) manifolds of standard Lagrange (Legendre) bundles to within Lagrange (Legendre) cobordisms. The classes of cobordant manifolds form groups. We compute these groups. For example, each wave front on the surface of a plane is Legendre cobordant with the front consisting of several bows similar to those shown in Fig. 1.

The present paper was written in connection with research conducted at the Moscow Institute of Electronic Machine Building on the shortwave asymptotics of electromagnetic fields. The author thanks the sponsor V. P. Maslov, as well as A. N. Varchenko, A. M. Gabriélov, A. G. Khovanskii, and Ya. M. Éliashberg for useful discussions.

1. Groups of Cylindrical and Stable Cobordism Classes

A. Cylindrical Cobordisms. A cobordism of Lagrange (Legendre) submanifolds of Lagrange (Legendre) bundle spaces is defined with the aid of a cobordism of the bases. We say that a Lagrange (Legendre) cobordism is cylindrical if the cobordism of bases corresponding to it is a cylinder (a direct product on a segment).

By the addition of immersed submanifolds of one manifold we shall mean the union of the immersions. This operation provides a set of classes of cylindrically cobordant immersions with fixed base of the structure of a commutative semigroup. In all the cases examined further on these semigroups prove to be groups. The results of the computation of these groups are presented below.

1. Lagrange Curves in $\mathbf{R}^2$. THEOREM. The group of classes of oriented Lagrange curves on a plane is isomorphic to $\mathbf{Z} + \mathbf{R}$.

An oriented curve on a phase plane (p, q) is defined to within Lagrange cobordism by two invariants, one continuous and the other discrete. The discrete invariant is the Maslov index and the continuous one is the integral of form p dq. Thus, all curves with a fixed value of the Maslov index bounding a fixed area are Lagrange cobordant.

*A maximal isotopic submanifold of a symplectic 2-form on a cotangent bundle space (see [1, 4]).

†An integral submanifold of maximal dimension of a contact distribution of hypersurfaces (see [1]).

Moscow State University. Translated from Funktsional'nyi Analiz i Ego Prilozheniya, Vol. 14, No. 3, pp. 1-13, July-September, 1980. Original article submitted February 26, 1980.

Originally publ. in: Funkts. Anal. Prilozh. 14(3), 1-13, © 1980 Russ. Acad. Sciences
English transl.: Funct. Anal. Appl. 14(3), 167-177, © 1980 Consultants Bureau, New York.

Fig. 1

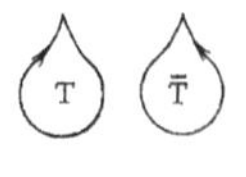

Fig. 2

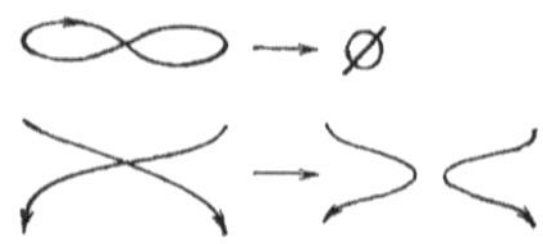

Fig. 3

2. Legendre Curves in Jet Space. We consider the space $J^1(\mathbf{R}, \mathbf{R})$ of 1-jets of functions of one variable x.

THEOREM. The group of classes of oriented Legendre curves in the space of 1-jets is a free cyclic group.

The generator is the class of a curve whose projection onto the plane (x, y) of 0-jets has two cusps and looks like a figure 8 with cusped vertices:

$$x = \cos t, \quad y = \sin 4t - 2 \sin 2t, \quad p = 8 \sin 3t. \tag{1}$$

The projection of curve (1) onto plane (x, y) is called a bow (Fig. 1).

THEOREM. The group of classes of unoriented Legendre curves in the space of 1-jets is trivial.

In particular, a bow [more precisely, the Legendre curve (1)] is drawn taut by an unorientable Legendre membrane in $J^1(\mathbf{R}^2, \mathbf{R})$, viz., a Möbius band.

3. Armed and Unarmed Fronts. A wave front is said to be armed* if it is equipped with a smooth field of normal vectors.

THEOREM. The group of classes of a Legendre cobordism of armed oriented wave fronts on a plane is a free cyclic group. The generator is the class of an armed bow (1) on plane (x, y).

Wave fronts of general position on a plane have no singularities other than points of self-intersection and nondegenerate (semicubic) cusps (see [1]).

We say that a nondegenerate cusp of an armed oriented front is positive if in circuiting the front as prescribed by its orièntation the passage from branch to branch at the cusp takes place on the side of the arming front of the normal.

Definition. The number of cusps, with due regard to the signs, is called the front's index.

THEOREM. The index of a front on a plane is even and is the single invariant of the Legendre cobordism of armed oriented fronts on the plane.

THEOREM. The group of classes of the Legendre cobordism of armed unoriented fronts on a plane is trivial. The group of classes of unarmed fronts on a plane is finite: $\mathbf{Z}_2$ for unoriented fronts and $\mathbf{Z}_2^2$ for oriented ones. As the generators we can take the classes of "teardrops"† (fronts with one cusp, Fig. 2); in the oriented case the two generators have different orientations.

4. Immersed Curves. The groups of classes of a cylindrical cobordism can be defined not just for Lagrange and Legendre submanifolds but also for other objects, for example, for the usual (neither Legendre nor Lagrange) immersions.

THEOREM. The groups of classes of immersed curves on a plane are finite: $\mathbf{Z}_2^2$ for armed oriented curves and $\mathbf{Z}_2$ for the remaining three cases.

As the generator we can take the class of the figure eight (an armed figure eight on an oriented plane has two nonequivalent orientations).

Henceforth we shall denote‡ armedness by the letter S and unarmedness by the letter P. The orientedness of curves and of their cobordisms will be denoted by the letter ω and unorientedness by the letter ν. The

*Translator's Note. The word "armed" here is not to be thought of as "having limbs" (arms). The picture that comes to my mind under the definition given here is that of a porcupine with quills distended.

†Translator's Note. The Russian word used here is simply "drop"; but I felt that in English the word "teardrop" more closely calls to mind the object shown in Fig. 2.

‡The letter S calls to mind the sphere formed by the armed contact elements at a point, while P calls to mind the projective space formed by the unarmed elements.

symbol Leg denotes the groups of classes of Legendre curves; the symbol Imm denotes the groups of classes of immersed curves.

The theorems in Paragraphs 3 and 4 are summarized in the following table of groups of classes with base $\mathbf{R}^2$:

	S Leg	*P* Leg	*S* Imm	*P* Imm
ω	$\mathbf{Z}$ B	$\mathbf{Z}_2^2$ T, $\bar{\text{T}}$	$\mathbf{Z}_2^2$ $\infty, \overline{\infty}$	$\mathbf{Z}_2$ ∞
ν	0 —	$\mathbf{Z}_2$ T	$\mathbf{Z}_2$ ∞	$\mathbf{Z}_2$ ∞

[B denotes bow (1), T denotes the teardrop, the overbar denotes a change of orientation.]

5. Other Bases. We compute the semigroups of classes of cobordism of Lagrange (Legendre) curves over bases other than the Euclidean straight line (plane). They all turn out to be groups. The map into a group of one-dimensional homologies of the base defines a homomorphism onto. The kernel is isomorphic to the group of classes over the Euclidean base, but the group of classes does not always decompose into a direct product.

At first we consider wave fronts on a projective plane.

THEOREM. The group of classes of the Legendre cylindrical cobordism of oriented armed fronts on a projective plane is a free cyclic group. As the generator we can take the class of curve $x^2 = y^3$.

The single invariant of the cobordism of armed oriented fronts in $\mathbf{RP}^2$ is the index, equal to the number of cusps with due regard to signs. This number may be odd.

THEOREM. The index of an oriented armed front of general position on a projective plane equals the number of points of inflection of the front, with due regard to signs: a point of inflection is reckoned to be positive if the front after the point of inflection turns aside from the arming normal.

THEOREM. Projectively dual curves have opposite indices.

Now let the base be an arbitrary two-dimensional compact manifold, b_0, b_1, b_2 be its Betti numbers, and c be the number of unorientable components with an even Euler characteristic.

THEOREM. The groups of classes of cylindrical cobordism of Legendre (Leg) or of simply immersed (Imm) oriented (ω) or unoriented (ν) and armed (S) or unarmed (P) curves are isomorphic to the groups in the following table:

	S Leg	*P* Leg	*S* Imm	*P* Imm
ω	$\mathbf{Z}^{b_0+b_1} + \mathbf{Z}_2^c$	$\mathbf{Z}^{b_1} + \mathbf{Z}_2^{2b_0}$	$\mathbf{Z}^{2b_1} + \mathbf{Z}_2^{b_0+b_2}$	$\mathbf{Z}^{b_1} + \mathbf{Z}_2^{2b_0-b_2}$
ν	$\mathbf{Z}_2^{b_0+b_1-b_2}$	$\mathbf{Z}_2^{2b_0+b_1-b_2}$	$\mathbf{Z}_2^{b_0+b_1}$	$\mathbf{Z}_2^{2b_0+b_1-b_2}$

Here the base may or may not be orientable, but the Legendre (respectively, immersed) surfaces defining the cobordisms are assumed to be oriented in the ω-cases and armed in the S-cases.

THEOREM. The group of classes of a Lagrange cylindrical cobordism of oriented curves in a cotangent bundle of circles is $\mathbf{Z} + \mathbf{Z} + \mathbf{R}$ (the invariants are. the index, the homology class, the integral of form p dq).

B. Stable Cobordisms. Two classes of a cylindrical cobordism of submanifolds over a base B are said to be stably equivalent if a cobordism of bases exists with borders (not necessarily cylindrical) diffeomorphic to B, over which the submanifolds of the two given classes are cobordant (in the sense of the theory being examined, i.e., the theory of Lagrange or Legendre cobordisms or of cobordisms of immersions, with due regard to orientations, arming, etc.).

The group of stable classes (of a cobordism of a fixed theory) over a given base is defined as the factor group of the group of cylindrical classes by the subgroup generated by classes stably equivalent to zero.

The results of the computation of the groups of stable classes are as follows.

Lagrange Cobordisms

THEOREM. The group of stable classes of oriented Lagrange curves over any one-dimensional manifold is $\mathbf{Z} + \mathbf{R}$.

Legendre Cobordism

a) The Bases and Their Cobordisms Are Oriented

THEOREM. The groups of stable classes of Legendre curves in the spaces of line elements of two-dimensional manifolds (Leg) and of immersions of curves in two-dimensional manifolds (Imm) are isomorphic to those in the following table:

	S Leg	*P* Leg	*S* Imm	*P* Imm
ω	$\mathbf{Z}$ B	$\mathbf{Z}_2^2$ T, $\bar{\text{T}}$	$\mathbf{Z}_2^2$ ∞, $\bar{\infty}$	$\mathbf{Z}_2$ ∞
ν	0 —	$\mathbf{Z}_2$ T	$\mathbf{Z}_2$ ∞	$\mathbf{Z}_2$ ∞

The generators too are shown in the table. B denotes the bow, T denotes the teardrop, ∞ denotes the figure eight (in the ω-case with orientation, in the case with arming the overbar denotes a change of orientation).

b) The Bases and Their Cobordisms Are Unoriented

THEOREM. The groups of stable classes of Legendre curves and of immersions of curves for orientable (Ω) and unorientable ($\mathfrak{N}$) bases are isomorphic to those in the following table:

	S Leg	*P* Leg	*S* Imm	*P* Imm
Ω, ω	$\mathbf{Z}$ B	$\mathbf{Z}_2$ T	$\mathbf{Z}_2$ ∞	$\mathbf{Z}_2$ ∞
Ω, ν	0 —	$\mathbf{Z}_2$ T	$\mathbf{Z}_2$ ∞	$\mathbf{Z}_2$ ∞
$\mathfrak{N}$, ω	$\mathbf{Z}$ β	$\mathbf{Z}_2^2$ T, δ	$\mathbf{Z}_2$ ∞	$\mathbf{Z}_2^2$ ∞, δ
$\mathfrak{N}$, ν	$\mathbf{Z}_2$ β	$\mathbf{Z}_2^2$ T, δ	$\mathbf{Z}_2$ ∞	$\mathbf{Z}_2^2$ ∞, δ

Here δ denotes the class of a disorienting smooth curve and β denotes the class of a disorienting curve with one cusp.

C. Fuks Groups. Fuks suggested the examination of the groups of classes of a cobordism with a variable base: addition of classes is defined with the aid of a nonintersecting union of bases. The Fuks group of classes of a cobordism Φ is the direct sum of the subgroup Φ_0 of classes of a cobordism over a base cobordant to zero and the group of classes of the cobordism of the bases.

THEOREM. The Fuks groups of classes of the cobordism of curves over a base cobordant to zero (Leg or Imm, S or P, ω or ν) are isomorphic to the corresponding stable groups with the base of a sphere if the bases and their cobordisms are treated as being oriented (Ω) and with the base of a Klein bottle if they are treated as being unoriented ($\mathfrak{N}$):

	S Leg	*P* Leg	*S* Imm	*P* Imm
Ω, ω	$\mathbf{Z}$ B	$\mathbf{Z}_2^2$ T, $\bar{\text{T}}$	$\mathbf{Z}_2^2$ ∞, $\bar{\infty}$	$\mathbf{Z}_2$ ∞
Ω, ν	0 —	$\mathbf{Z}_2$ T	$\mathbf{Z}_2$ ∞	$\mathbf{Z}_2$ ∞
$\mathfrak{N}$, ω	$\mathbf{Z}$ β	$\mathbf{Z}_2^2$ T, δ	$\mathbf{Z}_2$ ∞	$\mathbf{Z}_2^2$ ∞, δ
$\mathfrak{N}$, ν	$\mathbf{Z}_2$ β	$\mathbf{Z}_2^2$ T, δ	$\mathbf{Z}_2$ ∞	$\mathbf{Z}_2^2$ ∞, δ

Remark. The reason for the isomorphism of Fuks groups to stable groups is that the subgroup generated by the classes stably equivalent to zero turns out to be consisting of elements stably equivalent to zero for all the 16 theories listed.

With the cyclic class of curve γ over base B, we associate the class $[\gamma, B] - [\varnothing, B] \in \Phi_0$. This map is a homomorphism. Its kernel is generated by classes stably equivalent to zero. Consequently, the stable group of base B is identified with a subgroup of group Φ_0.

This subgroup is isomorphic to Φ_0 for a suitable choice of base B (for oriented bases and their cobordisms it is sufficient to take a sphere as the base; for unoriented ones, a Klein bottle).

D. Other Results. The proofs of the theorems presented yield more information than do the theorems themselves. For example, from them follow results such as the ones below.

THEOREM. Every oriented closed (not necessarily connected) curve, immersed in $\mathbf{R}^4$, with respect to which the integral of form p dq equals zero, is the border of an immersed oriented Lagrange surface.

The normalization* of an immersed curve in $\mathbf{R}^4$ by a normal vector antiorthogonal to the tangent vector in the sense of a symplectic structure, is called its Lagrange normalization.

For such a normalization of a closed oriented curve we can define the Maslov index (e.g., as the degree of the map $\gamma \to S^1$ obtained as the superposition of maps $\gamma \to \Lambda$ and $\Lambda \to S^1$, where $\Lambda = U(2)/O(2)$ is the manifold of all Lagrange planes in $\mathbf{R}^4$; its map onto a circle is induced by the square of the determinant $\det^2$: $U(2) \to S^1$, see [2]).

THEOREM. Every Lagrange normalization of a closed oriented curve with a zero integral of form p dq in $\mathbf{R}^4$, having an index 0, is a normal normalization of this curve on a suitable immersed oriented Lagrange surface.

THEOREM. Neither the projective plane nor any other closed surface with an odd Euler characteristic can be Lagrange immersed in $\mathbf{R}^4$.

THEOREM. A Klein bottle and, in general, every connected closed surface with an even Euler characteristic, admits of immersion in a five-dimensional space of 1-jets of functions on plane $J^1(\mathbf{R}^2, \mathbf{R})$ as a Legendre submanifold (and, consequently, admits of a Lagrange immersion into $\mathbf{R}^4$).

[Surfaces with an odd Euler characteristic do not admit of Legendre immersions in $J^1(\mathbf{R}^2, \mathbf{R})$ by the preceding theorem.]

We consider $\mathbf{R}^4$ as the space of quaternions $q_1 + ip_1 + jq_2 + kp_2$. We shall denote the operator of multiplication by quaternion h from the left ($a \mapsto ha$) by the same letter h as the quaternion itself.

Lagrange (more precisely, i-Lagrange) planes are defined as those passing into planes orthogonal to themselves under the action of operator i. i-Complex planes coincide with (j and k)-Lagrange planes (the Cauchy–Riemann equations).

THEOREM. The manifold of oriented Lagrange planes in $\mathbf{R}^4$ is diffeomorphic to the direct product $S^2 \times S^1$.

Quaternions permit an explicit indication of a diffeomorphism.

THEOREM. Each oriented Lagrange plane is uniquely representable in the form $e^{i\varphi}\lambda$, where λ is the j-complex plane and $0 \leq \varphi < \pi$.

Since the manifold of j-complex planes is $\mathbf{C}P^1 \approx S^2$, while $e^{i\varphi}/\pm 1$ ranges the circle S^1, the pair (λ, φ) is identified with a point from $S^2 \times S^1$.

COROLLARY. j-Complex surfaces are i-Lagrange; moreover, the Maslov index of each closed curve on such a surface equals zero.

THEOREM. The manifold of unoriented Lagrange planes in $\mathbf{R}^4$ is diffeomorphic to the factor space of $S^2 \times S^1$ by the antipodal involution $(x, y) \mapsto (-x, -y)$.

The Gauss map of a two-dimensional oriented Lagrange submanifold immersed in $\mathbf{R}^4$ into the manifold of oriented Lagrange planes in $\mathbf{R}^4$ enables us to determine a numerical invariant of a regular homotopy of the immersion: the degree of a composition onto S^2.

THEOREM. The degree mentioned equals the Euler characteristic of the Lagrange submanifold being examined (Fuks) and is not an invariant of the Lagrange cobordism (Varchenko).

A Euler class is not a stable class in the cohomologies of a Lagrange Grassmann manifold (see [5]); the characteristic numbers defined by stable classes are, clearly, invariants of Lagrange cobordisms.

E. Abstract Definitions. Above we have considered several different "cobordism theories" (e.g., $\mathfrak{N}S\omega$ Imm, viz., the theory of oriented armed immersions of closed curves in unoriented manifolds). Since the

*Translator's Note. Sometimes also called "rigging."

basic concepts (cylindrical, stable, and Fuks groups) are defined in uniform fashion for all theories, we present these definitions in an abstract form. We commence with the usual theory of cobordisms of manifolds (oriented or not) and we consider a category of cobordisms, whose objects are manifolds and whose morphisms are cobordisms: $\partial\beta = B_1 - B_0$ is treated as a morphism β from B_0 into B_1.

Definition. A functor F from the category of cobordisms into the category of Abelian correspondences is called a cobordism theory. Functor F associates an Abelian group F(B) with each manifold B and associates a correspondence, i.e., a subgroup $F(\beta) \subset F(B_0) \times F(B_1)$, with each cobordism ($\partial\beta = B_1 - B_0$).

Example. If $F = \mathfrak{R}S\omega$ Imm, then F(B) is a cylindrical group of classes of oriented normalized immersions of curves into B and β is an unoriented manifold with borders B_1 and B_0; (x_0, x_1) belongs to $F(\beta)$ if there exists an armed immersion of an oriented surface z into β, transversal to borders B_1 and B_0 and cutting out x_0 in B_0 and x_1 in B_1 (with due regard to orientations and armings).

The fulfillment of the following three axioms is assumed.

1) If β is the cylinder B × [0, 1], then $F(\beta)$ contains the diagonal $\{(x, 0), (x, 1)\} \subset F(B \times 0) \times F(B \times 1)$.

2) $F(B_1 \amalg B_2) = F(B_1) \oplus F(B_2)$.

3) $F(\beta_1 \amalg \beta_2) = \{(x_1^0, x_2^0), (x_1^1, x_2^1)$, where $(x_k^0, x_k^1) \in F(\beta_k)\}$.

Definition. The correspondence

$$NF(B_0, B_1) = \sum (F(\beta) \mid \partial\beta = B_1 - B_0)$$

is called the total correspondence from B_0 into B_1 of theory F. This is either a subgroup in $F(B_0) \times F(B_1)$ or an empty set.

Definition. The group $NF(B) \subset F(B)$, defined by the condition

$$NF(B) \times 0 = NF(B, B) \cap (F(B) \times 0)$$

is called the group of classes cobordant to zero of theory F over B.

Definition. The group St F(B) = F(B) / NF(B) is called the group of stable classes of theory F over B.

Definition. The group generated by the semigroup $\amalg F(B)$, in which the sum of $\varphi_1 \in F(B_1)$ and $\varphi_2 \in F(B_2)$ is $\varphi_1 + \varphi_2 \in F(B_1 \amalg B_2)$, is called the monster group Monstr F of theory F.

Definition. The subgroup Cbd F of elements cobordant to zero of the monster group of theory F is defined as the subgroup generated by sums $x_0 \amalg x_1$, where $(x_0, x_1) \in F(\beta)$, $x_0 \in F(B_0)$, $x_1 \in F(B_1)$.

Definition. The factor group Monstr F/Cbd F is called the Fuks group $\Phi(F)$ of theory F.

Definition. The kernel of homomorphism α: $\Phi(F) \to \Omega$ (or $\mathfrak{R}$) associating with the coset $[x]$, $x \in F(B)$, the class [B] in the usual group Ω (or $\mathfrak{R}$) of cobordisms of manifolds is called the Fuks group $\Phi_0(F)$ of theory F.

Obviously, $\Phi(F) \simeq \Phi_0(F) \oplus \Omega$ (or $\mathfrak{R}$) (we set $\alpha^{-1}([B]) = [0 \in F(B)]$).

Definition. Theory F is called a Fuks theory if the total correspondence $NF(B_0, B_1)$ is not a sum for any B_0 and B_1, but is the union of correspondences $F(\beta)$ such that $\partial\beta = B_1 - B_0$.

It is easy to prove the

THEOREM. For any Fuks theory F there exists a homomorphism with zero kernel St F(B) $\to \Phi_0(F)$; this homomorphism is induced by the map φ: F(B) $\to$ Monstr F associating with class $x \in F(B)$ the class [x] − [0], where [y] is the class of element y from F(B) in group Monstr F.

Examples. The theory of Lagrange immersions of curves Lag_1 [in which F(B) is the group of cylindrical classes of oriented Lagrange curves immersed in T*B] is not a Fuks theory. All 16 theories Leg and Imm of immersions of curves (S or P, ω or ν, Ω or $\mathfrak{R}$) are Fuks theories.

At my request Fuks computed the first groups $\Phi_0(F)$ for the case when F is a homology theory.

For q-dimensional homologies of oriented n-dimensional manifolds he obtained the following groups:

$n-q$	q					
	1	2		3	4	
1	0	0	0	0	$\mathbf{Z}$	$CP^2 \subset CP^2 \times S^1$
2	0	Z	$CP^1 \subset CP^2$	0	$\mathbf{Z}^2$	$CP^2 \subset CP^2 \times S^2$, $CP^2 \subset CP^3$
$\geqslant 3$	0	Z_2	$CP^2 \subset CP^2 \times S^{n-4}$	0	$\mathbf{Z} \oplus \mathbf{Z}_2$	?

The table shows the generators as well: e.g., $CP^1 \subset CP^2$ signifies the class in Φ_0 of the difference of elements $[CP^1] \in H_2(CP_1^2, \mathbf{Z})$ and $[0] \in H_2(CP_0^2, \mathbf{Z})$ of the homology groups of two different complex projective planes.

For small q Fuks has also computed the groups Φ_0 for the theory of q-dimensional homologies of unoriented n-dimensional manifolds:

q	Group	Generator
1	$\mathbf{Z}_2$	Disorienting path on Kl x S^{n-2}
2	$\mathbf{Z}_2$	$CP^1 \subset CP^2 \times S^{n-4}$

here Kl is the Klein bottle.

2. Lagrange Cobordisms

A. Lagrange Border. The border of the cotangent bundle space of a manifold is fibered over the cotangent bundle space of the border, with fiber $\mathbf{R}$: the map $\pi: \partial T^*B \to T^*\partial B$ is defined as the contraction of 1-forms.

Definition. The Lagrange border of a Lagrange submanifold L of space T*B, transversal to ∂T^*B, is defined as the projection of its intersection with the border into the border's cotangent bundle space,

$$\underline{\partial} L = \pi(\partial L \cap \partial T^*B) \subset T^*\partial B.$$

A Lagrange border has been defined only for Lagrange manifolds transversal to ∂T^*B and is a submanifold in $T^*\partial B$, but, in general, this is only an immersed submanifold even if L has been embedded in T*B. Further on the words Lagrange manifold everywhere signify the Lagrange immersion of a borderless compact manifold.

B. Lagrange Cobordisms. We shall denote the usual cobordism β between manifolds B_1 and B_0 thus: $\partial\beta = B_1 - B_0$.

Definition. A Lagrange submanifold λ in $T^*\beta$ with a Lagrange border $\underline{\partial}\lambda = L_1 - L_0$ is called a Lagrange cobordism over β between Lagrange manifolds L_1 in T^*B_1 and L_0 in T^*B_0.

A cobordism is said to be cylindrical if $\beta = B \times [0, 1]$ (λ may or may not be a cylinder).

THEOREM. Lagrange cobordantness is an equivalence relation.

Proof. Difficulty arises only with transitivity: if $\underline{\partial}\lambda_1 = L_1 - L_0$ and $\underline{\partial}\lambda_2 = L_2 - L_1$ are Lagrange cobordisms over $\partial\beta_1 = B_1 - B_0$ and $\partial\beta_2 = B_2 - B_1$, then the union of λ_1 and λ_2 is, in general, not a Lagrange manifold since a discontinuity appears over B_1. This discontinuity is annihilated by the insertion between λ_1 and λ_2 of a vertical Lagrange manifold formed by segments of fibers of bundle π and by smoothing the corners formed. This is formally realized by the following

LEMMA. Let $\Pi = B \times \mathbf{R}$, $\Pi_\pm = B \times \mathbf{R}_\pm$; let $\lambda_\pm$ be Lagrange submanifolds over $\Pi_\pm$ with a common Lagrange border L over $B \times 0$. We prescribe the map f: $\Pi \to \Pi$ by the formula $f(b, t) = (b, t\varphi(t))$, where the smooth function φ equals 0 when $|t| \le \varepsilon$ and 1 when $|t| \ge 2\varepsilon$ and is contained between 0 and 1 when $\varepsilon < |t| < 2\varepsilon$. If ε is sufficiently small, the map $f^*: T^*\Pi \to T^*\Pi$ takes $\lambda_+ \cup \lambda_-$ into an immersed Lagrange manifold coinciding with $\lambda_+ \cup \lambda_-$ when $|t| \ge 2\varepsilon$; its parts over $\Pi_\pm$ have the same Lagrange border as $\lambda_\pm$.

The lemma's proof is trivial.

Remark. The smoothing lemma provides each Lagrange cobordism with a "collar" possessing the two properties indicated below.

In a neighborhood of border B of base β we introduce the direct product coordinates: $q \in B$ and $q_0 \in \mathbf{R}$. By p and p_0 we denote the corresponding momenta. Then: 1) the collar is a cylinder with a generator, viz., axis q_0; 2) the coordinate p_0 along the collar equals 0 everywhere.

In the definition of Lagrange cobordism we could have assumed that λ is provided with a collar with properties 1) and 2). Then we would not have required smoothing.

Definition. The product of the Lagrange cobordisms $\partial\lambda_1 = L_1 - L_0$ and $\partial\lambda_2 = L_2 - L_1$ is defined as the Lagrange cobordism $\partial\lambda = L_2 - L_0$ obtained by smoothing the union $\lambda_1 \cup \lambda_2$ along the common border L_1.

The product of cylindrical cobordisms is, obviously, a cylindrical cobordism.

C. Group of Classes of a Cylindrical Cobordism. The union of Lagrange immersions over a common base is once again a Lagrange immersion. This allows us to define the addition of classes of cylindrical cobordantness ($\approx$): we have the trivial

THEOREM. If $L_1 \approx L_1'$ and $L_2 \approx L_2'$, then $L_1 \cup L_2 \approx L_1' \cup L_2'$.

By the same token the set of classes of a Lagrange cylindrical cobordism over a fixed base B is turned into a commutative semigroup Lag(B) with the operation $[L_1] + [L_2] = [L_1 \cup L_2]$.

Remark. The theorem's assertion is false for noncylindrical cobordism classes.

THEOREM. The semigroup of classes of a cylindrical Lagrange cobordism over $B = \mathbf{R}^n$ is a group: $[L] + [-L] = [\emptyset]$, where the immersion $-L$ is obtained from immersion L by reversing the orientation.

Proof. Let $(q_1, \ldots, q_n)$ be coordinates in $\mathbf{R}^n$, $q_0 \geq 0$ in $\mathbf{R}_+$. By g_φ we denote the rotation of plane (q_0, q_1) through an angle φ. The symplectic transformation, induced by it, of space $T^*\mathbf{R}^{n+1}$ is denoted G_φ. Let L be a Lagrange manifold over $\mathbf{R}^n$, on which $q_1 < 0$. Setting $p_0 = 0$, we construct over L an isotopic manifold L' in $T^*\mathbf{R}^{n+1}$. We consider the union of the turned manifolds $G_\varphi L'$, where $0 \leq \varphi \leq \pi$. This union λ is a Lagrange submanifold over $\mathbf{R}^{n+1}$. It specifies the Lagrange cobordism $\partial\lambda = L - L''$, where L" is obtained from L by the involution σ: $(p_1, q_1) \mapsto (-p_1, -q_1)$. Thus, $[L''] + [L] = 0$, and our semigroup is a group. It is easily seen that $[L''] = [-L]$ [since σ is included in the family of rotations of plane (p_1, q_1)]. Finally, $[L] = [-L] = 0$, etc.

D. Lagrange Covering Homotopy. A family of time-dependent Lagrange manifolds cannot always be united into one Lagrange manifold over space–time. For example, curves on a phase plane must cover equal areas in order that a Lagrange surface can be made up from them.

Let $B \times \mathbf{R}$ be space–time, π: $T^*(B \times \mathbf{R}) \to (T^*B) \times \mathbf{R}$, τ: $(T^*B) \times \mathbf{R} \to \mathbf{R}$ be natural projections, $I = [0, 1]$.

Definition. A Lagrange immersion over space–time F: $M \times I \to T^*(B \times \mathbf{R})$ is said to be 1) fib ered if $\tau \circ \pi \circ F(m, t) \equiv t$, 2) a covering immersion f: $M \times I \to (T^*B) \times \mathbf{R}$ if $\pi \circ F = f$.

By α we denote a canonic 1-form "p dq" on T*B and by φ: $(T^*B) \times \mathbf{R} \to T^*B$ we denote a canonic projection.

Definition. The immersion f: $M \times I \to (T^*B) \times \mathbf{R}$ is said to be 1) fibered if $\tau \circ f(m, t) \equiv t$, 2) admissible if the integrals of form $t^*\varphi^*\alpha$ with respect to any cycles lying in $M \times t_1$ and $M \times t_2$ and homologic in $M \times I$ coincide.

Remark. A fibered immersion f defines a family of immersions f_t: $M \to T^*B$ $(f(m, t) = (f_t(m), t))$, and vice versa. If a fibered immersion f is admissible, then all the immersions f_t are Lagrange; however, in general, the inverse is false.

THEOREM. Every admissible fibered immersion f is covered by a Lagrange fibered immersion F.

Proof. The condition for the Lagrangeness of F defines in terms of f an increment s of the momentum h conjugate to time $\theta = \tau \circ \pi$ along any path lying on $M \times t$. The admissibility of f guarantees the independence of this increment from a path with fixed beginning and end. This leads to the following formula for F. By g^s we denote the transformation in time s under the action of a phase flow with Hamiltonian θ: $T^*(B \times \mathbf{R}) \to \mathbf{R}$. Let i: $(T^*B) \times \mathbf{R} \to T^*(B \times \mathbf{R})$ be the zero section. Then $F(m, t) = g^{s(m,t)} i f(m, t)$, where $s(m, t) = \frac{d}{dt} \int_{m_0}^{m} f_t^* \alpha$.

Remark. We cannot ensure the fulfillment of the initial condition for F when $t = 0$, but if $f(m, t)$ is independent of t for small t, then for these t the momentum h equals 0. In this case immersion F is furnished a collar.

E. Surgery. We now consider a not necessarily fibered Lagrange manifold in space–time. As time changes the instantaneous Lagrange manifold will be reorganized at isolated critical instants. For example, the birth of a Lagrange curve from emptiness is possible.

The usual newly born curve has the form of an ellipse.

For a Lagrange curve this is impossible since the cobordism's invariants (area and index) are nonzero for a circle. It is not difficult to grasp that a newly born Lagrange curve has the form of a figure eight. This is the appearance, e.g., of the track of a Lagrange torus corresponding to motion in a central field on a plane over a straight line close to the tangent at the apocenter. By examining the pericenter as well, we arrive at the following conclusion.

THEOREM. The Lagrange cobordisms shown in Fig. 3 exist.

Proof. We consider the generating family

$$\Phi(x, q) = x^3 - x(q_2 \pm q_1^2).$$

We define a Lagrange surface in $\mathbf{R}^4$ by the formulas $\partial\Phi/\partial x = 0$, $p = \partial\Phi/\partial q$. The cobordisms needed are determined by a part of this surface above the plane $|q_2| \leq \delta$.

It is useful to note that two like directed curves are obtained after the surgery in Fig. 3.

F. Generators and Relations. THEOREM. Every oriented Lagrange curve on a plane is Lagrange orientedly cobordant with a curve from a nonintersecting circle.

Proof. A curve remains cobordant with itself under the following operations: 1) regular homotopies of immersions, preserving the areas of the immersion's components; 2) birth and annihilation of figure eights of zero area; 3) annihilation of an intersection, under which the total area of the participating component is preserved (the inverse surgery is the birth of an intersection from like-directed close parallel parts of a curve).

This follows from the theorems proved [the combination of homotopies (D) and of surgeries (E) is carried out by a smoothing analogous to the construction of the collar in (B); we do not here dwell on the details on this smoothing].

It is clear that transformations 1) and 3) can turn any curve into a collection of nonintersecting circles.

By $[s]$ we denote the class of circles of area s. We have already proved (C) that $[-s] = -[s]$.

THEOREM. Let $s_1 + s_2 = s_3 + s_4$, $s_k > 0$. Then $[s_1] + [s_2] = [s_3] + [s_4]$.

The proof consists of the surgeries −3, 1, 3 (Fig. 4).

By $\langle t \rangle$ we denote a function on the semiaxis $s > 0$, equal to 1 at point t and 0 at the remaining points. Let Γ be the additive group of finite integral functions on the semiaxis, i.e., the group with generators $\langle t \rangle$.

THEOREM. Let $\Delta \subset \Gamma$ be a subgroup with generators $\langle s_1 \rangle + \langle s_2 \rangle - \langle s_3 \rangle - \langle s_4 \rangle$, where $s_1 + s_2 = s_3 + s_4$. Then $\Gamma/\Delta \approx \mathbf{Z} + \mathbf{R}$.

Proof. Let m be a function from Γ. Then $\sum m(s)$ and $\sum s m(s)$ for all $s > 0$ define the homomorphism $\alpha: \Gamma \to \mathbf{Z} + \mathbf{R}$. Obviously, $\Delta \subset \operatorname{Ker} \alpha$.

Conversely, let us prove that $\operatorname{Ker} \alpha \subset \Delta$. Any function m from Γ can be decomposed into a sum of its own positive part $p \geq 0$ and negative part $n \leq 0$. Any positive function γ from Γ is representable as a sum of a function from Δ and of a function $\mu(\gamma)$ uniquely defined by γ and nonzero at no more than one point. If $m \in \operatorname{Ker} \alpha$, then $\mu(p) = \mu(n)$, i.e., $m \in \Delta$.

G. Groups ω Lag. From the three theorems in the preceding paragraph and from the invariance of the index and of the integral relative to cobordisms of orientable Lagrange curves follows the

THEOREM. The group ω Lag $(\mathbf{R})$ of classes of an oriented Lagrange cobordism of curves on the phase plane $T^*\mathbf{R}$ is isomorphic to $\mathbf{Z} + \mathbf{R}$.

Remark. As a matter of fact, the groups are isomorphic also as topological groups.

Let us now consider Lagrange curves on cylinder T^*S^1.

THEOREM. $\omega \operatorname{Lag}(S^1) \approx \mathbf{Z} + \mathbf{Z} + \mathbf{R}$.

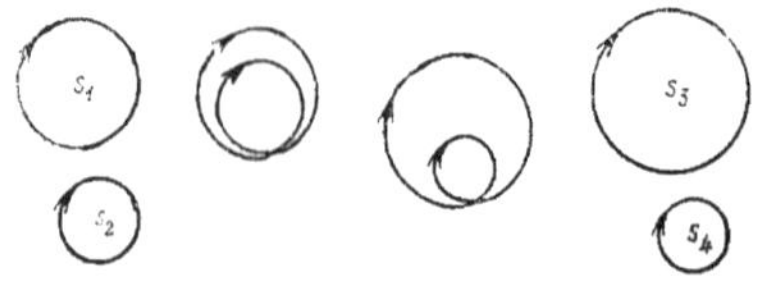

Fig. 4

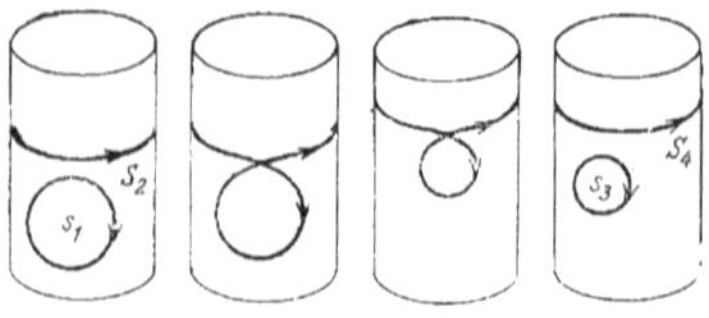

Fig. 5

Here the invariants of the cobordism are the index, the homology class, and the integral; the fact that the semigroup is a group was not obvious in advance.

For the proof we first annihilate the intersections as in Paragraph E and we replace the original curve by standard curves of the following two types.

By [s] we denote the class of a connected closed embedded curve, homologic to zero, of area $s > 0$ and by [S] we denote the class of curve $p = S$ on the cylinder $T^*S^1 = \{p, q \bmod 1\}$, oriented to the side of growth of q. By the symbol $-[S]$ we denote the class of the oppositely oriented curve. It is easy to prove

LEMMA 1. $[S] + (-[S]) = [\varnothing]$, and the semigroup ω Lag (S^1) is a group with generators [s], $s > 0$, and [S], $S \in \mathbf{R}$.

The relations supply the following

LEMMA 2. Let $s_1 + s_2 = s_3 + s_4$, $S_1 + S_2 = S_3 + S_4$, $s_1 + S_2 = s_3 + S_4$. Then $[s_1] + [s_2] = [s_3] + [s_4]$, $[S_1] + [S_2] = [S_3] + [S_4]$, $[s_1] + [S_2] = [s_3] + [S_4]$.

The proof consists of the surgeries -3, 1, 3 (Fig. 5).

By $\langle S \rangle$ we denote a function on the real line, equal to 1 at point S and to 0 at the remaining points. The functions $\langle s \rangle$, $s > 0$ and $\langle S \rangle$, $S \in \mathbf{R}$, generate an additive group Γ (the group of finite functions on $\mathbf{R}_+ \amalg \mathbf{R}$). It is easy to prove

LEMMA 3. Let $\Delta \subset \Gamma$ be a subgroup generated by functions $\langle s_1 \rangle + \langle s_2 \rangle - \langle s_3 \rangle - \langle s_4 \rangle$, $\langle S_1 \rangle + \langle S_2 \rangle - \langle S_3 \rangle - \langle S_4 \rangle$, $\langle s_1 \rangle + \langle S_2 \rangle - \langle s_3 \rangle - \langle S_4 \rangle$, where $s_1 + s_2 = s_3 + s_4$, $S_1 + S_2 = S_3 + S_4$, $s_1 + S_2 = s_3 + S_4$. Then $\Gamma/\Delta \approx \mathbf{Z} + \mathbf{Z} + \mathbf{R}$.

The isomorphism is defined by the invariants taking on a function m from Γ the values $\operatorname{ind} = \sum m(s)$, $\deg = \sum m(S)$, $I = \sum s m(s) + \sum S m(S)$ (everywhere $s \in \mathbf{R}_+$, $S \in \mathbf{R}$).

The theorem follows from Lemmas 1-3.

H. Other Groups of Classes of a Lagrange Cobordism. The following results are easily derived from the preceding.

THEOREM. The classes of an unoriented cylindrical Lagrange cobordism of curves in $T^*\mathbf{R}$ form a trivial group.

The classes of an oriented cylindrical Lagrange cobordism of unoriented curves in $T^*\mathbf{R}$ form group $\mathbf{Z}_2$.

The invariant of an oriented cobordism of an unoriented Lagrange curve is the index reduced modulo 4: it does not depend upon the choice of the orientations of the curve's components. Indeed, the index of each component is even and, hence, changes by a number that is a multiple of 4 when the orientation changes.

The classes of an unoriented cobordism of curves over a circle (i.e., in T^*S^1) form a group $\mathbf{Z}_2$ isomorphic to the group of one-dimensional homologies.

The groups of stable classes of the Lagrange cobordism of curves over closed manifolds B consisting of any number of components (i.e., the groups of classes of curves in T^*B) are isomorphic to the groups of cylindrical classes over the straight line: $\mathrm{St}\,\omega\,\mathrm{Lag}(B) \approx \mathbf{Z} + \mathbf{R}$, $\mathrm{St}\,\nu\,\mathrm{Lag}(B) \approx 0$ (the invariants in the oriented case are the index and the area).

The article is concluded in the next issue.

LITERATURE CITED

1. V. I. Arnol'd, Mathematical Methods of Classical Mechanics [in Russian], Nauka, Moscow (1974).

2. V. I. Arnol'd, "On characteristic class entering in quantization condition," Funkts. Anal. Prilozhen., 1, No. 1, 1-14 (1967).
3. V. I. Arnol'd [Arnold], "Wave fronts evolution and equivalent Morse lemma," Commun. Pure Appl. Math., 29, No. 6, 557-582 (1976) [errata: 30, No. 6, 823 (1977)].
4. V. P. Maslov, Perturbation Theory and Asymptotic Methods [in Russian], Moscow State Univ. (1965).
5. D. B. Fuks, "On characteristic Maslov–Arnol'd classes," Dokl. Akad. Nauk SSSR, 178, No. 2, 303-306 (1968).
6. L. Hörmander, "Fourier integral operators. I," Acta Math., 127, 79-183 (1971).
7. Ya. M. Éliashberg, "Estimate of the number of fixed points of area-preserving transformations," Syktyvkar (1978), RZhMat [manuscript deposited in VINITI, IV 374-379, 5A 491 (1979)].
8. C. T. C. Wall, "Cobordism of pairs," Commun. Math. Helv., 35, 136-145 (1961).
9. R. Wells, "Cobordism groups of immersions," Topology, 5, 281-294 (1966).

LAGRANGE AND LEGENDRE COBORDISMS. II

V. I. Arnol'd

UDC 513.83

Legendre cobordisms describe the relation between the traces of a wave front extended through the air on the surface of the earth at different moments of time. Here we give the proofs of the theorems on Legendre cobordisms formulated in the first part of the paper, published in the preceding issue of this journal, where one can also find all the necessary definitions.

3. Legendre Cobordisms

A. Legendre Boundary. We consider the manifold PT^*B of contact elements on the manifold B with boundary ∂B (i.e., the projectivization of the space of the cotangent bundle). The bundle $\pi: PT^*B \to B$ is Legendre, i.e., its fiber is a Legendre submanifold of the contact manifold PT^*B. Let $\xi \in \pi^{-1}(\partial B)$ be a contact element, transversal to the boundary. The intersection of ξ with the tangent plane of the boundary defines a contact element $\rho\xi \in PT^*(\partial B)$.

Let λ be a Legendre submanifold of PT^*B, transverse to $\pi^{-1}(\partial B)$. Then $\rho(\lambda)$ will be a Legendre immersed submanifold of $PT^*(\partial B)$; it is also called the Legendre boundary of the manifold λ.

An analogous construction is applicable to the bundle of armed contact elements $ST^*B \to B$ and to the bundle of 1-jets $J^1(V, \mathbf{R}) \to J^0(V, \mathbf{R})$: we associate the 1-jet of a function on V at a point of ∂V with the 1-jet of the restriction of the function to ∂V.

The rest of the definitions are introduced by the standard scheme of Sec. 1.

B. Legendre Immersions in $J^1(\mathbf{R}, \mathbf{R})$. We need some general facts about Legendre manifolds (cf. [1]).

The projection of a Legendre submanifold of the space of the Legendre bundle $J^1(V, \mathbf{R}) \to V \times \mathbf{R}$ onto the base is called a front. For almost any Legendre submanifold (those excluded constitute a set of infinite codimension) the front is a hypersurface in the base, having a tangent plane everywhere. Such a hypersurface uniquely determines the Legendre manifold, so we can pass from Legendre manifolds and cobordisms to fronts and their cobordisms.

The front of a Legendre curve in general position in $J^1(\mathbf{R}, \mathbf{R})$ is a planar curve, without singularities, other than points of self-intersection and the simplest cusps: a front is everywhere nonvertical, i.e., its tangent at no point contains the fiber of the bundle $V \times \mathbf{R} \to V$.

Conversely, any curve with the properties indicated is the front of a Legendre manifold uniquely determined by it.

In exactly the same way, fronts of Legendre surfaces in general position in $J^1(\mathbf{R}^2, \mathbf{R})$ are surfaces in $\mathbf{R}^3$, having nonvertical tangent planes everywhere, and having no singularities other than a cuspidal edge of semicubical type, swallow tails, and transversal self-intersections; conversely, any surface with these properties is the front of a uniquely determined Legendre manifold. This allows us to consider instead of Legendre curves and their cobordisms, fronts and their cobordisms (a cobordism of fronts – of curves defined as fronts – is a surface, transversal to the boundary of the base as a stratified manifold).

Proposition 1. The semigroup of cylindrical cobordism classes of Legendre immersions (oriented or not) in $J^1(\mathbf{R}^n, \mathbf{R})$ is a group.

Moscow State University. Translated from Funktsional'nyi Analiz i Ego Prilozheniya, Vol. 14, No. 4, pp. 8-17, October-December, 1980. Original article submitted February 26, 1980.

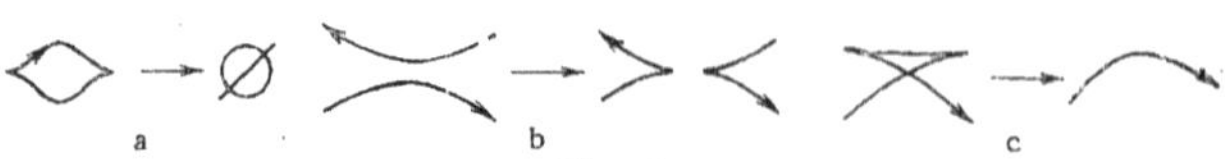

Fig. 6

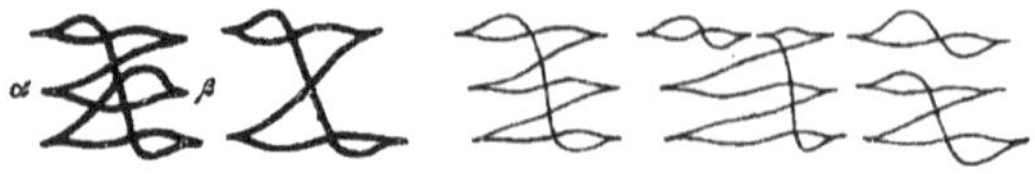

Fig. 7

In fact, we denote by $(x_1,\dots,x_n)$ the coordinates in $\mathbf{R}^n$, by x_0 that in $\mathbf{R}_+$, by y that in $\mathbf{R}$. Rotations by angles $0 \le \varphi \le \pi$ in the plane (x_0, x_1) induce a rotation $G\varphi$ of the space $\mathbf{R}^{n+2}$ with coordinates (x, y). Let Φ be a front in $\mathbf{R}^n \times \mathbf{R}$, on which $x_1 < 0$. Then $\cup G\varphi\Phi$ defines a front in $\mathbf{R}^n \times \mathbf{R} \times \mathbf{R}_+$, whose boundary is $\Phi - \Phi'$ (where Φ' is the reflection of Φ in a hyperplane and minus denotes change of orientation). Thus, $[\Phi] + [-\Phi'] = 0$, which is what had to be proved.

Proposition 2. Let $\{g_t\}$ be a one-parameter family of diffeomorphisms of the base, $V \times \mathbf{R} \to V \times \mathbf{R}$; $g_0 =$ id; $\varphi: M \to V \times \mathbf{R}$ be an immersion of a front. Let us assume that the tangent plane to $g_t \circ \varphi M$ for $0 \le t \le 1$ remains nonvertical. Then $g_1 \circ \varphi : M \to V \times \mathbf{R}$ is an immersion of a front, cobordant with the original.

Let us assume now that $M = M_1 \cup M_2$ and that on $\varphi(M_1 \cap M_2)$ the diffeomorphisms g_t are the identity. Then φ on M_1 together with $g_1 \circ \varphi$ on M_2 determine an immersion $M \to V \times \mathbf{R}$, cobordant with the original.

Proposition 3. The reconstructions pictured in Fig. 6, carry fronts of immersions in $J^1(\mathbf{R}, \mathbf{R})$ into cobordant ones (cf. [3, 7]).

It is useful to note that reconstruction b is applicable to two close parallel oppositely directed parts of a front.

THEOREM. Each oriented front of an immersion in $J^1(\mathbf{R}, \mathbf{R})$ is oriented cobordant with a sum of bows (Fig. 1).

Proof. Cusps (ears) of a front are divided into four types: positive or negative and left or right (reconstruction a of Fig. 6 annihilates a positive left and negative right ear, reconstruction b generates a right positive and a left negative).

With a right ear there always follows a left, so the number of ears on each component of a front is even.

Proposition 2 allows one to change each front of a curve situated in the strip $|x_1| \le 1$ so that the successive ears are found at the points (1, 0), (−1, 0), (1, 1), (−1, 1),..., (1, k), (−1, k) (Fig. 7).

If in the sequence there occur two ears of different signs in succession (α, β in Fig. 7a), then they can be annihilated by reconstruction c.

Repeating this procedure, we arrive at a curve, all of whose ears are of one sign. Using Corollary 2 again, we can position this curve in the standard way in the form of a zigzag, crossing itself obliquely (Fig. 7b left). Separating out a bow by reconstructions b, c, as shown in Fig. 7b, we decrease the number of ears, until we arrive at a sum of bows.

THEOREM. The group of classes of oriented Legendre immersions in $J^1(\mathbf{R}, \mathbf{R})$ is isomorphic with the group of even numbers; the isomorphism associates with an immersion the number of cusps of the front, taking signs into account.

Proof. We consider the map $J^1(\mathbf{R}, \mathbf{R}) \to T^*\mathbf{R}$. It carries a Legendre immersion into a Lagrange one and Legendre cobordant immersions into Lagrange cobordant ones. The index of the Lagrange curve obtained is equal, as is easy to calculate, to the number of cusps of the front (assumed to be general) taking account of signs.

Since each front is cobordant with a sum of bows, and the sum of a positive (index 2) and a negative (index −2) bow is cobordant to 0 (cf. above Corollary 1), the theorem is proved.

THEOREM. A bow is nonoriented cobordant to zero.

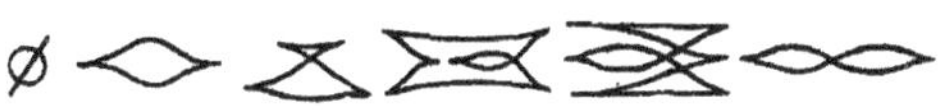

Fig. 8

Proof. In Fig. 8 there is pictured a sequence of reconstructions pulling a bow into a Möbius strip.

COROLLARY. The group of classes of nonoriented Legendre immersion in $J^1(\mathbf{R}, \mathbf{R})$ is trivial.

COROLLARY. A Klein bottle (and any connected closed two-dimensional manifold of even Euler characteristic) can be Legendre immersed in $J^1(\mathbf{R}^2, \mathbf{R})$.

To construct an immersion of a Klein bottle, it suffices to join two examples of the cobordism of Fig. 8 along the boundary (a bow). Handles can be attached to the bottle by reconstructions b and $-$b.

THEOREM. The projective plane cannot be Legendre immersed in $J^1(\mathbf{R}^2, \mathbf{R})$, and moreover $\mathbf{R}P^2$ cannot be Lagrange immersed in $\mathbf{R}^4$.

In fact, the normal bundle of a Legendre manifold is isomorphic with the cotangent bundle. Hence, the Whitney sum of the tangent bundle of a Legendre manifold immersed in $\mathbf{R}^{2n}$ with itself is trivial. The Whitney sum of the tangent bundle of $\mathbf{R}P^2$ with itself is nontrivial (the Stiefel class is $(1+a+a^2)^2 = 1+a^2 \bmod 2$).

An elementary proof can be given as follows. 1°. Any surface Lagrange immersed in $\mathbf{R}^4$ can be approximated by immersions in general position with respect to the projection $(p, q) \mapsto (q_1, q_2) \mapsto q_2$.

An immersion in general position can be described by a sequence of standard reconstructions of Lagrange curves in the plane (p_1, q_1) (traces over the lines $q_2 = c$): for each reconstruction, to the part of the surface where $q_2 \leq c$, one glues a handle of index 0, 1, or 2.

2°. A reconstruction is called disorienting if a handle of index 1 is glued to a component of the boundary along two opposite sides, where the handle and the half of the component of the boundary between the ends of the handle form a disorienting path.

3°. The number of disorienting reconstructions is congruent with the Euler characteristic modulo 2. In fact, the number of components of the boundary decreases by 1 upon gluing any nondisorienting handle and is unchanged upon gluing a disorienting one. In varying c from $-\infty$ to $+\infty$ the number of components of the boundary is unchanged, so the number of disorienting reconstructions is always even. Now more generally the number of reconstructions is equal modulo 2 to the Euler characteristic (by the Poincaré formula $\chi = M_0 - M_1 + M_2$).

4°. The Maslov index of the boundary of the surface over $q_2 \leq c$ reduced modulo 4 is unchanged by nondisorienting reconstructions and changes by 2 under disorienting ones. This is the central moment of the proof.

We choose on each component of each boundary any orientation.

If a reconstruction is nondisorienting, then, under compatible choice of orientation of the components, the index is unchanged. This means that modulo 4 it is unchanged under any choice of orientations.

We consider the disorienting reconstruction (I $\to$ II in Fig. 9).†

We shall calculate the index as the number of passages of the direction of the tangent through the horizontal direction.

In order to observe its change under reconstruction, we make another disorienting reconstruction II $\to$ III. In the passage I $\to$ III the orientation of the branch db changes (this leads to the change of the total index by twice the index of the branch bdb of the curve III) and two points (u, v) of identical index on curve I disappear. Consequently, ind (mod 4) changes by 2.

5°. The number of disorienting reconstructions is even. In fact, upon varying c from $-\infty$ to $+\infty$ ind (mod 4) is unchanged, and for each disorienting reconstruction, it changes by 2.

From 3° and 5° follows the impossibility of the Lagrange immersion of a manifold with odd Euler characteristic in $\mathbf{R}^4$, and hence, also of a Legendre immersion in $J^1(\mathbf{R}^2, \mathbf{R})$.

Remark. It follows from the theorem proved that not any isotropic mapping into a symplectic manifold can be approximated by isotropic immersions.

†In Fig. 9 (III) the arrow should be on the opposite side (from d to b).

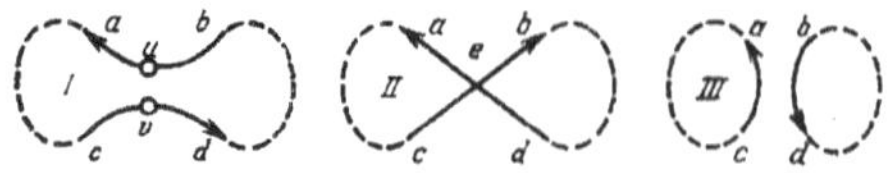

Fig. 9

Fig. 10

Nevertheless, Lagrange (Legendre) immersions can be approximated by Lagrange (Legendre) immersions in general position with respect to projections.

C. Fronts on the Plane. We consider oriented Legendre submanifolds of the space $ST^*\mathbf{R}^2$ transverse to oriented contact elements of the plane. The study of the group of cylindrical cobordisms $S\omega$ Leg $(\mathbf{R}^2)$ reduces to the study of cobordisms of fronts in general position (curves with cusps on the plane). The difference from the case considered above is the discarding of the condition of nonverticality of fronts (and the addition of armament, which is chosen canonically in the space of jets).

Proposition. The semigroup $S\omega$ Leg $(\mathbf{R}^n)$ of classes of oriented Legendre cobordisms in $ST^*\mathbf{R}^n$ is a group.

The proof is by rotation by π as in Subsection B.

This proof shows that the opposite of each front is the front subjected to reflection in a hyperplane and change of orientation. An armed bow goes into the opposite under change of orientation.

A front can, while remaining in its cobordism class, be freely moved in the plane, as in Proposition 2 of Subsection B (without being concerned any more about nonverticality). The list of reconstructions in passing from jets of functions to contact elements is extended (cf. [3, 7]):

Proposition. The reconstructions pictured in Fig. 10 carry fronts of Lagrange immersions in $ST^*\mathbf{R}^2$ into ones which are oriented Legendre cobordant to them.

It is useful to note that the Morse reconstruction c is applicable to two close parallel parts of a front, opposite in orientation and in armament, and consequently, orienting the plane identically.

On the contrary, the reconstruction d is applicable to parts which are opposite in orientation but identical in armament, and consequently, orienting the plane differently.

THEOREM. Each armed oriented front in the plane is cobordant to a sum of bows.

The proofs of the following seven lemmas are pictured in Fig. 11.

LEMMA 1. A front with the intersection of an identically orienting plane of sections is cobordant with a front without this intersection.

LEMMA 2. A front with the intersection of a variously orienting plane of sections is cobordant with the sum of a bow and a front without this intersection, but with two new cusps.

From Lemmas 1 and 2 follow

LEMMA 3. Each front is cobordant with a sum of bows and fronts without points of self-intersection.

LEMMA 4. A front with two cusps of different signs in succession is cobordant with a front without such a pair.

From Lemmas 3 and 4 follow

LEMMA 5. Each front is cobordant with a sum of bows and fronts, diffeomorphic with a saw with k teeth, pictured in Fig. 11 for $k = 4$.

LEMMA 6. A saw with k teeth is cobordant with k single-toothed saws.

LEMMA 7. A one-toothed saw is cobordant with a bow.

The theorem follows from Lemmas 1-7.

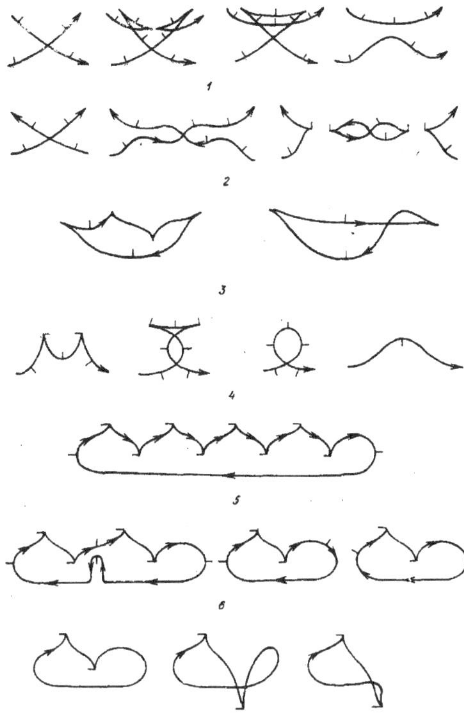

Fig. 11

COROLLARY. Upon reversing orientation, a Legendre curve in $ST^*\mathbf{R}^2$ changes its cobordism class into the opposite one.

In fact, for a bow this was proved above, any curve can be reduced to bows.

THEOREM. A bow is not cobordant to zero.

The proof is based on the invariance of the index of Legendre curves with respect to cobordisms. This invariance is easy to deduce from the preservation of the index under all reconstructions in general position (i.e., the reconstructions of Fig. 10).

One can also use the invariance of the index of Lagrange curves.

We consider an oriented armed front in general position in the plane, having no singularities other than simple cusps and points of self-intersection.

This front determines a Legendre curve in $ST^*R^2 \subset T^*R^2$. We consider the action of the multiplicative group $\mathbf{R}^*$ on $T^*\mathbf{R}^2$. Under the action of this group our curve goes into a "conical" surface in $T^*\mathbf{R}^2$.

Proposition. The surface obtained is, in a neighborhood of the curve considered, an immersed Lagrange manifold.

Proof. Let ξ denote a vector from $\mathbf{R}^4$ with components (q, p) and let the given curve have the form $q = r(t)$, $p = n(t)$ (n is the unit normal to $\dot{r}$). Our surface $\xi = \xi(t, s)$ is defined by the formulas $q = r(t)$, $p = (1+s)n(t)$, $|s| < \varepsilon$. We shall denote the derivative with respect to t by a dot and that with respect to s by a prime. Then $\dot{\xi} = (\dot{r}, (1+s)\dot{n})$, $\xi' = (0, n)$. These vectors are everywhere skew-orthogonal, since $\langle \dot{r}, n \rangle = 0$. For $s = 0$, $\dot{r} \neq 0$, they obviously are independent. At cusps, $\dot{r} = 0$, but $\dot{n}$ and n are independent. Hence $\dot{\xi}$ and ξ' are everywhere independent.

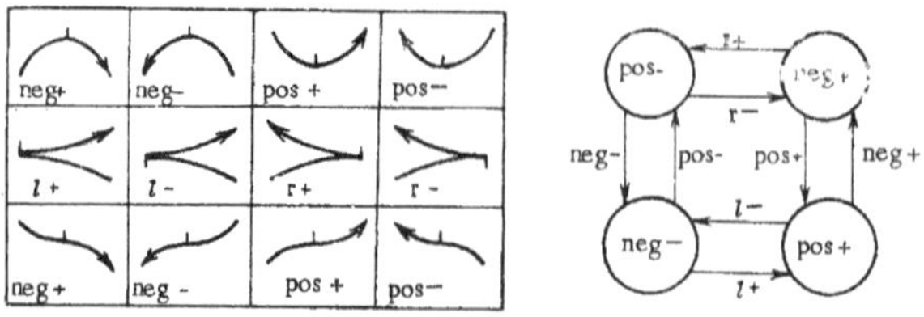

Fig. 12

Definition. By the index of an oriented armed front in the plane is meant the index of the corresponding Lagrange curve in the conical Lagrange manifold constructed.

THEOREM. The indices of fronts of Legendre cobordant curves are identical.

Proof. The conical Lagrange surfaces of Legendre cobordant curves are Lagrange cobordant, and the index of a curve in a Lagrange surface is an invariant of cobordism.

THEOREM. The index of a front in general position is equal to the number of cusps taking account of sign (+ denotes that a motion orienting the front carries a point with a branch onto a branch in the direction of the arming normal).

Proof. 1°. We define a surface $\xi = \xi\varphi(t, s)$, $|s| < \varepsilon$, by the formulas $q = r(t) + s\cos\varphi\cdot n(t)$, $p = n(t)(1 + s\sin\varphi)$.

2°. The map ξ_φ is a Lagrange immersion for each φ.

In fact the skew-scalar product of the vectors

$$\dot{\xi} = (\dot{r} + s\cos\varphi\cdot\dot{n},\quad \dot{n}(1 + s\sin\varphi)),\qquad \xi' = (\cos\varphi\cdot n,\quad \sin\varphi\cdot n)$$

is equal to zero, since $\langle\dot{r}, n\rangle = \langle\dot{n}, n\rangle = 0$. We shall prove the independence of $\dot{\xi}$ and ξ' for $s = 0$.

Let $a\dot{\xi} + b\xi' = 0$, i.e., $a\dot{n} + b\sin\varphi\cdot n = 0$, $a\dot{r} + b\cos\varphi\cdot n = 0$. If $a \neq 0$, then multiplying the first by $\dot{n}$ and the second by $\dot{r}$, we find $\dot{n} = \dot{r} = 0$, in spite of the nondegeneracy of the cusp. If $b \neq 0$, then multiplication by n gives $\sin\varphi = \cos\varphi = 0$.

3°. The index of the axial curve $s = 0$ on ξ_φ is independent of φ. This follows immediately from the homotopy definition of the index as the degree of a map [2].

4°. The surface $\xi_{\pi/2}$ is the conical Lagrange surface of the original front. This means that the index of the front is also equal to the index of the axial curve on the surface $\xi_0 : q = r(t) + sn(t)$, $p = n(t)$.

5°. The singularities of the projection of the surface ξ_0 along the fibers of $T^*\mathbf{R}^2$ onto the plane q are determined from the condition of linear independence of the vectors $\dot{q} = \dot{r} + s\dot{n}$, $q' = n$. Each cusp of the front determines a point of transversal intersection of the Legendre curve with the curve of singularities of the projection.

The index of intersection under the usual [2] transversal orientation of the line of singuliarities is equal to 1 at cusps of positive type.

The theorem is proved. Since the index of a bow is equal to 2, this completes the calculation $S\omega\,\mathrm{Leg}(\mathbf{R}^2) \approx \mathbf{Z}$.

Remark 1. The definition of the index of a front as well as its calculation in terms of cusps is suitable for armed oriented fronts on any surface (orientable or not). The proofs are also almost unchanged. In essence, everything is in the transversal orientability of the stratum A_2 in the space of functions.

Remark 2. Projecting the surface $\xi_{\pi/2}$ onto a p-space, we get the expression for the index indicated in Sec. 1 in terms of the number of points of inflection of the front. A nondegenerate front on an oriented plane divides the cusps and points of inflection into parts of four types, pictured in Fig. 12.

Cusps of an oriented curve in an oriented surface can be divided into left (the orientation given by rotation from the incoming to the outgoing branch is positive) and right; they can be armed positive and negative. Points of inflection also occur in four types, pictured in Fig. 12 (we number them by successive arcs).

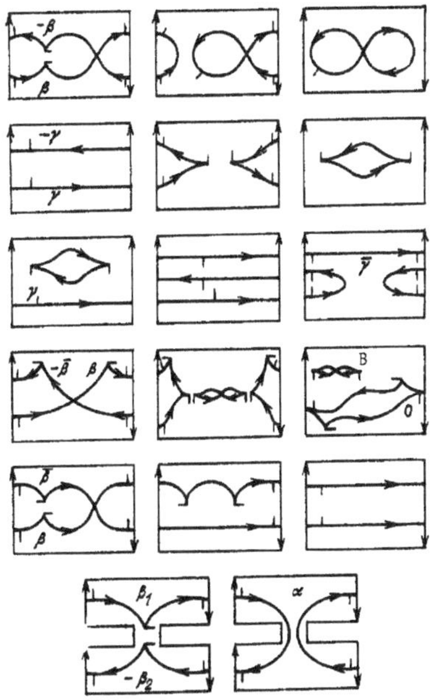

Fig. 13

We construct an oriented graph, the four vertices of which correspond to the four types of arcs, and the eight edges of which correspond to transitions from arc to arc at cusps and points of inflection. To each closed curve corresponds a closed path in the graph. From the calculation of the degree of the combinatorial mapping of the circle onto the circle which arises, we get the following formulas for the number of cusps and points of inflection of various types:

$$(l+) - (l-) = (r+) - (r-) = (\text{neg}-) - (\text{pos}-) = (\text{neg}+) - (\text{pos}+). \tag{1}$$

We put together the number of cusps $(+) = (l+) + (r+)$, $(-) = (l-) + (r-)$, and points of inflection $(\text{neg}) = (\text{neg}+) + (\text{neg}-)$, $(\text{pos}) = (\text{pos}+) + (\text{pos}-)$. From (1) follows

$$(\text{neg}) - (\text{pos}) = (+) - (-) = \text{ind}. \tag{2}$$

On the projective plane the difference between left and right cusps and positive and negative arcs vanishes, but (2), as before, follows from the calculation of the degree.

D. Fronts on Surfaces. 1°. The homology class of an oriented front on B is an invariant of cylindrical cobordism with values in $H_1(B, Z)$. The map which arises of the (semi)group of cylindrical cobordism classes $S\omega\,\text{Leg}(B) \to H_1(B, Z)$ is a homomorphism onto.

In fact, each homology class can be realized by an imbedded curve; the two-sided components of this curve are armed at once, and on the one-sided ones it is necessary first to create a cusp (cf. Example 3.3.4 of [6]).

2°. Each armed oriented front on B is cobordant with a sum of bows and pairwise disjoint two-sided imbedded circles γ and one-sided imbedded curves β with one cusp.

This is proved by the same lemmas as in Subsection C.

3°. The semigroup $S\omega\,\text{Leg}(B)$ is a group. In fact, it is proved in Fig. 13, for the generators β and γ defined above, that $[\beta] + [-\beta] = [\varnothing], [\gamma] + [-\gamma] = [\varnothing]$ (the minus sign denotes change of orientation while preserving armament).

4°. Let a dash denote change of armament while preserving orientation, B denote a bow, and Bo denote the boundary of a Möbius strip (consisting of a circle of the front β). In Fig. 13 the relations $[\gamma] = [\gamma]$, $[\beta] - [\bar{\beta}] = [B]$, $[\beta] + [\bar{\beta}] = [\text{Bo}]$ are proved.

TABLE 1

B	$S\omega$ Leg B	Generators	Relations
$\mathbf{R}P^2$	$\mathbf{Z}$	$[\beta]$	$2[\beta] = [B]$
Mob	$\mathbf{Z}^2$	$[\beta]$, $[\bar{\beta}]$	$[\beta] - [\bar{\beta}] = [B]$, $[\beta] + [\bar{\beta}] = [Bo]$
Kl	$\mathbf{Z}^2 + \mathbf{Z}_2$	$[\beta]$, $[\bar{\beta}]$, $[\alpha]$	$[\beta] - [\bar{\beta}] = [B]$, $2[\alpha] = 0$

5°. A smoothly imbedded front which is homologous to zero is cobordant to zero. Hence, on a connected orientable surface B, the intersection of the kernels of the homomorphisms ind: $S\omega\mathrm{Leg}(B) \to \mathbf{Z}$, $h: S\omega\mathrm{Leg}(B) \to H_1(B, \mathbf{Z})$ is trivial. Consequently, for an orientable connected surface $S\omega\,\mathrm{Leg}(B) \approx \mathbf{Z} \oplus H_1(B, \mathbf{Z})$.

6°. Let β_1 and β_2 be fronts with one positive cusp in the connected sum of two Möbius strips. It is proved in Fig. 13 that $[\beta_1] - [\beta_2] = [\alpha]$, where α is a two-sided imbedded curve.

7°. Thus, each front on a connected surface is cobordant to the sum of smoothly imbedded two-sided components, a certain number of bows, and possibly, a β-curve chosen once and for all. Whence it follows that the intersection of the kernels of the homomorphisms ind and h is always trivial.

In fact, each front from the intersection of kernels is cobordant to a null-homologous sum of smoothly imbedded two-sided components, and consequently is cobordant to zero.

From everything said follows the list of generators and relations in Table 1 for the group $S\omega$ Leg(B) for the projective plane, Möbius strip, and Klein bottle.

Upon connected summation of B with k tori, one adds $\mathbf{Z}^{2k}$ with the standard generators a_i, b_i.

From this it is easy to deduce the tables given in Sec. 1 of cylindrical and stable groups S Leg and P Leg.

E. Groups of Classes of Immersions. The cobordism groups of armed oriented immersions of curves can be calculated on the basis of reconstructions a and c of Fig. 10. These reconstructions do not affect intersections of the differently named components of the curve and allow one to replace each point of intersection of components of the same name by a figure eight of the same sign. The evenness of the number of points of intersection of components of each sign is invariant under reconstruction, and consequently under cobordisms of the theory $S\omega$ Imm.

Immersions in $\mathbf{R}^n$ form a group; the opposite of an immersion is constructed by reflection in a hyperplane. In particular, a figure eight of any sign is opposite to itself. Thus, $S\omega\,\mathrm{Imm}\mathbf{R}^2 \approx \mathbf{Z}_2^2$ (cf. [8, 9]).

In the case of an arbitrary connected surface B, it is useful to consider the oriented covering B'. An armed oriented curve lifts uniquely to B' (as a positive curve). The homology class of the lift in $H_1(B', \mathbf{Z})$ together with the evenness of the number of points of intersection (and in the case of an orientable surface, even more the evenness of the number of points of self-intersection of positive components) also determines the cobordism class. This means the subgroup of cobordism classes is a group: $S\omega\,\mathrm{Imm}\,(B) \approx H_1(B', \mathbf{Z}) \oplus \mathbf{Z}_2^{b_0+b_1}$. This leads to the tables of groups of classes of immersions from Sec. 1.

Supplement of June 30, 1980. Upon becoming acquainted with the manuscript of the present paper, Ya. M. Éliashberg got the following result.

We denote by $L = \oplus L_n$ the graded ring of cobordisms of oriented Legendre manifolds of spaces of 1-jets of Euclidean spaces.

THEOREM. $L_n \approx \pi_{2n}(T\lambda_n)$, where λ_n is the tautological n-dimensional vector bundle over the manifold of oriented Lagrange planes in $\mathbf{R}^{2n}$, T is the Thom space.

COROLLARY. $L \otimes \mathbf{Q}$ is an exterior algebra on generators of degree $4k+1$.

The group of cylindrical cobordisms with any bases, the group of cobordisms of fronts, and the group of Lagrange cobordisms are also expressed in homotopy terms.

LITERATURE CITED

1. V. I. Arnol'd, Ergodic Problems of Classical Mechanics, W. A. Benjamin (1968).
2. V. I. Arnol'd, "On a characteristic class appearing in a quantization condition," Funkts. Anal. Prilozhen., 1, No. 1, 1-14 (1967).

3. V. I. Arnol'd, "Wave fronts evolution and equivariant Morse lemma," Commun. Pure Appl. Math., 28, No. 6, 557–582 (1976).
4. V. P. Maslov, Theory of Perturbations and Asymptotic Methods [in Russian], Moscow State Univ. (1965).
5. D. B. Fuks, "On Maslov–Arnol'd characteristic classes," Dokl. Akad. Nauk SSSR, 179, No. 2, 303–306 (1968).
6. L. Hörmander, "Fourier integral operators. I," Matematika, 16, No. 2, 67–136 (1972), especially Example 3.34 (p. 128).
7. Ya. M. Éliashberg, "Estimate of the number of fixed points of transformations preserving area," Syktyvkar, 1978 (manuscript deposited in VINITI, IV, 374-70, 5A 491, 1979).
8. C. T. C. Wall, "Cobordism of pairs," Commun. Math. Helv., 35, 136–145 (1961).
9. R. Wells, "Cobordism groups of immersions," Topology, 5, 281–294 (1966).

SWEEPING A CAUSTIC BY A CASPIDAL EDGE OF A MOVING FRONT

V.I. ARNOLD

(Joint sessions of the Petrovskiĭ seminar on Differential equations and mathematical problems of physics and the Moscow Mathematical Society; the morning meeting on January 22, 1981, section 2.[1])

Conjecture. *There exists a theory of Lagrangian varieties (manifolds with singularities).*

Example. The surface Π defined below is swept by the caspidal edge of the moving front in the bifurcation A_4 in space-time $\mathbb{R}^4$. One can obtain Π in the same way as a family of curves in $\mathbb{R}^3$ dual to smooth curves with a bi-flattening decomposition.

Let $P = x^5 + \lambda_1 x^3 + \lambda_2 x^2 + \lambda_3 x + \lambda_4$, $\Sigma = \{\lambda \in \mathbb{R}^4 : \exists x : P' = P'' = 0\}$.

Theorem 1. *Any generic symplectic structure on $\mathbb{R}^4$ in a neighborhood of 0 can be reduced to the normal form $\mathrm{d}\lambda_1 \wedge \mathrm{d}\lambda_4 + \mathrm{d}\lambda_3 \wedge \mathrm{d}\lambda_2$ by a formal diffeomorphism preserving Σ.*

Let Q be a surface in $\mathbb{R}^3$ covered by a pencil of geodesics, let M^4 be the symplectic manifold of oriented straight lines in $\mathbb{R}^3$, and let $\tau : Q \to M$ associate to each point the tangent to the corresponding geodesic. Let the geodesic at 0 be doubly tangent to the asymptotic.

Theorem 2. *A generic germ τQ at 0 is formally diffeomorphic to the Lagrangian variety*

$$\Pi = \{\lambda \in \mathbb{R}^4 : \exists x : P = P' = P'' = 0\}.$$

Let $\mathcal{F}$ be a function in $\mathbb{R}^6$; $\omega = \sum \mathrm{d}p \wedge \mathrm{d}q$; $N^4 : \mathcal{F} = 0 = q_0$; at 0 one has $\mathcal{F} = v\mathcal{F} = v^2\mathcal{F} = v^3\mathcal{F} = 0$ $(v = \partial/\partial p_0)$.

Theorem 3. *Such a generic N is formally symplectically diffeomorphic to $p_0^4 + p_1 p_0^2 + q_2 p_0 + p_2 = 0 = q_0$ while preserving the plane $q_0 = 0$.*

Let $K \subset N : v\mathcal{F} = 0$; let $\pi : \mathbb{R}^5 \to \mathbb{R}^4$ be the p_0-forgetful map for $q_0 = 0$. Let γ be the preimage on K of an integral curve on πK of the direction field skew-orthogonal to the tangent planes to πK.

Theorem 4. *Along γ one has a root of p_0-multiplicity 3 for*

$$\Phi = p_0^5/5 + p_1 p_0^3/3 + q_2 p_0^2/2 + p_2 p_0 + q_1/2, \quad p_1 = \text{const}.$$

[1]Translation of Uspekhi Mat. Nauk, **36** (1981), no. 4, p. 233.

Originally publ. in: Usp. Mat. Nauk 36(4), 233,
Translated by Boris Khesin and Andrey Khesin
Springer-Verlag Berlin Heidelberg 2018 with kind permission

Singularities of Legendre varieties, of evolvents and of fronts at an obstacle

V. I. ARNOL'D

Department of Mathematics and Mechanics, Moscow State University, Lenin Hills 117234, *Moscow, USSR*

(*Received* 25 *May* 1982)

To the memory of V. M. Alexeyev

Abstract. The Legendre manifolds are the maximal integral submanifolds of a contact manifold. The shortest path problem on a manifold with boundary leads to Legendre varieties. We find normal forms of their generic singularities in terms of binary forms invariants theory.

The fronts are the hypersurfaces, equidistant from a given one. We consider fronts in a medium containing an obstacle (i.e. in a manifold with a boundary). For instance, if the obstacle is bounded by a plane curve, the fronts are its evolvents.

The Legendre manifold of a front is the set of all its tangent planes. A smooth front may acquire singularities while travelling through a smooth medium, but its Legendre manifold remains non-singular. At an obstacle, however, even the Legendre manifold may become singular. In this paper we classify generic singularities of Legendre varieties thus defined. We will describe below the normal forms for the singularities in terms of irreducible SL $(2, R)$-modules. More precisely, we define natural contact structures in the manifolds of odd degree 0-dimensional projective hypersurfaces and in some manifolds of polynomials in one variable. We will prove that the singularities of the Legendre varieties at generic obstacles are diffeomorphic to those of the varieties of polynomials admitting high multiplicity roots.

The simplicity of the final results is rather misleading; the polynomials, and even their degrees, are hidden. Even though they are known to exist it is still difficult to find them from geometrical considerations.

The relation of the obstacle problem to SL $(2, R)$-modules was discovered as a result of a series of works [**1–5**] featuring geometrical observations based on the resemblance of bifurcation diagrams occurring in different theories, strange cancellations of many terms in long calculations due to some properties of the varieties of polynomials with multiple roots, which seem to be new for the algebraists and new concepts in symplectic and contact geometry, namely the 'triads' of Guivental, describing families of rays and of fronts at obstacle points.

Originally publ. in: Ergodic Theory and Dynamical Systems 2(3-4), 301-309,

The future general theory of the singularities for Legendre varieties seems to be an extension of the classification of the triads presented below, rather than a theory of Legendre ideals or of the generating families hierarchy (both of which are drastically different from the triad classification).

1. *The obstacle problem*

Let Γ be a smooth hypersurface in Euclidean space R^n. Let us consider the length S of the shortest path from a fixed set to a variable end point, avoiding the obstacle bounded by Γ. The study of the singularities of S as a function of the end point leads to the following problem. Let us consider a family of geodesics on Γ orthogonal to some hypersurface on Γ. The straight lines, tangent to these geodesics, define an $(n-1)$-parameter family of rays in R^n, namely the family of all normals to some front hypersurfaces in R^n. The problem is to study the singularities of these fronts.

Example 1. Let the obstacle be bounded by a generic plane curve Γ ($n=2$). The fronts are the evolvents of Γ. They have singularities of order $3/2$ at generic points of Γ (Huyghens). A generic curve may have some inflection points. A calculation shows that the fronts have singularities of order $5/2$ at inflectional tangent points.

Example 2. Let the obstacle be bounded by a generic surface Γ in the 3-space ($n=3$). The fronts are surfaces with cuspidal edges. These edges are of order $3/2$ at generic points of Γ. Our one-parameter family of geodesics covers a domain in Γ. The geodesic direction may become asymptotic along some curve in this domain. The rays tangent to the geodesics at the points of this curve have asymptotic directions. The fronts' singularities at the points of an asymptotic ray are edges of order $5/2$ (see [**1**], [**2**]), unless the ray is 'bi-asymptotic' (this may happen at some points of our curve) in which case the singularity is more complicated. The main object of our study is the contact geometry of these complicated singularities and of their higher dimensional counterparts.

2. *The contact geometry*

We describe here some basic notions we need from contact geometry. For the proofs and for more details see [**6**], [**7**].

DEFINITION. *A contact structure on a manifold is a field of tangent hyperplanes (called contact planes), which is 'maximally non-integrable' that is, if the field of contact planes is locally defined as the zeros of an 1-form α, then the 2-form $d\alpha$ is non-degenerate on each contact plane.*

Example 1. A contact element with contact point P on a manifold is a linear hyperplane in the tangent space at P. Let us consider the manifold of all contact elements on V. We denote it by PT^*V (the projectivized cotangent bundle of V).

Let us define a field of tangent hyperplanes on PT^*V; the element velocity belongs to the field hyperplane if and only if its contact point velocity belongs to the element.

This field is a contact structure on PT^*V, called the *natural contact structure* of the contact elements bundle.

Example 2. Now let us consider a hypersurface in a contact manifold, transversal to the contact planes so that the intersections of tangent planes to the hypersurface with contact ones have codimension one in both. The $d\alpha$ form defines a symplectic structure in the contact plane $\alpha = 0$ which is unique up to a non-zero multiple, as is α itself. The $d\alpha$-skeworthocomplement to the intersection with the tangent space does not depend on this multiple; it is called the hypersurface *characteristic direction*.

The integral curves of the characteristic directions field are called the hypersurface *characteristics*. Thus a hypersurface in a contact space transversal to the contact planes is locally fibred in characteristics.

The base space of this fibration has a natural contact structure. This structure is given by the projections to the base along the characteristics of the intersections of the contact planes with tangent planes (use the Stokes formula).

Thus the characteristics space of a hypersurface in a contact manifold inherits its contact structure from the latter.

Example 3. Let us consider the pseudo-Euclidean space–time bundle of contact elements and the hypersurface formed by all light elements. The characteristics space is well defined globally in this case. One may identify it with the bundle of all transversally oriented contact elements over the physical space (the isochrone).

Indeed, a characteristic is defined by a light velocity motion. So it is defined by its point of intersection with one of the isochrones (say, $t = 0$) and by its physical space velocity vector. The length of the latter vector is fixed, and the vector is defined by the isochrone contact element transversally oriented, orthogonal to this vector.

PROPOSITION. *The identification thus defined identifies the contact structure of the space of characteristics (inherited from the space–time contact elements bundle) with the natural contact structure of the isochrone contact elements bundle.*

Proof. The space–time contact element at an isochrone's point is generated by the corresponding isochrone's contact element and by the direction of the projection of the characteristics line in the space–time. The isochrone's contact element velocity lies in the contact plane of the contact structure inherited from the space–time's contact structure, if and only if its contact point moves along the isochrone, being tangent to the contact element in the space–time. This occurs exactly when the contact point velocity belongs to a contact element of the isochrone. This means that both contact structures coincide.

Remark. The same reasoning works in a more general setting; we may substitute for the space–time isochrone any hypersurface in a manifold M, transversal to the projections of the characteristics of a fixed hypersurface in its contact elements bundle PT^*M (for instance the result is still true in the pseudo-Riemannian case); we also may substitute for PT^*M the bundle of transversally oriented contact elements (the spheres bundle, ST^*M).

DEFINITION. *A Legendre manifold in a contact manifold is an integral manifold of the contact planes field having maximal dimension (that is dimension $n-1$ in a contact manifold of dimension $2n-1$).*

Example. The set of all contact elements, tangent to a given submanifold (of arbitrary dimension) is a Legendre submanifold of the contact elements bundle of the ambient manifold.

For instance, fibres of the projective cotangent bundle are Legendre manifolds. The Legendre fibrations are fibrations of contact manifolds into Legendre fibres.

The *front* variety of a Legendre submanifold in a contact manifold with a Legendre fibration is its image under the projection to the base space of the fibration. Isotopies of a Legendre manifold generate motions and even metamorphoses of its front variety. These singularities of the fronts come from the projection. They are present even on the fronts travelling through smooth media. But if the medium contains an obstacle the Legendre submanifold itself may acquire singularities.

Example. Let us assume that the obstacle is bounded by a plane curve with a simple inflection point. The corresponding front (the curve evolvent) acquires a singularity of order $5/2$. Thus the corresponding variety of plane contact elements, i.e. the Legendre variety, consisting of all plane contact elements tangent to the front, is singular with a singularity of order $3/2$. In other words, the Legendre variety is locally diffeomorphic to a semicubical parabola in the 3-space.

3. *The contact triads*

The obstacle problem leads to the following definition motivated by the examples and theorems below.

DEFINITION. *A contact triad* (H, L, l) *consists of*

(1) *a hypersurface* H *in a contact manifold transversal to the contact planes*;

(2) *a Legendre manifold* L *in the same contact manifold*;

(3) *a smooth hypersurface* l *in* L,

such that the hypersurface H *is tangent to the Legendre manifold* L *with first order of tangency at every point of* l.

We shall study the germ of a triad at a point 0 of l.

DEFINITION. *The Legendre variety*, generated by the triad (H, L, l) *at* 0, *is the image of the germ of* l *at* 0 *by the projection of* H *onto its space of characteristics.*

Example 1. Let us extend the Euclidean space R^n to the space–time $R^n \times R = \{q, t\}$. We call a space–time contact element *non-vertical*, if it may be defined by $dt = p\,dq$, and we call it a *light* contact element, if $p^2 = 1$. All light contact elements form a hypersurface H in the bundle of contact elements over the space–time. This hypersurface H is the first element in a triad defined by the family of geodesics on a hypersurface $\Gamma \subset R^n$, consisting of all geodesics, normal to a surface of codimension 1 in Γ.

Let $s: \Gamma \to R$ be the time function, defining the family of geodesics on Γ (the geodesics are orthogonal to the surfaces $s = \text{const}$, and $(\nabla s)^2 = 1$ on Γ). The graph of s is a codimension-2 submanifold of space–time. Let us consider the set of all space–time contact elements, tangent to this graph. This set is a Legendre manifold, and it will serve as L of the triad.

PROPOSITION 1. *The hypersurface H is (first order) tangent to the Legendre manifold L along a submanifold l of codimension 1 in L, consisting of all contact elements, belonging to L, which contain the normal to $\Gamma \times t$ in $R^n \times t$.*

Proof. The elements of L at a point of the graph form a linear (one-parameter) family. Let us consider an element of L defined by

$$dt = p\,dx.$$

By the definition of L, the restriction of pdx to $T_x\Gamma$ is ds. But

$$(\nabla s)^2 = 1.$$

Hence the function p^2, restricted to the above-defined one-parameter family, has a non-degenerate minimum, equal to 1; it is attained for such a p, that

$$p\,dx = 0$$

for all the vectors normal to Γ in R^n.

Proposition 1 is proved: (H, L, l) is a triad. □

PROPOSITION 2. *The Legendre variety generated by the triad constructed above, is the Legendre variety formed by those contact elements of R^n which are tangent to the same front for the obstacle problem with boundary Γ.*

Proof. The manifold of H-characteristics was identified in § 2 with the bundle of transversally oriented contact elements of an isochrone $t = c$, that is with ST^*R^n. The characteristics intersecting l are identified with those elements, normal to the tangent rays of geodesics of the family on Γ, which correspond to the isochrone $t = c$. An element corresponds to $t = c$, if and only if the oriented distance along the ray, from the point x of contact of the ray with a geodesic to the contact point of the element, is equal to $c - s(x)$. These elements are precisely all elements tangent to the level variety $t = c$ of the (multi-valued) time function in R^n, defined by the family of geodesics on the obstacle surface. This proves proposition 2. □

We will now construct a series of triads examples, providing normal forms for the germs of generic triads at all their points. This implies, for instance, normal forms of singularities of the Legendre varieties, consisting of all contact elements tangent to a front for a generic obstacle in an Euclidean or Riemannian space.

4. *Contact structures on polynomials spaces*

Let $V = \{a_0x^d + \cdots + a_dy^d\}$ be the linear space of all real binary forms of degree d. The group SL $(2, R)$ of linear unimodular transformations of the (x, y)-plane acts on V. If d is odd, the dimension of V, which is equal to $d + 1$, is even and (as A. B. Guivental has remarked) V carries a linear symplectic structure (a non-degenerate skew symmetrical bilinear form), invariant under the SL $(2, R)$ action; this form Ω is unique up to a non-zero multiple. The form Ω defines on V a differential 2-form ω with 'constant coefficients' which is a symplectic structure.

This structure may be written as $\sum dp_i \wedge dq_i$ using suitable 'Darboux coordinates' in which the binary form $\varphi \in V$ with coordinates (p_i, q_i) is defined by

$$\varphi(x, 1) = q_1 \frac{x^d}{d!} + \cdots + q_n \frac{x^n}{n!} - p_n \frac{x^{n-1}}{(n-1)!} + \cdots + (-1)^n p_1,$$

where $d = 2n - 1$ and the signs before the p_k's alternate.

Let $v = \sum a_i \, \partial/\partial a_i$ be the Euler vector field on V. It is $\mathrm{SL}(2, \boldsymbol{R})$-invariant (every linear transformation sends v to v). The interior product of v with ω defines on V an $\mathrm{SL}(2, \boldsymbol{R})$-invariant differential 1-form $i_v\omega$. This 1-form is equal to zero on v. Hence it defines an $\mathrm{SL}(2, \boldsymbol{R})$-invariant hyperplane field in the projective space $\boldsymbol{RP}^d$ of dimension 0 hypersurfaces of degree $d = 2n - 1$. We obtain:

Proposition 1. *The hyperplane field defined above is an* $\mathrm{SL}(2, \boldsymbol{R})$*-invariant contact structure on* $\boldsymbol{RP}^d$

For instance, in the domain consisting of those polynomials $\varphi(x, 1)$ with $q_1 = 1$;

$$\{f: f(x) = \frac{x^{2n-1}}{(2n-1)!} + q_2 \frac{x^{2n-2}}{(2n-2)!} + \cdots + q_n \frac{x^n}{n!} - p_n \frac{x^{n-1}}{(n-1)!} + \cdots \pm p_1 \qquad (*)$$

the contact structure on the projective space is represented by the following 1-form

$$\alpha = p' dq' - q' dp' - dp_1,$$

where $q' = (q_2, \ldots, q_n)$, $p' = (p_2, \ldots, p_n)$.

Let us consider a one-parameter group of contact diffeomorphisms. Points, where the velocity field vector lies in the contact plane, form a hypersurface given by the equation $\alpha | \xi = 0$ if the velocity field is ξ and the contact structure is $\alpha = 0$. We will call this hypersurface the *neutral surface* of the group. Let us suppose, that the neutral hypersurface is transversal to the contact planes.

Proposition 2. *The orbits of the group on the neutral surface are its characteristics.*

Proof. By the homotopy formula,

$$i_\xi d\alpha + d i_\xi \alpha = L_\xi \alpha.$$

The field ξ being contact,

$$L_\xi \alpha = 0 \quad \text{if } \alpha = 0.$$

On the neutral surface

$$i_\xi \alpha = 0.$$

Hence

$$d i_\xi \alpha = 0$$

on its tangent vectors. Thus

$$i_\xi d\alpha = 0$$

on the intersection of the contact plane with the tangent plane to the neutral surface, as required.

The group of translations of polynomials along the x-axis acts on the space of polynomials of the form (*), and preserves its contact structure:

PROPOSITION 3. *The neutral surface is defined by* $K=0$, *where*

$$K=p_2+p_3q_2+\cdots+p_nq_{n-1}+q_n^2/2.$$

Indeed, the velocity field is

$$\xi=\sum q_{k-1}\partial/\partial q_k-p_{k+1}\partial/\partial p_k \quad (p_{n+1}=q_n, q_1=1).$$

PROPOSITION 4. *The triple* $H(K=0)$; $L(p=0)$, $l(p=q_n=0)$ *is a contact triad.*

This follows directly from the formula for K.

DFINITION. *We call the* standard variety of dimension m *and denote* Σ^m *the Legendre variety generated by the triad of proposition* 4 (*here* $m=n-2$).

COROLLARY. *The standard Legendre variety* Σ^{n-2} *consists of those polynomials of the form*:

$$\frac{x^{2n-1}}{(2n-1)!}+q_3\frac{x^{2n-3}}{(2n-3)!}+\cdots+q_n\frac{x^n}{n!}-p_n\frac{x^{n-1}}{(n-1)!}+\cdots+(-1)^np_1 \qquad (**)$$

where p_2 *means* $-(p_4q_3+\cdots+p_nq_{n-1}+q_n^2/2)$, *which have a root of multiplicity greater than* n; *the contact structure is defined by the* 1-*form*

$$\alpha=p''dq''-q''dp''-dp_1,$$

where $q''=(q_3,\ldots,q_n)$, $p''=(p_3,\ldots,p_n)$.

From this corollary and from [5] follows:

COROLLARY. *The standard Legendre variety* Σ^m *is diffeomorphic to the open swallow-tail of dimension* m, *that is to the variety of all polynomials*:

$$x^{2m+1}+A_1x^{2m-1}+\cdots+A_{2m} \qquad (***)$$

which have a root of multiplicity at least $m+1$.

Example. Let $n=3$. Formula (**) implies

$$f=\frac{x^5}{120}+q_3\frac{x^3}{6}-p_3\frac{x^2}{2}-\frac{q_3^2}{2}x-p_1,$$

$$\alpha=p_3\,dq_3-q_3\,dp_3-dp_1.$$

The polynomials which have roots of multiplicity greater than 3 form a Legendre curve

$$\Sigma^1=\{f=(x-t)^4(x+4t)/120\},$$

that is

$$q_3=-t^2/2, \qquad p_3=-t^3/3, \qquad p_1=-t^5/30.$$

This curve is diffeomorphic to a cubic parabola. Its singularity is responsible for the singulariites of order 5/2 of evolvents at the inflections tangent.

Remark. One sees from (**), that the variety Σ^{n-2} is quasi-homogeneous, for the coordinates weights

$$\deg q_i = i-1, \quad \deg p_i = 2n-i$$
$$(\deg K = 2n-2, \quad \deg \alpha = 2n-1).$$

The standard variety Σ^{n-2} is the image of a space with coordinates of weights $(1, \ldots, n-2)$ under a quasi-homogeneous mapping.

Example. For $n=4$ the coordinates $(q_3, q_4, p_4, p_3, p_1)$ have weights $(2, 3, 4, 5, 7)$. The Legendre variety Σ^2 is the image of a plane (t, r), where $\deg t = 1$, $\deg r = 2$:

$$\Sigma^2 = \{f = (x^2 + 5tx + r)(x-t)^5/7!\}.$$

5. *Contact triads normal forms*

Let us consider the space R^{2k+1} equipped with the contact structure $\alpha = 0$,

$$\alpha = p'\,dq' - q'\,dp' - dp_1,$$

where $p' = (p_2, \ldots, p_k)$; $q' = (q_2, \ldots q_k)$.

DEFINITION. *The standard contact triad* $\tau_{k,n}$ $(n \le k)$ *is the triad* (H, L, l) *in this space, defined by the equations*

$$H: \quad p_2 + p_3 q_2 + \cdots + p_n q_{n-1} + q_n^2/2 = 0;$$
$$L: \quad p = 0; \qquad l: \quad p = 0, q_n = 0.$$

Remark. The triad $\tau_{k,n}$ is a trivial extension (by addition of new p's and q's) of $\tau_{n,n}$, the triad studied in § 4.

THEOREM 1. *The germs at* 0 *of the standard triads* $\tau_{k,n}$ *are stable, as germs of contact triads considered up to the contact equivalence.*

THEOREM 2. *The germs of generic contact triad at all points are contact equivalent to the germs of standard triads at* 0.

COROLLARY. *The germs of the Legendre variety, formed by the contact elements tangent to a front in the generic obstacle problem (see* §§ 3 *and* 4) *are contact stable and contact equivalent to the germs of standard Legendre varieties (may be multiplied by the linear spaces) — and hence are diffeomorphic to the germs of open swallowtails.*

Proof. We need a simple lemma:

LEMMA 1. *In a neighbourhood of a point, where a hypersurface in a contact space is transversal to the contact plane, one may choose an equation* $\alpha = 0$ *for the contact planes in such a way, that the kernel of the* 2*-form* $d\alpha$ *is tangent to the hypersurface.*

Proof. One may reduce the equation of the hypersurface germ to the form $P_2 = 0$, the contact structure being $P'\,dQ' - dP_1$. This follows, for instance, from the Guivental–Darboux theorem [**2**], which says that in contact geometry there is no extrinsic curvatures: the traces of the contact planes on the germ of a submanifold define the germ up to a contact equivalence, and define the restriction of a contact form up to a contact form preserving equivalence.

We apply the lemma to the H of a contact triad. The field of kernels of $d\alpha$ defines (locally) a fibration of the contact space into curves (in terms of the above coordinates: $(P', Q', P_1) \mapsto (P', Q')$). The base B of this fibration π has a natural symplectic structure defined by the projection of the 2-form $d\alpha$ along its kernels' integral curves. The hypersurface H is fibred into these integral curves. Hence this hypersurface is projected onto a hypersurface πH in B.

The kernel of the 2-form $d\alpha$ is not tangent to the Legendre manifold L, because its tangent plane belongs to the contact plane $\alpha = 0$, which is transversal to the kernel of $d\alpha$ by the definition of a contact structure.

Hence the projection of the triad (H, L, l) to the symplectic manifold B defined above is the triple $(\pi H, \pi L, \pi l)$, where H is a hypersurface in B, first order tangent to the submanifold πL along its hypersurface πl. Note, that germs of L and of l are projected diffeomorphically.

πL is a Lagrange manifold because L is a Legendre manifold. Thus the triple is a symplectic triad in B and the symplectic case of our theorem is known (A. B. Guivental). Thus we obtain normal forms of the projections of our triads to the (P', Q') space.

The normal form for a contact triad follows because $P_1 = \int P' \, dQ'$ (integrals along paths on the Lagrange manifold πL, starting at 0). So we have derived the classification of the contact triads from that of symplectic triads.

REFERENCES

[1] O. A. Platonova. Singularities in the shortest way problem at an obstacle. *Funct. Anal.* **15** (1981), 86–87.

[2] V. I Arnol'd. *Catastrophes theory*. Znanie: Moscow, 1981. (In Russian.)

[3] V. I. Arnol'd. Evolution of singularities of potential flows in collision free media and caustics metamorphosae in 3-space. *Trudy semin. im. I. G. Petrovskogo* **8** (1982), 21–57. (In Russian.)

[4] V. I. Arnol'd. Lagrange varieties, asymptotical rays and the open swallowtail. *Funct. Anal.* **15** (1981), 1–14.

[5] A. B. Guivental. Varieties of polynomials having a prescribed multiplicity root and the generalized Newton equation. *Funct. Anal.* **16** (1982), 8–13.

[6] V. I. Arnol'd. *Mathematical Methods of Classical Mechanics*. Springer: New York, 1978.

[7] V. I. Arnol'd, A. N. Varchenko & S. M. Gusein-Zade. *Singularities of Differentiable Mappings*. Nauka: Moscow, 1982. (In Russian.)

ELEMENTS OF THE LARGE-SCALE STRUCTURE OF THE UNIVERSE

V.I. ARNOLD, YA.B. ZELDOVICH, S.F. SHANDARIN

(The Petrovskiĭ seminar on Differential equations and mathematical problems of physics; meeting on November 26, 1980.[1])

The simplest model of a collisionless medium in the absence of forces is described by the motion $x \mapsto x + v(x)t$, where v is the initial velocity field. As t increases the trajectories of the particles begin to intersect and caustics are formed, where the medium density becomes infinite. The problem of an increase in perturbations turns out to be equivalent, where v is potential [1], [2].

According to the theory of the hot Universe at an early stage of expansion, when the size of the Universe was roughly 1000 times smaller than at present, the distribution of matter was practically homogeneous ($\delta\rho/\rho \sim 10^{-3}$). At present there are great inhomogeneities (stars, galaxies, ...) of size less than 1/30 of the size of the domain of the Universe that is observable in principle. There are no inhomogeneities of a larger scale. The formation of inhomogeneities is related to gravitational instability.

As a model of the formation of inhomogeneities of the greatest size (superclusters of galaxies) we suggest the formation of caustics of potential flows of collisionless particles. For the Universe, whose main mass is contained in protons and neutrons, the model is applicable up to the formation of density singularities. On the other hand, if the main mass is contained in massive neutrinos (which has been indicated by recent experiments) then the model that includes the intersection of flows is applicable to later stages as well.

From the mathematical point of view, it is a question of singularities of a one-parameter family of Lagrangian mappings. (Consequently, in the motion of a collisionless medium in any potential field, for example, in its own gravitational field, the singularities are the same as for motion by inertia, provided that the initial velocity field is potential.)

The critical values of the family form a big caustic in the space-time, its sections by the isochrones being instantaneous caustics. Their singularities have been classified in [3], [4] (the types $A_3, \dots, D_5$) and depicted in [5]. The ratio of the mass lying in an ε-neighborhood of the point to the volume of that neighborhood will be called the averaged density. The asymptotics of the averaged density is of the order of $\varepsilon^{-2/3}$ for A_3, $\varepsilon^{-3/4}$ for A_4, $\varepsilon^{-4/5}$ for A_5, $1/\varepsilon$ for D_4, and $\ln\varepsilon/\varepsilon$ for D_5. The family of instantaneous caustics in the conventional space is described as the family of

[1]Translation of Uspekhi Mat. Nauk, **36** (1981), no. 3, 244–245.

Originally publ. in: Usp. Mat. Nauk 36(3), 244-245,
Translated by Gerald Gould and Boris Khesin
Springer-Verlag Berlin Heidelberg 2018 with kind permission

projections, along the world lines from the space-time into the conventional space, of the sections of the big caustic by the isochrones. The theorem on the family of cuspidal edges of caustics also describes the motion of a cuspidal edge of a wave front along a caustic in optics.

Theorem A_4. *Let Σ be the variety of all polynomials $x^5+\lambda_1x^3+\lambda_2x^2+\lambda_3x+\lambda_4$ having a root of multiplicity $\geqslant 3$. Let $t(\lambda)$ be a generic function in a neighborhood of $\lambda=0$. Then the restriction of t to Σ can be reduced in a neighborhood of the point 0 to the normal form $t'=\pm\lambda'_1\pm\lambda'_4+c$ by formal[2] diffeomorphisms $\lambda'(\lambda)$, $t'(t)$ preserving Σ and the field of directions of the world lines $\partial/\partial\lambda_4$.*

The following theorem describes sweeping the space by moving caustics.

Theorem A_3. *Let Σ be the surface $\lambda_2^2=\lambda_1^3$ in $\mathbb{C}^n$ and $t(\lambda)$ a function whose restriction to the cuspidal edge $\lambda_1=\lambda_2=0$ has a critical point at 0 and is otherwise generic. Then the restriction of t to Σ can be reduced in a neighborhood of the point 0 to the normal form $t'=\pm\lambda'_1\pm\lambda'_2\pm{\lambda'_3}^2\pm\cdots\pm{\lambda'_n}^2+c$ by formal[2] diffeomorphisms $\lambda'(\lambda)$, $t'(t)$ keeping Σ and the field of directions of the world lines $\partial/\partial\lambda_2$.*

Definition. The *bicaustic* is the projection of the variety of singularities A_3 of the big caustic along the world lines (in $\mathbb{R}^3$ this is the surface swept out by the cuspidal edges of a moving caustic).

Singularities of generic bicaustics in $\mathbb{R}^3$. A_3: $x^2y=z^2$ (umbrella); A_4: $u^2=v^3$, $u^2=v^3w^2$, $u^2w=v^3w$; A_5: swallowtail; D_4: $z(z-y^2)(z-Ay^2)=0$, $A=a+x$ or $a\pm x^2$; D_5.

The following theorem describes sweeping the umbrella by the cuspidal edges of caustics.

Theorem. *Let $t(u,v)$ be a generic function in a neighborhood of the point $u=v=0$ and let $\varphi(u,v)=(u,v^2,u^2v)$. Then in a neighborhood of the point 0 one can reduce t to the normal form $t=v+a(v^2)$ by a local diffeomorphism of the (u,v)-plane preserving φ ($\varphi\circ g_2=g_3\circ\varphi$).*

References

[1] Ya.B. Zeldovich. Disintegration of homogeneous matter into parts under the action of gravity. *Astrofizika* (a journal of the Armenian SSR Academy of Sciences), 6 (1970), no. 2, 319–335.

[2] A.G. Doroshkevich, V.S. Ryabenkiĭ, S.F. Shandarin. Nonlinear theory of the development of potential perturbations. *Astrofizika* (a journal of the Armenian SSR Academy of Sciences), 9 (1973), no. 2, 257–272.

[3] V.I. Arnold. Normal forms of functions near degenerate critical points, the Weyl groups A_k, D_k, E_k, and Lagrangian singularities. *Funktsional. Anal. i Prilozhen.*, 6 (1972), no. 4, 3–25. English translation: *Funct. Anal. Appl.*, 6 (1972), no. 4, 254–272.

[4] V.I. Arnold. Wave front evolution and equivariant Morse lemma. *Commun. Pure Appl. Math.*, 29 (1976), no. 6, 557–582. Correction: *Commun. Pure Appl. Math.*, 30 (1977), no. 6, 823.

[5] V.I. Arnold. Catastrophe theory. *Priroda*, 1979, no. 10, 54–63. English translation: V.I. Arnold. Collected Works. Volume III. Singularity Theory (1972–1979). Springer, Berlin, 2016, 441–456.

[2] As a rule, the transformation series diverge, even if the function is holomorphic.

The Large Scale Structure of the Universe I. General Properties. One- and Two-Dimensional Models

V. I. ARNOLD
Moscow State University, U.S.S.R.

and

S. F. SHANDARIN and YA. B. ZELDOVICH
Institute of Applied Mathematics, Moscow, U.S.S.R.

(*Received August* 11, 1981)

Evolution of initially smooth perturbations in a cold self-gravitating medium in a Friedmann Universe gives rise to the formation of singularities in the distribution of density in a manner similar to that of catastrophe theory. Using the approximate nonlinear theory of gravitational instability the objects formed were found to possess a very oblate shape—a "pancake" (Zeldovich, 1970). This result is shown in this paper to be a general feature of the evolution of so-called Lagrangian systems (Arnold, 1980). Pancakes are one of the several kinds of generic singularity formed at the nonlinear stage of evolution of such a system. Some of the others are a cusp, a beak-to-beak, a swallow-tail. In this paper we present the full list of singularities for the one- and two-dimensional cases (Figures 1–9). The three dimensional singularities will be discussed in Part II of the paper. We discuss the geometrical and some dynamical properties of each kind of singularity. We give also asymptotic laws for the growth of the density near each kind of singularity. This list of singularities gives the elements from which the large scale structure of the Universe is constructed.

Our scheme naturally explains the existence of the flattened superclusters, as well as the rich clusters of galaxies by themselves and their chains. It is useful to note that the process of formation of smaller objects (galaxies) must be considered against the background of structures described in this paper.

1. INTRODUCTION

There are indications, observational and theoretical, suggesting a rather unique and definite scenario for the build up of the structure of the Universe. The scenario begins with a phase of small perturbations superimposed on the regular isotropic expansion of the uniform Friedmann Universe. In an earlier discussion of the problem of a cosmological

Originally publ. as: Preprint No. 100, Keldysh Inst. Prikl. Mat. (Keldysh Inst. Appl. Math.), USSR Acad. Sciences, Moscow, 32 pp., © 1981 Russ. Acad. Sciences
English transl.: Geophys. Astrophys. Fluid Dynamics 20(1-2), 111-130, © 1982 Taylor and Francis.

singularity in the General Relativity Lifshitz, Khalatnikov and Sudakov (see the review by Lifshitz and Khalatnikov (1963) and references therein) found that a coordinate singularity arises in the time orthogonal frame of reference and, what is more, this is due to one of the eigen values of the tensor g_{ik}. They explained this singularity as a caustic and have pointed out that, if the medium is purely pressureless, what seems to be a non-realistic singularity in the density distribution arises.

In an earlier paper one of the authors (Zeldovich, 1970) showed that the evolution of adiabatic perturbations gives rise to such a relation between the density perturbation spectrum after decoupling and the gas temperature when the pressure is negligible. An approximate theory for the evolution of perturbations predicting the formation of very oblate objects of high density was developed in the framework of Newtonian theory. There the subsequent evolution of the singularity was considered. However at present it is clear that this singularity is only one of several generic singularities.

A very general mathematical analysis of the motion of sets of particles and the classification of all possible types of singularity was carried out by another of the authors (Arnold, 1972).

The theoretical analysis and numerical calculations made by the third author (Shandarin) and his collaborators (Doroshkevich *et al.*, 1980b) confirm the applicability of the approximate theory, including the formation of singularities and points, to specific patterns formed in the subsequent evolution.

The actual astronomical observations and their theoretical discussion proceed on two distinctly different lines. One course, due mainly to Peebles and his collaborators, consists of the calculation of correlation functions as the most general characteristic of the spatial distribution of galaxies. It seems to be most "impartial" and done without any prejudices i.e. without any previous assumptions about the scenario. The most important results are: (i) the determination of the characteristic length and mass dividing large ($\xi > 1$) and small ($\xi < 1$) correlations and (ii) the power law for correlation dependence on distance: $\xi \propto r^{-n}$. The first result gives—at least by order of magnitude—the characteristic scale of perturbations. Its value confirms the possibility of the pressureless description mentioned above.

However, concerning the second result, the universal slope ($n \approx 1.8$) of the correlation function for the two regions [large and small amplitude perturbations ($\xi > 1$ and $\xi < 1$) corresponding to small and large distance scales ($r < 5h^{-1}$ and $r > 5h^{-1}$ Mpc)] seems to be devoid of any direct physical meaning. In the large scale domain the linear growth of small amplitude perturbations is involved. In the small scale domain, where the correlation function is large, one has to deal with gravitationally bound objects. Quite different factors could be involved in the formation of the density distribution and correlation function on small scales: cooling-rate, the problem how to get the required rotation of galaxies etc. The equality of *n* in the two regions is a miracle?!

Another line of observational research, influenced partly by the pancake theory, investigates the particular patterns in the large structure. The detailed study of the three-dimensional distribution of galaxies has produced convincing arguments favouring the existence of sheets and chains of galaxies and clusters of galaxies (de Vaucouleurs, 1978; Chincarini and Rood, 1976, 1979; Joeveer *et al.*, 1978; Einasto *et al.*, 1980; Tarenghi *et al.*, 1979, 1980; Gregory and Thompson, 1978; Tully and Fisher, 1978; Gregory *et al.*, 1980). Another piece of evidence came from Oort (1981) who proposed the existence of flat gas clouds in order to explain the numerous absorption lines in the spectra of distant quasars.

Given so much theoretical and observational work concerning the structure of the Universe, it seems necessary to make more precise statements about the typical patterns and singularities which arise due to the growth of perturbations and their astrophysical counterparts.

Some technical remarks are needed before we proceed to a description of the singularities.

The initial pattern of density perturbations and particular velocities could be of an arbitrary type at the moment of decoupling. In principle it is possible that the velocity field possessed a vorticity. However the vorticity decreased during the expansion of the Universe, so it was probably not significant in the formation of the structure. In addition it causes non-Friedmann expansion at a stage of nucleosynthesis.

On the contrary, arbitrary perturbations of the density and velocity field of a potential type contain both decreasing and increasing modes. Observations unambiguously show that, at the decoupling, perturbations were small.† Therefore the time period, when the increasing mode was growing and the decreasing mode was diminishing, must have been long. Thus, without loss of generality, one can consider the increasing mode only.

Here we are assuming only primordial perturbations of the so-called adiabatic type. Generally speaking, there could exist isothermal primordial perturbations. However, an assumption about isothermal initial conditions results in a quite different scenario of structure formation (Gott and Rees, 1975; Jones, 1976; Rees, 1978; Peebles, 1980). Giving no detailed

†Thus the adiabatic perturbations are consistent with an almost Friedmann Universe at the stage of nucleosynthesis.

comparison of the two scenarios, we point out only that present developments in particle physics favour primordial perturbations of the adiabatic type (Zeldovich, 1978; Weinberg, 1979; Turner and Schramm, 1979; Press, 1980). The formation of the structure in the neutrino dominated Universe is very similar to the adiabatic scenario in the baryon dominated Universe. During the time perturbations grow, the kinetic energy and pressure of the massive neutrino gas are negligible (Doroshkevich *et al.* 1980a).

Thus it is specific to the adiabatic scenario that, just after decoupling at $z \sim 10^3$, the spectrum of inhomogeneities did not contain short waves. Therefore the distribution of density as well as the velocity field were smooth at that time.

As was mentioned above, the growth of perturbations under the conditions in question results in the formation of pancakes, so the adiabatic theory of galaxy formation is often referred to as the pancake theory. At the non-linear stage when pancakes are formed, there are singularities in the distribution of density and therefore the spectrum contains short waves. However, the standard spectrum approach is useless at this stage because the phases are highly correlated.

In the ordinary nucleon dominated Universe the creation of a pancake implies the formation of shock fronts which are the boundaries of the pancake. The gas involved in the pancakes is heated, ionized and compressed by these shock waves. The fragmentation of pancakes into galaxies is shown to be rather complicated and involves many physical processes like cooling, thermal and hydrodynamical instabilities, turbulence etc.: (Sunyaev and Zeldovich, 1972; Doroshkevich and Shandarin, 1974; Doroshkevich, Shandarin and Saar, 1978).

However, pure collisionless particles are found to form pancake-like structures as well as ordinary gas. There is of course a great difference in the inner structure of the pancakes in these models. For instance, in a collisionless medium shock waves are not formed, but a surface of infinite density instead—caustics are the boundaries of the pancakes.

It is important to note that the mass of each singularity is infinitesimal. The density reaches infinity only in the case of a cold ($T=0$) continuous medium. Discreteness and velocity dispersion of particles remove singularities. However, at low temperature and small initial pressure density contrasts are large.

In this paper we give a general description of the formation of the large scale structure of the Universe in the adiabatic scenario. The observed structure is assumed to be described as a process of the formation of singularities (not only pancakes!) in the medium, interacting due to gravitation only. We present the description of all kinds of singularities arising in one- and two-dimensional models provided no special kind of symmetry is imposed on the initial conditions. We do not use the so-called standard forms (Arnold, 1972) which produce too refined patterns but instead operate in the frame of the approximate solution for gravitational instability, that in general gives the same patterns but having a more real appearance. For every kind of singularity we provide the asymptotic law of density growth.

The three dimensional case is now under way and will be published in Part II of the paper together with further astronomical discussion.

2. FORMATION OF THE STRUCTURE AND LAGRANGIAN SINGULARITIES. GENERAL PROPERTIES

The increasing mode of perturbations is well described by the approximate solution (Zeldovich, 1970)

$$R_i(\mathbf{q}, t) = a(t)q_i - b(t)\partial S(\mathbf{q})/\partial q_i, \tag{1}$$

where R_i are Eulerian coordinates, at time t, of the particle with Lagrangian coordinates q_i; $a(t)$ is a scale factor of the Friedmann Universe and $b(t)$ is a known function of time only, describing the evolution of the increasing mode of perturbations, and $\partial S/\partial q_i$ is a vector field determined by the initial perturbations of the density and the velocity field. At the linear stage

$$(\delta\rho/\rho)_{\mathrm{incr}} \propto \partial^2 S/\partial q_i^2. \tag{2}$$

It is easy to show that the motion in accordance with (1) is potential.†

In mathematical language we consider this process as a one parameter family of mappings of Lagrangian space (corresponding to the so-called unperturbed state) into an Eulerian one, parameterized by time. These mappings are Lagrangian (see Arnold (1980) for the definition and discussion of properties of Lagrangian maps). Thus we can use the classification of the Lagrangian mappings' singularities (Arnold, 1972) and the metamorphoses of these singularities (Arnold, 1976). Singularities are categorized in these mathematical works in classes A_k, D_k and so on; we adopt this notation here for the singularities and for their metamorphoses (in which case the notation is that of the corresponding Lagrangian singularities of the mapping of Lagrangian space-time to the Eulerian one). The Lagrangian mappings' singularities metamorphoses theory is a generalisation of the theory by Zeldovich (1970).

†One can speak about potential motion until multi-stream flows occur. In gasdynamics shock waves violate potential motion.

Equation (1) describes the motion of any particle with rather high precision up to the intersection of its trajectory with that of another particle. Besides being the exact solution of the problem in the one dimensional case, it provides a good approximation also in the general case up to the intersection of trajectories (Zeldovich, 1970; Doroshkevich, Ryabenki and Shandarin, 1973).

In fact in this paper we shall use the approximate solution (1) but the results concerning the structure of the singularities are more general than the approximate solution (Arnold, 1980). Indeed, for their validity one needs only motion of a potential type. For instance, the same kinds of singularities arise in a collisionless, medium of noninteracting particles each moving at a constant speed. (In the latter model non-expanding space is considered because otherwise the perturbations do not grow enough if one starts from small perturbations no singularities will develop.)

It is convenient to transform **R** into proper coordinates and introduce a new time θ, i.e.

$$\mathbf{r}=\mathbf{R}/a(t), \qquad \theta=b(t)/a(t), \tag{3}$$

so that (1) becomes

$$r_i=q_i-\theta\,\partial S/\partial q_i. \tag{4}$$

This confirms the previous statement and one may interpret $\partial S/\partial q_i$ as the velocity of the particle marked by the Lagrangian coordinate q_i. However, one should remember that in the expanding Universe small inhomogeneities grow due to gravitational attraction only.

Knowing the law of motion (4) for every particle, it is easy to calculate the density at any time:

$$\rho(\theta,\mathbf{q})=|\delta_{ik}-\theta\;(\partial^2 S/\partial q_i\partial q_k)|^{-1}. \tag{5}$$

Here the density ρ is supposed to be expressed in terms of the mean density. The tensor $\partial^2 S/\partial q_i\partial q_k$ is symmetric and can be reduced to the principle axes at any point. If the eigenvalues of the tensor $\partial^2 S/\partial q_i\partial q_k$ are denoted by $\alpha(\mathbf{q})$, $\beta(\mathbf{q})$ and $\gamma(\mathbf{q})$, where $\alpha\geqq\beta\geqq\gamma$, one can deduce that

$$\rho(\theta,\mathbf{q})=1/\{(1-\theta\alpha(\mathbf{q}))(1-\theta\beta(\mathbf{q}))(1-\theta\gamma(\mathbf{q}))\}. \tag{6}$$

Equation (6) makes it clear that a singularity arises at some point **q** when

$$1-\theta\alpha(\mathbf{q})=0. \tag{7}$$

Then the density at the point marked by the Lagrangian radius vector $\mathbf{q}$ becomes infinite and a set of such points forms a caustic, a surface (or a line in two-dimensional space) of infinite density in Eulerian space.

To study the local properties of caustics at any chosen point, let us expand $S(\mathbf{q})$ in a Taylor series in the vicinity of this point. In addition let this point be at the origin of the coordinates and the basis be produced by the principal axes of the tensor $\partial^2 S/\partial q_i \partial q_k$ at the point. Thus

$$S(\mathbf{q}) = S_0 + \sigma_i q_i + \tfrac{1}{2} A_{ik} q_i q_k + \tfrac{1}{6} B_{ikl} q_i q_k q_l + \tfrac{1}{24} C_{iklm} q_i q_k q_l q_m + \cdots, \tag{8}$$

where

$$\left.\begin{aligned} &S_0 = S(0); \qquad \sigma_i = \partial S/\partial q_i|_{\mathbf{q}=0}; \\ &A_{ik} = \left.\frac{\partial^2 S}{\partial q_i \partial q_k}\right|_{\mathbf{q}=0} = \begin{pmatrix} \alpha_0 & 0 & 0 \\ 0 & \beta_0 & 0 \\ 0 & 0 & \gamma_0 \end{pmatrix}; \\ &B_{ikl} = \left.\frac{\partial^3 S}{\partial q_i \partial q_k \partial q_l}\right|_{\mathbf{q}=0}; \text{ etc.} \end{aligned}\right\} \tag{9}$$

It should be stressed that the vector field is smooth and regular. So the expansion (8) is valid at each point. However, the evaluation of the eigen values $\alpha(\mathbf{q})$, $\beta(\mathbf{q})$ and $\gamma(\mathbf{q})$ is a nonlinear operation so it can, in principle, give rise to singularities. But it is easy to show that $\alpha(\mathbf{q})$, $\beta(\mathbf{q})$ and $\gamma(\mathbf{q})$ are continuous functions, although their derivatives have beaks on the lines where $\beta = \alpha$ or $\beta = \gamma$.

It is quite clear from (6) that the singularities connected with the maxima of the largest eigenvalue α arise first of all. By its definition $\theta(t)$ is a monotonically increasing function of time and, as we start from small perturbations, the condition (7) is satisfied first by the particles with the largest magnitude of $\alpha(=\alpha_m)$ at time $\theta_m = 1/\alpha_m$. When $\theta - \theta_m$ is positive but small pancakes form in the vicinity of the maxima of α. In due course they grow and interact with each other forming a well-pronounced cellular structure. Assuming a statistical character for the initial inhomogeneities, it is rather difficult to evaluate the significance of all kinds of singularities for the evolution of the structure. For example, all particles with negative α (and therefore negative β and γ) will never reach a singularity in the sense we use.

However, the theory of Lagrangian singularities predicts that in three-dimensional space, along with oblate structures (pancakes), there must exist very prolate as well as compact structures. In addition the density grows more steeply near prolate objects than near oblate ones. In turn the growth of the density near compact structures is steeper than near prolate ones. Nevertheless pancakes are formed earlier and are more massive.

3. TYPES OF ONE-DIMENSIONAL SINGULARITIES

The expression (1), or its equivalent (4), is said above to provide the exact solution to the problem of the evolution of an arbitrary one-dimensional perturbation. It is exact up to the intersections of trajectories, but one can also infer qualitative features of the motion for a finite time—after the intersections in the case of a collisionless medium. In the one-dimensional case the density can be found from an equation similar to (6) (Zeldovich and Myshkis, 1973), viz:

$$\rho(\theta, q) = [1 - \theta\, \alpha(q)]^{-1},$$

where $\alpha(q) = d^2S/dq^2$. So it is evident that the singularity arises at the point under study when equation (7) is satisfied. Moreover, the singularity initially arises just at the point of the highest maxima of $\alpha(q) = \alpha_m$. At this time there is only one point of infinite density but at the next instant the density at this point becomes finite. Instead the infinite density now arises in two neighboring particles possessing a slightly smaller value of α. There is a triple stream flow between the singular points. Two are the boundaries of the pancake; they move away from each other making the pancake thicker. At $\theta - \theta_m \ll \theta_m$ the thickness of the pancake grows as $\alpha \propto (\theta - \theta_m)^{3/2}$. During this time another pancake may arise in a different region and so on. An example of α and dS/dq is shown in Figure 1. There are two maxima of different magnitudes. The density ρ as a function of r (the Eulerian coordinate not the Lagrangian one) is plotted for the instant when the next pancake in the second maximum of $\alpha = \alpha_{m2} < \alpha_{m1}$ is formed. Near the boundary points the density goes to infinity only inside a pancake, $\rho \propto r^{-1/2}$. At the instant a pancake is created the density goes to infinity, $\rho \propto r^{-2/3}$, (see Zeldovich, 1970). The former is a singularity of type A_2 and the latter of type A_3 in accord with the general classification (Arnold, 1972; 1976; 1980). The law of density growth is determined by the type of singularity and is the same regardless of the dimensions of the space. These two kinds of singularity are all there are in the one-dimensional case.

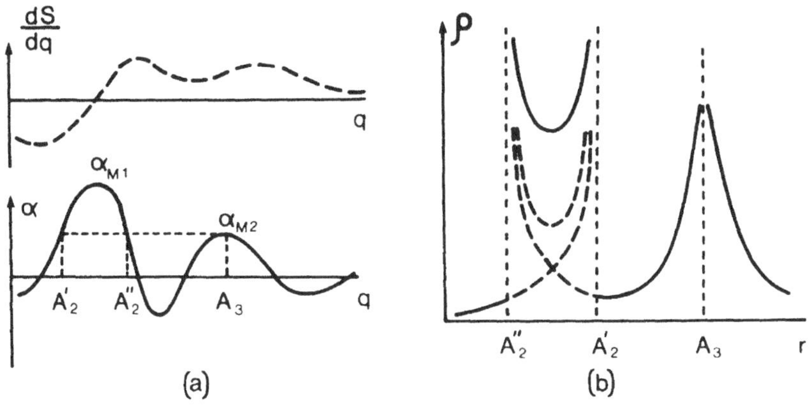

FIGURE 1 (a) An example of the function dS/dq—dashed line. Its derivative $\alpha(q)$ is the solid line. (b) The distribution of density at the time of birth of the second "pancake", connected with the smallest of the two maxima of α_{m2}. The vertical dotted lines are asymptotes for the density curve. There are two points of type A_2 (the boundary of a pancake) and a point of type A_3 (the point of origin of a pancake). The dashed lines between the two points of type A_2 represent the density in each of the three flows.

4. TYPES OF SINGULARITIES IN TWO-DIMENSIONAL SPACE

In this section we describe all the generic singularities in two-dimensional motion of a potential type. As above we use the proper coordinates $\mathbf{r}$ and time θ introduced by (3). In addition it is convenient to assume that, at the point under consideration,

$$\alpha_0 = 1, \tag{10}$$

thus the time singularity develops is always

$$\theta_0 = 1/\alpha_0 = 1. \tag{11}$$

The description of singularities is arranged in order of increasing complexity of their structure. Chronologically the process of the formation of the structure starts from the creation of pancakes—those of the A_3 type, which however, are more complicated, will be described later.

4.1. A point of a boundary line—the singularity of type A_2

The boundary of a pancake is easily seen from (7) to be the set of points having equal values of α. Thus in Lagrangian space it is a component of a level line of α. The Lagrangian and Eulerian patterns for an arbitrary point of this set are shown in Figure 2a, b. The local level lines in the vicinity of the chosen point are straight lines (this means that the curvature of the line is unimportant)

$$B_{11i}q_i = 1 - \theta. \tag{12}$$

In Figure 2a there are three level lines corresponding to three instants of time when they consecutively become caustics. The point under consideration is plotted by a closed circle. Figure 2b shows the evolution of the caustics in Eulerian space. Dots show the region of three stream flow. The chosen point approaches the caustic from within the pancake and then departs, moving into pancake again. When one approaches the caustic from inside the pancake, the density goes to infinity as $\rho \propto r^{-1/2}$ as in the one-dimensional case. The particles of the third stream come to the region of a caustic from outside but their density is finite.

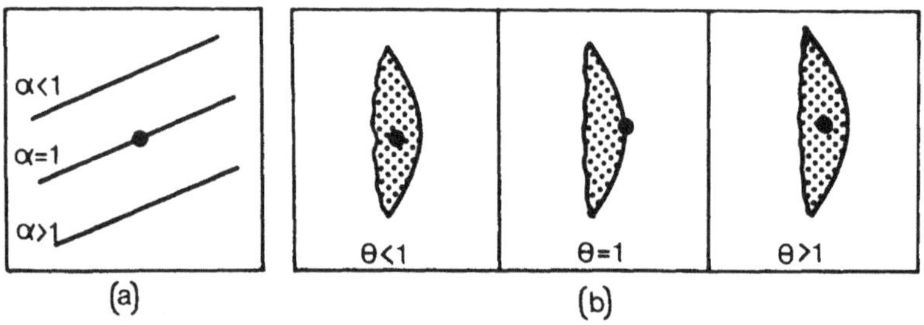

FIGURE 2 (a) The vicinity of a point of type A_2 on the Lagrangian plane. The solid lines are the three closed level lines of $\alpha(\mathbf{q})$. The density becomes infinite at the points situated on these lines at three "successive" instants. At $\theta = 1$ the density becomes infinite at the level line on which the point under study is situated. (*b*) Three pictures of a singularity of type A_2 in three "successive" moments on the Eulerian plane. At each moment only one fragment exists. The solid lines are images of the solid lines from (a). The wavy lines are the edges of the picture. In the regions filled with points there are three stream flows. The closed circle is the point under investigation.

4.2. The edge of a pancake, points of birth and surgery of pancakes—the singularities of type A_3 (cusp, lips, beak-to-beak)

If a closed level line is mapped from Lagrangian to Eulerian space at the time of the singularity $\theta = 1/\alpha$ (or, remembering (10) and (11), at $\theta = 1$) one observes that the image curve has degenerated and has cusps. These cusps correspond to the singularities of type A_3, and we shall also call them edge points, (Figure 3b). Singular points of type A_3 are points where the principal vector $\mathbf{m}(\mathbf{q})$, corresponding to the eigenvalue α, is tangential to the level line of α. A set of these points for all θ makes up the lines. The condition of tangency is

$$m_i(\mathbf{q})\,\partial\alpha/\partial q_i=0. \tag{13}$$

This is possible if, in our notation,

$$B_{111}=0. \tag{14}$$

The level lines in the vicinity of the edge point are given by the equation

$$B_{112}q_2+[\tfrac{1}{2}C_{1111}-B_{112}^2(1-\beta_0)^{-1}]q_1^2=1-\theta \tag{15}$$

and plotted in Lagrangian coordinates in Figure 3a for the three times $\theta<1$, $\theta=1$ and $\theta>1$. As above the points under consideration are depicted as a closed circle. Figure 3b shows the time evolution of caustics. The cusp is determined by a semicubical parabola $y\propto x^{3/2}$.

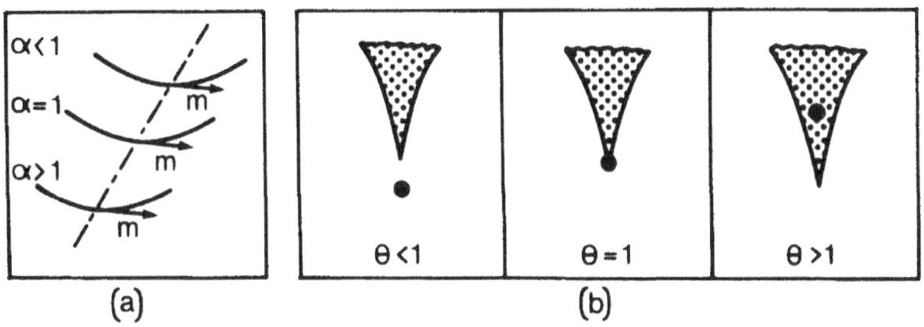

FIGURE 3 An edge of a pancake ("cusp" type A_3). The same designations as in Figure 2. The dash-dot line is formed by the edge points realized at different times.

Another kind of singularity of type A_3 are the points of the extrema of the function $\alpha(\mathbf{q})$. The condition of an extremum is

$$B_{111}=0 \qquad \text{and} \qquad B_{112}=0. \tag{16}$$

Every extremum point lies on the lines (13), for $\partial\alpha/\partial q_i$ is certainly zero at these points. The level lines in the vicinity of these points are practically ellipses or hyperbolas

$$\tfrac{1}{2}C_{1111}q_1^2+C_{1112}q_1q_2+[\tfrac{1}{2}C_{1122}+B_{122}^2(1-\beta_0)^{-1}]q_2^2=1-\theta. \tag{17}$$

If an extremum is a maximum of α a pancake is created. In Eulerian space a new-born pancake looks like a sickle but having two points of inflection on the convex part (i.e. the greater part) of its boundary (lips, Figures 4a, b). The width of a new-born pancake grows as $(\theta-1)^{3/2}$ and its length as $(\theta-1),^{1/2}$ thus the ratio of its width to its length is proportional to $(\theta-1)$.

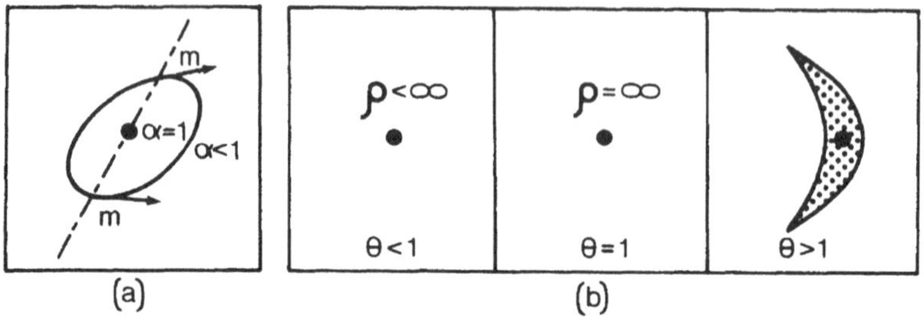

FIGURE 4 The birth of a pancake ("lips", type A_3). The same designations as in Figure 3.

Saddle points are subdivided into two versions depending on the relative positions of the edge line and level lines. If the edge line intersects higher level lines one can observe the joining of two pancakes at the edge points (Figures 5a, b; beak-to-beak). Otherwise the pancakes join at points lying on the boundary lines (Figures 6a, b; the other kind of beak-to-beak). In both cases the cusps move as $|\theta-1|^{1/2}$.

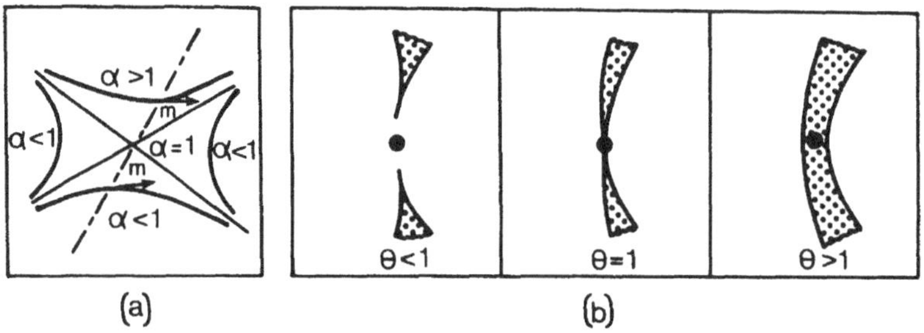

FIGURE 5 The point of junction of two pancakes through edge points ("beak-to-beak", type A_3). The same designations as in Figure 3.

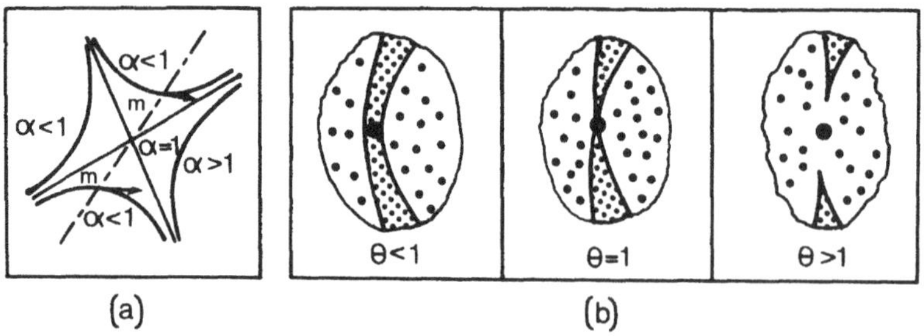

FIGURE 6 The point of junction of two pancakes through boundary points (the other kind of "beak-to-beak", type A_3). The same designations as in Figure 3. The regions of higher density of points shows the regions of five stream flow.

Every point mentioned is a singularity of type A_3 and at the time of infinite density ($\theta=1$) the mean density in the region with characteristic dimensions of r is $\rho \propto r^{-2/3}$

4.3. A break of the boundary line—a singularity of type A_4 (swallow-tail in Eulerian space-time)

The singularity discussed in the previous section results from the tangency of the principal vector $\mathbf{m}(\mathbf{q})$ to level lines of α. In Lagrangian space these points form a line on which birth points and join points lie. However, the points may happen to occur where the eigen vector $\mathbf{m}$ is a tangent to these lines (Arnold, 1976). These points are singularities of type A_4 and are realized when two conditions are fulfilled, viz:

$$B_{111}=0 \quad \text{and} \quad (1-\beta_0)C_{1111}+3B_{112}^2=0. \tag{18}$$

Figures 7a and 7b show the level lines and their images at three successive instants. The caustic looks like $y \propto x^{4/3}$ at the moment of the metamorphosis ($\theta=1$) and then, inside the pancake, a new closed line of infinite density is formed possessing two semicubical parabolic cusps ($\theta>1$). Inside the triangle the flow consists of five streams. At the moment of the metamorphosis the mean density in the r-neighbourhood of the point goes to infinity as $\rho \propto r^{-3/4}$.

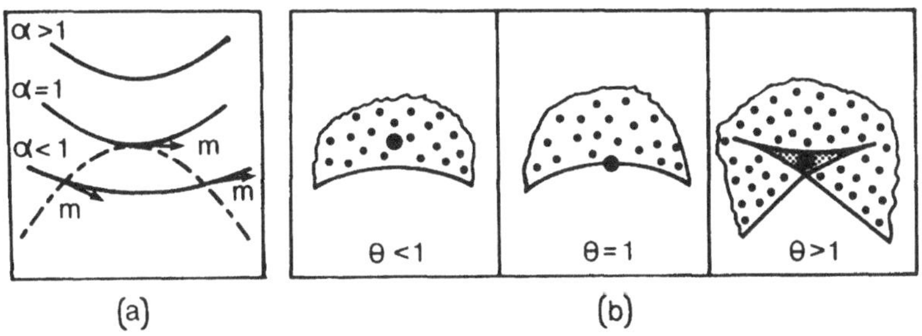

FIGURE 7 The singularity "swallow-tail" (type A_4). The same designations as in Figure 3. The region of higher density of points represents the region of five stream flow.

4.4. Points where $\alpha=\beta$ are singularities of type D_4 (pyramid and purse in Eulerian space-time)

Points where both eigenvalues are equal ($\alpha=\beta$), as Eq. (6) for the density suggests, should be peculiar. Compression of matter near the particles where $\alpha>\beta$ is shown to have a one-dimensional character in accord with the fact that only one term in the denominator of (6) tends to zero. If $\alpha=\beta$ the singularity in the density is stronger.

At such a point the surfaces of $\alpha=\alpha(\mathbf{q})$ and $\beta=\beta(\mathbf{q})$ look like cones. The top cone is a surface of α, the lower one a surface of β (Figures 8a and 9a). Depending on the position of the cone, a singularity D_4 can be one of two types: "purse" or "pyramid". These got their names from their shapes in Eulerian space-time.

Figure 8 shows the first one—the "purse", in space-time. In Lagrangian space there are two families of curves, the first one connected with α and the second one with β. The edge line also consists of two parts divided by the point where $\alpha=\beta$. The first family of curves is formed by the points where **m** is a tangent to the level lines of α and the second one is formed by those where **n** is a tangent to the level lines of β. According to what has been said, the caustic also consists of two branches. The first curve is smooth and the second has a cusp. After the metamorphosis, they interchange their appearance (Figure 8b).

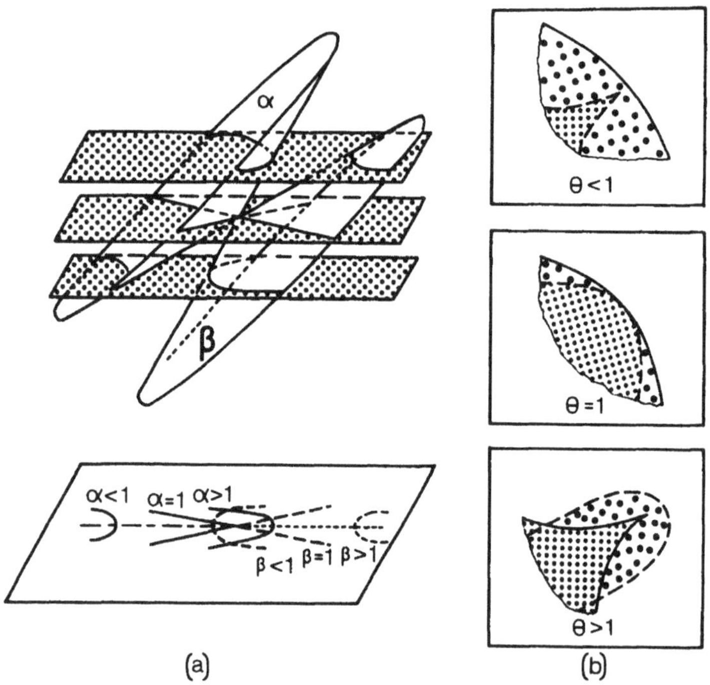

FIGURE 8 The structure of functions $\alpha=\alpha(\mathbf{q})$ and $\beta=\beta(\mathbf{q})$ near the point D_4 where $\alpha=\beta$. The top cone is $\alpha(\mathbf{q})$, the lower one is $\beta(\mathbf{q})$. The three intersecting planes show the three levels of eigenvalues. The dashed lines are the level lines of β in (a) and their images in (b) at corresponding instances. The dotted line is similar to the dash-dot line (Figure 3) but constructed for the smallest eigenvalue β.

The second version of the caustic D_4 (the "pyramid" in space-time) is depicted in Figure 9. In this case the level lines are closed with the point of equilibrium $\alpha=\beta$ being inside them. In Eulerian space the caustic forms a curved triangle having 3 cusps of type A_3 at the vertices. At the moment of the metamorphosis, the caustic degenerates into a point.

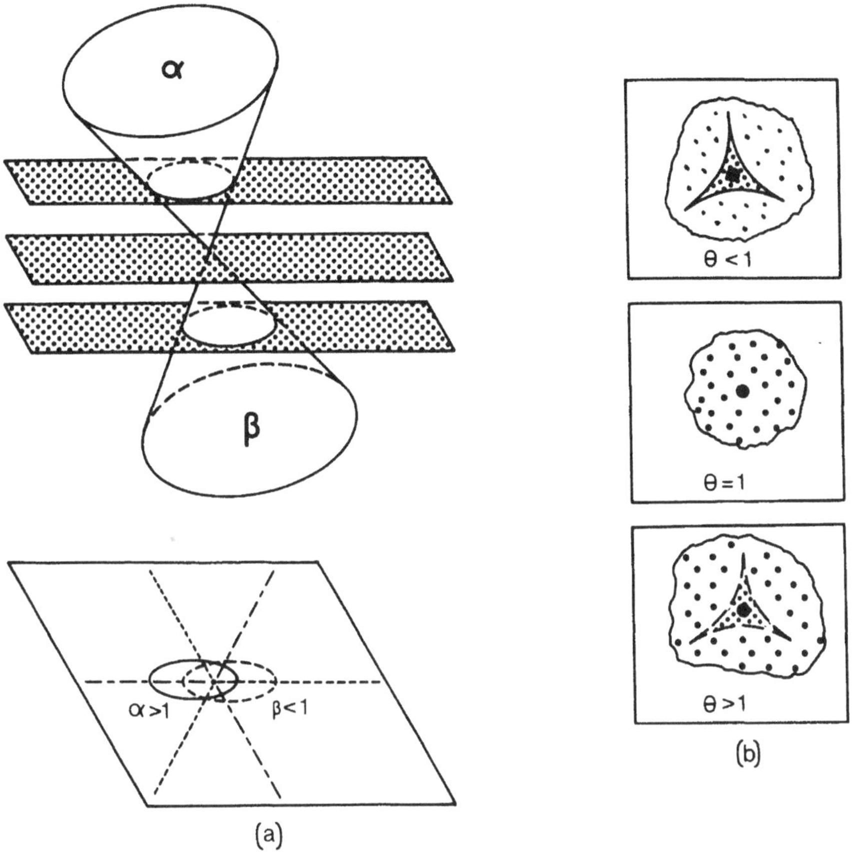

FIGURE 9 The other kind of points of equality $\alpha=\beta$ ("pyramid", type D_4). The same designations as in Figure 8.

At the moment of the metamorphosis $\theta=1$ there is the very steep density profile ($\rho\propto r^{-1}$) mentioned before. It should be noted that the contraction of matter near a point of equality is highly anisotropic.

4.5. The general picture of a two-dimensional structure

The singularities described above form the list of structure elements from which a two-dimensional structure is built. However, deciphering the real structure, arising for instance as a result of numerical simulation under some sort of random initial conditions, represents certain difficulties. To illustrate this statement we give only two examples.

When doing a numerical simulation (Doroshkevich *et al.*, 1980b) one can often observe "triple knots"—groups of particles forming structures like the one shown in Figure 10a. The detailed analysis of many pictures of this kind shows that they are commonly connected with a singularity of type A_4 (Figure 10b). Initially an ordinary swallow-tail is formed (Figure 7b) and then one of the edge points (of type A_3) goes out from the pancake† and joins an edge point belonging to another pancake. At the same time the second edge point stays inside the pancake, so the exterior form resembles the triple knot.

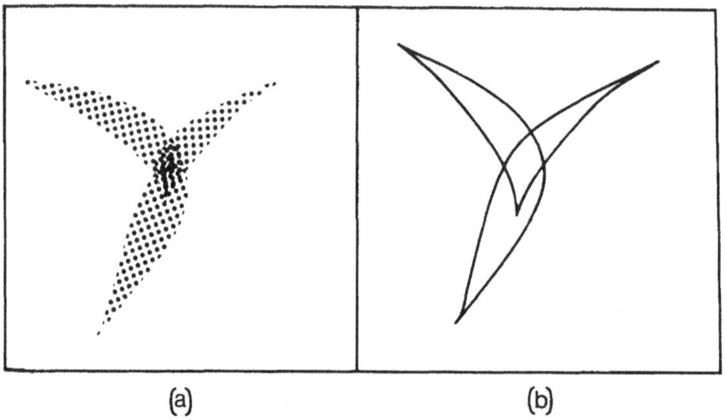

FIGURE 10 (a) An ordinary kind of structure arising in numerical simulations with statistical initial conditions. The points have been preserved only in the region of multi-stream flow. (b) The deciphering of (a). This is a typical swallow-tail structure (see Figure 3) but one of the two edge points has left the initial pancake. This does not imply any metamorphosis of the flow.

The second difficulty is the complicated character of a real picture. Usually in a small region there are several singular points. The following example illustrates the situation. Figure 11a shows a region in Lagrangian space taken from a model with random initial conditions. It is about a tenth by area of the whole pattern. There are two maxima of α, a saddle point, a maximum of β, two points of equality $\alpha=\beta$ and a point of type A_4. Figure 11b shows the caustics (in Eulerian space at some time) connected with both the level lines of α and those of β. The events have occurred as follows: at first the pancake connected with the largest

†It should be noted that it is *not* another type of singularity, for in this case there is no metamorphosis of flows.

Lagrangian maximum α_{m1}, which is situated inside the level line marked by "7", is formed. After that another pancake, connected with the smaller maximum α_{m2} which is situated inside the level line marked by "6", was formed. They merged and soon the pancake connected with the second eigenvalue β was born. In Figure 11b the Eulerian picture is presented for the time when the caustics are formed by the level lines marked by "3". This is the moment of the D_4 metamorphosis. One can observe 3-, 5-, 7-stream flows which are marked by dots of progressively increased density.

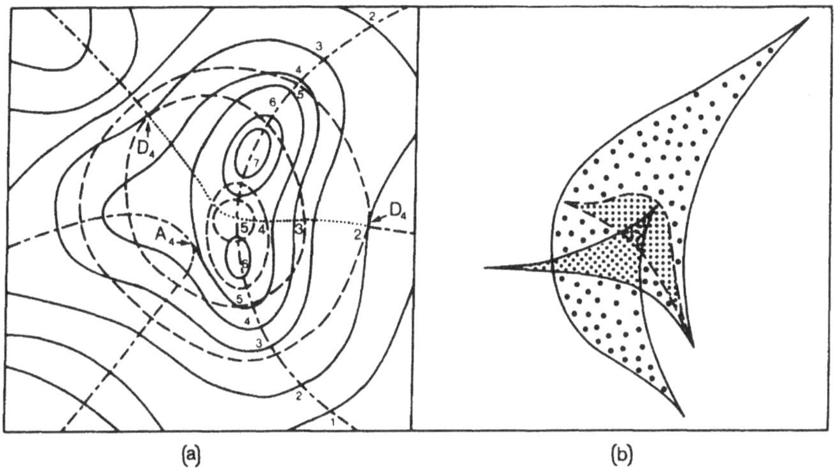

FIGURE 11 An example of the real structure arising in numerical simulations with statistical initial conditions. The designations are the same as in previous figures. (b) represents the Eulerian plane at the time of realization of one of two singularities of type D_4 ("purse").

5. DISCUSSION

The results presented in this paper give a general answer to the question as to what structure is formed in a two-dimensional collisionless self-gravitating medium in the non-linear stage. Every singularity has an analogue in three-dimensional space but, in addition to the above singularities, there are two singularities A_5 and D_5 which first appear in the three dimensional case. They have a much more complicated geometrical structure (see also Arnold, 1976; 1981) and will be discussed in Part II of this paper. To continue the discussion of the similarity of the two- and three-dimensional singularities, it should be noted that every law of density growth near singular points is the same in three dimensional space.

Let us characterize a singularity by its mean density $\rho(r)$, being the ratio of the mass in a sphere of radius r centred at the point in question to the volume of the sphere. The full set of density singularities is as follows (see Vasiljev, 1977):

type of singularity	A_2	A_3	A_4	D_4	A_5	D_5
$\rho(r)$	$r^{-1/2}$	$r^{-2/3}$	$r^{-3/4}$	r^{-1}	$r^{-4/5}$	$r^{-1}\ln(1/r)$

Points of type A_2 form the surface of the pancake, A_3-lines, A_4 and D_4 are only realized at isolated points and A_5 and D_5 at isolated moments.

One should note that the stronger singularities like A_4 and D_4 are realized on less extensive sets, therefore in the two dimensional picture patches of high density are more pronounced than line structures. The same is true in the three-dimensional case. Line structures are more pronounced than surface ones and patches are more pronounced than lines of high density.

The generality of the results presented above is exceedingly important. In spite of the fact that in this paper an approximate solution has been used, all the results will be the same for the exact potential motion in a collisionless self-gravitating medium.

The above singularities describe directly the geometrical structure of the neutrino component in the neutrino dominated Universe. However, galaxies are formed from the small residual fraction of mean density-baryons and atoms, which are not collisionless. So inside the neutrino pancakes ordinary pancakes are formed. They consist of matter which moved with the neutrino before it collided with the matter flowing in the opposite direction. Shock waves are formed, then some portion of matter is cooled and fragments into dense clouds of 10^4 K. Further thermal evolution is likely to proceed rather similarly to that in a matter dominated Universe (Doroshkevich *et al.*, 1978). However, the question arises: how will the large inhomogeneities in the neutrino density distribution, and hence gravitational potential, influence the motion of ordinary matter?

Another question concerns the influence of the initial thermal velocity dispersion on the neutrino component. It is clear that the velocity dispersion removes the singularities so that the maximum density and its width are determined by the extent of the initial dispersion. However, this question needs a special quantitative study.

It seems reasonable to speculate that the different kinds of singularities may be connected with specific differences in the real large-scale structure (different types of rich clusters, quasars and so on) but at present the question is open. The possible role of early star formation and explosions may be very important as pointed out by Ostriker and Cowie (1981).

References

Arnold, V. I., "Normal forms for functions near degenerate critical points, the Weyl groups of A_k, D_k, E_k and Lagrangian singularities," *Functional Anal. Appl.* **6,** 254 (1972).

Arnold, V. I., "Wave front evolution and equilibrium Morse lemma," *Commun. Pure Appl. Math.* **29,** 557 (1976).

Arnold, V. I., *Mathematical Methods of Classical Mechanics*, Springer-Verlag, New York, Heidelberg, Berlin (1980).

Arnold, V. I., "Surgery of singularities in potential collisionless media and caustic metamorphoses in 3-space," *Trudy Seminara im. I. G. Petrovski*, **8** (1981).

Chincarini, G. and Rood H. J., "The Coma supercluster: analysis of Zwicky–Herzog cluster 16 in field," *Astrophys. J.* **206,** 30 (1976).

Chincarini, G. and Rood, H. J., "Space distribution of S_c galaxies: clues to the large-scale structure," *Astrophys. J.* **230,** 648 (1978).

Doroshkevich, A. G., Ryabenki, V. S. and Shandarin, S. F., "Non-linear Theory of development of potential perturbations," *Astrophysics* **9,** 144 (1973).

Doroshkevich, A. G., Shandarin, S. F. and Saar, E., "Spatial structure of protoclusters and the formation of galaxies," *Mon. Not. R. Astron. Soc.* **184,** 643 (1978).

Doroshkevich, A. G., Khlopov, M. Yu., Sunyaev, R. A., Szalay, A. S. and Zeldovich, Ya. B., "Cosmological impact of the neutrino rest mass," *Proc. Xth Texas Symp. on Relativistic Astrophysics*, Baltimore (1980a).

Doroshkevich, A. G., Kotok, E. V., Novikov, I. D., Polyudov, A. N., Shandarin, S. F., and Sigov, Yu. S. "Two-dimensional simulation of the gravitational system dynamics and formation of large-scale structure of the Universe," *Mon. Not. R. Astron. Soc.* **192,** 321 (1980b).

Doroshkevich, A. G. and Shandarin, S. F., "Galaxy formation in nonlinear gravitational-instability theory," *Sov. Astron.* **18,** 24 (1974).

Einasto, J., Joeveer, M. and Saar, E., "Structure of superclusters, and supercluster formation," *Mon. Not. R. Astron. Soc.* **193,** 353 (1980).

Gott, R. and Rees, M., "A theory of galaxy formation and clustering," *Astron. and Astrophys.* **45,** 365 (1975).

Gregory, S. A. and Thompson, L. A., "The Coma/A1367 supercluster and its environs," *Astrophys. J.* **222,** 784 (1978).

Gregory, S. A., Thompson, L. A. and Tifft, W. C., "The Perseus supercluster," Preprint (1980).

Joeveer, N., Einasto, J. and Tago, E., "Spatial distribution of galaxies in the southern galactic hemisphere", *Mon. Not. R. Astron. Soc.* **185**, 357 (1978).

Jones, B. J. T., "The origin of galaxies: A review of recent theoretical development and their confrontation with observation," *Rev. Mod. Phys.* **48,** 107 (1976).

Lifshitz, E. M. and Khalatnikov, I. M., "Investigations in relativistic cosmology", *Adv. Phys.* **12,** 185 (1963).

Oort, J. H., "Superclusters and Lyman-α absorption lines in quasars," *Astron. and Astrophys.* **94,** 359 (1981).

Ostriker, J. P. and Cowie, L. L., Galaxy formation in an intergalactic medium dominated by explosions", *Astrophys. J.* **243,** L127 (1981).

Peebles, P. J. E., *The Large-Scale Structure of the Universe*, Princeton: Princeton University Press (1980).

Press, W. H., "Spontaneous production of the Zeldovich spectrum of cosmological fluctuations," *Phys. Scr.* **21,** 702 (1980).

Rees, M. J., "Growth and fate of inhomogeneities in a "Big Bang" cosmology," *Observational Cosmology*, p. 261, by J. E. Gunn, M. S. Longair and M. J. Rees, Saas-Fee (1978).

Sunyaev, R. A. and Zeldovich, Ya. B., "Formation of clusters of galaxies: protocluster fragmentation and intergalactic gas heating," *Astron. and Astrophys.* **20,** 189 (1972).

Tarenghi, M., Tifft, W. C., Chincarini, G., Rood, H. J. and Thompson, L. A., "The Hercules supercluster I. Basic data," *Astrophys. J.* **234,** 793 (1979).

Tarenghi, M., Chincarini, G., Rood, H. J. and Thompson, L. A., "The Hercules supercluster II. Analysis," *Astrophys. J.* **235,** 724 (1980).

Tully, R. B. and Fisher, J. R., *Proc. IAU Symposium* **79,** 31 (1978).

Turner, M. S. and Schramm, D. N., "The origin of baryons in the Universe," *Nature,* **279,** 303 (1979).

Vasiljev, V. A., "Asymptotics of exponential integrals, Newtoni diagrams and minima points classification," *Funct. Anal. Appl.* **11,** No. 3, 5 (1977).

De Vaucouleurs, G., "The local supercluster," in *The Large Scale Structure of the Universe,* eds. M. S. Longair and J. Einasto, p. 205, D. Reidel Publ. Comp., Dordrecht, Holland/Boston, USA (1978).

Weinberg, S. A., "Cosmological production of baryons," *Phys. Rev. Lett.* **42,** 850 (1979).

Zeldovich, Ya. B., "Gravitational instability: An approximate theory for large density perturbations," *Astron. and Astrophys.* **5,** 84 (1970).

Zeldovich, Ya. B., "*The Theory of the Large Scale Structure of the Universe*, eds. M. S. Longair, J. Einasto, p. 409, D. Reidel Publ. Comp. Dordrecht: Holland/Boston, U.S.A. (1978).

Zeldovich, Ya. B. and Myshkis, A. D., *Elements of Mathematical Physics. Medium of Non-Interacting Particles,*" Nauka, Moscow (1973).

EVOLUTION OF SINGULARITIES OF POTENTIAL FLOWS IN COLLISION-FREE MEDIA AND THE METAMORPHOSIS OF CAUSTICS IN THREE-DIMENSIONAL SPACE

V. I. Arnol'd

UDC 517.5

Following Zel'dovich [8, 9], we consider a medium of particles moving under the action of a force field on them. Let us assume that at the initial time the velocities of the particles form a smooth vector field on the space. In time t, each particle moves into a new position. Thus, there arises a one-parameter family of differentiable maps: the map at time t takes the initial point into its terminal position.

For sufficiently small t the map at time t is one-to-one but as time goes on the particles start to outdistance one another and singularities arise for the map at time t.

If at the initial time the particles were distributed over the space with positive density, then as a result of the motion the density will change: at time t it can become infinite at points which are critical values of the map at time t (i.e., at positions of collision of infinitely close particles).

In the present paper we describe the critical values of the maps at time t and their evolution as t changes for potential initial velocity fields in general position under the assumption that the force field is potential. In particular, the results are applicable to the case of media of noninteracting particles (moving under inertia) and to the case of media of particles moving under the action of a gravitational field on them.

From the mathematical point of view we are concerned with the structure and evolution of caustics of a general one-parameter family of Lagrangian maps of manifolds of dimension not exceeding three. The normal forms for the evolution of singularities of this form are described in [1], and the evolution of caustics corresponding to them are illustrated in [2] (Figs. 13 and 16). In the present paper, for each type of evolution we give a detailed geometric description of the structure of the singularity. We get a number of new normal forms (for example, for singularities of bicaustics, swept out by the cuspidal edges of instantaneous caustics, and for families of cuspidal edges on bicaustics). The investigation of these questions required new algebraic information about the manifold of polynomials with multiple roots, which is also given in the paper.

The author thanks Ya. B. Zel'dovich and S. F. Shandarin for numerous discussions, stimulating the writing of this paper.

1. Characteristics of Singularities

Here we introduce the geometric concepts in terms of which the singularities are described later.

A. Lagrangian Mappings. The map $x \to \partial f/\partial x$ is called the *gradient*. The gradient map is a special case of Lagrangian mappings (cf. [3, 4]).

Example 1. The mappings at time t for the motion of noninteracting particles with potential initial field of velocities have the form $x \to x + t\nabla S$. Consequently, they form a one-parameter family of Lagrangian maps.

A submanifold of dimension n of phase space $\mathbb{R}^{2n}$ with canonical coordinates (p, q) is called *Lagrangian* if the form pdq is a (locally) exact differential. We enumerate some basic properties of Lagrangian submanifolds.

A submanifold which projects uniquely to configuration space [i.e., which is the graph of a map $p = f(q)$] is Lagrangian if and only if the map f is a gradient: $p = \partial S/\partial q$.

Under the action of the transformation of any Hamiltonian phase flow (even with time-dependent Hamiltonian) a Lagrangian manifold goes into a Lagrangian one. The property of

Translated from Trudy Seminara imeni I. G. Petrovskogo, No. 8, pp. 21-57, 1982. Original article submitted August 29, 1980.

Originally publ. in: Tr. Semin. Im. I.G. Petrovskogo 8, 21-57, © 1982 Moscow State University Press
English transl.: J. Soviet Math. 32(3), 229-258, © 1986 Consultants Bureau, New York.

unique projectability is not preserved here: the projection of the transformed Lagrangian manifold to configuration space along the p-direction generally has singularities (critical points).

The projection of a Lagrangian submanifold to the configuration space is called a *Lagrangian mapping*, and its singularities are called *Lagrangian singularities*.

Example 2. We consider a medium of particles moving in a potential field (possibly depending on time, for example, created by the particles themselves). Let us assume that the initial distribution of momenta of particles is potential: $p = \partial S/\partial q$. A Lagrangian submanifold of phase space corresponds to this initial distribution. Under the action of the transformation of the phase flow at time t this manifold goes into a new Lagrangian manifold, no longer necessarily projecting uniquely to configuration space.

Thus, the singularities of the mapping which associates the terminal positions of the particles to the initial ones are Lagrangian singularities.

The motion of particles corresponds to a one-parameter family of Lagrangian mappings. Under change of parameter (time) the corresponding Lagrangian singularities evolve.

Everything said is applicable, in particular, to the case when the potential field mentioned is zero (the particles move according to inertia) or is the gravitational field, created by the medium itself. Thus, the singularities of the map associating the terminal position of the particles to the initial one are Lagrangian in both these cases.

B. Caustics. The set of critical values of a Lagrangian mapping is called a *caustic*. Normal forms are described for the singularities of Lagrangian mappings in general position for spaces of small dimension (cf. [3, 4]). From these formulas there follows a classification of singularities of caustics in general position in spaces of a small number of dimensions. For example, general caustics in three-dimensional space do not have singularities other than cuspidal edges and point singularities of three types: swallow's tail, pyramid, and purse (Fig. 1).

C. Big Caustics. In order to get information about evolutions of caustics in general one-parameter families from the description of caustics, we consider a caustic in space–time.

We shall call the mapped space (preimage) *Lagrangian*, and the image space *Eulerian* [since the map at time t is the passage from the Lagrangian coordinates of the particles, x, to the Eulerian ones, y: for example, for motion by inertia, $y = x + t\nabla S(x)$].

We further introduce *Lagrangian* and *Eulerian space–time* $\{(x, t)\}$ and $\{(y, t)\}$. All the motions of particles can be described by one mapping of the Lagrangian space–time into the Eulerian, $(x, t) \to (y, t)$. We shall call it the *big map*.

The critical points and critical values of the big map are made up of the critical points and values of the maps at time t for all values of t.

It turns out that the big map can be considered as Lagrangian (for more details, cf. [1]). The set of critical values of the big map is hence a caustic in Eulerian space–time. We shall call this set a big caustic.

Thus, a big caustic in Eulerian space–time is composed of instantaneous caustics corresponding to different times, situated in different instantaneous Eulerian spaces.

A big caustic of a general one-parameter family of Lagrangian maps has only the singularities which occur for caustics in general position in Eulerian space–time. Thus, the first step in the description of general evolutions of singularities of caustics consists of investigating general singularities of caustics in space of the next dimension (for example, the singularities enumerated above of caustics in three-dimensional space allow us to investigate evolutions in the plane.

The normal forms of general Lagrangian singularities of mappings of spaces whose dimension does not exceed four are given below (the forms A_2, A_3, A_4, A_5, D_4, D_5, Sec. 2A).

D. Isochrones. In transforming a big caustic in space–time to normal form we do not pay attention to the difference between spatial and temporal coordinates.

The second step in the investigation of evolution is the reduction to normal form of the time function. Here we do not want to spoil the big caustic, which has already been reduced to normal form. Hence, we reduce time to normal form by diffeomorphisms of space–time preserving the big caustic.

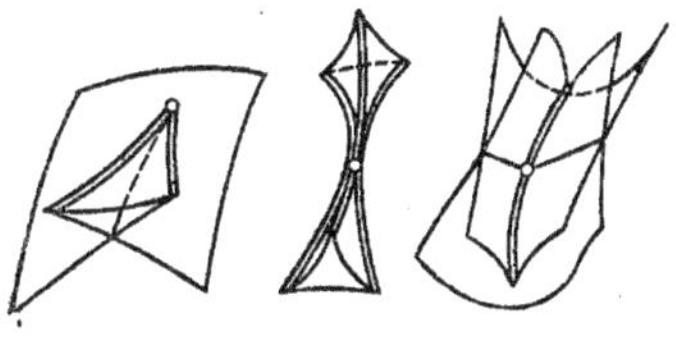

Fig. 1

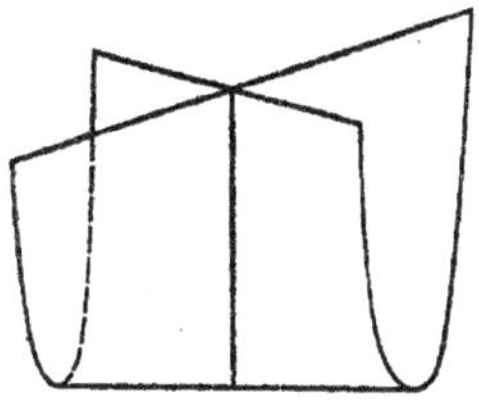

Fig. 2

Time is a function on space–time which has no critical points (points where the total differential of the function is equal to zero).

Normal forms having no critical points of functions in general position in neighborhoods of points of caustics A and D are given in [1]. We reproduce this list of normal forms below, in Sec. 2A.

For singularities of the series A the reduction of time to normal form can be accomplished by a local diffeomorphism preserving the caustic and adding constants. For singularities of the D series one also uses passage to a new coordinate on the time $t' = t'(t)$.

The level sets of the time function are called *isochrones*. Each isochrone is an instantaneous Euler space and is diffeomorphic to the Euler space $\{y\}$. The intersections of a big caustic (reduced to normal form) with isochrones of normalized time form a family of small caustics, evolving in the course of time. These evolutions, up to a smooth time-dependent diffeomorphism and a smooth change of time, coincide with the evolutions of instantaneous caustics of the initial family of maps as time t passes through bifurcation values.

For each normal form of a time function, we give below, in Sec. 2B, three pictures of caustics: before, at the time, and after evolution (Figs. 3, 5, 6). These pictures are found as follows:

We fill out the normalized time to a local coordinate system in space–time, choosing any additional coordinates in the space of normalized caustics.

The values of the additional coordinates at points of intersection of a (normalized) big caustic with the normalized isochrones form instantaneous caustics in the space of additional coordinates. They are also illustrated in the figures.

E. Traces of Singularities. Now we consider singularities of instantaneous caustics. (If the initial space is three-dimensional, then this is a cuspidal edge, i.e., a line in three-dimensional Eulerian space.)

In the course of time the instantaneous caustic moves. Its singularities move in Euler space and fill a certain *trace* (in the three-dimensional case, a surface, formed by the instantaneous cuspidal edges of caustics in one three-dimensional Euler space).

We study the singularities of the trace of the singularities of a moving caustic in Euler space for a general one-parameter family of Lagrangian mappings.

We consider the set of singularities of a big caustic in Euler space–time. The trace studied is found from this set by projecting it onto Euler space along world lines.

By a *bicaustic* (of a given caustic) we mean the projection of the set of singularities of the caustic (not counting self-intersections) onto a space whose dimension is equal to the dimension of the initial caustic. Thus, *the trace of the singularities of instantaneous caustics is a bicaustic of a big caustic.*

Example 1. If the initial (small) space is two-dimensional, then a bicaustic is a line in the plane. The singularities of a bicaustic in general position in this case are singularities of general projections of a cuspidal edge, swallow's tail, pyramid, and purse to the plane. A list of normal forms of traces (or bicaustics) in the plane obtained in this way is given in Sec. 4A.

Example 2. If the initial caustic in $\mathbf{R}^k$ has a singularity A_{k+1}, then its general bicaustic has the same kind of singularity as a caustic A_k in $\mathbf{R}^{k-1}$.

In the case of a three-dimensional initial (small) space, to singularities A_5 on a big caustic there corresponds a trace with singularity, diffeomorphic to a swallow's tail.

Example 3. Singularities of a *one-dimensional* bicaustic in general position in the plane all arise from *singularities* of the set of singularities of the initial caustic. For a *two-dimensional* bicaustic in general position in three-dimensional space there are also possible singular points whose preimages on the original caustic are *regular* points of the set of its singularities.

In the neighborhood of such a point the bicaustic is diffeomorphic to a Whitney umbrella (Fig. 2).

Thus, a general bicaustic in three-dimensional space has in addition to self-intersections only the following singularities: 1) cuspidal edge A_4, 2) swallow's tail A_5, 3) Whitney umbrella, 4) lines D_4 with singular points D_5, 5) intersections of lines A_4 and D_4 with lines of self-intersection (for more details cf. Sec. 4).

F. Covering of the Trace of Instantaneous Singularities. The trace detected by manifolds of singularities of instantaneous caustics does not decompose into these manifolds (since they can intersect), but covers them. In Sec. 5 we study this covering up to a local diffeomorphism of the ambient space.

Example 1. If the bicaustic is a surface, diffeomorphic to a Whitney umbrella, its covering turns out to be locally diffeomorphic to the covering of the umbrella by its rectilinear generators.

Example 2. If the bicaustic is a surface, diffeomorphic to a swallow's tail, in Sec. 5B there is given a formal normal form for the covering (composed of the covering of the curves evolving from the spatial curves with two cusps without points of self-intersection with respect to nonsingular ones). Apparently reduction to this normal form is also possible in the class of infinitely differentiable changes of variables, but not in the class of analytic changes: with respect to the latter, coverings of a swallow's tail have moduli (even functional moduli), analogous to those found by Voronin in [7].

The investigations of normal forms of bicaustics and their coverings are the basic new mathematical results of the present paper. They are based on new information on the manifold of singularities A_3 of caustics A_5 and on vector fields preserving it.

Example. Each vector field, holomorphic in a neighborhood of a singular point of a "swallow's tail" on a bicaustic in $\mathbf{C}^3$, tangent to the bicaustic, can be lifted to a vector field in $\mathbf{C}^4$, holomorphic in a neighborhood of a singular point A_5 of the caustic and tangent to the manifold of A_3 singularities of the caustic, other than self-intersections.

The information indicated is found as a result of unexpected reductions in the lengths of calculations with vague a priori outcome; it would be very desirable to understand the reason for these phenomena in order to transfer the results at least to all caustics A_k.

G. Lagrangian Description: Critical Sets. In Lagrangian space—time the set of critical points of a big map in general position is a hypersurface.

Example 1. If the initial (small) space is two-dimensional, then a big set of critical points in three-dimensional space—time is a two-dimensional surface with conical singular points (in a neighborhood of each of which the surface is diffeomorphic to a circular cone).

Example 2. If the initial (small) space is three-dimensional, then a big set of critical points is a three-dimensional submanifold of four-dimensional space—time with singular lines. In a neighborhood of each point of a singular line this three-dimensional manifold is locally diffeomorphic with the direct product of a circular cone by a line.

A big set of critical points in Lagrangian space—time is the union of instantaneous (small) sets of critical points (which are the sections of the big set by the isochrones t = const).

Example 3. For a medium of noninteracting particles, moving according to the law $y = x + t\partial S/\partial x$, the instantaneous set of critical points at time t is determined by the condition $\det(E + t\partial^2 S/\partial x^2) = 0$, i.e., by the condition $-1/t = \lambda(x)$ = an eigenvalue of the Hessian, $\partial^2 S/\partial x^2$. In this case the big critical set is the graph of an n-valued function, where n is the dimension of the initial space. In the case of a two-dimensional (small) initial space, the instantaneous set of critical points at a general moment of time is a smooth curve. As time changes this curve is transformed at moments corresponding to singularities of the big critical set or its tangent to an isochrone.

Example 4. For the two-dimensional medium of Example 3 the transformation of a small critical set in general position occurs at moments of time corresponding to the junction of branches and critical values of the two-valued function $t = -1/\lambda(x)$.

In the three-dimensional case a small critical set at a general moment of time is a surface with isolated conical singular points. The transformations at different moments of time correspond to tangency of the big critical set to an isochrone (this is the ordinary Morse reconstruction) and the tangency of a line of singular points of the big critical set to an isochrone (fusion of two conical singular points).

Example 5. For the three-dimensional medium of Example 3, the transformations of a small critical set in general position occur for $t = -1/\lambda$, where λ is either a critical value of one of the branches of the three-valued function $\lambda(x)$, or a critical value of the restriction of this function to the line of coincidence of two roots of the characteristic equation $\det(\partial^2 S/\partial x^2 - \partial E) = 0$.

H. Field of Kernels. At a critical point the derivative of a Lagrangian mapping has a nonzero kernel.

Example 1. For the instantaneous map $y = x + t\partial S/\partial x$ of Example 3, the kernel of the derivative coincides with the eigenspace of the operator $\partial^2 S/\partial x^2$, corresponding to the eigenvalue $\lambda = -1/t$.

The kernel of the derivative of the big map (of Lagrangian space–time into Eulerian space–time) at each point of Lagrangian space–time is tangent to the isochrone and coincides in it with the kernel of the derivative of the corresponding instantaneous map of Lagrangian space into Eulerian.

Example 2. In the case of a two-dimensional or three-dimensional initial (small) space the kernel of the derivative of a big Lagrangian map in general position is one-dimensional at all points of smoothness of the big critical set and two-dimensional at its conical points.

Thus, both big and small critical sets of a general big Lagrangian mapping are provided with the field of one-dimensional kernels of the derivative (which dengerate a singular points of the big critical set).

I. Secondary Singularities. Points where the kernel is tangent to the critical set form secondary singularities. In the situation of general position the set of such points is a smooth hypersurface of the critical set. Points of tantency of the kernel of this hypersurface form, in general, a manifold of one dimension lower, etc. Thus, there arises the series of singularities A_k (A_1 are nonsingular points, A_2 is the simplest singularity, A_3 is tangency of the kernel with A_2, A_4 that with A_3, etc.).

Knowing the normal form of a big Lagrangian map in general position and the time function near singularities, we get, up to a diffeomorphism, normal forms of the critical manifold, the field of kernels, and the family of isochrones in Lagrangian space–time, and consequently, also the normal forms of the sets A_k in it.

Example 1. In the case of a two-dimensional initial (small) space in three-dimensional Lagrangian space–time, there is a surface of critical points A_2 with lines A_3 and points A_4. The transformations of the instantaneous maps occur at times of tangency of isochrones with the surface A_2 and the lines A_3, and also at the times that they pass through the points A_4 and through the conical points of the critical set (D_4).

Example 2. In the case of three-dimensional initial (small) space in four-dimensional space–time there is a three-dimensional manifold of critical points A_2 with surface A_3, lines A_4, and points A_5 (in addition to the line of conical points D_4 already mentioned).

The transformations of the instantaneous maps occur at times of tangency of isochrones with the manifold A_2, the surface A_3, and the lines A_4 and D_4, and also at times when the isochrones pass through points A_5 (and in addition through certain points D_5 on the line D_4).

In Figs. 4, 7, 8, and 9 in Sec. 2, for each transformation there is given the form of the manifolds A_k, D_k in instantaneous Lagrangian space up to the time, at the time, and after the transformation, and also the field of kernels on the critical set.

J. Sheets of Critical Sets. In the case of an n-dimensional medium of noninteracting particles ($y = x + t\partial S/\partial x$) the big critical set consists of n single-valued sheets extended over x-space (corresponding to the n real eigenvalues of the symmetric matrix $\partial^2 S/\partial x^2$). We order these sheets in the order of their appearance ($t_1 \leqslant \ldots \leqslant t_n$).

We divide each instantaneous critical set into parts, corresponding to these sheets (first,...,n-th), and we call these parts the critical sets on the first,...,n-th sheet.

Parts of different instantaneous critical sets corresponding to one sheet (but different moments of time) in Lagrangian space do not intersect.* The instantaneous critical sets of one sheet form a partition of the Lagrangian space.

We can also provide the Lagrangian space with the field of kernels made up of the fields of kernels of the instantaneous critical sets of one sheet. (This is the field of characteristic directions of the Hessian $\partial^2 S/\partial x^2$.)

In Figs. 4, 7, 8, 9, and 14 we make use of this method for picturing the transformations of the instantaneous manifolds of singularities and fields of kernels in a general one-parameter family of Lagrangian mappings.

The instantaneous critical sets in Lagrangian space, corresponding to different values of time, illustrated in one figure should move apart in different isochrones of Lagrangian space–time. Then the picture obtained in Lagrangian space–time will describe the corresponding evolution in a general one-parameter family of Lagrangian maps up to diffeomorphisms of Lagrangian space–time, preserving the partition into isochrones $t = \text{const}$ (but, in general, not preserving the lines $x = \text{const}$).

The objects formed by the dotted lines in the figures corresponding to singularities D involve the second sheet.

2. Normal Forms

Here we write out the normal forms to which one can reduce a general function without critical points in a neighborhood of each point of a caustic of a general Lagrangian map in a space of dimension $\leqslant 4$.

A. Generating Functions. We shall define a big caustic in $\mathbf{R}^{n+1}$ with the help of a generating function

$$F(x, \lambda),\ x \in \mathbf{R}^k,\ \lambda \in \mathbf{R}^l,\ k + l = n + 1.$$

A caustic is determined as the set of critical values of the map† $(x, \lambda) \to (y, \lambda)$, where $y = \partial F/\partial x$.

Example. A singularity A_μ is defined by the generating function

$$F = x^{\mu+1} + \lambda_1 x^{\mu-1} + \lambda_2 x^{\mu-2} + \ldots + \lambda_{\mu-2} x^2.$$

In the tables given below we indicate the normal forms of the time function $t(y, \lambda)$ in a neighborhood of a caustic reduced to normal form for all singularities of caustics of general Lagrangian mappings of spaces of dimension $k + l = 3$ and 4 (according to [1]). All the normal forms act in a neighborhood of the point 0.

Remarks. 1. In the cases D_4 it is sometimes more convenient to use the normal form with $\lambda_1 x_2^2$ instead of $\lambda_1(x_1^2 \mp x_2^2)$.

2. In the cases A_k reduction of t to normal form can be effected by a diffeomorphism preserving the caustic of the space (y, λ) containing it and translation of the origin for reading t, and in the cases D_k one also uses a smooth change of time.

3. For t in general position the parameter a in cases D_k assumes general values.

4. In the formula for D_4 in [1] there is a misprint: τ_1^2 is missing.

B. Two-Dimensional Medium, Euler Space. In Fig. 3 for each case of Table 1 we have pictured:

1) the big caustic with isochrone $t = 0$;

2) instantaneous caustics up to, at the time of, and after transformation;

*Here one makes essential use of the property of the medium of noninteracting particles -- for general one-parameter families of Lagrangian maps the situation is different.
†The corresponding Lagrangian map is the projection of the Lagrangian manifold $y = \partial F/\partial x$, $\varkappa = -\partial F/\partial \lambda$ of the space with symplectic structure $dx \wedge dy + d\varkappa \wedge d\lambda$, $(x, y, \varkappa, \lambda) \mapsto (y, \lambda)$.

TABLE 1

Name	F	k	t
A_3	$x^4 + \lambda_1 x^2$	1	τ_1 or $\pm\lambda_1 \pm \tau_1^2$
A_4	$x^5 + \lambda_1 x^3 + \lambda_2 x^2$	1	$\pm\lambda_1$
$D_4^{\pm}$	$3x_1^2 x_2 \pm x_2^3 + 3\lambda_1 (x_1^2 \mp x_2^2)$	2	$\pm\lambda_1 + y_1 + ay_2$

Here the components of λ are denoted by λ_i and τ_j.

TABLE 2

Name	F	k	t
A_3	$x^4 + \lambda_1 x^2$	1	τ_1 or $\pm\lambda_1 \pm \tau_1^2 \pm \tau_2^2$
A_4	$x^5 + \lambda_1 x^3 + \lambda_2 x^2$	1	τ_1 or $\pm\lambda_1 \pm \tau_1^2$
A_5	$x^6 + \lambda_1 x^4 + \lambda_2 x^3 + \lambda_3 x^2$	1	$\pm\lambda_1$
$D_4^{\pm}$	$3x_1^2 x_2 \pm x_2^3 + 3\lambda_1 (x_1^2 \mp x_2^2)$	2	τ_1 or $\pm\lambda_1 + y_1 + ay_2 \pm \tau_1^2$
D_5	$x_1^2 x_2 + x_2^4 + \lambda_1 x_2^3 + \lambda_2 x_2^2$	2	$\pm\lambda_1 + y_1 + ay_2$

3) the bicaustic (trace of the cusps of instantaneous caustics in Euler space) with several caustics.

The case $A_3(+)$ corresponds to birth of a singularity at time $t = 0$ (birth of a pancake in the terminology of Ya. B. Zel'dovich). The length of the pancake grows at the first moment like $t^{1/2}$, and the thickness like $t^{3/2}$. (Thus, at the initial moment the rate of growth of the pancake is infinite, i.e., the newborn pancake quickly achieves finite size.) In the case of general position the curve bounding the newborn pancake has two points of inflection (there are four for the normal forms, but two vanish under a general diffeomorphism).

The case $A_3(-)$ corresponds to the combination of two pancakes into one. The rate of this process is also infinite (the spikes move along a bicaustic like $\pm\sqrt{-t}$). Thus, the spikes of the pancakes shoot out toward one another narrow, rapidly moving tongues which then join. Before joining there is, on each of the tongues, in general, one point of inflection.

All of this information is implicitly contained in the formulas of Table 1.

C. Two-Dimensional Medium, Lagrangian Space. In Fig. 4, for each case of Table 1 there are pictured:

1) a big critical set A_2 in three-dimensional Lagrangian space–time with isochrone $t = 0$, field of kernels and line of secondary singularities A_3, where the field of kernels is tangent to the critical set;

2) a family of instantaneous critical sets for a medium of noninteracting particles with field of kernels and line A_3.

One should keep in mind that the last picture is by no means defined by the normal form up to a diffeomorphism of the plane. The instantaneous critical sets and fields of kernels on them are only determined up to a diffeomorphism of the plane, depending smoothly on the time: the relative position of critical sets at different times on the plane can differ from that pictured in the figure. For example, even for a medium of noninteracting particles in the case of a singularity of pyramidal type (D_4^-) the line A_3 on the first sheet can be directed not along the angle smaller than π between the lines $t = 0$ on this sheet, but along the larger one.

In Fig. 4 everything involving the second sheet is dashed.

D. Three-Dimensional Medium, Euler Space. In Figs. 5 and 6, for each case of Table 2 we have pictured:

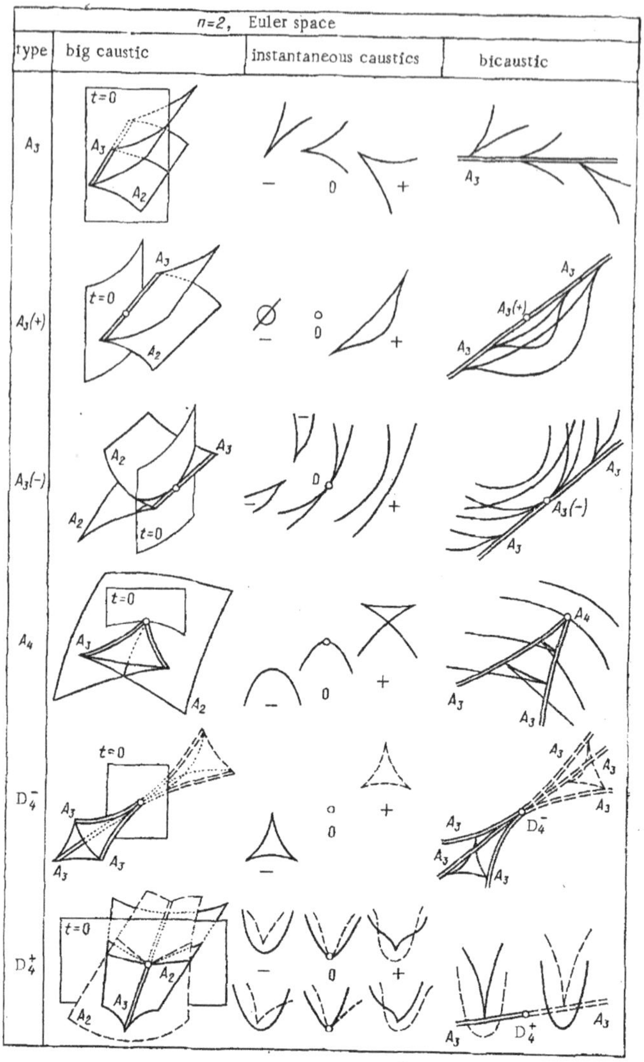

Fig. 3

1) instantaneous caustics up to, at the time of, and after transformation;

2) the bicaustic composed of cuspidal edges of the moving caustic in Euler space.

Although all these results are implicitly contained in Table 2, borrowed from [1], a drawing of instantaneous caustics apparently first appeared in [2],* and drawings of bicaustics have not yet been published.

Remark. In contrast with Fig. 3, the caustics are drawn in normal form. Under diffeomorphisms their curvature changes.

A singularity $A_3(+, +)$ describes the birth of a pancake with dimensions of order $(t^{1/2}, t^{1/2}, t^{3/2})$. In the case $A_3(+, -)$, two pancakes merge along their cuspidal edges, shoot out tongues toward one another, and merge into one pancake.

*In the French translation of [2] this drawing was omitted for some reason.

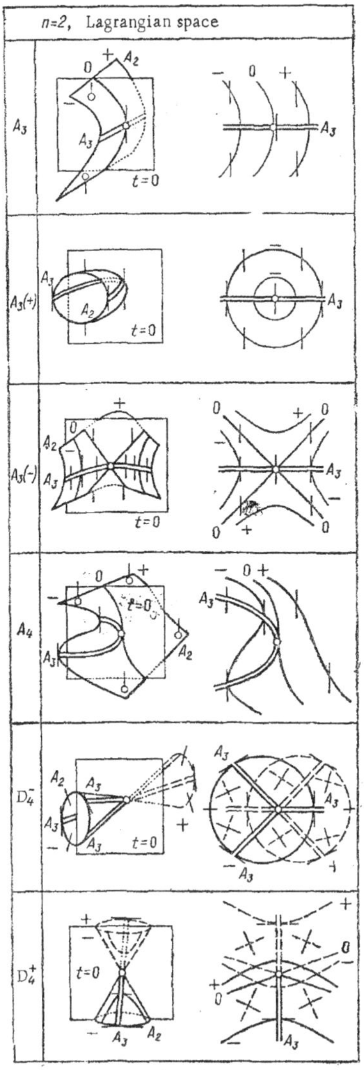

Fig. 4

In the case $A_3(-, -)$ the surfaces of the pancakes slowly approach one another (the distance between them varies as $|t|^{3/2}$, so that the rate of approach at the time of tangency vanishes), after which a constriction is formed between the pancakes with rapidly growing annular cuspidal edge (its length and width are of order $t^{1/2}$).

In the cases D (i.e., in Fig. 6) everything concerning the second sheet is denoted by dashed lines (the lines enclosed by the surfaces are denoted by series of dots).

E. Three-Dimensional Medium, Lagrangian Space. In Figs. 7 and 8, for each case of Table 2 we have pictured in Lagrangian space:

1) the family of instantaneous critical surfaces A_2 for a medium of noninteracting particles (provided with a field of kernels, curves A_3, and points A_4, D_4);

2) the surface formed by all instantaneous lines A_3.

As in the case of a two-dimensional medium, the normal form determines the instantaneous critical surfaces and fields of kernels on them only up to a diffeomorphism of Lagrangian

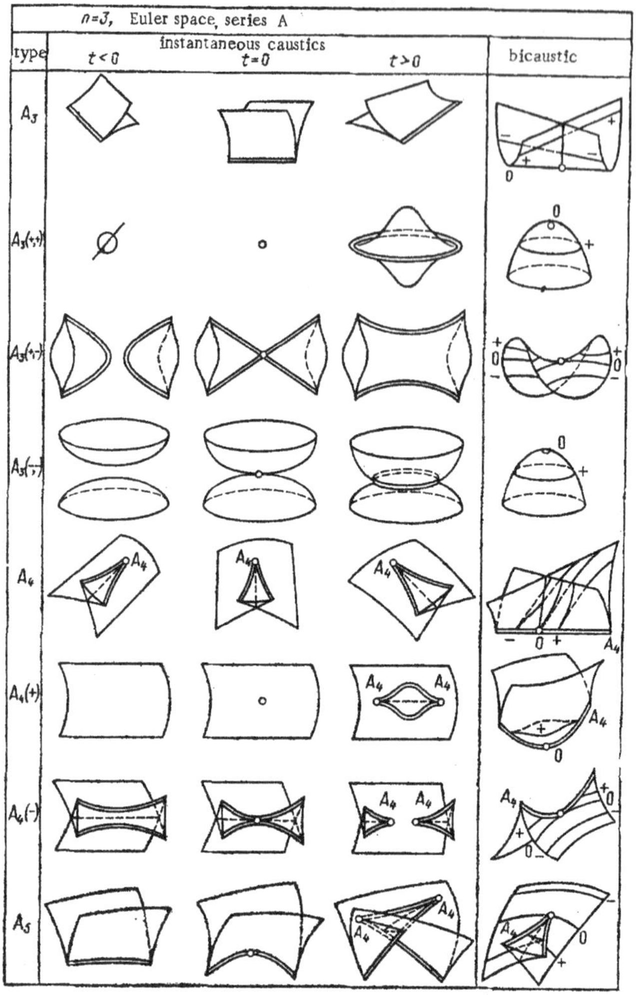

Fig. 5

space depending on time; the mutual positions of surfaces A_2 at different times can differ from that given in our figures.

For singularities of the series A we rectified the field of kernels in the Lagrangian space (in Fig. 7 it is everywhere vertical).

In the cases D in Fig. 8, we picture separately the families of instantaneous critical surfaces A_2 of the first sheet and the second sheet.

For the more complicated cases $D_4^+(+)$ and D_5, we give separately in Fig. 9 the combined transformations of both sheets of the critical surface A_2 and the transformation of the field of kernels on it as t passes through the bifurcation value 0. At a point of intersection of the kernel with a sheet of the surface A_2 only the half of the kernel situated above the sheet is indicated; at a point of tangency (i.e., on a line A_3) the kernel is pictured as a solid line if it is situated above the sheet and as a dashed line if not.

3. The Case of D_5

Here we describe a method, with the help of which one gets the results cited above, for the most complicated case, the evolution of D_5.

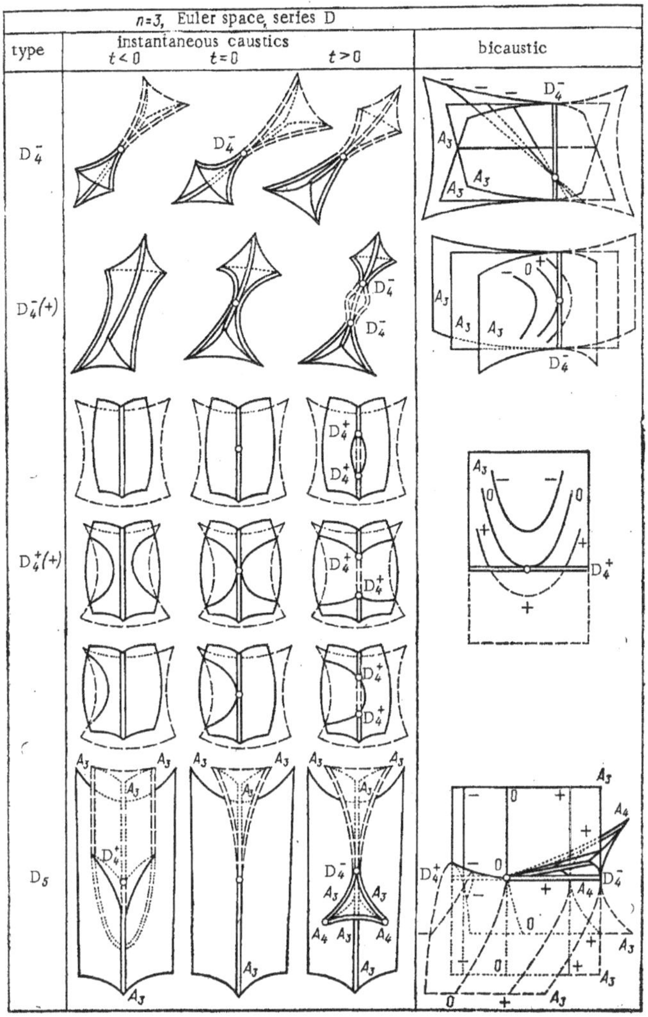

Fig. 6

A. <u>Normal Form.</u> We choose the generating function (equivalent to that given in Sec. 2),

$$F = 2x_1^2 x_2 + x_2^4 + \lambda_1 x_2^3 + \lambda_2 x_2^2.$$

The map of four-dimensional space–time being studied consequently has the form of a two-parameter family of mappings of the plane onto the plane

$$y_1 = 4x_1 x_2, \quad y_2 = 2x_1^2 + 4x_2^3 + 3\lambda_1 x_2^2 + 2\lambda_2 x_2.$$

We form the Hessian

$$\det(\partial^2 F/\partial x^2) = 4\begin{vmatrix} 2x_2 & 2x_1 \\ 2x_1 & 6x_2^2 + 3\lambda_1 x_2 + \lambda_2 \end{vmatrix}.$$

Consequently, the equation of the critical set has the form

$$2x_1^2 = 6x_2^3 + 3\lambda_1 x_2^2 + \lambda_2 x_2 \qquad (A_2)$$

and the field of kernels is generated for $(x_1, x_2) \neq (0, 0)$ by the vector field with components $(x_1, -x_2)$ in the plane (x_1, x_2).

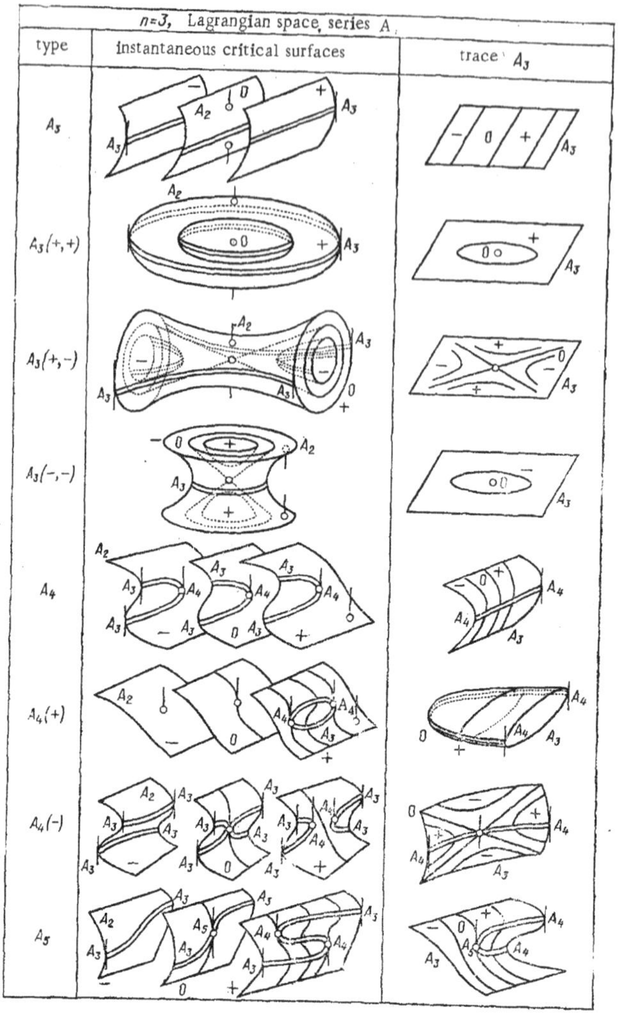

Fig. 7

The surface A_3 is cut out of A_2 by the additional equation of tangency

$$10x_2^3 + 4\lambda_1 x_2^2 + \lambda_2 x_2 = 0. \qquad (A_3)$$

The curve A_4 is determined by the condition of tangency of the field with A_3, from which it follows that:

$$\lambda_2 = \frac{2}{5}\lambda_1^2,\ x_2 = -\frac{1}{5}\lambda_1,\ x_1^2 = -\frac{1}{250}\lambda_1^3. \qquad (A_4)$$

The curve D_4 is the λ_1 axis:

$$x_1 = x_2 = \lambda_2 = 0.$$

We note that all the objects constructed are quasihomogeneous, with weights

$$\deg x_1 = 3,\ \deg x_2 = 2,\ \deg \lambda_1 = 2,\ \deg \lambda_2 = 4,$$
$$\deg y_1 = 5,\ \deg y_2 = 6,\ \deg F = 8.$$

B. Quasicones λ_1 = const. We shall study sections of the critical set A_2 by three-dimensional spaces λ_1 = const. These are two-dimensional surfaces in the (x_1, x_2, λ_2) space.

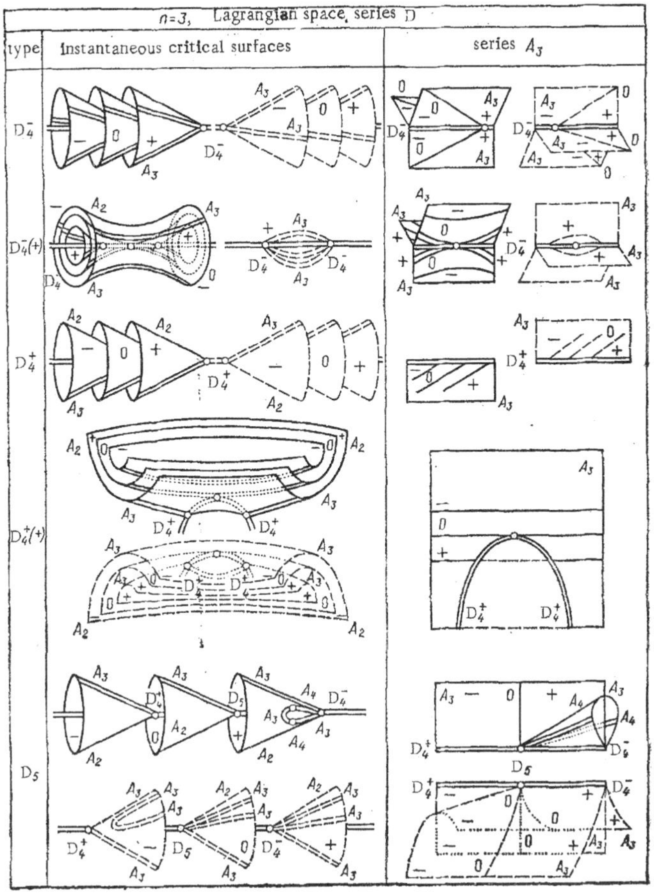

Fig. 8

It is clear from the equation (A2) that such a surface has a conical singular point at zero, so we shall call it a *quasicone*.

A quasicone is locally diffeomorphic with its tangent cone, defined by the equation $2x_1^2 = 3\lambda_1 x_2^2 + \lambda_2 x_2$; this diffeomorphism depends smoothly on the small parameter λ_1.

For $\lambda_1 = 0$ the tangent cone is situated in the space (x_1, x_2, λ_2) so that it encloses the line $(x_1 = 0, x_2 = \lambda_2)$ (Fig. 10). Consequently, for small $|\lambda_1|$ the functions x_1 and $\xi = x_2 - \lambda_2$ are local coordinates on each of the fields of the quasicone $\lambda_1 = \text{const}$ (in some neighborhood of the vertex, not contracting to zero as $|\lambda_1| \to 0$).

The rest of the investigation is made by drawing in the (x_1, ξ) plane, considered as a map of the field of the quasicone, various geometric objects.

C. Family of Elliptic Curves. We consider sections of the quasicone $\lambda_1 = \text{const}$ by the plane $\lambda_2 = \text{const}$. It is clear from formula (A2) that this section is an elliptic curve $2x_1^2 = p_3(x_2)$ in the (x_1, x_2) plane. This allows us to draw the level lines of the function λ_2 on the quasicone in the (x_1, ξ) plane – for this it suffices to shift the elliptic curve constructed by λ_2 along the x_2 axis.

The field of kernels $x_1\partial_1 - x_2\partial_2$ also lies in the plane $\lambda_1 = \text{const}$, $\lambda_2 = \text{const}$ and it can also be drawn in the (x_1, ξ) plane. The lines A_3 on the quasicone are pictured in this plane as lines of tangency of the field of kernels with the elliptic curves $\lambda_2 = \text{const}$.

In order to do all this we study first the family of elliptic curves (A2). As λ changes the form of the curve changes when the roots of the polynomial p_3 on the right side of formula

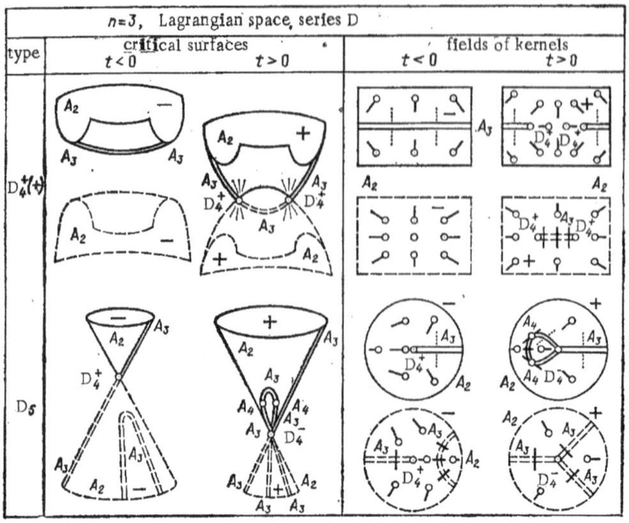

Fig. 9

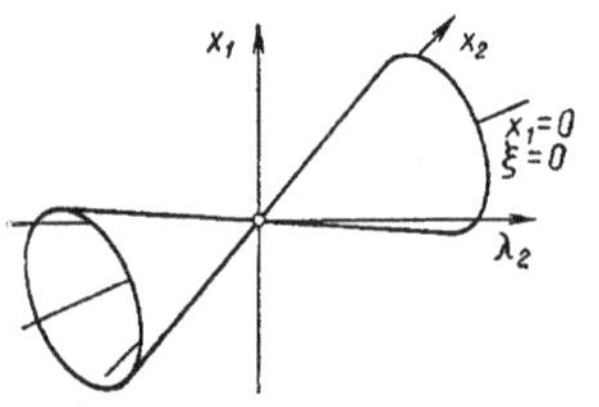

Fig. 10

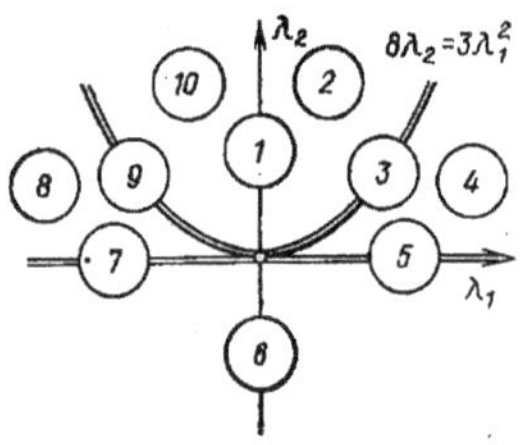

Fig. 11

(A_2) become multiple. This happens for $\lambda_2 = 0$ and for $8\lambda_2 = 3\lambda_1^2$. The λ plane is divided by the bifurcation diagram of the line and the parabola tangent to it into sectors (Fig. 11). The sequence of changes of the graph of the polynomial p_3 upon passage of λ from sector to sector is pictured in Fig. 12. Under each graph of p_3 is drawn the elliptic curve corresponding to it.

Equation (A_2) describes both fields of quasicones. On the first of them $x_2 \geqslant 0$, and on the second $x_2 \leqslant 0$. The elliptic curves in Fig. 12 consist of two parts. The part lying in the half-plane $x_2 \geqslant 0$ relates to the first field, and that in the half-plane $x_2 \leqslant 0$ to the second.

D. Construction of the Field of Kernels on the Quasicone. We fix λ_1 and we consider in the (x_1, x_2) plane, elliptic curves with different λ_2. It follows from Fig. 12 that as λ_2 changes we get, according to the sign of λ_1, one of three families of curves, pictured in Fig. 13.

The families in the right half-planes $x_2 \geqslant 0$ correspond to the first field of quasicones, those in the left, to the second. The dashed lines in Fig. 13 denote hyperbolas (integral curves of the field of kernels $x_1\partial_1 - x_2\partial_2$).

In order to get the curves $\lambda_2 = \text{const}$ and the field of kernels on the (x_1, ξ) plane which is a map of the quasicone, it suffices to shift each of the curves in the (x_1, x_2) plane constructed above by the λ_2 corresponding to it together with restriction to this curve of the field of kernels.

The result of the shift is given in Fig. 14. The points of tangency of the field of kernels with the curves $\lambda_2 = \text{const}$ form lines A_3; the points of tangency of these lines with the field are the points A_4. The upper part of Fig. 14 corresponds to the first field, the lower to the second.

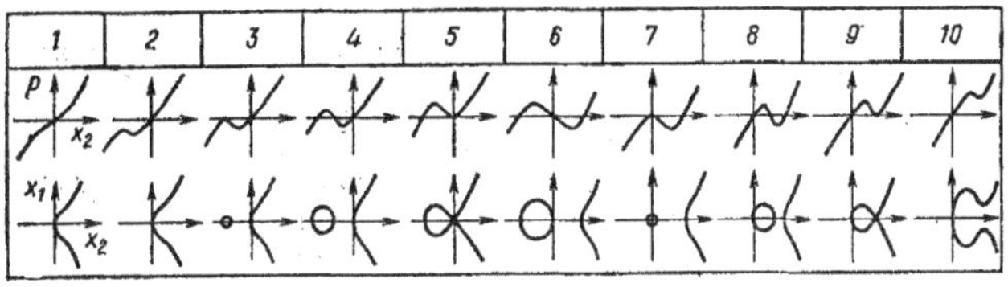

Fig. 12

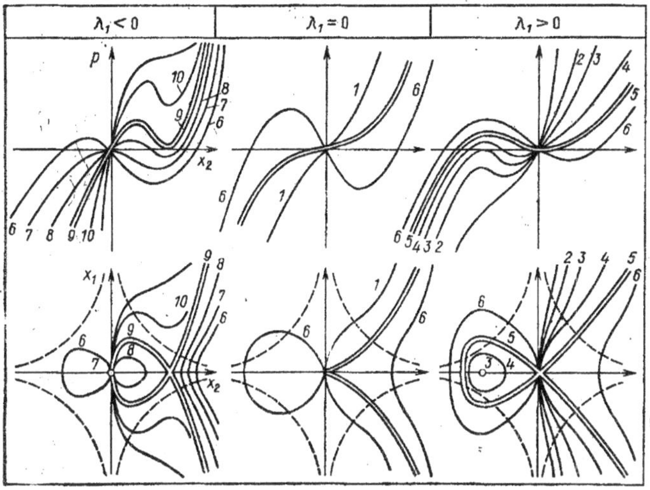

Fig. 13

E. Investigation of the Evolution. Above we studied the evolution of sections of the critical set A_2 in four-dimensional Lagrangian space-time by planes λ_1 = const as λ_1 passes the bifurcation value 0 (Fig. 14).

These sections are also pictured in Figs. 8 and 9 instead of the instantaneous critical sets up to, at the time of, and after bifurcation D_5.

Apparently the instantaneous critical surfaces t = const are transformed in the same way as the sections λ_1 = const. The justification of this assumption is the following.

1°. According to the normal form of the time function $t = \lambda_1 + y_1 + a y_2$ the family of isochrones is tangent to the family of surfaces λ_1 = const along the λ_1 axis (the quantities y_1 and y_2 are of the second order of smallness with respect to x and λ_2).

2°. None of the manifolds into which a big critical set splits (the manifolds A_2, A_3, D_4, etc.) is tangent to the plane $\lambda_1 = 0$ at zero.

This follows from quasihomogeneity considerations, since λ_1 (along with x_1) has the smallest weight of the coordinates (x_1, x_2, λ_1, λ_2) and in the plane $\lambda_1 = 0$, besides the singularity D_5 at zero, there are no superfluous degenerations.

From 1°, 2°, and the sufficiently good abutment to one another of the manifolds mentioned above, it apparently follows that the transformations of the sections of the big critical set by the family of planes λ_1 = const and by the close family t = const are equivalent (at least topologically and combinatorially).

Unfortunately I have not been able to discover a suitable theorem on the topology of the transformations in both families of sections of quasihomogeneous hypersurfaces. Thus, the transformations of the critical set under bifurcation D_5 pictured in Figs. 8 and 9 must more cautiously be considered hypothetical.

In exactly the same way as for transformations of an instantaneous caustic under a bifurcation D_5 we have pictured in Fig. 6 the transformation of a section of a big caustic in

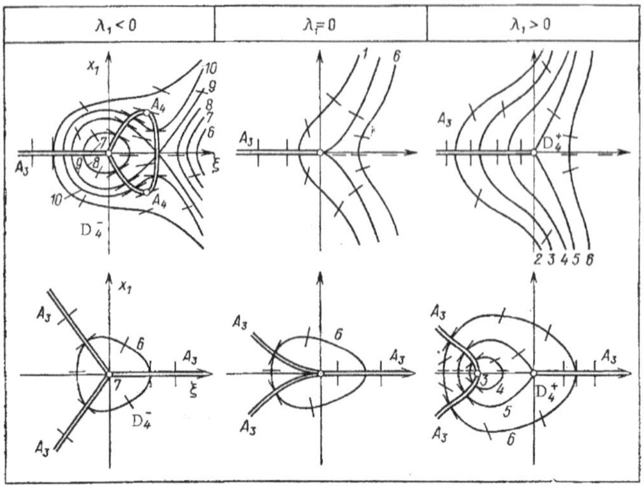

Fig. 14

the $(y_1, y_2, \lambda_1, \lambda_2)$ space by the planes λ_1 = const. (Justification: the variable λ_1 has strictly smaller weight.)

We shall not dwell on the details of the caustics in three-dimensional space pictured in Fig. 6, since their sections by two-dimensional planes λ = const have been described more than once (cf., e.g., the dust jacket of [6]). That the transformations D_5 of instantaneous caustics are homeomorphic to the transformations pictured in Fig. 6 of the sections λ_1 = const must carefully be considered conjectural.

4. Bicaustics

Here we describe the traces of the singularities of instantaneous caustics moving in Euler space.

A. Two-Dimensional Case. We consider the case of a two-dimensional medium, i.e., of three-dimensional space–time. The trace left by cusp points of instantaneous caustics is a bicaustic, i.e., the projection of the cuspidal edge of a big caustic from three-dimensional space–time on the plane.

THEOREM. A general bicaustic in the plane in a neighborhood of each of its points is locally diffeomorphic with one of the following curves, given by the equation $f(u, v) = 0$ in a neighborhood of zero:

Type	A_3, $A_3(+)$, D_4^+	$2A_3$	A_4	D_4^-
f	u	uv	$u^3 - v^2$	$u^3 + au^2v^2 + v^6$, $4a^3 + 27 \neq 0$

These singularities are stable (under perturbation only the parameter a in the normal form can change).

Proof. A cuspidal edge of a general caustic in three-dimensional space has singularities only at points of a swallow's tail (A_4) and pyramidal points (D_4^-). From the normal forms for a swallow's tail and a pyramid it is easy to see that: 1) the cuspidal edge of a swallow's tail lies on a smooth surface (tangent to the caustic at the vertex) and is diffeomorphic to a plane semicubic curve; 2) the three cuspidal edges of a pyramid are smooth curves with first-order tangency at the vertex of the pyramid.

We consider projection of three-dimensional space onto a plane (i.e., a smooth map of rank 2). The preimages of points of the plane under the projection are called world lines. The tangent to a world line is the kernel of the projection.

World lines of a general projection are nowhere tangent to a cuspidal edge. Consequently, for a general projection the nonsingular points of a cuspidal edge go into nonsingular points of the projection or into points of the transverse intersection of two of its branches.

At points A_4 the direction of the kernel of a general projection is not tangent to the cuspidal edge, so the projections are smooth plane curves. For a projection in general position these three plane curves have first-order tangency pairwise.

Three plane curves with first-order tangency can be reduced by a diffeomorphism of the plane to three parabolas $u = c_i v^2$, and such a triple of parabolas can be reduced to the normal form indicated in the table (cf. [3]).

The parameter a in the normal form is a modulus with respect to the diffeomorphisms. Thus, for different general projections of one caustic one can get bicaustics with nondiffeomorphic singularities.

Remark. In Fig. 3 near the bicaustic the instantaneous caustics are drawn. The position of the instantaneous caustics near a point in general position on the bicaustic is standard: in the neighborhood of such a point the family of instantaneous caustics can be reduced by a diffeomorphism to the family of parallel semicubical parabolas $u^3 = (v - t)^2$.

At individual regular points of the bicaustic the family of instantaneous caustics is transformed. Besides points of the type $A_3(+)$ and $A_3(-)$ transformation can occur at ordinary points of type A_3, as is indicated in Fig. 3. This transformation corresponds to points of the cuspidal edge of a big caustic in space-time, where the kernel of the projection lands in the tangent plane of the big caustic.

The family of instantaneous caustics has other singularities in addition to the bicaustic: they correspond to singularities of the projection of part of the big caustic different from the cuspidal edge onto the plane. The folds and gathers of this projection appear as enveloping families of instantaneous caustics and singularities of these envelopes (cf. [1]).

In the remainder of this section the dimension of the medium is assumed to be equal to three.

B. Whitney Umbrella. In the case of a three-dimensional medium the bicaustic is a surface, swept out by the cuspidal edges of the instantaneous caustics. In four-dimensional space-time the cuspidal edges of the instantaneous caustics form a two-dimensional surface. The bicaustic is the projection of this surface from four-dimensional space to three-dimensional. The singularities of the bicaustic occur due to the singularities of the indicated surface in space-time and due to the singularities introduced by the projection. In the case of a two-dimensional medium a general projection did not introduce singularities (excluding the transverse intersection of two branches of the bicaustic). In the three-dimensional case this is no longer so.

THEOREM. In the neighborhood of each regular point of a surface of singularities A_3 of a general caustic in four-dimensional space, a projection in general position of four-dimensional space to three-dimensional space restricted to the surface A_3 is diffeomorphic either to an imbedding of a plane in three-dimensional space or to the map of a plane onto a Whitney umbrella $(u, v) \to (u, uv, v^2)$ in the neighborhood of the origin.

In other words, a bicaustic in general position in three-dimensional space has in addition to transverse self-intersections, only singularities diffeomorphic to a Whitney umbrella and singularities on lines swept out by point singularities of instantaneous caustics. As is well known, an umbrella is stable, i.e., a singularity of this type is preserved by small perperturbations of the indicated map of the plane into three dimensional space.

Proof. We consider a fixed linear projection of four-dimensional to three-dimensional space. The set of two-dimensional subspaces of a four-dimensional linear space containing a fixed one-dimensional subspace is two-dimensional. The set of directions of tangent spaces to a surface A_3 at all of its points is also two-dimensional. The whole manifold of two-dimensional subspaces of a four-dimensional space is four-dimensional. Two two-dimensional submanifolds of four-dimensional space can intersect transversely at isolated points.

Hence the tangent space to a general smooth two-dimensional surface in four-dimensional space-time is tangent at isolated points to the direction of world lines, and this tangency is not removable by small perturbations (in the case of a two-dimensional medium the corresponding tangency is removable).

The restriction of the projection to the surface has rank 1 at points of tangency.

The rest of the investigation of the singularities of a general projection does not differ from the investigation of singularities of a general map of a two-dimensional surface into three-dimensional space; Whitney [11] showed that such maps have only the singularities indicated in the theorem.

C. Singularity A_4. On a surface A_3, swept out in four-dimensional space–time by the cuspidal edges of instantaneous caustics, the vertices of swallow tails of instantaneous caustics sweep out a line. This line is a cuspidal edge A_4 of a two-dimensional surface A_3 in four-dimensional space. On this line there are other singular points of types A_5 and D_5.

The projection of a general smooth curve from four-dimensional space to three-dimensional is a smooth curve so for a general projection the projection of a curve A_5 into Euler space is a smooth curve without singularities and self-intersections.

THEOREM. A general bicaustic in three-dimensional space in a neighborhood of each point of the projection of a line A_4 (excluding the points A_5, D_5) is diffeomorphic to one of the three surfaces

$$u^2=v^3, \qquad u^2=v^3w^2, \qquad u^2w=v^3w$$

in a neighborhood of the point 0 of three-dimensional space.

In other words, the projection of lines A_4 is a semicubical cuspidal edge of a bicaustic, except for isolated points, close to which both parts of the bicaustic separated by the cuspidal edge intersect (cf. Fig. 5).

Proof. A surface A_3 in four-dimensional space–time has along A_4 a semicubical cuspidal edge: in a neighborhood of each point of the line A_4 the surface A_3 goes into a smooth three-dimensional manifold and is diffeomorphic to the surface in it defined by the equation $u^2 = v^3$.

We consider the tangent space to the three-dimensional manifold indicated at a point of the line A_4. It is independent of the choice of three-dimensional manifold; we call it the *accompanying* three-dimensional space.

The directions of the accompanying spaces at all points of the line A_4 form a one-dimensional subset of three-dimensional manifold of all three-dimensional subspaces of four-dimensional linear space. The three-dimensional spaces containing the "vertical" direction form a two-dimensional submanifold in the same three-dimensional manifold. For a big caustic in general position the curve of directions of accompanying spaces intersects the surface of spaces containing verticals transversely, i.e., at isolated points.

If at the initial point of the line A_4 the accompanying space does not contain verticals, then the projection to the three-dimensional space indicated above is a diffeomorphism. In a neighborhood of the projection of such a point the bicaustic has the projection of the line A_4 by the usual cuspidal edge.

We investigate the points of (transverse) intersection.

The restriction of the projection to the three-dimensional manifold described above has at the point considered a line A_4 of singularities. For a big caustic in general position this is the simplest singularity, i.e., is a fold of three-dimensional space: the projection is given by the formula $(x, y, z) \to (x, y, z^2)$ in a suitable local coordinate system with origin at the point considered, and the line A_4 intersects the surface of the fold ($z = 0$) transversely. The form of the bicaustic in this case follows from the following proposition.

LEMMA. The image of a surface in general position with semicubical cuspidal edge under a folding map of three-dimensional space is locally diffeomorphic (in a neighborhood of the image of a point of intersection of the cuspidal edge with the fold) to the surface $u^2 = v^3w^2$ in a neighborhood of the origin.

Proof. Let us assume that the cuspidal edge intersects the fold $z = 0$ transversely at the origin. We try to reduce the surface in a neighborhood of the point of intersection to simplest form, preserving the formula of folding by diffeomorphisms. Such diffeomorphisms are even, i.e., have the form $(x, y, z) \to (X, Y, z\Phi)$, where the functions X, Y, Φ are even with respect to z.

By diffeomorphisms of this form it is easy to locally reduce the cuspidal edge to the form $x = z$, $y = 0$ (it is assumed that the tangent to it at zero is not invariant with respect to the involution changing the sign of z). One can also assume that the tangent plane of the surface at zero is the plane x = z. In this case the sections of the surface by the planes z = const can be given by the equation $x = z + a_2(z)t^2 + a_3(z, t)t^3$, where $a_3(0, 0) \neq 0$ (t denotes $\sqrt{y}$).

We represent a_3 in the form $A(z, t^2) + tB(z, t^2)$ and then we set $a_2 + Bt^2 = zC(z^2, t^2) + D(z^2, t^2)$. Now our equation assumes the form

$$x = z[1 + yC(z^2, y)] + yD(z^2, y) + t^3A(z, y).$$

The substitution $x' = x - yD(z^2, y)$, $z' = z[1 + yC(z^2, y)]$ defines an even diffeomorphism, after which the equation of our surface assumes the form $x = z + t^3H(z, y)$, $H(0, 0) \neq 0$. We represent H in the form $E(z^2, y) + zF(z^2, y)$. Obviously $E(0, 0) \neq 0$. The substitution $y' = yE(z^2, y)^{2/3}$ defines an even diffeomorphism, after which the equation of the surface can be written in the form $x = z + t^3[1 + zG(z^2, y)]$, whence $z[1 + t^3G(z^2, y)] = x - t^3$. The image of this surface after the folding $(x, y, z) \to (x, y, \sigma = z^2)$ has the equation

$$\sigma[1 + t^3G(\sigma, y)]^2 = x^2 - 2xt^3 + y^3.$$

We choose new coordinates $(x, y, s = \sigma - x^2 - y^3)$. In the new coordinates the equation of the image assumes the form

$$(s + x^2 + y^3)(1 + 2t^3G + y^3G^2) = x^2 - 2xt^3 + y^3,$$

i.e.,

$$s + 2\sigma t^3G + \sigma y^3G^2 = -2xt^3,$$

or

$$s + \sigma y^3G^2 = -2t^3(x + \sigma G).$$

We introduce new variables $u = s + \sigma y^3G^2$, $w = x + \sigma G$, which form with $v = 4^{1/3}y$ a coordinate system. In this coordinate system the equation of the image assumes the form $u^2 = v^3w^2$, which is what had to be proved.

D. Three-Dimensional Case: Singularities D_4. The singularities of type D_4 form in four-dimensional space–time smooth curves including the points D_5.

THEOREM. For a general projection to three-dimensional space, the projection of the line of singularities D_4 is a smooth curve (not excepting the points D_5).

Because a general projection of a smooth curve from four-dimensional space to three-dimensional is nonsingular (cf. above the case of A_4).

We consider the structure of a bicaustic in a neighborhood of the curve D_4.

THEOREM. In a neighborhood of each point of the projection of the curve D_4, excluding the points D_5, the projection of a neighborhood of the curve D_4 to the surface A_3 under a general projection of a general big caustic from four-dimensional space to three-dimensional is diffeomorphic to one of the following surfaces

D_4^+ (purse)	D_4^- (pyramid)
$w = 0$	$w(w - v^2)(w - Av^2) = 0$, $A = a + u$ или $a \pm u^2$

in a neighborhood of the point 0 of the u axis of the (u, v, w) space.

Proof. Under a general projection the tangent plane to the surface A_3 at each point of the curve D_4 is mapped isomorphically (cf. case A_4 above). Hence the projection of the surface A_3 consists of one (in the case of D_4^+) or three (in the case of D_4^-) smooth surfaces passing through the projection of the curve D_4, pairwise tangent at all points of the curve in the case of D_4^-.

We choose coordinates so that the curve is the u axis, and the first of the three surfaces is the plane w = 0. Then the equations of the other two will have the form $w = a_i(u, v)v^2$. In the case of general position $a_i(0, 0) \neq 0$, but at isolated points of the u axis one of the functions a_i can vanish (to the first order in u). Without loss of generality we assume $a_i(0, 0) \neq 0$. Then by the choice of the coordinate $w' = w/a_1$ we reduce the equation of the second surface to the form $w = v^2$. Now the equation of the third surface has the form $w = b(u, v)v^2$. We consider the restriction of b to the axis v = 0. In the case of general position the restriction has only simple critical points. In a neighborhood of a noncritical point the restriction can be reduced by a smooth change of the u coordinate to the form a + u, and in the case of a nondegenerate critical point, to the form $a \pm u^2$ [where $a = b(0, 0)$].

Remark. The case $a = 0$ in $A = a + u$, although exceptional, can occur at isolated points of the curve considered. In this case the equation of the third surface can be reduced to the form $w = uv^2$, i.e., the second and third surfaces intersect along a line which is transverse to our curve (cf. the bicaustic D_4^- in Fig. 6).

The case $A = a \pm u^2$ can be realized at isolated points of the curve, and for a general bicaustic at these points $a \neq 0, 1$.

E. Singularity A_5. We consider on a general caustic in four-dimensional space a singularity A_5. This singularity occurs at isolated points.

THEOREM. In a neighborhood of the projection of a point A_5 a general bicaustic in three-dimensional space is diffeomorphic to a swallow's tail.

Proof. We consider the family of world lines of the projection. For a general projection the direction of a world line at a point A_5 is transverse to the tangent plane to the caustic. According to [5, 10], a vector field with this property can be straightened by a diffeomorphism preserving the caustic.

A caustic with singularity A_{n+1} in $\mathbf{R}^n$ is locally diffeomorphic with a surface {λ: the polynomial $p = x^{n+1} + \lambda_1 x^{n-1} + \ldots + \lambda_n$ has a multiple root}. Its tangent plane at zero has the equation $d\lambda_n = 0$. A transverse vector field can be reduced by a diffeomorphism preserving the caustic in a neighborhood of the point $\lambda = 0$ to normal form $\partial/\partial\lambda_n$.

Thus, a general projection of the caustic A_{n+1} is locally equivalent to the standard projection $(\lambda_1,\ldots,\lambda_n) \to (\lambda_1,\ldots,\lambda_{n-1})$.

The manifold of singularities A_3 is defined by the condition: p has a root of multiplicity greater than 2.

The bicaustic of the standard projection is hence defined by the condition

$$\{(\lambda_1, \ldots, \lambda_{n-1}) : p' \text{ has a multiple root}\}$$

This surface is obviously diffeomorphic to the standard caustic A_n in $\mathbf{R}^{n+1}$.

COROLLARY. For a general bicaustic in three-dimensional space the projection of the line A_4 in a neighborhood of points A_5 is diffeomorphic to a plane curve, namely a semicubical parabola.

F. Singularity D_5. A bicaustic in general position in three-dimensional space has in addition isolated singular points of type D_5. The question of the normal form of a bicaustic in a neighborhood of such a point is unsolved. In Fig. 6 as the bicaustic we have pictured the projection of the set A_3 of singularities of the caustic D_5, defined by the normal form in the space $(y_1, y_2, \lambda_1, \lambda_2)$ to the space $(y_1, \lambda_1, \lambda_2)$ along the λ_2 axis.

Apparently a bicaustic in general position has the same kind of singularities (i.e., at least topologically and combinatorially equivalent ones).

The justification for this is that the tangent plane to the caustic at zero has the equation $dy_2 = 0$ [since of the coordinates $(\lambda_1, \lambda_2, y_1, y_2)$ the last has strictly maximal weight]. Due to this y_2 along the caustic is small in relation to the other coordinates and the calculation of it apparently has little influence on the form of the projection.

However this is not proved so that the homeomorphism of a general bicaustic in a neighborhood of a point D_5 with the special bicaustic illustrated in Fig. 6 should strictly be considered conjectural.

Remark. One can verify that close to our projection $(y_1, y_2, \lambda_1, \lambda_2) \to (y_1 + \varepsilon y_2, \lambda_1, \lambda_2)$ leads to a bicaustic which is *not locally diffeomorphic* to ours (obtained for $\varepsilon = 0$).

In fact, calculations show that the configuration formed by the plane $y_1 = 0$ (constituting part of our bicaustic), the cuspidal edge (A_4) of the second component, and the two lines of intersection of the components upon passage from $\varepsilon = 0$ to $\varepsilon \neq 0$ is replaced by a nondiffeomorphic configuration.

Of course for all small ε except zero one gets configurations which are diffeomorphic to one another, so that it is not impossible that the bicaustics of all general projections in a neighborhood of a point D_5 are diffeomorphic (or that the nonsmooth components of these bicaustics are diffeomorphic for all projections whose kernels are transverse to the tangent plane of the caustic D_5).

5. Motion of Singularities on a Bicaustic

In this section we investigate the bicaustic swept out by the manifolds of singularities of the instantaneous caustics.

A. Two-Dimensional Case. We consider a singular point of a general bicaustic on the plane which is a cusp A_4.

THEOREM. For a general bicaustic in the plane there exists a diffeomorphism of the plane after which the bicaustic is reduced in the neighborhood of a cusp to the normal form $y_1^3 + y_2^2 = 0$, and the time of passage of a point of the bicaustic by the cusp of the instantaneous caustic assumes the form $t = \pm y_1 + \text{const}$.

Proof. A caustic A_{n+1} in $\mathbf{R}^n$ is diffeomorphic to the surface of the swallow's tail

$$\{\lambda : x^{n+1} + \lambda_1 x^{n-1} + \ldots + \lambda_n \text{ has a multiple root}\}.$$

By a diffeomorphism preserving the swallow's tail one can reduce a function t without critical points with $\partial t/\partial\lambda_1|_0 \neq 0$ to the form $t = \pm\lambda_1 + \text{const}$ (cf. [1]).

In the case of interest to us, $\mathbf{R}^n$ is three-dimensional space-time, the bicaustic is the projection of the cuspidal edge of a swallow's tail in $\mathbf{R}^3$. This cuspidal edge $\lambda_1 = -6x^2$, $\lambda_2 = 8x^3$, $\lambda_3 = -3x^4$ lies on the smooth surface $12\lambda_3 + \lambda_1^2 = 0$. On this surface λ_1 and λ_2 are coordinates. The world lines of a general projection intersect the surface transversely. We shall index the world lines by the coordinates (λ_1, λ_2) of the point of intersection with the surface. On the cuspidal edge $t = \pm\lambda_1 + \text{const}$ and $\lambda_1^3 + c^2\lambda_2^2 = 0$, so by a dilatation we get the surface of coordinates sought $y_1 = \lambda_1$, $y_2 = c\lambda_2$.

Remark. The family of instantaneous caustics close to a cusp A_4 of a bicaustic cannot be reduced to finite normal form, even if one admits a time change along with diffeomorphisms of the plane. The fact is that such a family has moduli (invariants) which are functions of two variables; these invariants are even topological.

In fact, it is easy to see that through each point of the domain inside a cusp of a bicaustic there pass three instantaneous caustics, corresponding to three moments of time t_i. Two coordinates (t_1, t_2) determine a point of the plane, so there arises a function of two variables $t_3(t_1, t_2)$, determined by the geometry of the family of instantaneous caustics invariantly (up to an arbitrary function of one variable, if one permits a time change).

The same kind of argument shows that a family of instantaneous caustics has functional moduli in a neighborhood of points D_4, in a neighborhood of points where the tangent to the instantaneous caustic at a cusp is tangent to the bicaustic, and also in a neighborhood (not lying on the bicaustic) of a cusp of an enveloping family of caustics (this is a critical value of the projection of a big caustic to the plane at gathering points). In general, functional moduli in the form of functions of n variables appear in a one-parameter family of instantaneous caustics in n-dimensional space whenever the local multiplicity of the interaction of a world line with a big caustic becomes greater than n (in our case $n = 2$).

Example. In a neighborhood of the transformations $A_3(+)$ and $A_3(-)$ (birth of a pancake or junction of pancakes) the multiplicity is equal to two, so one can seek a normal form for the family of caustics. We shall give a family of caustics expressing the time at which the caustic passes through Euler coordinates.

THEOREM 1. In a neighborhood of a transformation $A_3(\pm)$ the whole family of instantaneous caustics can be reduced to the normal form

$$t = \varphi(\lambda) + \lambda^{3/2} \pm \tau^2,$$

where φ is a smooth function, $\varphi'(0) \neq 0$, by a diffeomorphism of the Euler plane (τ, λ).

THEOREM 2. By a formal diffeomorphism of the Euler plane and a formal time change the same family can be reduced to the normal form

$$t = \lambda + \lambda^{3/2} \pm \tau^2.$$

Proof of Theorem 1. A big caustic and its projection to the Euler plane along the family of world lines can be reduced to the normal form $\{\lambda_1, \lambda_2, \tau : \lambda_1^3 = \lambda_2^2\}$, $(\lambda_1, \lambda_2, \tau) \to (\lambda_1, \lambda_2)$

(cf. [5, 10]). It remains to reduce the time function on the big caustic to normal form with the help of a diffeomorphism preserving the projection. On the big caustic we introduce coordinates (s, τ), where $\lambda_1 = s^2$, $\lambda_2 = s^3$.

Diffeomorphisms of the (s, τ) plane commuting with the involution $(s, \tau) \to (-s, \tau)$ give diffeomorphisms of the big caustic preserving the projection. We shall reduce the time function to normal form by these diffeomorphisms of the plane $(s, \tau) \to (sa(s^2, \tau), b(s^2, \tau))$.

In the cases $A_3(\pm)$ the time function $t(y_1, y_2, y_3)$ has a critical point on the cuspidal edge of the big caustic, but is otherwise a function in general position. Consequently, in (s, τ) coordinates its restriction to the big caustic can be written in the form

$$t = p(\tau)\tau^2 + s^2 q(s^2, \tau) + s^3 r(s^2, \tau) + \text{const},$$

where p, q, and r are different from zero at zero.

We reduce time by dilatations of the form $\tau' = \sqrt{|p|}\tau$, $s' = \sqrt[3]{rs}$ to the form

$$t = \pm\tau^2 + s^2 Q(s^2, \tau) + s^3.$$

By Morse's lemma the function $\pm\tau^2 + \tau B(s^2, \tau)$ can be reduced to the form $\pm\tau^2 + \psi(s^2)$ by a substitution for τ depending smoothly on s^2. Theorem 1 is proved.

Remark 1. It is impossible to get rid of the arbitrary function of one variable in the normal form if one does not transform time.

In fact, we consider a point of the Euler plane at which the instantaneous caustics passing at different moments of time are tangent. All such points form a smooth curve. We associate with each point of this curve the pair of times of passing. We get a curve in the (t_1, t_2) plane which is a functional modulus.

Remark 2. Theorem 1 extends directly to the multidimensional case of the evolution $A_3(\pm, \ldots)$, but $\pm\tau^2$ is replaced by $\pm\tau_1^2 \pm \tau_2^2 \pm \ldots$. Theorem 2 carries over to the multidimensional case in the same way.

Proof of Theorem 2. We shall transform the function t of Theorem 1. By dilatations and a translation of the time origin we achieve $\varphi'(0) = 1$. Now we normalize t for $\tau = 0$.

LEMMA. The series $t = s^2 + s^3 + s^4 f(s)$ can be reduced to normal form $t = s^2 + s^3$ by the formal substitution $s = s'u(s'^2)$ and the formal time change $t' = v(t)$.

Proof. In $t = s^2 + s^3$ we substitute $s = s'(1 + as'^{2k-2})$, $t' = t + bt^k$ $(k \geqslant 2)$. We get $t' = s'^2 + s'^3 + As'^{2k} + Bs'^{2k+1} + \ldots$, where $A = 2a + b$, $B = 3a + kb$. The determinant of this system of equations for a and b is nonzero. Consequently, in the series $t = s^2 + s^3 + s^{2k}(A + sB + \ldots)$ with any A and B one can eliminate terms of degrees 2k and 2k + 1.

Successively eliminating terms of degrees 2 and 3, 4 and 5, etc., we get the substitution needed. The lemma is proved.

To prove Theorem 2 first we choose a change of time which (with an even change of s) reduces the restriction of the function t to the line $\tau = 0$ to the normal form $s^2 + s^3$.

By the lemma there exists such a formal substitution. After the substitution we get

$$t' = s'^2 + s'^3 + \tau^2[u(s'^2, \tau^2) + s'^3 v(s'^2, \tau^2)],$$

where $u(0, 0) \neq 0$. We make the substitution $s'[1 + \tau^2 v(s'^2, \tau^2)]^{1/3} = s''$. After this substitution we get

$$t' = s''^2 + s''^3 + \tau^2 w(s''^2, \tau^2),$$

while $w(0, 0) \neq 0$. Replacing τ by $\tau\sqrt{|w|}$ completes the proof.

Remark. Theorem 2 can be carried over to the case of infinitely differentiable functions but is false in the class of analytic ones: with respect to analytic substitutions there is even a functional modulus. In fact, on the preimage of the line of tangency of caustics already mentioned (i.e., $\tau = 0$ in our coordinates) there are defined two holomorphic involutions: one interchanges the preimages of one point of the Euler plane $(s \to -s)$, and the other interchanges the preimages of one caustic (points with identical t). As Voronin proved [7], pairs of involutions have functional moduli with respect to holomorphic variable changes. Thus, the formal series of Theorem 2 and the lemma diverge.

In the rest of this section we consider the motion of the cuspidal edge of a caustic along the bicaustic in three-dimensional space.

B. Sweeping Out of a Whitney Umbrella. A Whitney umbrella on a bicaustic is the image of a smooth map of a surface into three-dimensional space which can be written in normal form $\varphi(p, q) = (p, pq, q^2)$. The cuspidal edges of the instantaneous caustics form a family of lines on the umbrella; this family is obtained from a general family of lines in the (p, q) plane under the map φ.

THEOREM 1. For a big caustic in general position the family of cuspidal edges of instantaneous caustics in a neighborhood of a singular point of the bicaustic of umbrella type can be reduced to the normal form $t = q + a(q^2)$ by a diffeomorphism of three-dimensional space (preserving the umbrella).

In particular, a general family of smooth curves sweeping out an umbrella can be transformed into a family of straight lines preserving the umbrella by a diffeomorphism of three-dimensional space.

Definition. A diffeomorphism $g:A \to A$ is said to be a *lowering* of a smooth map $\varphi:A\to B$, if there exists a diffeomorphism $h:B \to B$ such that $\varphi g = h\varphi$.

Theorem 1 follows from the following proposition.

THEOREM 2. Let t be a smooth function in a neighborhood of the point $p = q = 0$ in the plane, $\partial t/\partial q|_0 \neq 0$. Then t can be reduced to normal form $q + a(q^2)$ by a local diffeomorphism of the plane which is a lowering of the Whitney map φ.

Proof. We consider a level line of the function t, passing through the origin. This curve is transverse to the q axis. Hence on it $q = p\alpha(p)$, where α is a smooth function.

LEMMA. There are diffeomorphisms $(p, q) \to (p, Q)$ which are φ-lowerings of functions Q of the following three types: $q - p\alpha(p)$, $q\beta(p, q^2)$, $q[1 + pq\gamma(p, q)]$.

Proof. For the coordinates in three-dimensional space we introduce the notation (u, v, w), so that the map φ is given by the formulas $u = p$, $v = pq$, $w = q^2$. To prove that Q can be φ-lowered it suffices to verify that pQ and Q^2 can be expressed smoothly in terms of u, v, and w, while

$$|\partial(pQ, Q^2)/\partial(v, w)| \neq 0.$$

In the first case this follows from the equations

$$pq - p^2\alpha(p) = v - u^2\alpha(u),$$
$$q^2 - 2pq\alpha(p) + p^2\alpha^2(p) = w - 2v\alpha(u) + u^2\alpha^2(u).$$

In the second case

$$pq\beta(p, q^2) = v\beta(u, w), \quad q^2\beta^2(p, q^2) = w\beta^2(u, w).$$

In the third case we set $\gamma(p, q) = A(p, q^2) + qB(p, q^2)$; then $pq\gamma(p, q) = vA(u, w) + uwB(u, w)$, so

$$pQ = pq(1 + pq\gamma) = v(1 + vA + uwB),$$
$$Q^2 = q^2(1 + pq\gamma)^2 = w(1 + vA + uwB)^2,$$

and the lemma is proved.

Using the first diffeomorphism of the lemma, we reduce the level line of t passing through 0 to the form $q = 0$. Now $t = \text{const} + q\beta(p) + q^2\varepsilon(p, q)$, where $\beta(0) \neq 0$. Using the second diffeomorphism we reduce β to one:

$$t = \text{const} + q + q^2\varkappa(p, q).$$

Now we consider the q coordinate of the point of intersection of the level line of t passing through (p, q) with the q axis, and we denote it by Q(p, q). Obviously $Q(p, q) = q + q^2\xi(p, q)$ and $Q(0, q) \equiv q$ so Q can be written in the form $Q = q + pq^2\gamma(p, q)$. The third diffeomorphism of the lemma reduces Q to the form q and hence t to the form $t = \text{const} + q + q^2\delta(q)$.

We represent δ in the form $\lambda(q^2) + q\mu(q^2)$ so that $t = \text{const} + q(1 + \mu(q^2)) + q^2\lambda(q^2)$. The second substitution of the lemma [where $\beta = 1 + \mu(q^2)$] reduces t to the form $\text{const} + q + q^2\nu(q^2)$, which is what had to be proved.

Remark. If we permit a change of time, then our family reduces simply to the normal form $t = q$.

Now if a change of time is not allowed, then the family has a functional modulus (in the form of a function of one variable).

In fact, we consider points of intersection on the umbrella. Through each of them there pass two lines of the family. The map of the time axis onto itself which associates one of the moments of passing with the other is a functional modulus.

In the remainder of the paper we will be concerned with the case when the big caustic has a singularity A_5. In this case the bicaustic has a swallow's tail. For the study of the sweeping out of the swallow's tail by the instantaneous cuspidal edges we need some information about the surface of singularities A_3 of the big caustic.

C. Investigation of the Surface A_3 in the Case A_5. We consider the four-dimensional space of polynomials of degree five

$$p = x^5 + \lambda_1 x^3 + \lambda_2 x^2 + \lambda_3 x + \lambda_4.$$

The big caustic of a singularity A_5 is diffeomorphic to the three-dimensional manifold formed in this space by the polynomials with multiple roots. The surface A_3 is formed by polynomials with roots of multiplicity higher than 2.

THEOREM 1. On the surface A_3 the quantity λ_4 is a rational function of λ_1, λ_2, λ_3, namely

$$\lambda_4 = \frac{\lambda_1\lambda_2}{25} - \frac{(9\lambda_1^2 - 20\lambda_3)\lambda_3}{75\lambda_2}.$$

THEOREM 2. On the surface A_3 one has relations of the form

$$\lambda_2\lambda_4 = a(\lambda_1, \lambda_2, \lambda_3), \tag{8}$$

$$(9\lambda_1^2 - 20\lambda_3)\lambda_4 = b(\lambda_1, \lambda_2, \lambda_3), \tag{9}$$

$$\lambda_4^2 = c(\lambda_1, \lambda_2, \lambda_3), \tag{10}$$

where a, b, and c are certain quasihomogeneous polynomials of quasihomogeneous degrees 8, 9, and 10, respectively (for $\deg\lambda_i = i + 1$).

The explicit form of these polynomials is given in the proof of the theorem.

THEOREM 3. Each power series $f(\lambda_1, \lambda_3, \lambda_3, \lambda_4)$ on the surface A_3 is equal to a series of the form $u(\lambda_1, \lambda_2, \lambda_3) + \lambda_4 v(\lambda_1)$. Moreover, there exists a linear operator

$$L: \mathbf{C}[\lambda_1, \lambda_2, \lambda_3, \lambda_4] \to \mathbf{C}[\lambda_1, \lambda_2, \lambda_3] \oplus \mathbf{C}[\lambda_1],$$

preserving the homogeneous graduation and such that if $Lf = (u, v)$, then $f = u + \lambda_4 v$ on A_3.

THEOREM 4. The projection of the surface A_3 along the λ_4 axis to the space with coordinates $(\lambda_1, \lambda_2, \lambda_3)$ is a swallow's tail. Each vector field in this three-dimensional space, which is holomorphic at the point 0 and tangent to the swallow's tail, is the projection of some vector field, holomorphic in a neighborhood of the point 0 in four-dimensional space, and tangent to the surface A_3.

In particular, quasihomogeneous fields of degrees 0, 1, and 2* with the components indicated in the following table are tangent to the surface A_3 and project along the λ_4 axis to the three-dimensional space:

Degree	$\dot\lambda_1$	$\dot\lambda_2$	$\dot\lambda_3$	$\dot\lambda_4$
0	$2\lambda_1$	$3\lambda_2$	$4\lambda_3$	$5\lambda_4$
1	$20\lambda_2$	$20\lambda_3 - 9\lambda_1^2$	$-6\lambda_1\lambda_2$	$-\lambda_2^2$
2	$20\lambda_3 - 9\lambda_1^2$	$-18\lambda_1\lambda_2$	$-9\lambda_2^2$	$-4\lambda_2\lambda_3$

*The degree of a quasihomogeneous field is defined so that differentiation in the direction of the field raises the degrees of homogeneous polynomials by the degree of the field: $\deg(\lambda^a \partial/\partial\lambda_i) = \Sigma(a_j \deg\lambda_j) - \deg\lambda_i$.

Proof of Theorem 1. The surface A_3 is determined as the set of λ, for which the system of equations $p = p' = p'' = 0$ is solvable. From these equations we express λ_2, λ_3, and λ_4 in terms of λ_1 and x (this is easy to do in the order indicated). We get the equations

$$\lambda_2 = -10x^3 - 3\lambda_1 x, \tag{1}$$
$$\lambda_3 = 15x^4 + 3\lambda_1 x^2, \tag{2}$$
$$\lambda_4 = -6x^5 - \lambda_1 x^3. \tag{3}$$

Eliminating x^5 from (2) and (3), we get

$$5\lambda_4 = \lambda_1 x^3 - 2\lambda_3 x. \tag{4}$$

Combining this with (1) and eliminating x^3, we get

$$x = -\frac{50\lambda_4 + \lambda_1\lambda_2}{3\lambda_1^2 + 20\lambda_3}. \tag{A}$$

Eliminating x^4 from (1) and (2), we get

$$2\lambda_3 = -3\lambda_1 x^2 - 3\lambda_2 x. \tag{5}$$

Combining (5) with (1), we get

$$3\lambda_1\lambda_2 = 30\lambda_2 x^2 + (20\lambda_3 - 9\lambda_1^2)\, x. \tag{6}$$

Eliminating x^4 from (4) and (2), we get

$$\lambda_1\lambda_3 = (30\lambda_3 + 3\lambda_1^2)\, x^2 + 75\lambda_4 x. \tag{7}$$

Combining (5) with (7) and eliminating x^2, we find

$$x = \frac{\lambda_3 (3\lambda_1^2 + 20\lambda_3)}{3\,(25\lambda_1\lambda_4 - 10\lambda_2\lambda_3 - \lambda_1^2\lambda_2)}. \tag{B}$$

Combining (6) with (5) and eliminating x^2, we find

$$x = \lambda_2 \frac{3\lambda_1^2 + 20\lambda_3}{20\lambda_1\lambda_3 - 9\lambda_1^3 - 30\lambda_2^2}. \tag{C}$$

Comparison of the relations (A), (B), and (C) gives equations for A_3 of quasihomogeneous degrees 11, 12, and 14. By "lucky chance" the numerators of (B) and (C) have the common factor $3\lambda_1^2 + 20\lambda_3$ (not identically equal to zero on the irreducible manifold A_3). After cancellation by it, (B) and (C) give an equation of degree 10. By another lucky chance, for $\lambda_1 = 0$ the fractions (B) and (C) coincide, so that the indicated polynomial of degree 10 is divisible by λ_1. After cancellation, we get the equation

$$75\lambda_2\lambda_4 - 3\lambda_1\lambda_2^2 + (9\lambda_1^2 - 20\lambda_3)\,\lambda_3 = 0, \tag{8}$$

which proves Theorem 1.

One can show that this equation for A_3 is of the least quasihomogeneous degree.

Proof of Theorem 2. We have just gotten (8). To get the remainder relations we note that for λ from A_3 the polynomial p' has a multiple root and hence discriminant zero. Hence on A_3 one has the following equation which does not contain λ_4:

$$27\,(\lambda_1^3 + 5\lambda_2^2 - 20\lambda_1\lambda_3)\,\lambda_2^2 = (9\lambda_1^2 - 20\lambda_3)^2\,\lambda_3 \tag{12}$$

which is of quasihomogeneous degree 12.

[One can also get this equation for a swallow's tail by comparing the expression for λ_4 found in Theorem 1 with the expression

$$\lambda_4 = \frac{\lambda_2\,(14\lambda_1^2\lambda_3 + 40\lambda_3^2 - 3\lambda_1\lambda_2^2)}{5\,(9\lambda_1^3 - 20\lambda_1\lambda_3 + 30\lambda_2^2)},$$

found by comparing (A) and (B); at the last moment another lucky cancellation by λ_1 occurs.]

We multiply by $9\lambda_1^2 - 20\lambda_3$ the expression for λ_4 given in Theorem 1. We get the difference of a polynomial and a fraction.

But on A_3 this fraction by "lucky chance" can be reduced due to (12). We get on A_3

$$5(9\lambda_1^2 - 20\lambda_3)\,\lambda_4 - 32\lambda_1\lambda_2\lambda_3 - 9\lambda_2^3 = 0. \tag{9}$$

In exactly the same way, squaring the expression for λ_4, we get a reducible fraction on A_3, from which we get on A_3

$$1875\lambda_4^2=3\lambda_2^2(\lambda_1^2+5\lambda_3)-5\lambda_1\lambda_3(3\lambda_1^2-4\lambda_3),$$

having the form (10).

Proof of Theorem 3. Relation (10) allows us to represent each function on A_3 in the form

$$f(\lambda_1,\lambda_2,\lambda_3,\lambda_4)=a(\lambda_1,\lambda_2,\lambda_3)+\lambda_4 b(\lambda_1,\lambda_2,\lambda_3).$$

Relation (8) allows us to get rid of all terms b, divisible by λ_2, in the summand which does not depend on λ_4: on A_3

$$f=c(\lambda_1,\lambda_2,\lambda_3)+\lambda_4 d(\lambda_1,\lambda_3).$$

LEMMA. On A_3 there exist decompositions

$$\lambda_3^k\lambda_4=C_k\lambda_1^{2k}\lambda_4+\alpha_k(\lambda_1,\lambda_2,\lambda_3). \qquad (*_k)$$

Proof. For k = 1, according to (9) we have on A_3

$$\lambda_3\lambda_4=C_1\lambda_1^2\lambda_4+\alpha_1(\lambda_1,\lambda_2,\lambda_3).$$

We multiply $(*_k)$ by λ_3. Then on A_3 we get

$$\lambda_3^{k+1}\lambda_4=C_k\lambda_1^{2k}\lambda_3\lambda_4+\lambda_3\alpha_k.$$

Replacing $\lambda_3\lambda_4$ by the preceding expression, we get the decomposition sought for $\lambda_3^{k+1}\lambda_4$.

Conclusion of the Proof of Theorem 3. Using the lemma, we represent $\lambda_4 d$ on A_3 in the form

$$\lambda_4 d(\lambda_1,\lambda_3)=\lambda_4 p(\lambda_1)+\beta(\lambda_1,\lambda_2,\lambda_3),$$

which gives the decomposition sought for f.

The linearity and quasihomogeneity of all steps of the construction are obvious.

Proof of Theorem 4. The first assertion is obvious, since p has, for suitable λ_4, a root of multiplicity 3, if and only if p' has a root of multiplicity 2.

The second assertion follows from the third, since the projections of the vector fields in three-dimensional space indicated in the table generate the module of all vector fields tangent to the swallow's tail over the ring of functions.

In fact, formulas for the vector fields in three-dimensional space, forming a (free) basis for the module of vector fields tangent to the standard swallow's tail, are given in [1]. We evaluate these formulas in the case of an expanded swallow's tail (defined by the discriminant of the polynomial $5x^4+3\lambda_1 x^2+2\lambda_2 x+\lambda_3$, instead of that of $x^4+\lambda_1 x^2+\lambda_2 x+\lambda_3$). Then we subtract the first field multiplied by λ_1 from the third, so as to the formulas following. As a result we get in three-dimensional space a basis for the module of fields tangent to the expanded swallow's tail, consisting of three fields, quasihomogeneous of degrees 0, 1, and 2, whose components are given by the rows in the first three columns of the table of components.

Now we calculate the derivative of the rational function $\lambda_4(\lambda_1,\lambda_2,\lambda_3)$ of Theorem 1 in the direction of each of these fields in three-dimensional space. We get three rational functions of $(\lambda_1,\lambda_2,\lambda_3)$, which by "lucky chance" coincide on A_3 with the polynomials indicated in the last column of the table. For the field of degree 0 the coincidence is obvious, since this is the Euler field of a quasihomogeneous dilatation. We prove the coincidence for the second and third fields. It suffices to verify the relation obtained by differentiation of the polynomial (8): on A_3,

$$-75(\dot\lambda_4\lambda_2+\dot\lambda_2\lambda_4)+(3\lambda_2^2-18\lambda_1\lambda_3)\dot\lambda_1+6\lambda_1\lambda_2\dot\lambda_2+(40\lambda_3-9\lambda_1^2)\dot\lambda_3=0.$$

Substituting the components of the field from the second row of the table, we get a quasihomogeneous polynomial of degree 9. Up to a numerical factor (equal to 15) it coincides by lucky chance with the polynomial (9) and hence vanishes on A_3.

Substituting the components of the third row, we get a quasihomogeneous polynomial of degree 10. Luckily it turns out to be equal to the product of $18\lambda_1$ by the polynomial (8) of the proof of Theorem 1 and hence also vanishes on A_3. Theorem 4 is proved.

D. Sweeping Out of a Swallow's Tail. We turn to the study of the sweeping out of a bicaustic by the cuspidal edges of instantaneous caustics. The cuspidal edges of instantaneous caustics in space–time are found from the intersection of the manifold A_3 with isochrones, i.e., they are level lines of the time function on the surface A_3.

The time function can be reduced in a neighborhood of a singular point A_5 by a diffeomorphism preserving the big caustic to the normal form $t = \pm\lambda_1 + \text{const}$. Thus we get a normal form for families of cuspidal edges of instantaneous caustics on the surface $A_3 : \lambda_1 = \text{const}$.

In order to study the sweeping out of the bicaustic by the cuspidal edges it remains to project the surface A_3 from four-dimensional space–time to three-dimensional Euler space along world lines.

The projection also reduces to normal form $(\lambda_1, \lambda_2, \lambda_3, \lambda_4) \to (\lambda_1, \lambda_2, \lambda_3)$ by a diffeomorphism preserving the big caustic. The difficulty is that this diffeomorphism in general destroys the normal form $t = \lambda_1 + \text{const}$ of the time function on A_3.

Thus, to study the sweeping out of a swallow's tail on a bicaustic by the instantaneous cuspidal edges it is necessary to reduce the time function on the surface A_3 to the simplest possible form by diffeomorphisms of four-dimensional space–time, which are lowerings of the standard projection; these diffeomorphisms must carry the surface A_3 into itself but do not necessarily carry the big caustic into itself.

Definition. Two functions defined in a neighborhood of the origin of the $(\lambda_1, \lambda_2, \lambda_3, \lambda_4)$ space are said to be *strongly equivalent* if their restrictions to A_3 are carried into one another by a diffeomorphism preserving the surface A_3 and carrying lines in the direction λ_4 into lines in the same direction.

Functions are said to be *weakly equivalent* if their restrictions to A_3 are carried into one another by a strong equivalence and diffeomorphism of the axis of values of the functions.

The preceding definition can be applied to the cases of holomorphic or smooth functions and mappings, and also (in the natural sense) to the case of formal power series. We shall use the terms *holomorphic equivalence,..., formal equivalence*, to differentiate these cases.

We show that a time function in general position can be reduced on A_3 in a neighborhood of zero by a weak formal equivalence to the normal form $\pm\lambda_1 + \lambda_4$. Thus, the family of instantaneous cuspidal edges can be reduced in the neighborhood of a swallow's tail of a bicaustic (at least at the level of formal power series) to the form $\pm\lambda_1 + \lambda_4 = \text{const}$ by a diffeomorphism of the three-dimensional Euler space containing the swallow's tail and a change of time.

We note that the function λ_4 is two-valued on the line of self-intersection of the swallow's tail (there $\lambda_4 = \pm c\lambda_1^{5/2}$). Hence the times when the parts of the cuspidal edge of the instantaneous caustics, sweeping out one or another self-intersecting part of the bicaustic, pass through a point of the line of self-intersection are different.

The normal form $\lambda_1 \pm \lambda_4$ shows that in time t after the bifurcation A_5 both times when the cuspidal edge of an instantaneous caustic pass through the same point of the Euler space are separated by an interval of order $t^{5/2}$.

Below we prove the following:

THEOREM 5. Let $f(0) = 0$, $\partial t/\partial\lambda_1|_0 \neq 0$. Then the function t is strongly formally equivalent to the function $\pm\lambda_1 + \lambda_4\varphi(\lambda_1)$. Here $\varphi(0) \neq 0$, if $\partial t/\partial\lambda_4|_0 \neq 0$.

THEOREM 6. Let $\partial t/\partial\lambda_1|_0 \neq 0$, $\partial t/\partial\lambda_4|_0 \neq 0$. Then the function t is weakly formally equivalent to the function $\pm\lambda_1 + \lambda_4$.

Remark 1. Apparently the reduction to normal form of Theorem 5 is also possible in the smooth and holomorphic cases.

An arbitrary function of one variable must necessarily participate in the normal form with respect to strong equivalence. In fact, the function which expresses one of the times when the cuspidal edge passes through a point of self-intersection of the swallow's tail in terms of the other is invariantly defined with respect to strong equivalences.

Remark 2. The reduction to normal form of Theorem 6 is apparently possible in the smooth case also, but is trivially impossible in the class of holomorphic diffeomorphisms.

In fact, we consider the preimage of the line of self-intersection of the bicaustic on the surface A_3. On this line two involutions are defined: in one case we interchange a point with its general projection on the bicaustic, and in the other case with the point on the same isochrone.

Such a pair of involutions reduces to simple normal form in the class of formal power series, but in the holomorphic case these series diverge as a rule: pairs of involutions even have functional moduli as was shown by Voronin [7].

E. Proof of Theorem 5. By hypothesis, $t = a\lambda_1 + \ldots$, where the dots denote terms of quasihomogeneous degree higher than 2 (since $\deg \lambda_i = i + 1$). By dilatations we can achieve $a = \pm 1$. Expanding t by Theorem 3, we get

$$t = \pm\lambda_1 + u(\lambda_1, \lambda_2, \lambda_3) + \lambda_4 w(\lambda_1). \qquad (*)$$

We shall denote the *quasihomogeneous order of zero* of functions (the smallest quasihomogeneous degree of monomials of the series) by ord. We have $\operatorname{ord} u \geqslant 3$, $\operatorname{ord} \lambda_4 w \geqslant 5$.

We construct an equivalence of t with a function of the same form with bigger $\operatorname{ord} u$. We say that a vector field has *a zero of quasihomogeneous order* $\geqslant\delta$ at zero if upon differentiating in the direction of the field the quasihomogeneous order of zero of any function increases by at least δ:

$$\operatorname{ord} L_v f \geqslant \operatorname{ord} f + \delta$$

(here and below L_v denotes the derivative along v).

LEMMA 1. Each power series in $(\lambda_1, \lambda_2, \lambda_3)$ without free term can be represented as a combination

$$f = 2\lambda_1 h_0 + 20\lambda_2 h_1 + (20\lambda_3 - 9\lambda_1^2) h_2$$

of the first components of the three fields v_i of Theorem 4 with the coefficients of $h_i(\lambda_1, \lambda_2, \lambda_3)$, where $\operatorname{ord} h_i \geqslant \operatorname{ord} f - 2 - i$.

Proof. The first assertion is Hadamard's lemma, the second follows from the fact that the components are quasihomogeneous of degrees 2, 3, 4.

LEMMA 2. Let v be a vector field with zero of positive order δ. We denote by g the transformation of the phase flow in time 1 and by g^* its action on functions:

$$(g^*F)(x) = F(g(x)).$$

Then

$$g^*F = F + L_v F + R, \text{ where } \operatorname{ord} R > \operatorname{ord} F + \delta.$$

Proof. By Taylor's formula $g^* = \exp(L_v)$. Since $\operatorname{ord} L_v^n \geqslant n\delta > \delta$, one has $\operatorname{ord}(g^*F - F - L_v F) \geqslant \operatorname{ord} F + 2\delta$, which is what was required.

We conclude the proof of the theorem. By Lemma 1, there exists a field h in four-dimensional space, for which

1) $L_h\lambda_1 = -u$,

2) $h = h_0 v_0 + h_1 v_1 + h_2 v_3$, $h_i = h_i(\lambda_1, \lambda_2, \lambda_3)$,

3) $\operatorname{ord} h \geqslant \operatorname{ord} u - 2$.

By Theorem 4 the transitions of the phase flow of this field preserve the surface A_3 and carry lines in the direction λ_4 into lines in the same direction.

By Lemma 2, the transformation in time 1 carries the function (*) into the function

$$g^*t = \pm\lambda_1 + [L_h\lambda_1 + u] + \lambda_4 w(\lambda_1) + \ldots,$$

where the dots denote the remainder of order *strictly* higher than $\operatorname{ord} u$.

Expanding this remainder by Theorem 3, we again get an expansion of the form (*), but with higher order of zero of the function u.

Repeating this process an infinite number of times, we completely eliminate the summand u, which is what was required.

(The terms of any fixed degree in the Taylor series stabilizes after a finite number of steps; the whole procedure really determines a formal change of variables.)

It remains to verify that $\varphi(0)\neq 0$ for $\partial t/\partial\lambda_4|_0 \neq 0$. For this we note that the diffeomorphisms which we used carry the λ_4 axis into itself, so under the diffeomorphisms the condition $\partial t/\partial\lambda_4|_0 \neq 0$ is not violated. It is also not violated under replacement of a function by another which agrees with the given on A_3:

LEMMA 3. If a function is equal to zero on A_3, then its derivative with respect to λ_4 at zero is equal to zero.

Proof. In view of the quasihomogeneity of the surface A_3, the terms of degree 5 in the series must vanish on A_3. These terms have the form $a\lambda_4 + b\lambda_1\lambda_2$. But such a combination is equal to zero on the whole surface A_3 only if $a = b = 0$ (by Theorem 1).

Theorem 5 is proved.

Remark. The diffeomorphism g *has quasihomogeneous order d* if for each function f one has $\operatorname{ord}(g^*f - f) \geqslant \operatorname{ord} f + d$. The diffeomorphisms of strictly positive order form a group. We reduced the time function on A_3 to the normal form of Theorem 5 by diffeomorphisms of positive order not less than $\operatorname{ord} u - 2$.

F. Proof of Theorem 6. 1°. We translate the time origin to achieve $t(0) = 0$. By Theorem 5 we reduce t to the form $\pm\lambda_1+\lambda_4\varphi(\lambda_1)$, $\varphi(0)\neq 0$.

2°. We denote $\varphi(0)$ by a^3. A quasihomogeneous dilatation $t' = a^2 t$, $\lambda_i' = a^{i+1}\lambda_i$ turns $\varphi(0)$ into one.

3°. Let $t = \pm\lambda_1 + \lambda_4 + c\lambda_4\lambda_1^k + \dots$.

LEMMA. The term $c\lambda_4\lambda_1^k$ can be eliminated by an admissible diffeomorphism (preserving the surface A_3 and carrying lines in the direction of the λ_4 axis into lines of the same direction) of order $\geqslant 2k$ and the time change $t' = t + At^{k+1}$.

The *proof of the lemma* is in points 4°–7°.

4°. Let $v_0 = 2\lambda_1\partial/\partial\lambda_1 + 3\lambda_2\partial/\partial\lambda_2 + 4\lambda_3\partial/\partial\lambda_3 + 5\lambda_4\partial/\partial\lambda_4$ be the first field of Theorem 4. We shall seek a diffeomorphism in the form of a transformation of the phase flow of the field $\alpha(\lambda_1)v_0$, where $\operatorname{ord}\alpha \geqslant 2k$.

It is clear that such a transformation is admissible. We denote it by g. The diffeomorphism g has order $\geqslant 2k$.

5°. We let $t_1 = g^*t$, $t_2 = t_1 + At_1^{k+1}$. We calculate the smallest terms of the expansions of t_1 and t_2. It will be necessary for us to analyze the cases $k = 1$ and $k > 1$ separately.

6°. If $k = 1$, we choose $\alpha = p\lambda_1 + q\lambda_1^2$ and we calculate, neglecting summands of order higher than seven. We get

$$t_1 = \pm\lambda_1 \pm 2p\lambda_1^2 + \lambda_4 + (2q \pm 4p^2)\lambda_1^3 + (c + 5p)\lambda_1\lambda_4 + \dots,$$
$$t_2 = \pm\lambda_1 + (A \pm 2p)\lambda_1^2 + \lambda_4 + (2q \pm 4p^2 + 4Ap)\lambda_1^3 + (c + 5p \pm 2A)\lambda_1\lambda_4 + \dots.$$

We choose the parameters A, p, and q, so that all three terms in parentheses in the expression for t_2 become zero. The vanishing of the first and third parenthetical terms determines A and p, after which we find q from the second parenthetical term.

7°. If $k > 1$ we choose $\alpha = p\lambda_1^k$ and we calculate, neglecting summands of order higher than $2k + 5$.

We get

$$t_1 = \pm\lambda_1 + \lambda_4 \pm 2p\lambda_1^{k+1} + (c + 5p)\lambda_1^k\lambda_4 + \dots,$$
$$t_2 = \pm\lambda_1 + \lambda_4 + P\lambda_1^{k+1} + Q\lambda_1^k\lambda_4 + \dots,$$

where

$$P = A(\pm 1)^{k+1} \pm 2p,\quad Q = c + 5p + A(k+1)(\pm 1)^k.$$

The equations $P = Q = 0$ determine A and p. The lemma is proved.

8°. Applying the lemma, we reduce our function to the form $\pm\lambda_1 + \lambda_4 + R$ with remainder term of higher order, $\operatorname{ord} R > \operatorname{ord}\lambda_1^k\lambda_4 = 2k + 5$.

9°. Using Theorem 5 again, we get a function of the form $\pm\lambda_1 + \lambda_4 + b\lambda_4\lambda_1^{k+1} + \dots$, equivalent to the original function.

Repeating this process, we can shove it aside arbitrarily far.

10°. All the diffeomorphisms and time changes of the lemma and of Theorem 5 are of strictly positive order, increasing with the order of the shove aside, k. Hence the terms of any fixed degree in the Taylor series of the substitution effecting the equivalence stabilize as $k \to \infty$. The formal substitutions found as $k \to \infty$ establish the equivalence required in Theorem 6.

LITERATURE CITED

1. V. I. Arnold, "Wave front evolution and equivariant Morse lemma," Commun. Pure Appl. Math., 39, No. 6, 557-582 (1976).
2. V. I. Arnol'd, "Theory of catastrophes," Priroda, No. 10, 54-63 (1979).
3. V. I. Arnol'd, "Normal forms of functions near degenerate critical points, Weyl groups, and Lagrangian singularities," Funkts. Analiz, 6, No. 4, 3-25 (1972).
4. V. I. Arnol'd, Mathematical Methods of Classical Mechanics [in Russian], Moscow (1974).
5. V. I. Arnol'd, "Indices of singular points of 1-forms on a manifold with boundary, displacement of invariants of groups generated by reflections, and singular projections of smooth surfaces," Usp. Mat. Nauk, 34, No. 2, 3-38 (1979).
6. T. Brecker and L. Lander, Differentiable Germs and Catastrophes [Russian translation], Mir, Moscow (1977).
7. S. M. Voronin, "Analytic classification of germs of conformal maps $(C, 0) \to (C, 0)$ with linear part the identity," Funkts. Analiz, 15, No. 2, 10-30 (1981).
8. Ya. B. Zel'dovich, "Disintegration of homogeneous matter into parts under the action of gravity," Astrofizika, 6, No. 2, 319-335 (1970).
9. Ya. B. Zeldovich, "Gravitational instability: an approximate theory for large density perturbations," Astron. Astrophys., 5, 84-89 (1970).
10. O. V. Lyashko, "Geometry of bifurcation diagrams," Usp. Mat. Nauk, 34, No. 4, 205-206 (1970).
11. H. Whitney, "The general type of singularity of a set of $2n - 1$ smooth functions of n variables," Duke Math. J., 45, 220-283 (1944).

On the Newtonian Potential of Hyperbolic Layers*

V. I. Arnold

According to a famous theorem of Newton, a homogeneous spherical shell does not attract interior points. Here we shall formulate and prove two generalizations of this theorem: to the case (I) of layers bounded by hypersurfaces of the second degree; (II) of level hypersurfaces of hyperbolic polynomials of arbitrary degree in a space of any number of dimensions (in n-space the force of attraction is inversely proportional to the $(n-1)$th power of the distance).

The simplest of the generalizations is as follows. In the (x, y) plane consider the hyperbola $xy = 1$. We charge one of its branches positively and the other negatively, with density proportional to the thickness of the layer between the given hyperbola and an infinitely close level line of the function xy. Then the hyperbola does not attract points within one of its branches (i.e., in the domain $xy > 1$).

1. The charge associated with a point

Let A be a point on a given hypersurface in $\mathbf{R}^n$.

Definition. The *index* of A with respect to the point O is the number of points of the hypersurface that are strictly interior to the interval (OA). Notation: $\operatorname{ind}(O,A)$.

We consider a hypersurface defined by the equation $f = 0$ in the Euclidean space $\mathbf{R}^n$. On it we define the differential form $\omega = (dx_1 \wedge \cdots \wedge dx_n)/df$ (the density of the layer $0 < f < df$ of infinitesimal thickness).

Definition. The charge on the hypersurface $f = 0$, *associated with the point* O, is the $(n-1)$-form defined at each point A belonging to the hypersurface by the formula $\omega_0 = (-1)^{\operatorname{ind}(O,A)}|\omega|$.

Originally publ. in: Tr. Tbiliss. Univ., Ser. Mat., Mekh., Astron. 232/233(13-14), 23-29,

English transl.: Selecta Math. Sov. 4(2), 103-106,

Example. Let $f = 0$ be the usual equation of a circle. If the point O is inside the circle then ind $\equiv 0$ and the charge associated with O is distributed uniformly over the circle. If O is outside the circle, the charge is positive on the arc inverse to O and negative on the complementary arc where ind $= 1$. The charge associated with the point $(1, 1)$ on the hyperbola $xy = 1$, oriented by the condition $dx > 0$, has density dx/x.

2. Hyperbolic polynomials

Let O be a point not on an algebraic hypersurface in affine space.

Definition. The point O is *hyperbolic* (with respect to the given hypersurface) if all the intersections of each complexified real line through O with the (complexified) surface are real.

Example. A point inside a disk is hyperbolic with respect to its circumference; the point $(1, 1)$ is hyperbolic with respect to the hyperbola $xy = 1$. If a curve of degree $2m$ consists of m ovals, one inside another, then the domain bounded by the innermost oval consists completely of hyperbolic points.

A hypersurface, and its defining polynomial, are called hyperbolic if they possess one or more hyperbolic points.

3. Statement of results

Consider a nonsingular surface of *second* order, defined in $\mathbf{R}^n$ by the equation $f = 0$, where f is a polynomial of degree 2.

Theorem 1. *A point not on the surface of second order is not attracted by its associated charge on the surface.*

Now consider a singular *hyperbolic* surface in $\mathbf{R}^n$.

Theorem 2. *A point that is hyperbolic with respect to a surface is not attracted by its associated charge.*

Example. Consider a curve of degree $2m$ consisting of m ovals successively contained in one another. Distribute the charge on the curves with density $(-1)^i|\mathrm{grad}\, f|$, where i is the number of the oval. Then the curve does not attract points inside the innermost oval.

4. Proofs

The proof of Theorem 1 is in essence contained in Newton's *Principia*. Consider an infinitesimal solid angle with vertex at O and containing the line OA. The force of attraction of the part of the layer intercepted by this angle, in the neighborhood of an intersection of OA with the surface, say A_i, is equal to $dF_i = -e_{A_i} R_i^{1-n}\, dm_i$, where e_{A_i} is a unit vector along OA_i, R_i is the length of OA_i, and dm_i is the differential charge. The volume of a prism is equal to the product of the length of an edge and the area of its projection on a plane orthogonal to an edge. Therefore $|dm| = R^{n-1}|d\Omega|\,|dR|$, where $|d\Omega|$ is the measure of the solid angle, and $|dR|$ is the thickness of the layer along OA. Thus,

$$dF_i = -e_{A_i}|d\Omega|\,|dR_i|(-1)^{\mathrm{ind}(O,A_i)}.$$

However, the line segments intercepted by the layer between two surfaces of second order, $f = a$ and $f = b$, near both points A_1, A_2 of intersection of a straight line with the surfaces, have the same length: $|dR_1| = |dR_2|$. The directions of the unit vectors e_{A_i} are the same when the signs $(-1)^{\mathrm{ind}(O,A_i)}$ are different and opposite when the signs are the same. Hence the sum of the two terms is zero, and it follows that the total force of attraction is zero (after a simple investigation of the convergence of the integrals).

The proof of Theorem 2 is similar, except that instead of two terms we obtain $d = \deg f$ terms of the form

$$dF_i = -|d\Omega|e_{A_i}|dR_i|(-1)^{\mathrm{ind}(O,A_i)},$$

where A_i run through the d points of intersection of OA with the surface.

The sum of the d terms is zero by the formula

$$\sum \frac{1}{p'(x_i)} = 0$$

(sum over all d simple roots of a polynomial p of degree d). This is the simplest special case of the Euler–Kronecker–Jacobi formula; applying the formula to the restriction of f to the line OA, we obtain $\sum dF_i = 0$, which establishes Theorem 2 (in which the density of the charge may be multiplied by any polynomial of degree less than $d - 1$).

5. Remarks

1. The author has not been able to find Theorem 1 in the literature, although it is hinted at in N. E. Zhukovsky's *Theoretical Mechanics*.

2. Theorem 2 was suggested by a paper of I. G. Petrovsky on lacunas for hyperbolic equations, although there is no obvious direct connection between potentials and the Herglotz–Petrovsky formula.

3. The theorem on the attraction of a sphere on an exterior point can also be generalized: hypersurfaces of second order can be replaced by confocal ones without changing the field in the corresponding domains; that is, Ivory's theorem on ellipsoidal layers can be extended to hyperbolic layers [**1**].

4. The results presented above arose as a by-product of an investigation of Ya. B. Zeldovich's model of the clustering of matter. In this model one considers potential flows of dust particles moving in their own gravitational field. The velocities of the particles form a Lagrangian manifold in phase space, evolving over time. Clustering occurs when the projection of the Lagrangian manifold on configuration space has singularities. Such singularities produce singularities of the density of particles and consequently of the gravitational field. A singularity of the field can make an originally smooth Lagrangian manifold become unsmooth, or at least less smooth. Newton's theorem shows that restriction of the gravitational field to the Lagrangian manifold produced by a Lagrangian singularity D_4 of elliptic type is nonsingular. The singularity D_4 has another "hyperbolic" real form. Investigation of the singularities of the corresponding field leads to the necessity of considering hyperbolic layers instead of elliptic ones; the present paper is a first step in this direction.

Reference

[1] V. I. Arnold, *Magnetic field analogues of Newton's and Ivory's theorems* (in Russian), Uspehi Mat. Nauk **38**:5 (1983), 145–146.

ON THE NEWTONIAN ATTRACTION OF CLUSTERS OF DUST-LIKE PARTICLES

V.I. ARNOLD

(Joint sessions of the Petrovskiĭ seminar on Differential equations and mathematical problems of physics and the Moscow Mathematical Society; the evening meeting on January 20, 1982, section 1.[1])

Potential motions of a dust-like medium lead to the formation of density singularities. To study the singularity D_4, one has to generalize to the hyperbolic case Newton's theorems on the attraction of spherical layers (in $\mathbb{R}^n$ the attraction force is proportional to r^{1-n}).

Theorem. *A non-singular hypersurface $f = 0$ of degree 2, charged with a density of $|\nabla f|^{-1}$ with the sign "+" at the points visible from P and the sign "−" at the points invisible from P, does not attract the point P.*

For instance, the hyperbola $xy = 1$ does not attract the points where $xy > 1$.

Theorem. *A non-singular hypersurface $f = 0$ of degree n, hyperbolic with respect to a point P (i.e. such that all its points of intersection with straight lines passing through P are real), does not attract the point P if the charge is distributed with a density of $|\nabla f|^{-1}$ with the positive sign on the component nearest to P, with the negative sign on the next component, with the positive sign on the third component, and so on. This result is still valid if the density is multiplied by any polynomial of degree less than $n-1$.*

This theorem, suggested by I.G. Petrovskiĭ's work on lacunas, is proven in the same way as Newton's theorem on the attraction inside a sphere.

Example. A curve of degree $2m$ with m ovals inside one another does not attract points inside the most interior oval provided that the density is $dx \wedge dy/df$ while the integration chain is oriented by Rokhlin's "complex orientation".

---

[1]Translation of Uspekhi Mat. Nauk, **37** (1982), no. 4, p. 125.

Originally publ. in: Usp. Mat. Nauk 37(4), 125,
Translated by Boris Khesin and Andrey Khesin
Springer-Verlag Berlin Heidelberg 2018 with kind permission

Some Algebro-Geometrical Aspects of the Newton Attraction Theory*

V. I. Arnold

To I. R. Shafarevich

According to the Zeldovich theory[1], the observed large scale structure of the universe (the drastically non-uniform distribution of galaxy clusters) is explained by the geometry of caustics of a mapping of a Lagrange submanifold of the symplectic total space of the cotangent bundle to its base space. This Lagrange submanifold is formed by the particle velocities. Contemporary theory of the hot universe predicts a smooth potential velocity field at an early stage (when the universe was about 1000 times "smaller" than now). At this stage the Lagrange manifold is a cotangent bundle section. Then it evolves according to Hamiltonian equations of motion, and hence continues to be Lagrangian. However, it does not need to be a section at all times. The set of critical values of its projection on the base space is called the caustic. At the caustic the particle density becomes infinite (mathematically); the caustic is the place where clustering occurs (generation of galaxies and so on).

The singularities of caustics and their metamorphoses are classified in a way usual for the Lagrangian singularity theory: $A_k, D_k, E_k, \ldots$. This is true for non-interacting particles or for particles in any potential field. But in cases where the field is generated by the particles, a new difficulty occurs. After the caustic has been formed the force field is no longer smooth (because of density singularities). Hence our Lagrange manifold may acquire singularities. Thus we are led to the problem: to generalize the Lagrange mappings singularity theory to the case where the Lagrange

*from a letter to M. Artin, May 10, 1982

[1]Arnold, V.I., Shandarin, S.F., Zeldovich, Ya. B., the Large Scale Structure of the Universe I, Geophys, Astrophys. Fluid Dynamics 1982.

Arnold, V.I., Surgery of singularities in potential collisionless media and caustics metamorphoses in 3-space, Trudy Seminary, 3, (1982), 22–58.

Originally publ. in: Arithmetic and Geometry. Papers Dedicated to I.R. Shafarevich. Vol. II: Geometry (Progr. Math. 36), 1-3,
Reprint by Springer-Verlag Berlin Heidelberg 2018 with kind permission

source manifold becomes a Lagrange variety.

[Lagrange varieties occur also in other situations, for instance in the study of the shortest length function on a manifold with boundary (see V.I. Arnold, Lagrange varieties singularities, asympototical rays and the open swallowtail, Funkt. Anal. 15 (1982) 1–14). In this case typical singularities of Lagrange varieties are those of the set of odd degree polynomials having a root of multiplicity greater than one half of the degree. (The symplectic structure of the space on polynomials is inherited from the invariant symplectic structure on the space of binary forms.) But Lagrange variety singularities arising from the Newton attraction theory seem to be different (and in any case they are not known).]

The first step in the study of such Lagrange variety singularities is the study of the singularities of force fields, generated by clustering of free particles. Among typical clusterings in the plane, elliptic and hyperbolic D_4 singularities occur at some (exceptional) moments in time. At these singularities, the particle density is inversely proportional to a quadratic form (in suitable local coordinates). Thus constant density lines are homothetic ellipses or hyperbolas. We are led to the problem of Newtonian attraction by an elliptic or hyperbolic layer.

The theory of attraction by elliptic layers is classic. The first results are due to Newton: a uniform spherical layer does not attract interior points, and attracts exterior ones as if the mass were concentrated at its center. These results were extended to the case of ellipsoids by Ivory, but the hyperbolic case seems not to be settled by classical authors.

Two generalizations of the Newton interior points theorem are formulated below. Let us consider a level hypersurface of a real polynomial in a euclidean space of dimension h, and a point not on the hypersurface. We call natural the density inversely proportional to the gradient length. Let us distribute a Coulomb charge (force inversely proportional to the $(n-1)$st power of the distance) along the hypersurface, with natural density but with a sign, depending on the chosen point: + for points of the hypersurface which one can see from the chosen one, − for those obstructed once, + for those obstructed twice, and so on. We call such a charge "associated" to the chosen point. Then the following generalized Newton theorem holds:

(1) The charge on a second degree hypersurface, associated to a point, does not attract this point.

For instance, let us consider a plane hyperbola with natural density and component charges of different signs. Such a hyperbola does not attract points of the convex parts of the plane it bounds.

Now let us consider any real algebraic hypersurface. We call a point p hyperbolic with respect to the hypersurface if all real lines through p meet the hypersurface at real points only. The following generalization of Newton's theorem holds:

(2) The charge associated to a hyperbolic point does not attract this point.

For instance, let us consider a plane algebraic curve of even degree, consisting of a sequence of ovals one inside other (their number being equal to one half of the degree). Let the charges change signs from each oval to the next one and let the density be natural. Such a curve does not attract points inside the inner oval.

The theorem on the attraction of exterior points leads to results in the geometry of confocal quadrics which seem to be new. The generalization to nonconnected hyperboloids is direct; for instance, the attraction of points between the components of the hyperbola charged as above does not change if we substitute for the hyperbola a larger confocal one. For quadratic forms of other signatures one needs to consider differential forms of appropriate degree instead of charges and instead of the Newton-Coulomb attraction law one needs to consider the Biot-Savart law and its higher dimensional analogues.

Received June 2, 1982

Professor V. I. Arnold
Mathematics Department
Moscow State University
Moscow 117 234
USSR

MAGNETIC ANALOGUES OF NEWTON'S AND IVORY'S THEOREMS

V.I. ARNOLD

(Joint sessions of the Petrovskiĭ seminar and the Moscow Mathematical Society; the evening meeting on January 19, 1983, section 2.[1])

The interior (simply connected) domain bounded by a one-sheet hyperboloid in $\mathbb{R}^3$ is foliated into the meridians of the corresponding elliptic coordinate system, while the exterior domain is foliated into the parallels (diffeomorphic to circles).

Theorem. *The magnetic field of the current of an appropriate density, flowing along the surface of a hyperboloid along the meridians (respectively, the parallels), vanishes in the interior domain and is directed along the parallels in the exterior one (respectively, it vanishes in the exterior domain and is directed along the meridians in the interior one).*

The density of the meridian current is determined by the condition that its flow through any curve on the hyperboloid is proportional to the volume of an infinitesimally thin homeoidal neighborhood of the projection of this curve to the focal ellipse along the two-sheet hyperboloids. The density of the current flowing along the parallels is determined by the corresponding projection to the focal hyperbola.

The magnetic field in the interior domain outside of a charged conductive ellipsoid confocal to the hyperboloid coincides up to a sign with the electric field of the ellipsoid. For the exterior domain, the ellipsoid is replaced by a two-sheet hyperboloid $f = 0$ on which an infinite charge is distributed with the homeoidal density $\mathrm{d}x\,\mathrm{d}y\,\mathrm{d}z/\mathrm{d}f$, whose electrostatic meaning has been clarified by O.P. Shcherbak.

B.Z. Shapiro and A.D. Vaĭnshteĭn have carried over all these results to a hyperboloid in $\mathbb{R}^n$, diffeomorphic to $S^k \times \mathbb{R}^l$, by constructing a harmonic k-form in the exterior domain and a harmonic l-form in the interior domain.

The vanishing of the corresponding Biot–Savart integrals of algebraic differential forms is a non-trivial identity similar to Petrovskiĭ's lacunas, while the discovered harmonic forms in algebraic non-compact domains suggest "a mixed de Rham structure".

The constructed fields also provide exact solutions of the problems on a potential flow of an incompressible fluid through the interior domain of a triaxial hyperboloid and on an irrotational flow past the hyperboloid.

[1]Translation of Uspekhi Mat. Nauk, **38** (1983), no. 5, 145–146.

Originally publ. in: Usp. Mat. Nauk 38(5), 145-146,
Translated by Boris Khesin and Andrey Khesin
Springer-Verlag Berlin Heidelberg 2018 with kind permission

SOME REMARKS ON ELLIPTIC COORDINATES

V. I. Arnol'd

UDC 519.4

A coordinate-free exposition of the theory of multidimensional Jacobi coordinates, which are used to integrate the equations of geodesics on an ellipsoid as well as other equations, is given. Generalizations of the theorems of Newton and Ivory, concerning the attraction field produced by an ellipsoid, are given.

It is well known that with every ellipsoid in finite-dimensional Euclidean space are related Jacobi elliptic coordinates, in terms of which one can integrate the equations of geodesics on this ellipsoid, as well as other equations, for example, the equations of motion of a point mass on the sphere under the action of a force field with quadratic potential.

This suggests the idea that in the infinite-dimensional, Hilbert space too with every symmetric operator there should be connected its own class of integrable systems. To investigate these systems one must carry over the theory of elliptic coordinates to the infinite-dimensional case. To this end, in turn, one must first of all give an exposition of the usual finite-dimensional theory of confocal second-degree surfaces in coordinate-free language.

Such a coordinate-free exposition of the theory of Jacobi's elliptic coordinates was given in the report [1] and is reproduced here. To pass to the infinite-dimensional case one must replace everywhere symmetric operators in finite-dimensional Euclidean space by self-adjoint operators in Hilbert space. Note that since elliptic coordinates are related not to the operator itself, but rather to its resolvent, the fact that the given operator (which can be, for example, differential) might be unbounded is not a very serious obstruction.

In certain cases the elliptic coordinates obtained in this way in Hilbert space form a countable collection. However, the case of continuous spectrum may arise, in which the resulting collection of coordinates is continual. In this case the passage from a given point of the Hilbert (say, functional) space to the continual set of elliptic coordinates of this point may be viewed as a linear transformation of the functional space. By analogy with the Fourier transform, this transformation may be referred to as the Jacobi transform: it sends a given function into the function that expresses the dependence of the continual elliptic coordinate on its index number (i.e., a measure on the axes of the spectral parameter). Hoping that the investigation of functional-elliptic properties of the direct and inverse Jacobi transforms is not a matter of too distant future, I confine myself here to a preliminary preparation and give a geometric, noncomputational description of the finite-dimensional theory of elliptic coordinates (a description in terms of formulas can be found in Jacobi [2] and Moser [3]).

Translated from Zapiski Nauchnykh Seminarov Leningradskogo Otdeleniya Matematicheskogo Instituta im. V. A. Steklova AN SSSR, Vol. 133, pp. 38-50, 1984.

Originally publ. in: Zap. Nauchn. Semin. Leningr. Otd. Mat. Inst. Steklova 133, 38-50,

English transl.: J. Soviet Math. 31(6), 3280-3289,

1. Elliptic Coordinates and Confocal Quadrics

Elliptic coordinates in Euclidean space are defined by means of confocal quadrics (surfaces of degree two). The geometry of confocal quadrics is in turn obtained from the geometry of pencils of quadratic forms in Euclidean space (i.e., from the theory of principal axes of ellipsoids or the theory of small oscillations) by passing to the conjugate (dual) space.

Definition 1. An Euclidean pencil of quadrics (respectively, quadratic forms) in Euclidean space V is a one-parameter family of second-degree surfaces

$$\tfrac{1}{2}(A_\lambda x, x) = 1$$

(respectively, quadratic forms A_λ), where

$$A_\lambda = A - \lambda E$$

and A is a symmetric operator

$$A: V \to V^*, \quad A^* = A .$$

Definition 2. A confocal family of quadratics in Euclidean space is a family of quadrics conjugate in the quadrics of an Euclidean pencil (of quadrics in the space conjugate to the given one):

$$\tfrac{1}{2}(A_\lambda^{-1}\xi, \xi) = 1 .$$

Thus, mutually confocal quadrics form a one-parameter family, but the quadratic form of this family depends on the parameter in a nonlinear way.

Example. The plane curves confocal to a fixed ellipse are all ellipses and hyperbolas with the same foci.

In Fig. 1 we see below the curves forming a confocal family and above — the curves of the corresponding Euclidean pencil.

The elliptic coordinates of a point relative to a fixed family of confocal quadrics are defined as the values of the parameter λ for which the quadrics in this family pass through the given point.

Fix in Euclidean space an ellipsoid whose axes have all unequal lengths.

THEOREM 1 (Jacobi). Through each point of the n-dimensional Euclidean space are passing n quadrics confocal to the fixed ellipsoid. Smooth confocal quadrics intersect under straight angles.

Proof. To every nonzero point of the given space there corresponds in the conjugate space an affine hyperplane consisting of all linear forms which take the value 1 on this point. In terms of the conjugate space Theorem 1 asserts that every hyperplane which does not pass through 0 in n -dimensional Euclidean space is tangent to exactly n quadrics of a given Euclidean pencil, and that the vectors drawn from 0 to the tangency points are pairwise orthogonal (see Fig. 1).

The proof of the indicated property of Euclidean pencils is based on the fact that the aforementioned vectors determine the principal axes of a suitable quadratic form, namely $B = \frac{1}{2}(Ax,x) - \frac{1}{2}(\ell,x)^2$, where $(\ell,x) = 1$ is the equation of the hyperplane in question.

In fact, if B is an arbitrary quadratic form and λ is one of its eigenvalues, then both form $B - \lambda E$ and its gradient vanish on the principal axis corresponding to 0. The fact that the form B itself vanishes at the point of intersection of this principal axis with the given hyperplane means that the intersection point lies on the quadratic $\frac{1}{2}(A_\lambda x, x) = 1$, whereas the vanishing of the gradient means that at this point the quadratic is tangent to the hyperplane.

THEOREM 2 (Chasles). A general position straight line in n-dimensional Euclidean space is tangent to $n-1$ distinct quadrics; moreover, the planes tangent to these quadrics at the points where they intersect the given line are pairwise orthogonal.

Proof. Let us project the quadrics of the confocal family parallel to a pencil of (parallel) straight lines on the hyperplane perpendicular to the pencil. Every quadric determines an apparent contour (the set of critical values of the projection mapping of the quadric). For directions of projection in general position the apparent contours of the quadrics are surfaces of degree two in the target-hyperplane of the projection.

LEMMA. The apparent contours of the quadrics of a confocal family form themselves a confocal family of quadrics.

Proof. The passage to dual objects takes sections into projections and projections into sections. Therefore, the apparent contours of projections of confocal quadrics along a pencil of parallel straight lines are dual to sections of the dual quadrics by a hyperplane passing through zero.

But the sections of the quadrics of an Euclidean pencil by a hyperplane passing through 0 form an Euclidean pencil of quadrics in this hyperplane. The lemma follows by duality.

Now let us apply the lemma to the projection along the straight line of Theorem 2. By the lemma, the apparent contours of the projection mapping of the quadrics of this theorem form a conformal family of quadrics in a hyperplane. By Theorem 1, these apparent contours intersect under straight angles, which proves Theorem 2.

2. Geodesics on Ellipsoids

THEOREM 3 (Jacobi and Chasles). The tangent lines to a geodesic curve of a quadric in n-dimensional space, taken to all points of the given geodesic, are tangent, besides this quadric, to other $n-2$ quadrics confocal to the given one, which are the same for all points of the geodesic.

Beginning of the Proof. Consider the manifold of oriented straight lines in Euclidean space. This manifold is endowed with a natural symplectic structure, being the manifold of characteristics of the hypersurface: $p^2 = 1$ in the phase space of a free particle in inertial motion in our Euclidean space.

A characteristic on a hypersurface in a symplectic manifold is, by definition, an integral curve of the field of characteristic directions, i.e., of the field of skew-orthogonal complements to the tangent spaces to the hypersurface. In other words, it is a characteristic on the hypersurface in a phase curve (lying on this hypersurface) of the Hamilton equations with a Hamiltonian having a first-order zero on this hypersurface.

The symplectic structure of the manifold of characteristics of a hypersurface in a symplectic manifold is specified by the condition that the skew-scalar product of any two vectors tangent to the hypersurface in the ambient symplectic manifold must be equal to the skew-scalar product of their projections on the manifold of characteristics.

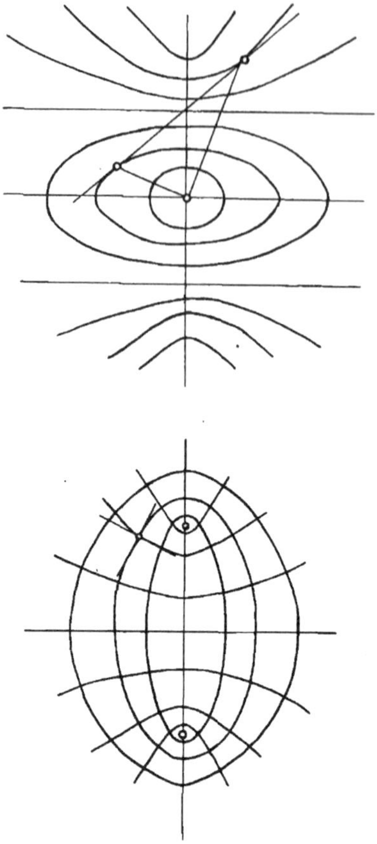

Fig. 1

LEMMA A. Every characteristic of the manifold of all straight lines tangent to a given hypersurface in Euclidean space consists of the tangents to a single geodesic of the hypersurface (taken to all the points of this geodesic).

Proof of Lemma A. For the sake of brevity, here we identify cotangent and tangent vectors in Euclidean space using the Euclidean inner product. Accordingly, we regard the original phase space as the space of vectors emanating from the points of Euclidean space (we identify momenta with velocities). The unit vectors with origin at the points of the given hypersurface and tangent to it form in the phase space a submanifold of odd codimension (equal to 3). The characteristics of this submanifold determine the geodesic flow on the considered hypersurface.

The mapping sending each vector into the line on which it lies takes the indicated submanifold of codimension 3 into the manifold of lines tangent to the hypersurface. This mapping takes characteristics into characteristics (this follows from the definition of the symplectic structure on the space of lines), which proves the lemma.

Remark. The foregoing arguments are readily generalized to the following, more general situation, considered first by Melrose [4] (see also [5, 6]).

Let Y, Z be a pair of hypersurfaces in the symplectic manifold X, which intersect transversally along submanifold W. Consider the manifolds B and C of characteristics of hypersurfaces Y and Z, respectively, together with the canonical bundles with characteristics as fibers, $Y \twoheadrightarrow B$ and $Z \twoheadrightarrow C$. Manifolds B and C inherit symplectic structures from X.

In the intersection W we single out the hypersurface Σ (which thus has codimension 3 in X), at whose point the restriction of the symplectic structures X on W degenerates. Hypersurface Σ in W can be alternately described as the set of critical points of the composite mapping $W \hookrightarrow Y \twoheadrightarrow B$ (or, equivalently, $W \hookrightarrow Z \twoheadrightarrow C$). The objects just introduced form the following commutative diagram

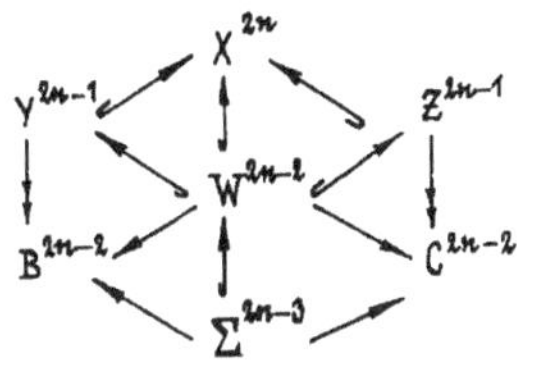

In these circumstances the analog of Lemma A asserts that the characteristics on the images of the maps $\Sigma \to B$ and $\Sigma \to C$ are images of exactly the same curves on Σ (namely, of the characteristics of submanifold Σ of the symplectic manifold X).

Lemma A itself follows from this assertion in the particular case where $X = \mathbb{R}^{2n}$ is the phase space of a free particle moving in $\mathbb{R}^n$, hypersurface Y consists of unit vectors (it is given by the equation $p^2 = 1$, i.e., is a level surface of the free-particle Hamiltonian), and hypersurface Z consists of all vectors with origin on the surface in $\mathbb{R}^n$ under study. In this case B is the manifold of all oriented straight lines in Euclidean space, while Σ is the manifold of unit tangent vectors. The mapping $\Sigma \to B$ takes each unit tangent vector into the line on which it lies. Manifold C is the total space of the (co)tangent bundle to the surface under study. Finally, $\Sigma \to C$ is the embedding in this total space of the total space of the unit sphere bundle (or, in other terms, the embedding of the level hypersurface of the kinetic energy, i.e., of the Hamiltonian of the constrained motion).

It is useful to keep in mind the commutative diagram given above whenever one studies constraints in symplectic geometry.

Continuation of the Proof of Theorem 3. Suppose that in the Euclidean (configuration) space one is given a smooth function and that the restriction of this function to some straight line has a nondegenerate critical point. Under this assumption a similar critical point (i.e., a point at which the line is tangent to a level surface of the given function) exists on each straight line close enough to the given one. The value of the function at the critical point is thus a function of lines. We shall refer to this function of lines as the function induced (by the given function of points).

LEMMA B. Suppose given two functions of points of Euclidean space with the property that the planes tangent to their level surfaces at the points where these surfaces are tangent to some fixed line* are orthogonal. Then the Poisson bracket of the induced functions vanishes at the point representing the given line.

Proof of LEMMA B. Let us compute the derivative of the second induced function along the phase flow defined by the first induced function, taken as a Hamiltonian. The phase curves determined by the first induced function on any of its level surfaces are characteristics of this surface. But each level surface of the first induced function consists of all lines tangent to a fixed level surface of the first function of points. Also, by Lemma A each characteristic of this manifold of lines consists of all tangents to some fixed geodesic on the level surface of the first function of points.

At an infinitesimal shift of the point along the geodesic on the surface the tangent to the geodesic rotates (modulo higher-order infinitesimals) in the plane determined by the original tangent and the normal to the surface. By assumption, the tangent plane to the level surface of the second function at the point where this surface is tangent to our line is perpendicular to the tangent plane to the level surface of the second function. Consequently, at the indicated infinitesimal rotation the line remains tangent (modulo higher-order infinitesimals) to the same level surface of the second function. Thus, the speed at which the second induced function changes under the action of the phase flow given by the first one vanishes at the point in question of the space of lines. This proves Lemma B.

Completion of the Proof of Theorem 3. Fix a line in general position in $\mathbb{R}^n$. By Theorem 2, this line is tangent to $n-1$ quadrics of a confocal family at $n-1$ points. We construct in the neighborhood of each such point a smooth function with no critical points, whose surface levels are the quadrics of our confocal family.

Fix one of these quadrics (label it the "first") and consider the Hamilton equations in the space of lines with Hamiltonian given by the first induced function of lines. Each phase curve on a fixed level surface of this Hamiltonian consists of the tangents to a single geodesic on the first quadric (Lemma A). By Lemma B, the remaining induced functions have zero Poisson bracket with the first (because the planes tangent to mutually confocal surfaces at the points of one fixed line are orthogonal in view of Theorem 2).

Thus, all induced functions are first integrals of the Hamiltonian system defined by any of them. Since each phase curve of such a system consists of the tangents to a single geodesic on the first surface, it follows that all induced functions assume on all these tangents constant (i.e., not depending on the point of the geodesic) values. This proves Theorem 3 as well as the following result.

THEOREM 4. The geodesic flow on a second-degree central surface in Euclidean space is completely integrable in the sense of Liouville (i.e., it has as many independent integrals in involution as degrees of freedom).

Remark. Strictly speaking, we proved Theorem 3 only for lines in general position, but the result is easily extended by continuity to exceptional cases (in particular, to the

*The tangency point is, generally speaking, different for each function.

asymptotic lines of our quadrics). Similarly, Theorem 4 was proved for surfaces with non-equal principal axes, but passing to the limit one can extend it to more symmetric rotation quadrics (as well as to noncentral "paraboloids").

3. Magnetic Analogs of the Theorems of Newton and Ivory

Elliptic coordinates permit us to extend the well known theorems of Newton on attraction by spheres to the case of attraction by ellipsoids.

Definition. We call homeoid density on the surface of an ellipsoid the density of the layer between the given ellipsoid and an infinitesimally close homothetic ellipsoid with the same center. The following theorem is well-known.

THEOREM (Ivory). A finite mass distributed with homeoid density on the surface of an elliposid does not attract interior points and attracts exterior ones as does the same mass, but distributed with homeoid density on the surface of a smaller confocal ellipsoid.

Here attraction is governed by Newton or Coulomb law: in n-dimensional space the force is proportional to r^{1-n} (as prescribed by the fundamental solution of Laplace's equation).

The generalizations of Newton's theorem (on the attraction of interior points) to the case of hyperbolic homeoid layers and to the case of attraction by a mass distributed on the level hypersurface of a hyperbolic polynomial of arbitrary degree are proved in [7] (see also [8]).

A polynomial $f(x_1, \ldots, x_n)$ of degree m is called hyperbolic (with respect to the point zero) if its restriction to any line passing through 0 has only real roots.

The homeoid charge density on the hypersurface $f=0$ is defined as the density on a infinitely thin homogeneous layer enclosed between the hypersurfaces $f=0$ and $f=\varepsilon$, $\varepsilon \to 0$ (the signs of charges are taken so that successive ovaloids carry opposite charges).

A theorem given in [7] asserts that a homeoid charge does not attract the point 0 (nor any other point inside the most inner ovaloid), and that this property of the charge is preserved upon multiplying its density by any polynomial of degree at most $m-2$.

Generalization: Upon multiplying the homeoid charge density by an arbitrary polynomial of degree $m-2+r$ one obtains a charge whose potential inside the most inner ovaloid is a harmonic polynomial of degree r (A. B. Givental', 1983). The attempt of generalizing Ivory's theorem on the atrraction by confocal ellipsoidal layers to hyperboloids revealed the essential role played by the topology of the hyperboloid. When hyperboloids of various signatures are considered, homeoid densities must be replaced by harmonic differential forms of various degrees on the hyperboloid while replacing the Newtonian or Coulombian potential by suitably generalized Biot–Savart potentials.

In the simplest nontrivial case of a one-sheeted hyperboloid in three-dimensional Euclidean space the results can be described as follows [10].

The hyperboloid divides the space into the "inner" and "outer" regions (the latter is not simply connected). Consider the lines of the elliptic coordinates, the surface levels of which are quadrics confocal to the given hyperboloid.

We call parallels of our hyperboloids the lines of the elliptic coordinates on our hyperboloid obtained by intersecting it with ellipsoids (i.e., the closed curvature lines of the hyperboloid). Also, we call meridians the lines, perpendicular to the parallels, of intersection with two-sheeted hyperboloids.

Although elliptic coordinates do have singularities (on all symmetry planes of the quadrics of our family), the hyperboloid is smoothly foliated by parallels (diffeomorphic to circles) and meridians (differmorphic to lines).

The region inside the hyperboloid tube is also smoothly foliated by meridians (orthogonal to the confocal ellipsoids of the family), whereas the annular outer region is foliated by parallels (orthogonal to one-sheeted hyperboloids).

THEOREM. A current of appropriate density, flowing along the meridians of the hyperboloid, generates a magnetic field which is equal to zero inside the tube of the hyperboloid while being directed along parallels in the outer annular region. A current of appropriate density, flowing along the parallels of the hyperboloid, generates a magnetic field which is equal to zero in the outer annular region, while being directed along the meridians inside the tube of the hyperboloid.

The density of the currents generating such fields generalize the homeoid densities on ellipsoid surfaces and may be described as follows.

Two "focal curves," an ellipse and a hyperbola, are associated with a family of confocal quadrics in three-dimensional Euclidean space. The focal ellipse is the boundary of the limit ellipsoid of the family, the small axis of which is shrunk to zero (the focal hyperbola is obtained in the same manner from the two-sheeted hyperboloid).

We define the homeoid density on the focal ellipse as follows. First consider some nonplanar parallel, defined as the nonplanar intersection of a mutually confocal ellipsoid and one-sheeted hyperboloid. The homeoid density on this parallel is defined as the density of an infinitesimally thin layer obtained by intersecting the layer between the given ellipsoid and an infinitesimally close homothetic ellipsoid with the same center, and the layer between the given hyperboloid and an infinitesimally close homothetic hyperboloid with the same center. We normalize this homeoid density on the parallel so that the mass of the whole parallel becomes equal to 1.

Now regard the focal ellipse as the limit of non-planar parallels. Then it turns out that the normalized homeoid densities on these parallels tend to a definite limit as the parallels tend to focal ellipse. It is this limit that will be called the homeoid density on the focal ellipse.

The homeoid density on the focal hyperbola is defined similarly.

We can now describe the densities of the currents which generate the magnetic fields described in the theorem. The surface of the one-sheeted hyperboloid is fibered over the focal ellipse (the fiber over a point is the meridian lying on the same two-sheeted hyperboloid as the point itself).

For the meridian current described in the theorem, the flux through any curve on the hyperboloid equals the integral of the homeoid density form on the focal ellipse over the projections of the given curve on the focal ellipse (along two-sheeted hyperboloids).

The density of the current flowing along parallels is similarly derived from the homeoid density on the focal hyperbola.

Remark. The magnetic field generated by the parallel current with the indicated density inside the tube of the hyperboloid is identical (modulo signs), in the outer domain of each confocal ellipsoid, with the Newtonian or Coulombian field of the charge distributed on the surface of this ellipsoid with homeoid density.*

Just as in the previous case, the magnetic field generalized by the meridian current in the annular outer domain of a one-sheeted hyperboloid is identical (modulo signs), in the region between the sheets of each two-sheeted confocal hyperboloid, with the Coulombian field produced by two equal charges of opposite signs distributed with homeoid density throughout the two sheets of this two-sheeted hyperboloid (O. P. Shcherbak).

The results formulated above were recently carried over to hyperboloids in Euclidean spaces of arbitrary dimension by B. Z. Shapiro and A. D. Vainshtein. Thus, for a hyperboloid in $\mathbb{R}^n$, differomorphic to $S^k \times \mathbb{R}^\ell$, one constructs a harmonic k-form in the outer domain (diffeomorphic to the product of S^k on a half space) and a harmonic ℓ-form in the inner domain.

The corresponding homeoid densities are defined on a k-dimensional focal ellipsoid and an ℓ-dimensional focal two-sheeted hyperboloid and are derived from densities of intersections of layers between infinitesimally close homothetic quadrics by passing to the limit, in the same way as we did above for $k=\ell=1$.

Noncomputational proofs of these geometric theorems are not known, not even in the case of a magnetic field in three-dimensional space.

Remark. The presence of distinguished harmonic forms on hyperboloids and their complementary domains prompts the idea that on noncompact (and possibly also on singular) algebraic and semialgebraic varieties in spaces of differential forms one should try to find filtrations similar to those arising in the theory of mixed Hodge structures. This would allow to restore the bijective correspondence between cohomology classes and harmonic forms existing on smooth compact varieties (however, in the noncompact or singular case the cohomology classes of a fixed dimension may be realized by forms of various degrees).

LITERATURE CITED

1. V. I. Arnol'd, "Integrable Hamiltonian systems connected with quadrics (according to Moser)," Usp. Mat. Nauk, 34, No. 5, 214-215 (1979).
2. C. G. J. Jacobi, Vorlesungen über Dynamik, Verlag G. Reimer, Berlin (1884).
3. J. Moser, "Various aspects of integrable Hamiltonian systems," in: Dynamical Systems, J. Guckehheimer, J. Moser, and E. S. Newhouse (eds.), C.I.M.E. Lectures, Bressanone, Italy, 1978, Progress in Math., Vol. 8, Birkhäuser, Boston (1980).
4. R. B. Melrose, "Equivalence of glancing hypersurfaces," Invent. Math., 37, 165-191 (1976).
5. V. I. Arnol'd, "Lagrangian manifolds with singularities, asymptotic rays, and the open swallow-tail," Funkts. Anal. Prilozhen., 15, No. 4, 1-14 (1981).
6. V. I. Arnol'd, "Singularities in variational calculus," J. Sov. Math., 27, No. 3 (1984).
7. V. I. Arnol'd, "On the Newtonian potential of hyperbolic layers," Tr. Tbil. Univ., Mat. Mekh. Astron., 232-233, 23-28 (1982).
8. V. I. Arnol'd, "On the Newtonian attraction of accumulations of dustlike particles," Usp. Mat. Nauk, 37, No. 4, 125 (1982).

*This is exactly the density with which a charge distributed itself on the surface of a conducting ellipsoid.

9. V. I. Arnold, "Some algebrogeometrical aspects of the Newton attraction theory," in: Arithmetics and Geometry, Papers Dedicated to I. R. Shafarevich on the Occasion of his Sixtieth Birthday, Vol. II, Geometry, Progress in Math., Vol. 36, Birkhäuser, Boston (1983), pp. 1-4.
10. V. I. Arnol'd, "Magnetic analogs of Newton's and Ivory's theorems," Usp. Mat. Nauk, 38, No. 5, 145-146 (1983).

LAGRANGIAN MANIFOLDS WITH SINGULARITIES, ASYMPTOTIC RAYS, AND THE OPEN SWALLOWTAIL

V. I. Arnol'd

UDC 513.83

The Lagrangian manifold* of the phase space of a variational problem is the description in the language of phase space of the field of extremals on configuration space (possibly, multivalued).

In the description of fields of extremals and time functions in problems with constraints (e.g., in the problem of bypassing an obstruction) there also arise Lagrangian manifolds with singularities. In the present paper we study one of the simplest singularities of this form; we call it an open swallowtail.

The ordinary swallowtail is the surface in $\mathbf{R}^3$, consisting of all polynomials $x^4 + ax^2 + bx + c$, having multiple roots.

The open swallowtail is the surface in $\mathbf{R}^4$, consisting of all polynomials $x^5 + Ax^3 + Bx^2 + Cx + D$, having roots of multiplicity ≥ 3.

Differentiation of polynomials maps $\mathbf{R}^4$ into $\mathbf{R}^3$, and the open tail into the ordinary one. For the open swallowtail the cuspidal edge is preserved, and the self-intersection disappears.

The open swallowtail appears in the problem of bypassing an obstruction as a singularity of the Lagrangian manifold of rays, falling from the surface of the obstruction.

We consider a smooth surface in Euclidean space $\mathbf{R}^3$, covered by a one-parameter bundle of geodesics. We associate with each point of the surface a ray falling on it (an oriented line, tangent to the geodesic bundle at the point considered).

All the falling rays form in the four-dimensional manifold of lines in $\mathbf{R}^3$ a two-dimensional set (depending on the surface of the obstruction and on the bundle of geodesics on it).

The manifold of all lines in $\mathbf{R}^3$ has a natural symplectic structure. The set of falling rays turns out to be a Lagrangian submanifold with singularities in this symplectic manifold.

The singular points of the manifold of falling rays are the rays falling in asymptotic directions. Points of the surface where the direction of the geodesic bundle is asymptotic form the line of asymptoticity of the bundle. Rays, falling on the line of asymptoticity, form (in the case of general position) the cuspidal edge of the Lagrangian manifold of rays.

At its individual points, the line of asymptoticity itself has an asymptotic direction. Rays falling in this direction we call biasymptotic.

The basic result of the present paper is that for a surface and bundle in general position the Lagrangian manifold of falling rays has close to a biasymptotic ray a singularity diffeomorphic with the singularity of the open swallowtail at zero (at least at the level of formal power series).

In passing we arrive at the normal form of several more objects, e.g., (1°) the symplectic structure in the neighborhood of the direct product of the ordinary swallowtail by a line; (2°) the pair, consisting of a smooth hypersurface $H = 0$ in a symplectic space and a submanifold $H = F = 0$ lying in it in the neighborhood of a point at which $\{H, F\} = \{H, \{H, F\}\} - \{H, \{H, \{H, F\}\}\} = 0$ (all objects are in general position, all normal forms are in the class of formal power series).

We note that in the case of a nonasymptotic direction of falling (i.e., $\{H, \{H, F\}\} \neq 0 \neq \{F, \{H, F\}\}$) Melrose [1] obtained a normal form for a pair of hypersurfaces $H = 0$, $F = 0$ in the neighborhood of a point, where $\{H, F\} = 0$. In the case of an asymptotic direction such a pair has functional moduli. Our result (2°) shows that

*Maximal isotropic submanifold of a symplectic manifold.

Moscow State University. Translated from Funktsional'nyi Analiz i Ego Prilozheniya, Vol. 15, No. 4, pp. 1-14, October-December, 1981. Original article submitted March 9, 1981.

Originally publ. in: Funkts. Anal. Prilozh. 15(4), 1-14, © 1981 Russ. Acad. Sciences
English transl.: Funct. Anal. Appl. 15(4), 235-246, © 1981 Consultants Bureau, New York.

without formal moduli one can advance by two more steps (to consider asymptotic and biasymptotic rays), if one normalizes instead of the pair ($H = 0$, $F = 0$) the pair ($H = 0$, trace of the hypersurface $F = 0$ on $H = 0$).

The open swallowtail supplies normal forms in several mutually dissimilar problems (e.g., in the problem of covering caustics by the cuspidal edge of a wave front moving in three-dimensional space and in the problem of decomposition of singularities of spatial curves by projectively dual nonsingular ones). Apparently this surface is one of the basic examples of a future theory of Lagrangian and Legendrian manifolds with singularities.

The present work was done in connection with the studies done at MIEM of shortwave asymptotics of electromagnetic fields. The author thanks their initiator V. P. Maslov and also O. A. Platonova, A. B. Givental', and O. P. Shcherbak for helpful discussions.*

1. Definitions

1°. Symplectization of the Hierarchy of Tangents. We consider a smooth surface in Euclidean space $\mathbf{R}^3$. A line tangent to it is called asymptotic if the order of tangency is higher than the first. We shall describe below the hierarchy of asymptotic tangents of various orders in terms of symplectic geometry. Thus, we carry it over to a more general situation than contact of extremals of a variational problem with a submanifold of configuration space. The opposite passage from general concepts to the usual asymptotics in $\mathbf{R}^n$ is carried out in Paragraph 7°, motivating the definitions of 2°-5°.

2°. Manifold of Rays. Let (X^{2n}, ω) be a symplectic manifold, Y^{2n-1} be a smooth hypersurface in it. The points of this hypersurface we shall call unit vectors. The skew-orthogonal complement to the tangent space to Y is called the characteristic direction. The integral curves of the field of characteristic directions on Y we shall call rays. Each unit vector determines a ray passing through it.

The manifold of unit vectors Y is locally fibered by rays. The base of this bundle $\pi: Y \to M$ we shall call the manifold of rays M. The manifold of rays has dimension $2n-2$. It has a natural symplectic structure. A symplectic 2-form on the manifold of rays is obtained from the 2-form ω on X by restriction to Y and projection by π.

If $H = 0$ is the equation of the surface Y (and $dH \neq 0$), then the rays are phase curves of the Hamilton equations with Hamiltonian function H, lying on the surface Y.

3°. Tangent Rays. Let Z be another hypersurface in X. We shall call its points boundary points. Points of intersection of Z with Y we shall call boundary unit vectors. We shall assume that Z intersects Y transversely in the $(2n-2)$-dimensional manifold of boundary unit vectors $W = Y \cap Z$.

We consider the restriction of the projection $\pi: Y \to M$ to the manifold of boundary unit vectors W. This smooth mapping of manifolds of identical dimension $2n-2$ we shall denote by $\rho: W \to M$. The map ρ assigns to each boundary unit vector its ray.

Definition. The critical points of the map ρ are called tangent unit vectors, and the critical values are called tangent rays.

Definition. The set of all critical points of the mapping ρ is called the manifold of tangent unit vectors and is denoted by Ω.

It is a hypersurface in the manifold of all boundary unit vectors W.

Proposition. Let us assume that Y and Z are given in X by the equations $H = 0$ and $F = 0$, respectively. Then Ω is singled out in W by the equation: the Poisson bracket $\{H, F\} = 0$.

Proof. The characteristic direction on Y is tangent to W. This means that at points of Ω the derivative of F along a field with Hamiltonian H is equal to 0.

Remark. The equation of Ω is symmetric with respect to permutation of F and H.

---

*After seeing the manuscript of the present paper, A. B. Givental' proved that the symplectic type of the germ of a submanifold of any dimension in a symplectic space is determined by the restriction of the symplectic form to the tangent bundle to the submanifold. Together with the results of Martine, Roussari, Golubitsky, and Tyshler on degenerate symplectic structures, it leads to the classification of submanifolds in general position in a symplectic space. He has an analogous result in the contact case also. The results of the present paper are carried over to the contact case by E. E. Landis.

COROLLARY. On Ω both characteristic directions of the hypersurfaces Y and Z are tangent to the intersection W of these hypersurfaces.

Definition. The set of all critical values of the map ρ is called the manifold of tangent rays and is denoted by Σ.

It is a hypersurface in the manifold of all rays.

4°. Geodesic Flows of Rays and Unit Vectors. At nonsingular points of the hypersurface Σ of tangent rays in the symplectic manifold M of all rays there is a characteristic direction (skew-orthogonal to the tangent space to Σ).

Definition. The integral curves of the field of characteristic directions on Σ are called trajectories of the geodesic flow of rays.

On the manifold Ω of tangent unit vectors there is also a distinguished field of directions. We consider the skew-orthogonal complement δ to the tangent space to W at the point 0 from Ω.

Proposition 1. Let us assume that 0 is a nonsingular point of the restriction of ρ to Ω. Then the intersection of δ with the tangent plane to Ω at 0 is one-dimensional.

Proof. The plane δ is generated by the characteristic directions of the hypersurfaces Y and Z at 0. According to Paragraph 3°, $\delta \subset T_0W$. Consequently, $\dim \delta \cap T_0\Omega \geqslant 1$. But $T_0\Omega$ does not contain the characteristic direction of the hypersurface Y (this direction is the kernel of $d\rho$). Consequently, $\dim \delta \cap T_0\Omega < 2$.

Definition. The direction $\delta \cap T_0\Omega$ is called the geodesic direction on Ω; the integral curves of the field of geodesic directions on Ω are called trajectories of the geodesic flow of unit vectors.

Remark. The manifold Ω and the field of geodesic directions on it are defined by the manifold W of codimension 2 in the symplectic manifold X and are independent of the choice of hypersurfaces Y and Z. In particular, the field of geodesic directions is unchanged under permutation of Y and Z.

The geodesic direction can be determined by the Hamiltonian $H\{F, \{H, F\}\} + F\{H, \{F, H\}\}$.

Proposition 2. The map ρ carries trajectories of the geodesic flow of unit vectors on Ω into trajectories of the geodesic flow of rays on Σ.

Proof. The geodesic direction is skew-orthogonal to $T_0\Omega$ in X. Hence its image under the action of ρ_* is skew-orthogonal to $T_{\rho 0}\Sigma$ in M. This means the image is the characteristic direction on Σ, which is what was required.

5°. Unital Singularities. The theory of singularities allows one to classify singularities of the map ρ: $W \to M$.

The kernel of the differential of π is one-dimensional, so $\dim \operatorname{Ker} \rho_* \leq 1$. This means (in the case of general position), the singularities of ρ are unital.

Definition. The Whitney projection $W \to Y \to M$ is defined by the inclusion of the manifold $W = \{y\colon F = 0\}$ and the projection $\pi\colon Y \to M$, which are defined by the formulas, where $k \leq N = \dim W = \dim M = \dim Y - 1$,

$$\pi(y_0, \ldots, y_N) = (y_1, \ldots, y_N), \quad F = y_0^{k+1} + y_1 y_0^{k-1} + \ldots + y_k. \tag{1}$$

By an equivalence of projections is meant a 3×2 commutative diagram, whose verticals are diffeomorphisms.

Example. $N = 2, k = 1$, π is the projection of the plane onto the y_1 axis, W is the parabola $y_0^2 + y_1 = 0$, ρ is its projection onto the y_1 axis, 0 is a fold point of the map ρ.

Let $\pi\colon Y \to M$ be any bundle with one-dimensional fibers.

Whitney's Theorem (cf. [2-4]). The projection of a hypersurface W in general position in the space of the bundle π onto the base of this bundle (i.e., a diagram $W \to Y \to M$) is locally equivalent to one of the normal forms (1) in a neighborhood of each point.

6°. Asymptotic Unit Vectors and Rays. Definition. A tangent unit vector (ray) is called asymptotic if the order of tangency of the ray with the manifold W is higher than the first, biasymptotic if it is higher than the second, etc.

Let Y and Z be given by equations $H = 0$ and $F = 0$, respectively. Then:

tangent unit vectors: $H = F = \{H, F\} = 0$;

asymptotic unit vectors: $H = F = \{H, F\} = \{H, \{H, F\}\} = 0$;

biasymptotic unit vectors: $H = F = \{H, F\} = \{H, \{H, F\}\} = \{H, \{H\{H, F\}\}\} = 0$ etc.

It is evident from Whitney's theorem that in the case of general position, the map ρ has at each point which is a tangent, but not asymptotic unit vector, a singularity of fold type, asymptotic but not biasymptotic, of gather type, etc.

In this case all tangent, asymptotic, etc. unit vectors form a flag of successively imbedded nonsingular manifolds. The corresponding manifolds of rays have singularities. Whitney's formula (1) shows that the manifold of tangent rays is locally diffeomorphic with the product of the swallowtail in $\mathbf{R}^k$ by $\mathbf{R}^{2n-2-k}$, the manifold of asymptotic rays with the product of the manifold of polynomials $y_0^{k+1} + y_1 y_0^{k-1} + \ldots + y_k$ with roots of multiplicity ≥ 3 by $\mathbf{R}^{2n-2-k}$, etc.

7°. Example: Hypersurface in Euclidean Space. Let $X = \mathbf{R}^{2n} = \{(x, p)\}$ be the phase space of a free particle x in Euclidean space $\mathbf{R}^n$, $H = (p^2 - 1)/2$ be its Hamiltonian function. Then:

1) Y: $H = 0$ is the ordinary manifold of unit vectors in $\mathbf{R}^n$;

2) M is the manifold of all oriented lines in $\mathbf{R}^n$.

Remark 1. In this case the manifold of rays M is defined in the large. The manifold M of all oriented lines in $\mathbf{R}^n$ is naturally diffeomorphic with the cotangent bundle of the sphere T^*S^{n-1} (a line $\to$ its unit vector). The symplectic structures of M and T^*S^{n-1} differ only by sign.

Remark 2. It follows from Remark 1, in particular, that the Gauss map is Lagrangian: the manifold of oriented normals to the hypersurface is a Lagrangian submanifold in the space of the Lagrangian bundle $M \simeq T^*S^{n-1} \to S^{n-1}$.

Let Π be a hypersurface (obstruction) in our Euclidean space $\mathbf{R}^n$. We define it by the equation $F(x) = 0$ and it will be assumed that the function F is also defined on the phase space: $F(x, p) \equiv F(x)$. Then

3) Z: $F = 0$ is the manifold of all vectors applied to Π;

4) W: $H = F = 0$ is the manifold of all unit vectors applied to Π.

Proposition 1. Ω: $H = F = \{H, F\} = 0$ is the manifold of all unit vectors tangent to Π (in the usual sense).

Proof. By definition Ω consists of unit vectors applied to Π such that a ray drawn through this unit vector is tangent to the manifold of applied unit vectors. For this the line representing the ray in $\mathbf{R}^n$ must be tangent to Π.

In exactly the same way asymptotic, biasymptotic, etc. unit vectors in the example considered become ordinary ones.

Proposition 2. The trajectories of the geodesic flow of unit vectors in the example considered are the phase curves of the usual geodesic flow on Π.

Proof. Temporarily we interchange the places of F and H. The corresponding objects we shall distinguish by dashes, e.g., $F' = H$, $H' = F$, $W' = W$, etc. We find $M' = T^*\Pi$. In fact, the phase flow with Hamiltonian F on the hypersurface $F = 0$ leaves in place the points of Π and adds to the vector ξ applied at x on Π a multiple of grad F at x. A trajectory of this flow is determined by the point of intersection of the trajectory of the vector ξ with plane at x tangent to Π, and thus by a vector from $T_x^*\Pi$. The symplectic quotient-structure of M' is compatible with the usual one on $T^*\Pi$.

The manifold $\Omega' = \Omega$ consists of unit vectors tangent to Π and under the map ρ': $W' \to M'$ goes into the manifold Σ' of vectors of length 1 from $T^*\Pi$. The restriction of ρ' to Ω' is a diffeomorphism $\Omega' \to \Sigma'$ (since the Hamiltonian H is strictly convex, the only singularities of ρ' are folds on Ω').

The characteristics on Σ' in M' are trajectories of the usual geodesic flow on Π, since Σ' is a level manifold of the kinetic energy of a point moving on Π. Consequently, the trajectories of the geodesic flow of unit vectors on Ω' are the usual trajectories of the geodesic flow of unit vectors (we apply Proposition 2 from Paragraph 4° to ρ'). But the trajectories Ω' and on Ω coincide (remark from Paragraph 4°). Consequently, the trajectories of the geodesic flow of unit vectors on Ω are the usual ones, which is what was required.

COROLLARY. The motion of a point on a characteristic of the manifold Σ in the example considered denotes the motion of a ray in $\mathbf{R}^n$, under which it is always tangent to one geodesic on II.

Proof. We apply Proposition 2 from Paragraph 4° to ρ and we use Proposition 2 from Paragraph 7°.

2. Results

1°. Theorem of Darboux + Whitney. By Darboux's theorem one can reduce to local normal form $\Sigma dp_i \wedge dq_i$ the symplectic structure of the manifold of rays M. By Whitney's theorem the manifold of tangent rays is also reduced by a diffeomorphism of M to local normal form: it is diffeomorphic to the product of a generalized swallowtail by a linear space [cf. (1)].

It turns out, if in (1) k = 1, 2, or 3 (i.e., in the cases of tangent, asymptotic, and biasymptotic rays in general position), that one can simultaneously reduce to local normal form both the symplectic structure of M and the manifold of tangent rays.

We give below the local normal forms to which one can reduce the generalized swallowtail (products of an ordinary swallowtail in $\mathbf{R}^3$ or a semicubical parabola in $\mathbf{R}^2$ by a linear space) in a symplectic manifold with the help of formal symplectic diffeomorphisms.

Equivalent formulation: we indicate the normal forms to which one can reduce symplectic 2-forms in general position in the neighborhood of a generalized swallowtail by means of a formal diffeomorphism preserving the tail.

2°. Symplectic Normal Form of the Manifold of Tangent Rays. We consider the space $\mathbf{R}^{2m}$ with the coordinate system (A, B, C, . . .).

Definition 1. By a standard hypersurface with cuspidal edge is meant a hypersurface Γ in $\mathbf{R}^{2m}$, given by an equation $A^3 = B^2$.

THEOREM 1. A symplectic structure in general position in the neighborhood of the point 0 in $\mathbf{R}^{2m}$ can be reduced by a formal diffeomorphism preserving Γ to the normal form

$$\omega = dA \wedge dB\,(+\,dC \wedge dD + \ldots \quad \text{for } m > 1)$$

(general position = nondegenerate on the plane A = B = 0 in $T_0\mathbf{R}^{2m}$

Definition 2. By the standard hypersurface with singularity of swallowtail type is meant the hypersurface Λ in $\mathbf{R}^{2m}$, given by the condition $x^4 + Ax^2 + Bx + C$ has a multiple root.

THEOREM 2. A symplectic structure in general position in a neighborhood of the point 0 in $\mathbf{R}^{2m}$ can be reduced by a formal diffeomorphism preserving Λ to the normal form

$$\omega = dA \wedge dD + dC \wedge dB\ (+\,dE \wedge dF + \ldots \quad \text{for } m > 2)$$

(general position = nondegenerate on the plane B = C = 0 in $T_0\mathbf{R}^{2m}$).

3°. Symplectic Normal Forms of Unital Projections. We consider in the symplectic manifold (X, ω) a pair, composed of a smooth hypersurface Y in X and a hypersurface W in Y. We form the local manifold of rays M and the local map π: Y → M, as described in Sec. 1, Paragraph 3°.

We consider the induced map ρ: W → M. Let us assume that the map ρ has at the point 0 the Whitney singularity (1) with k = 1, 2, or 3.

The normal form of the collection (X, ω, Y, W) is described by the following theorem. In the neighborhood of the point considered 0, one can choose a coordinate system, in which X, ω, and Y will have the form

$$X = \{(p_0, \ldots, q_m)\}, \quad \omega = dp_0 \wedge dq_0 + \ldots + dp_m \wedge dq_m, \quad Y\colon q_0 = 0.$$

Coordinates in which X, ω, and Y have such a form, we call Darboux coordinates.

THEOREM 3. In a neighborhood of the point 0 in general position with the properties indicated above, one can choose formal Darboux coordinates so that the manifold W will be singled out in Y by the condition F = 0, where

$$\begin{aligned}
F &= p_0^2 + p_1 && \text{for } k = 1,\\
F &= p_0^3 + p_1p_0 + q_1 && \text{for } k = 2,\\
F &= p_0^4 + p_1p_0^2 + q_2p_0 + p_2 && \text{for } k = 3.
\end{aligned}$$

Remark 1. The case $k = 1$ is considered by Melrose [1].

Remark 2. The collection (X, ω, Y, W) defines a projection $W \to Y \to M$ and a map $\rho: Y \to M$, which, thus, are also reduced to normal form by a symplectic formal diffeomorphism of the manifold X. Thus, the manifolds Ω, Σ and the trajectories of the geodesic flows on them are formally normalized.

4°. Normal Forms of Geodesics. In the cases $k = 1$ and 2 the form of the trajectories of the geodesic flows of unit vectors and rays in the normal coordinates of Theorem 3 are obvious: if we denote the parameter on the trajectory by ξ, then

$$q_0 = p_0 = 0, \qquad q_1 = \xi \qquad\qquad \text{for } k = 1;$$
$$q_0 = 0, \quad p_0 = \xi, \quad p_1 = -3\xi^2, \quad q_1 = -2\xi^3 \qquad \text{for } k = 2$$

(the coordinates not written are constant).

THEOREM 4. The trajectories of the geodesic flow of rays on the manifold of tangent rays Σ in the normal coordinates of Theorem 3 for $k = 3$ are obtained by translations along the coordinate axis q_1 (and also q_i, p_i, $i > 2$, if $m > 2$) from the curves in $\mathbf{R}^4$, given by the conditions:

1) the polynomial

$$\Phi = \frac{p_0^5}{5} + p_1 \frac{p_0^3}{3} + q_2 \frac{p_0^2}{2} + p_2 p_0 + \frac{q_1}{2} = \frac{q_1}{2} + \int F\, dp_0$$

has a root of multiplicity ≥ 3;

2) $p_1 = c$ (an arbitrary constant).

The trajectories of the geodesic flow of unit vectors on the manifold of tangent unit vectors Ω are obtained by adding the conditions $q_0 = 0$, $p_0 = \xi$ (root of Φ of multiplicity ≥ 3).

COROLLARY 1. Explicit expressions for the coordinates of points of trajectories in terms of the parameter ξ have the form

$$p_1 = c, \quad q_2 = -4\xi^3 - 2c\xi, \quad p_2 = 3\xi^4 + c\xi^2, \quad q_1 = -\frac{12}{5}\xi^5 - \frac{2}{3}c\xi^3 + \text{const.} \tag{2}$$

COROLLARY 2. A Lagrangian manifold in general position, lying on the hypersurface Σ studied in $\mathbf{R}^4$, can be locally reduced to normal form $\{(p_1, q_1, p_2, q_2) : \Phi \text{ has a root of multiplicity } \geq 3\}$ by a symplectic diffeomorphism of $\mathbf{R}^4$ preserving Σ.

Proof. A Lagrangian manifold on a hypersurface is composed of its characteristics. Any two-dimensional manifold composed of the characteristics of a three-dimensional hypersurface in a symplectic space is Lagrangian. Hence, for the proof it suffices to reduce to normal form a one-parameter subfamily in general position in the two-parameter family of Corollary 1.

For such a family in general position as the parameter one can take c, then const will be a smooth function $\varphi(c)$. The diffeomorphism sought has the form $(p_1, q_2, p_2, q_1) \mapsto (p_1, q_2, p_2, q_1 - \varphi(p_1))$.

COROLLARY 3 (cf. [9]). We consider in the Euclidean space $\mathbf{R}^3$ a surface Π and we assume that on Π there is given a one-parameter bundle of geodesics. Let us assume that the direction of the geodesic bundle at the point 0 is biasymptotic. For a surface and bundle in general position with the indicated properties the map $\tau: \Pi \to M$ of the surface into the manifold of rays in $\mathbf{R}^3$, associating to a point of the surface the tangent to the geodesic of the bundle passing through it, can be described in the form (2) after suitable choice of parameter ξ along the geodesic, parameter c, indexing the geodesics, and Darboux coordinates in M, at least at the level of power series.

In particular, the surface $\tau(\Pi)$ of all rays falling on the geodesic bundle is (formally) symplectically diffeomorphic with the standard open swallowtail $\{\Phi: \exists p_0: \ \Phi = \Phi' = \Phi'' = 0\}$, where the diffeomorphism reduces the whole manifold of tangent rays to the form $\{\Phi: \exists p_0: \ \Phi' = \Phi'' = 0\}$.

3. Proof of Theorems 1 and 2

1°. Preliminary Normalization. We consider the standard hypersurface Λ in $\mathbf{R}^{2m}$. In $T_0\mathbf{R}^{2n}$ we consider the flag of tangent spaces to the strata of the surface Λ: $\{C = 0\} \supset \{B = C = 0\} \supset \{A = B = C = 0\}$.

Definition. An exterior 2-form in $T_0\mathbf{R}^{2m}$ is called general in relation to Λ, if it is nondegenerate and its restriction to the space of codimension 2 $B = C = 0$ is nondegenerate.

Definition. A linear transformation $T_0\mathbf{R}^{2m} \to T_0\mathbf{R}^{2m}$ is called admissible if it is the derivative of a diffeomorphism preserving Λ.

LEMMA 1. A general exterior form ω can be reduced by an admissible transformation to the normal form $A \wedge D + C \wedge B$ $(+ E \wedge F + \ldots$, if $m > 2)$.

Proof. Let $m = 2$. The ω-skew-orthogonal complement $\varkappa$ to the plane $B = C = 0$ is transverse to this plane. Consequently, the projection of $\varkappa$ into (A, B, C)-space along the D axis is transverse to the A axis. A plane in $\mathbf{R}^3$, transverse to the A axis at 0, can be reduced by a diffeomorphism preserving the swallowtail in $\mathbf{R}^3$ to the normal form $A = 0$ (cf. [5]). The transformation

$$(A, B, C, D) \mapsto (A, B, C, \alpha A + \beta B + \gamma C + \delta D), \quad \delta \neq 0,$$

is admissible. Consequently, $\varkappa$ can be reduced by an admissible transformation to the plane $A = D = 0$. After this the transformation ω can be reduced to the form $\lambda A \wedge D + \mu C \wedge B$. The transformation $(A, B, C, D) \mapsto (\theta^2 A, \theta^3 B, \theta^4 C, \zeta D)$ for any $\theta \neq 0$, $\zeta \neq 0$ is admissible. Choosing θ and ζ, we can achieve $\lambda = 1$ and $\mu = 1$, which is what was required.

Let $m > 2$. We choose in the space $B = C = 0$ a subspace $\mathbf{R}^{2m-4}$, where ω is nondegenerate. Its ω-skew-orthogonal complement $\mathbf{R}^4$ in $\mathbf{R}^{2m}$ is transverse to it. The space $\mathbf{R}^4$ can be reduced by an admissible transformation to the form $E = F = \ldots = 0$. The restriction of ω to $\mathbf{R}^4$ is general with respect to the intersection of Λ with $\mathbf{R}^4$. Applying to $\mathbf{R}^4$ the lemma which has already been proved in the case $m = 2$ and reducing ω in $\mathbf{R}^{2m-4}$ to canonical form, we get the representation required.

Definition. We define the degrees of the monomials A, B, C, D by the conditions $\deg A = 2$, $\deg B = 3$, $\deg C = 4$, $\deg D = 5$. These conditions determine the degrees of monomial differential forms ($\deg dA = 2$, etc.) and vector fields ($\deg \partial/\partial A = -2$, etc.). The order of a formal series (form, field) is equal to the least of the degrees of monomials of the Taylor series.

Let us assume that the restriction of the closed nondegenerate 2-form ω in $\mathbf{R}^4$ to the plane $B = C = 0$ at 0 is nondegenerate.

LEMMA 2. ω can be reduced by a diffeomorphism of $\mathbf{R}^4$ preserving Λ and 0 to the form $dA \wedge dD + dC \wedge dB +$ (remainder of order greater than 7).

Proof. By Lemma 1 one can reduce ω to the indicated form with remainder vanishing at the point 0. The order of such a remainder is ≥ 7; the only monomial of degree 7 is $A\,dA \wedge dB$. But $dA \wedge dD + \lambda A\, dA \wedge dB = dA \wedge d(D + \lambda AB)$, and the diffeomorphism $(A, B, C, D) \mapsto (A, B, C, D + \lambda AB)$ preserves Λ.

2°. Solution of a Homological Equation. We consider in $\mathbf{R}^4$ with the symplectic structure $\omega = dA \wedge dD + dC \wedge dB$ the standard surface Λ and we denote by $\Delta = 0$ the equation of Λ (the discriminant Δ of the polynomial $x^4 + Ax^2 + Bx + C$ is equal to zero). The polynomial Δ is irreducible and has quasihomogeneous degree 12.

LEMMA 3. Let a formal vector field with Hamiltonian H be everywhere tangent to Λ. Then the series H can be represented in the form $a(A) + b(A, B, C, D)\Delta$, where this representation is unique.

Proof. The tangency condition $\{H, \Delta\} = c\Delta$ is quasihomogeneous relative to H, so it suffices to prove the lemma for quasihomogeneous polynomials H over $\mathbf{C}$. The tangency condition ($\{H, \Delta\} = 0$ on Λ) is equivalent with the condition ($\{\Delta, H\} = 0$ on Λ), and consequently means that H is constant along trajectories of the geodesic flow of rays on Λ. We use the explicit form (2) of these trajectories, given in Theorem 4 [the proof of (2) is independent of Theorem 2].

We consider a line l in the direction of the D axis, whose projection into (A, B, C)-space is a nonsingular point of the line of self-intersection of the swallowtail. A trajectory of the geodesic flow of rays on Λ, intersecting l, intersects l again (cf. Theorem 4), where the increment in D depends only on A.

Consequently, the polynomial H is constant on an arithmetic progression, and hence on the whole line l. This means on all trajectories intersecting l the value of H is identical. Consequently, H is constant on the entire surface in which Λ intersects the plane $A = \text{const}$.

But A is a parameter on the line of self-intersection of the tail in $\mathbf{R}^3$: the remaining coordinates along this line are polynomials in A. Consequently, there exists a polynomial $a(A)$, coinciding with H on this line. The difference $H - a$ vanishes on this line, and hence by what has been proved also on the whole surface Λ. This means this difference is divisible by Δ, which is what had to be proved (the uniqueness is obvious).

We denote by ω_7 the form $dA \wedge dD + dC \wedge dB$ in $\mathbf{R}^4$ and by i_v its displacement with the vector field v [i.e., $(i_v\omega)(\xi) \equiv \omega(v, \xi)$].

LEMMA 4 (Basic). Any Taylor series in a differential 1-form in $\mathbf{R}^4$ at 0 can be represented in the form $\alpha = i_v\omega_7 + df$, where v is a formal vector field tangent to Λ, f is a Taylor series of a function with null free term.

Proof. Vector fields tangent to Λ (formal, analytic, polynomial) can be represented uniquely in the form of linear combinations of the following basic fields, indexed by quasihomogeneous degree:

$$v_0 = 2A\frac{\partial}{\partial A} + 3B\frac{\partial}{\partial B} + 4C\frac{\partial}{\partial C},$$
$$v_1 = 3B\frac{\partial}{\partial A} + (4C - A^2)\frac{\partial}{\partial B} - \frac{1}{2}AB\frac{\partial}{\partial C},$$
$$v_2 = 4C\frac{\partial}{\partial A} - \frac{1}{2}AB\frac{\partial}{\partial B} + \left(2AC - \frac{3}{4}B^2\right)\frac{\partial}{\partial C},$$
$$v_{-5} = \frac{\partial}{\partial D},$$

with coefficients which are formal series (respectively, functions analytic at 0, polynomials) in A, B, C, D (cf. [5, 6]).

The Poincaré series of the graded space $\oplus M_k$ is defined as $\Sigma(\dim M_k)t^k$. The Poincaré series of the ring of polynomials in variables A, B, C, D of degrees 2, 3, 4, 5, graded by quasihomogeneous degree, is equal according to Euler [12] to $\Pi = [(1-t^2)(1-t^3)(1-t^4)(1-t^5)]^{-1}$.

We consider the Poincaré series of the following spaces graded by quasihomogeneous degree:

all 1-forms, $\{\alpha\}$: $p = (t^2 + t^3 + t^4 + t^5)\Pi$,

differentials of functions, $\{df\}$: $q = \Pi - 1$,

special forms, $\{i_v\omega_7\}$: $r = (t^2 + t^7 + t^8 + t^9)\Pi$,

$\{i_v\omega_7\} \cap \{df\}$: $s = t^2(1-t^2)^{-1} + t^{12}\Pi$

(we have used the form of the free generators v_k of the module $\{v\}$ and the fact that $\deg i_{v_k}\omega_7 = k + 7$, Lemma 3, and the fact that $\deg \Delta = 12$).

The basic fact on which the whole proof of the theorem is based is the identity $p + s = q + r$. From this identity it follows that for quasihomogeneous components of any degree

$$\dim(\{df\} + \{i_v\omega_7\}) = \dim\{df\} + \dim\{i_v\omega_7\} - \dim\{df\} \cap \{i_v\omega_7\} = \dim\{\alpha\},$$

i.e., $\{df\} + \{i_v\omega_7\} = \{\alpha\}$, which also proves Lemma 4.

3°. Proof of Theorem 2 for m = 2. We use a homotopy method. According to Lemma 2, the symplectic form can be reduced by a diffeomorphism preserving Λ to the form $\omega = \omega_7 + \delta$, where the closed form δ has order >7. We join ω with ω_7 by the segment $\omega^t = \omega_7 + t\delta$. We shall seek a one-parameter family of diffeomorphisms $\{g_t\}$ from the conditions:

$$1)\ g_t^*\omega^t \equiv \omega_7,\quad 2)\ g_0 = \mathrm{id},\quad 3)\ g_t(0) \equiv 0,\quad 4)\ g_t\Lambda \equiv \Lambda.$$

We denote by w^t the velocity field of the family g_t, so that

$$w^t|_{g_t(x)} \equiv \frac{\partial}{\partial t} g_t(x).$$

In terms of the field w^t, conditions 1)–4) assume the form:

$$1')\ L_{w^t}\omega^t + \delta \equiv 0,\quad 3')\ w^t|_0 \equiv 0,\quad 4')\ w^t \equiv \alpha_0 v_0 + \alpha_1 v_1 + \alpha_2 v_2 + \beta v_{-5}$$

with coefficients depending on A, B, C, D, t (L is the Lie derivative).

LEMMA 5. Equation 1') is solvable in the class of formal series w^t of positive order, satisfying 3'),4').

In fact, since $L = i\circ d + d\circ i$, and the form ω^t is closed, 1') can be rewritten in the form $d(i_{w^t}\omega^t) + \delta = 0$. We represent the closed form δ of order >7 in the form $d\gamma$, where γ is a 1-form of order >7. 1') assumes the form

$$i_{w^t}\omega^t = \gamma + df^t, \tag{1''}$$

where γ is a known 1-form of order >7, and w^t and f^t are unknown. We expand the unknown field w^t, the known forms ω^t and γ and the unknown function f^t by quasihomogeneous components: $w^t = w_1^t + w_2^t + \ldots$, $\omega^t = \omega_7 + \omega_8^t + \ldots$, $\gamma = \gamma_8 + \gamma_9 + \ldots$, $f^t = f_8^t + f_9^t + \ldots$. Comparing terms of degrees 8, 9, ... in (1''), we get the triangular system

$$\begin{cases} i_{w_1^t}\omega_7 = \gamma_8 + df_8^t & \Rightarrow w_1^t,\ f_8^t, \\ i_{w_2^t}\omega_7 = \gamma_9 - i_{w_1^t}\omega_8 + df_9^t & \Rightarrow w_2^t,\ f_9^t, \\ \dots\dots\dots\dots\dots\dots \end{cases} \qquad (*)$$

By Lemma 4, system (*) is solvable. We find a formal field w^t of positive order, tangent to Λ, satisfying (1') and (4'). (The dependence of the quasihomogeneous components on t is polynomial: w_1 and f_8 are independent of t, w_2 and f_9 are polynomials of the first degree in t, w_3 and f_{10} of the second, etc.). From the positivity of the order of the field it follows that $w^t|_0 \equiv 0$. Lemma 5 is proved.

The formal fields w^t determine a one-parameter family of formal diffeomorphisms g_t (the family is not formal in t, i.e., for any finite t there is defined a formal diffeomorphism g_t). Conditions (1)-(4) hold formally.

For the formal diffeomorphism g_1 we have $g_1^*\omega = \omega_7$, $g_1\Lambda = \Lambda$, $g_1(0) = 0$. Theorem 2 for m = 2 is proved.

4°. The Case m > 2. We assign to all 2k additional variables E, F, . . . the degree 7/2. After this the whole proof goes exactly as in the case m = 2. The new Poincaré series in Paragraph 2° have the form

$$\Pi_n = \Pi\,(1 - t^{7/2})^{-2k}, \quad p_n = (t^2 + t^3 + t^4 + t^5 + 2kt^{7/2})\,\Pi_n,$$
$$q_n = \Pi_n - 1, \qquad r_n = (t^2 + t^7 + t^8 + t^9 + 2kt^{7/2})\,\Pi_n,$$
$$s_n = (1 - t^2)^{-1}(1 - t^{7/2})^{-2k} - 1 + t^{12}\Pi_n.$$

Consequently, $p_n + s_n = q_n + r_n$. The rest of the proof goes as in the case m = 2 (with the obvious changes: $\omega^t = \omega_7^t + \omega_{7.5}^t + \omega_8^t + \dots$, $w^t = w_{0.5}^t + w_1^t + w_{1.5}^t + \dots$).

5°. Proof of Theorem 1. We set deg A = 2, deg B = 3, deg E = deg F = . . . = 5/2. Each formal vector field, tangent to the manifold $A^3 - B^2 \equiv \Delta = 0$, is representable uniquely in the form of a sum of products of formal series of functions by the fields $v_0 = 2A\partial/\partial A + 3B\partial/\partial B$, $v_1 = 2B\partial/\partial A + 3A^2\partial/\partial B$ and (for m > 1) $\partial/\partial C$, $\partial/\partial D$, etc. A Hamiltonian field is tangent to the manifold $\Delta = 0$ if and only if the Hamiltonian has the form $a(C, D, \dots) + b(A, B, C, \dots)\Delta$. Hence we get such Poincaré series:

$$\begin{aligned} \text{all functions, } \{f\}&: \Pi = (1 - t^2)^{-1}(1 - t^3)^{-1}(1 - t^{5/2})^{-2k}, \\ \text{all 1-forms, } \{\alpha\}&: p = (t^2 + t^3 + 2kt^{5/2})\,\Pi, \\ \text{differentials, } \{df\}&: q = \Pi - 1, \\ \text{special forms, } \{i_v\,\omega_5\}&: r = (t^5 + t^6 + 2kt^{5/2})\,\Pi, \\ \{df\} \cap \{i_v\omega_5\}&: s = t^6\,\Pi + (1 - t^{5/2})^{-2k} - 1. \end{aligned}$$

Thus, p + s = q + r, the rest is as in the proof of Theorem 2.

4. Proofs of Theorems 3 and 4

1°. Preliminary Lemmas. Let (X, ω, Y, and M) be just as in Sec. 1, 0 be a point in X and its image in M.

LEMMA 1. Any local symplectic diffeomorphism of (M, 0) is induced by a local symplectic diffeomorphism of (X, 0), preserving Y.

Proof. The symplectic manifold (X, ω) locally decomposes into the direct sum of symplectic manifolds (M, ω') and ($\mathbf{R}^2$, ω'') (cf. the proof of Darboux's theorem in [7]). The diffeomorphism sought is the sum of the given one and the identity.

We consider on X local coordinates ($p_0, \dots, q_m$), in which $\omega = dp_0 \wedge dq_0 + \dots + dp_m \wedge dq_m$. Y is the hypersurface $q_0 = 0$.

LEMMA 2. Let f: X → R be a function, f(0) = 0, $(\partial f/\partial p_0)_0 \neq 0$. Then there exists a local diffeomorphism h: X → X, preserving (ω, Y, 0) and carrying p_0 into f (i.e., $h^*f = p_0$).

Proof. We denote by $\{g^t\}$ the phase flow with Hamiltonian function f and by $p_0 = \varphi(\xi, \eta)$ a solution of the equation $f(p_0, 0, \eta)$ [where $\eta = (p_1, \dots, q_m)$]. The diffeomorphism sought: $h(p_0, q_0, \eta) = g^{q_0}(\varphi(p_0, \eta), 0, \eta)$ (cf. the proof of Darboux's theorem in [7]).

LEMMA 3. Let W → Y → M be the Whitney projection (1) and let ξ: M → M be a holomorphic diffeomorphism, preserving 0 and the set Σ of critical values of the map W → M. Then ξ is locally induced by a fiberwise holomorphic diffeomorphism χ: Y → Y preserving W.

Proof. This follows from two results of Lyashko [8]: 1) the group $\{\zeta\}$ is connected, so it suffices to prove the liftability of vector fields in M, tangent to Σ, preserving 0, to fields in Y, tangent to W, preserving 0; 2) liftability is guaranteed by the fact that the codimension in W of the submanifold, where the map $W \to M$ is not a fold, is equal to two.

Remark. Apparently Lemma 3 is also true in the formal and C^∞ cases. If ζ is a (formal) diffeomorphism, constructed by integration of combinations of basis Whitney-liftable fields v_i (with formal coefficients), tangent to Σ, then the liftability of ζ to χ is obvious.

2°. Proof of Theorem 3, k = 2 and 3. By Whitney's theorem, $W \to Y \to M$ can be reduced to normal form (1). The set of critical values Σ in M is also standard. By Theorems 1 and 2 in the case of general position, the symplectic structure in M can be normalized by a diffeomorphism ζ preserving Σ and 0. It is evident from the proof that this diffeomorphism is obtained by integration of a (formal) combination of basis Whitney-liftable fields. By Lemma 3 and the remark on it ζ is liftable in Y to a diffeomorphism preserving W.

Thus, we have reduced Σ to normal form by a (formal) symplectic diffeomorphism of (M, 0). By Lemma 1 it is induced by a symplectic diffeomorphism of (X, 0) preserving Y. The manifold W also reduces to standard form (e.g., for $k = 3 - y_0^4 + p_1 y_0^2 + q_2 y_0 + p_2 = 0$). In this formula y_0 is not the standard symplectic coordinate p_0 on the fiber of $\pi: Y \to M$, but a certain function on Y: $y_0 = y_0(p_0, \eta)$ [where η denotes $(p_1, \ldots, p_m)$]. At the point 0, $y_0 = 0$ and $(\partial y_0/\partial p_0) \neq 0$. We extend y_0 to X and we denote the extension by f. By Lemma 2 there exists a symplectic diffeomorphism h, preserving Y and the point 0 and carrying p_0 into f. The inverse diffeomorphism completes the normalization of W, and hence the proof of Theorem 3.

3°. Proof of Theorem 4. Suppose first m = 2. We consider the curve (2) (with const = 0). From the definition given in Theorem 4 it is evident that it lies on the surface Σ.

LEMMA. This curve γ is skew-orthogonal to the tangent fields to Σ, $v_0 = 2p_1\partial/\partial p_1 + 3q_2\partial/\partial q_2 + 4p_2\partial/\partial p_2$, and $v_{-5} = \partial/\partial q_1$.

Proof. The tangency is obvious (v_0 is a quasihomogeneous Euler field). We calculate the skew-scalar product of the tangent vectors: $\omega(\dot\gamma, v_0) = -2p_1\dot q_1 + 3q_2\dot p_2 - 4p_2\dot q_2$. We substitute the values of p and q and their derivatives with respect to ξ from (2):

$$\dot q_1 = -12\xi^4 - 2c\xi^2, \quad \dot p_2 = 12\xi^3 + 2c\xi, \quad \dot q_2 = -12\xi^2 - 2c.$$

We get zero. Since $\omega(\dot\gamma, v_{-5}) = \dot p_1 = 0$, the lemma is proved.

The vectors $\dot\gamma$, v_0, v_{-5} at a general point are linearly independent (if $c \neq 0$). All three are tangent to Σ. Consequently, $\dot\gamma$ is characteristic for $c \neq 0$, and hence also everywhere. Thus, γ is a trajectory of the geodesic flow of rays.

Translations along the axis A = B = C = 0 preserve Σ and ω. Hence the translation of a trajectory is a trajectory. (2) for m = 2 is proved. For m > 2 one now uses the skew-orthogonality of $\dot\gamma$ to the vectors $\partial/\partial p_3$, $\partial/\partial q_3, \ldots$.

The first assertion of Theorem 4 is proved. The second (about the geodesic flow of unit vectors) follows from the first, from Proposition 2 of Paragraph 4° of Sec. 1, and from the form of W in coordinates of Theorem 3.

5. Remarks

1. Curves, Dual to Smooth Ones. Let $\varphi: \mathbf{R} \to \mathbf{R}^3$ be a smooth mapping. We call it a curve of type (a, b, c) at 0, if the first nonzero derivative at 0 has number a, the first not collinear with it has number b, and the first not coplanar with them has number c.

Analogously we define curves of type $(a_1, \ldots, a_n)$ in $\mathbf{R}^n$, $\mathbf{RP}^n$, $\mathbf{C}^n$, and $\mathbf{CP}^n$.

The dual to $\varphi: \mathbf{R} \to \mathbf{RP}^3$ curve $\varphi^*: \mathbf{R} \to \mathbf{RP}^{3*}$ is defined as the manifold of osculating planes of the curve φ. Its type is $(c - b, c - a, c)$ [respectively, $(a_n - a_{n-1}, \ldots, a_n - a_1, a_n)$]. At an ordinary point in $\mathbf{R}^3$ the type of φ and φ^* is (1, 2, 3). Points of smoothness are distinguished by the condition $a_1 = 1$. It is easy to see that $\varphi^{**} = \varphi$.

The beginning of the hierarchy of points of smoothness of curves in $\mathbf{R}^3$ is the following:

(1, 2, 3) ordinary point ← (1, 2, 4) flattening ← (1, 2, 5) biflattening;
← (1, 3, 4) inflection.

The manifold of all tangent hyperplanes of the curve φ forms in the dual space a developable surface with cuspidal edge φ^*. We call it the front of the Legendre singularity of φ.

Example. Let φ have the simplest degeneracy of codimension 1 [flattening (1, 2, 4)]. Then φ^* has singularity (2, 3, 4) of the curve $x = t^2$, $y = t^3$, $z = t^4$. The tangent lines of this curve cover the ordinary swallowtail.

If φ is a curve in general position, then on it one can distinguish points of flattening, but there are no more complicated singularities. Consequently, the front of a curve in general position is a surface with cuspidal edge and swallowtails at isolated points.

In a general one-parameter family of curves $\{\varphi_s\}$ for isolated values of the parameter s there can occur points of biflattening. Let (x, y, z) be coordinates in the dual space, s be the parameter. The dual curves φ_s^* form in the four-dimensional space $\{(x, y, z; s)\}$ a surface V^2.

It is easy to prove

THEOREM 5. The surface V^2 in $\mathbf{R}^4$ is locally diffeomorphic to the open swallowtail, and its partition into the curves φ_s^* to its partition into the curves (2) with different c [for const = 0 in (2)].

2°. Covering a Caustic by the Cuspidal Edge of a Front. We consider a wave front in general position extended in $\mathbf{R}^3$. Its cuspidal edge covers a general caustic in $\mathbf{R}^3$. On a caustic in general position there occur singularities of swallowtail type. We consider the cuspidal edge of instantaneous fronts close to this singularity of the caustic. The surfaces corresponding to them in space–time cover a two-dimensional manifold V^2 in $\mathbf{R}^4$, lying on the direct product of a swallowtail by a line. One has

THEOREM 6 [10, 11]. The manifold V^2 in $\mathbf{R}^4$ is diffeomorphic with an open swallowtail. In space–time one can choose coordinates $(\lambda_1, \ldots, \lambda_4)$ such that: 1) the world-lines will have the direction of the axis λ_4 everywhere; 2) the manifold V^2 will be defined by the condition $V^2 = \{\lambda\colon \exists x\colon F = F' = F'' = 0\}$, where $F = x^5 + \lambda_1 x^3 + \lambda_2 x^2 + \lambda_3 x + \lambda_4$; 3) the partition of V^2 into instantaneous cuspidal edges on the level of formal series is equivalent to the partition of V^2 into lines $\pm\lambda_1 + \lambda_4 = \text{const}$ (these formal series in the holomorphic case, as a rule, diverge).

3°. Legendre Manifolds with Singularities and Their Fronts. A Lagrangian manifold with singularities in the space T^*B can be contactized to get a Legendre manifold with singularities in $J^1(B, \mathbf{R})$. This manifold defines a front which is a hypersurface in $J^0(B, \mathbf{R})$.

Example. In the situation of Corollary 3 of Sec. 2 such fronts are the level sets of the time function of the shortest circuit of the obstruction Π by the geodesic bundle and falling rays and also its graph.

THEOREM 7. We consider in $T^*\mathbf{R}$ a curve in general position with semicubic cusp. Then the corresponding front in $J^0(\mathbf{R}, \mathbf{R})$ is locally diffeomorphic with the curve $x^5 = y^2$.

Remark. This theorem explains the normal form of the time function in the neighborhood of an asymptotic ray, obtained by Platonova in [9]: $T = X + Z^{5/2}$.

Contactization of the singularity of the standard Lagrangian manifold of rays close to a general biasymptotic ray (Corollary 3 in Sec. 2) in the case of a Lagrangian fibration $(p_1, p_2, q_1, q_2) \mapsto (p_1, q_2)$ leads to a front, diffeomorphic to the surface $z = (150a^4 + 63a^2x + 7x^2)a^3$, $y = (10a^2 + 3x)a$ in (x, y, z) space.

LITERATURE CITED

1. R. B. Melrose, "Equivalence of glancing hypersurfaces," Invent. Math., 37, No. 3, 165–192 (1976).
2. H. Whitney, "On singularities of mappings of Euclidean spaces. I," Ann. Math., 62, 374-410 (1955).
3. B. Morin, "Formes canoniques des singularités d'une application differentiable," C. R. Acad. Sci., Paris, 260, 5662-5665, 6503-6506 (1965).
4. V. I. Arnol'd, "Indices of singular points of 1-forms on a manifold with boundary, displacement of invariants of groups, generated by reflections, and singular projections of smooth surfaces," Usp. Mat. Nauk, 34, No. 2, 3-38 (1979).
5. V. I. Arnold, "Wave fronts evolution and equivariant Morse lemma," Commun. Pure Appl. Math., 28, No. 6, 557-582 (1976).
6. L. Solomon, "Invariants of finite reflection groups," Nagoya Math. J., 22, 57-64 (1963).
7. V. I. Arnol'd, Ergodic Problems of Classical Mechanics, W. A. Benjamin (1968).
8. O. V. Lyashko, "Geometry of bifurcation diagrams," Usp. Mat. Nauk, 34, No. 4, 205-206 (1979).

9. O. A. Platonova, "Singularities in the problem of circuit of an obstruction," Funkts. Anal. Prilozhen., 15, No. 2, 86-87 (1981).
10. V. I. Arnol'd, Ya. B. Zel'dovich, and S. F. Shandarin, "Elements of the large-scale structure of the Universe," Usp. Mat. Nauk, 36, No. 3, 244-245 (1981).
11. V. I. Arnol'd, "Covering of a caustic by the cuspidal edge of a moving front," Usp. Mat. Nauk, 36, No. 4, 233 (1981).
12. L. Euler, Introductio in Analysin Infinitorum, Lausanne, Bosquet, No. 316 (1748).

Asymptotic rays in symplectic geometry (in Russian)

1. В. И. Арнольд «Асимптотические лучи в симплектической геометрии».

Каждое математическое понятие со временем симлектизируется. В 1976 г. Р. Мельроз дал симплектическую трактовку геометрий системы лучей, отражающихся от гладкой поверхности евклидова пространства. Симплектизацией этой бильярдной задачи оказалась задача о паре пересекающихся гиперповерхностей симплектического пространства.

Особенности расположения пары гиперповерхностей соответствуют асимптотическим направлениям отражающей поверхности. Простейшие особенности связанных с асимптотическими направлениями лагранжевых многообразий, «раскрытые ласточкины хвосты», встречались уже в работах Гильберта об инвариантах бинарных форм. Пространства форм и многочленов нечетной степени имеют естественные симплектические структуры. Многообразие многочленов с корнем кратности, большей половины степени, лагранжево.

С другой стороны, исследование пересечения пары гиперповерхностей приводит к локальной классификации гладких подмногообразий симплектических и контактных пространств. Например, Мельроз доказал, что двумерная поверхность общего положения в четырехмерном симплектическом пространстве приводится симплектическим диффеоморфизмом к одной из двух нормальных форм $y_1 = y_2 = 0$ либо $y_1 = 0, \ y_2 = x_1^2$ в окрестности каждой своей точки.

В докладе рассказано о теории отражающих лучей Мельроза, об особенностях лагранжевых многообразий и о локальной классификации гладких подмногообразий симплектических и контактных пространств, принадлежащей А. Б. Гивенталю.

Гивенталь доказал, что симплектический тип ростка подмногообразия определяется сужением симплектической 2-формы на его касательное пространство. Этим он свел классификацию ростков подмногообразий к теории вырождений симплектических структур Ж. Мартине, Р. Руссари, М. Голубицкого и Д. Тышлера, о которой также рассказано в докладе.

Originally publ. in: Usp. Mat. Nauk 37(2), 182-183,

Singularities of systems of rays

V.I. Arnol'd

CONTENTS

Introduction

A *system of rays* is what Hamilton called the system of normals to a hypersurface or, as we now say, a Lagrangian submanifold of the space of lines.

In many applications, along with smooth Lagrangian manifolds, Lagrangian manifolds with singularities are important. An account is given here of the first stages of a theory of singular Lagrangian manifolds.

*This translation incorporates some corrections and additions communicated to me by the author. Ed.

Originally publ. in: Usp. Mat. Nauk 38(2), 77-147, © 1983 Russ. Acad. Sciences
English transl.: Russ. Math. Surveys 38(2), 87-176, © 1983 London Mathematical Society.

The singularities of systems of rays have been studied for a long time (the theories of foci of Archimedes and Barrow, of evolutes and involutes of Huyghens, of caustics of Leibniz and Bernoulli, of conjugate points of Jacobi and Morse). Today these questions are a part of symplectic geometry, although we can refer them also to the calculus of variations, the theory of optimal control, the theory of partial differential equations, classical mechanics, optics, the theory of waves, or to the general theory of the singularities of smooth maps.

Many of the results about which we shall speak are related to such simple geometrical objects that it is astonishing that they have not been noticed before. For example, a local classification of the projections of surfaces in general position in three-dimensional space was found only in 1981. It turns out that there are finitely many, 14 to be precise, non-equivalent projection germs: that is the number of different ways in which we can see a neighbourhood of a point of a surface in general position if we look at it from different points of space.

The fact of the matter is apparently that the proofs are not simple: they are based on occasionally unexpected relationships with the theory of invariants, Lie algebras, groups of reflections, algebraic geometry, and mixed Hodge structures. Some of the results were stimulated by applications to the theory of perturbations of Hamiltonian systems and even to number theory, but an especially fruitful source of new points of view has been the problem of the quickest way round an obstacle bounded by a smooth surface in three-dimensional Euclidean space.

We consider the field of extremals joining the points of space to a fixed point and going round an obstacle (Fig. 1).

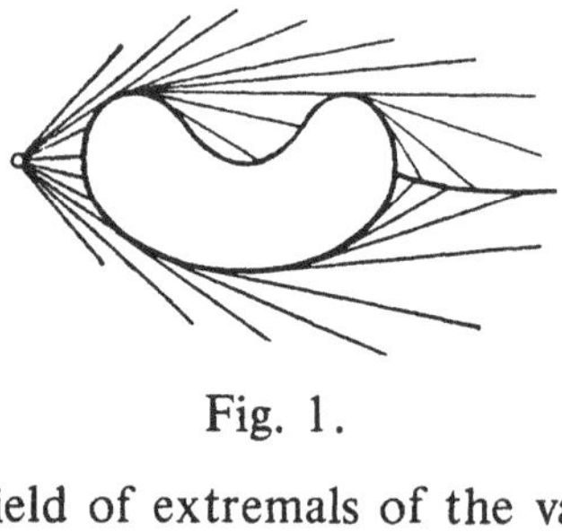

Fig. 1.

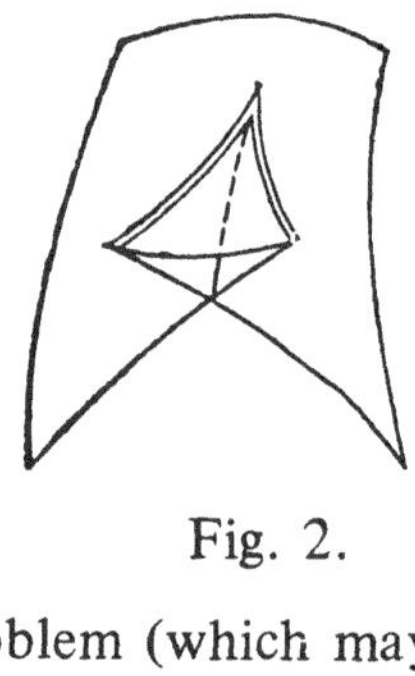

Fig. 2.

To the field of extremals of the variational problem (which may be many-valued) there corresponds in the symplectic geometry a Lagrangian manifold. The singularities of a caustic are those of the projection of a smooth Lagrangian manifold in the phase space into the configuration space. In the problem of going around an obstacle the Lagrangian manifolds have singularities. The simplest example is the two-dimensional surface in four-dimensional space that is called the open swallow-tail.

The ordinary swallow-tail is the surface in $\mathbf{R}^3$ (Fig. 2) consisting of all the polynomials x^4+ax^2+bx+c having a multiple root. The open swallow-tail is the surface in $\mathbf{R}^4$ consisting of all the polynomials $x^5+Ax^3+Bx^2+Cx+D$ having a root of multiplicity $\geqslant 3$. Differentiation of the polynomials maps the open tail to the ordinary one; for the open tail the edge of regression is preserved, but the self-intersection disappears.

The open swallow-tail turns up unexpectedly in the problem of going around an obstruction as a singularity of the manifold of rays breaking away from the obstruction surface in the symplectic space of lines. It appears that this variety is the first non-trivial example in a future theory of Lagrangian manifolds with singularities.

Another example is the appearance of the variety of irregular orbits of the Coxeter group H_3 (the group of symmetries of the icosahedron) in the calculus of variations. Taken by itself, the appearance of the discriminant varieties of finite groups generated by reflections, as normal forms of the singularities of graphs of solutions of variational problems, is quite striking: there are no clear a priori reasons for it. Following on the discovery in 1972 of the discriminant varieties of the Weyl groups A_k, D_k, E_k among the normal forms, and an understanding in 1978 of the variational meaning of the varieties of the irregular orbits of the remaining Weyl groups (B_k, C_k, F_4, G_2), there arose the conjecture that the non-crystallographic Coxeter groups were also bound up with singularities of the solutions of (more complicated) problems of the calculus of variations. In October 1982, as the result of joint efforts of Lyashko, the author, Givental', and Shcherbak, it became clear that the discriminant variety of the group H_3 is diffeomorphic to the graph of the (many-valued) time function in the problem of going around an obstacle bounded by a plane curve with a point of inflection. Therefore, with a point of cubic inflection of a plane curve there is associated an icosahedron whose group of symmetries controls the singularities of the involutes of this curve. (It is not easy to perceive this icosahedron in the neighbourhood of a point of inflection, even when we know of its existence.) It seems certain that the one remaining discriminant variety (H_4) will also be discovered in a deeper analysis of the singularities of systems of rays and wave fronts.

A significant part of the results presented in this survey was obtained in connection with research on the short-wave asymptotic behaviour of electromagnetic fields carried out at the Moscow Institute of Electronic Machine-building (MIEM). The author thanks their initiator V.I. Maslov, and also A.V. Alekseev, Ya.M. Eliashberg, D.B. Fuks, A.B. Givental', V.V. Goryunov, V.N. Karpushkin, A.G. Khovanskii, E.E. Landis, O.V. Lyashko, N.N. Nekhoroshev, O.A. Platonova, S.F. Shandarin, O.P. Shcherbak, A.N. Varchenko, V.A. Vasil'ev, V.M. Zakalyukin, and Ya.B. Zel'dovich for useful discussions.

Although the main aim of this survey is to present a number of recent results and the connections between them, it is appropriate to begin with things that have been known for a long time.

Chapter I

SYMPLECTIC GEOMETRY

Symplectic geometry is the geometry of the phase space. Along with the standard definitions and facts of symplectic geometry, this chapter contains some not altogether standard examples to be used in what follows (for example, the theory of normal forms of submanifolds of a symplectic manifold).

§1. Symplectic manifolds

A *symplectic structure* on a manifold is a closed non-degenerate 2-form.

Example 1 (Darboux coordinates). The 2-form $\omega = \sum dp_i \wedge dq_i$ in $\mathbf{R}^{2n}$ with coordinates (p_i, q_i) determines a symplectic structure. *All symplectic manifolds of the same dimension are locally symplectically diffeomorphic.* In other words, every closed non-degenerate 2-form is locally expressible by the above formula in suitable local coordinates (the so-called *Darboux coordinates*).

Two vectors tangent to a symplectic manifold at the same point are said to be *skew-orthogonal* if the value of the 2-form on them is zero. For example, the axes of the p-directions in the Darboux coordinates are skew-orthogonal.

Example 2 (phase space). *On the cotangent bundle space of any smooth manifold there is a natural symplectic structure* $\omega = d\alpha$ (the 1-form α on $M = T^*V$ is defined by the formula $\alpha(\xi) = \tau(\xi)\pi_*(\xi)$, where $\tau: TM \to M$ is the tangent bundle on M and $\pi: M \to V$ the cotangent bundle on V).

Vectors tangent to the fibre of the cotangent bundle at the same point are skew-orthogonal.

Example 3 (the space of extremals). We consider a hypersurface in a symplectic manifold. A direction on it is said to be *characteristic* if it is skew-orthogonal to a tangent plane. The integral curves of a field of characteristic directions are called *characteristics.* Thus, the hypersurface is locally fibered into curves (the characteristics). The base B of this fibration ρ has a natural symplectic structure (namely, $\omega_B(\xi, \eta) = \omega(\xi', \eta')$, where $\xi = \rho_*\xi', \eta = \rho_*\eta'$),

In particular, *on the space of extremals of a (one-dimensional) variational problem there is always a natural symplectic structure.* In some cases the manifold of extremals B is defined in the large.

Example 4 (the space of lines). We consider in the cotangent bundle space $T^*\mathbf{R}^n$ of a Euclidean space the hypersurface $p^2 = 1$. The characteristics of this hypersurface describe a uniform straight-line motion with fixed velocity squared. Consequently, *the manifold of all oriented lines in a Euclidean space has a natural symplectic structure.*

The manifold of lines B is diffeomorphic to the cotangent bundle space of the sphere S^{n-1}. For a line is defined by its direction vector and its point of intersection with the tangent hyperplane in $\mathbf{R}^n$ to the sphere at the endpoint of the direction vector (Fig. 3). It is not difficult to verify that the symplectic structure constructed in Example 3 on B differs from that induced from T^*S only in sign.

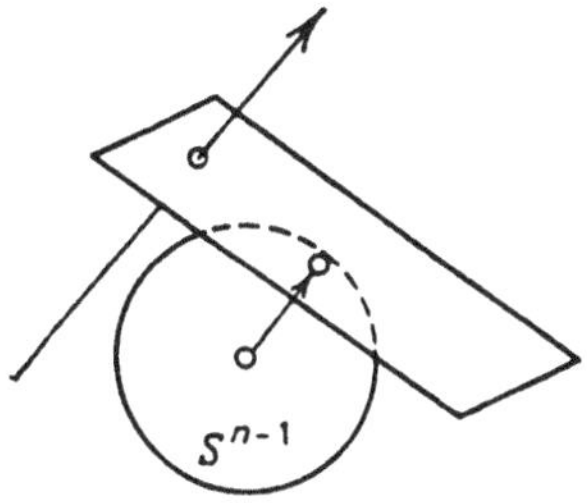

Fig. 3.

Example 5 (binary forms). We consider the linear space V of real forms of degree d in the variables x and y. The group $G = \mathrm{SL}(2, \mathbf{R})$ of unimodular (2×2)-matrices acts on this space in a natural way. If d is odd, then the space V is even-dimensional, and on it there is a G-invariant non-degenerate exterior 2-form (unique up to a real multiplier). Thus, *on the space of binary forms of odd degree there is* (up to a real multiplier) *a natural symplectic structure* (Givental').

In suitable Darboux coordinates a form φ of degree $d = 2n-1$ can be expressed as follows:

$$\varphi(x, 1) = q_1 x^d/d! + \ldots + q_n x^n/n! - p_n x^{n-1}/(n-1)! + \ldots + (-1)^n p_1.$$

Example 6 (polynomials of even degree). Forms like $x^d + \ldots$ constitute a hypersurface in the symplectic manifold V. We consider the manifold of its characteristics.

A form is determined by its value on the line $y = 1$, that is, by a polynomial $f = x^d + \ldots$ in x. Along a characteristic, forms differ by multiples of y^d. Therefore, the characteristic is defined by the derivative of f, consequently, by a polynomial like $x^{2m} + \ldots$ $(d = 2m+1)$. We obtain (up to a multiplier) *a natural symplectic structure on the space of monic polynomials of even degree in x.*

§2. Submanifolds of a symplectic space

Submanifolds of a Riemannian or Euclidean space differ not only in their intrinsic geometry, but also in the way they are situated (for example, for a surface in $\mathbf{R}^3$, besides the Gaussian curvature there is the mean curvature). In the symplectic case the matter is simpler: the inner geometry (locally) determines the outer.

Theorem (Givental').[(1)] *A germ of a submanifold of a symplectic space is determined, up to a symplectic diffeomorphism, by the restriction of its symplectic structure to its tangent bundle.*

Thus, the local classification of submanifolds of symplectic space reduces to that of the degeneracies of closed 2-forms. Using results of Martinet [1], Roussari [2], and Golubitsky and Tishler [3] we obtain the following corollary.

Corollary. *Every germ of a k-dimensional submanifold in general position in a 2n-dimensional symplectic space is symplectically diffeomorphic to one of the following germs at zero:*

k	$2n$	normal form
1	2	$p_1=0$
2	4	$p_1=q_1=0;\ q_1=p_1^2-p_2=0$
3	4	$p_1=0$
4	6	$p_1=q_1=0;\ q_1=p_2^2-p_2=0;\ p_3=p_2-p_1q_3\mp q_1q_2-q_3^3/6=0;$ (*)

Here $p_1, \ldots, q_n$ are the Darboux coordinates.

Remark 1. The second singularity for $k = 2$ or 4 is realized at points of a hypersurface (that is, on a curve of the 2-surface or on a three-dimensional surface in four-dimensional space). The third singularity for $k = 4$ is realized on a curve in four-dimensional space. The passage from "+" to "–" in this normal form (from the *elliptic* case to the *hyperbolic*) takes place at isolated parabolic points, whose existence is denoted in the table by the star *. The normal forms at parabolic points contain moduli. The remaining singularities in the table are stable and simple (do not have moduli).

Remark 2. The classification given refers to the C^∞-case. In the *analytic* case the question is open, as far as I know.

Remark 3. The simplest non-trivial normal form (the surface $p_1^2-p_2=q_1=0$ in $\mathbf{R}^4$) was apparently first found by Melrose [5]. His result refers, however, also to *the surface of codimension* 2 *in* $\mathbf{R}^{2n}$ given by the same equations: *the singularity of the germ at* 0 *of this surface is stable* (that is, every nearby surface has a symplectically equivalent germ at a suitable nearby point).

Remark 4. The germs of submanifolds of $\mathbf{R}^{2n}$ introduced above can be embedded in $\mathbf{R}^{2n}+\mathbf{R}^{2m}$ (with its standard symplectic structure). They are, therefore, normal forms of germs of submanifolds of symplectic manifolds of arbitrary dimension. For example:

---

(1) According to a theorem of Weinstein [4], a submanifold is determined by the restriction of ω to a larger bundle and this does not allow him to deduce the corollaries that follow.

a) *on four-dimensional submanifolds there occur curves of stable elliptic and hyperbolic singular points with the normal form*

$$p_2 = p_1 q_3 \pm q_1 q_2 + q_3^3/6, \quad p_3 = 0, \quad p_4 = q_4 = p_5 = q_5 = \ldots = 0;$$

b) *the germ of the generalized Melrose singularity*

$$p_2 = p_1^2, \quad q_1 = 0, \quad p_{n+1} = q_{n+1} = \ldots = p_{n+m} = q_{n+m} = 0$$

is stable for any $m \geqslant 0$, $n \geqslant 2$;

c) *a two-dimensional surface in general position in a symplectic space of arbitrary dimension has only singularities of the preceding type*:

$$p_2 = p_1^2, \quad q_1 = 0, \quad p_3 = q_3 = \ldots = 0.$$

We now describe briefly the geometry of the elliptic and hyperbolic singularities of the situation of a four-dimensional manifold in a symplectic space of large dimension. The restriction of the symplectic form to a four-dimensional manifold is closed, but need not be a non-degenerate 2-form on it. A generic 2-form on a four-dimensional manifold (no matter whether it is closed or not) degenerates on a three-dimensional submanifold. At points of this submanifold the form has a two-dimensional kernel (the subspace skew-orthogonal to the whole space with respect to this form).

Therefore, on a four-dimensional submanifold in general position in a symplectic manifold there is invariantly defined a three-dimensional submanifold and a field of 2-planes on it. There are two independent conditions for the 2-plane in the four-dimensional space to be tangent to the three-dimensional space. Two conditions on a point of the three-dimensional space determine a curve (smooth, if the original four-dimensional submanifold is in general position).

Thus, on the three-dimensional manifold we have constructed there is a distinguished curve, where the kernel is tangent to the three-dimensional manifold. At the remaining points of the three-dimensional manifold the kernel intersects the tangent plane to the manifold along a line. Therefore, away from the curve in question there is a field of directions invariantly defined on the three-dimensional manifold.

This field of directions is the field of kernels of a closed 2-form on the three-dimensional manifold (namely, the restriction of the original symplectic form). In local coordinates we can identify this 2-form with a vector field of divergence zero. On the curve mentioned above this field degenerates. Note that this vector field is not invariantly connected with the original problem, but its field of directions is.

Thus, we arrive at a strongly divergent vector field of divergence 0 on the three-dimensional space: its singular points form a whole curve. It turns out that the problem of reducing such an infinitely degenerate field to a normal form (by diffeomorphisms of the ambient space and multiplication by an appropriate function) is easier than the (practically hopeless) problem of the normal form of a non-degenerate field of divergence 0 at a singular point.

We consider to begin with the linear part of the field at the origin. One of the eigenvalues and the sum of all three is zero. Therefore, the remaining two are either real or pure imaginary. At a generic point of the curve of degeneracy they are different from 0. Then the linear part of the field reduces either to the form $x\partial_x - y\partial_y$ or to the form $x\partial_y - y\partial_x$ (both at 0 and at nearby points of the z-axis) for a suitable choice of coordinates and after multiplication by a suitable function. In the first case the point is said to be hyperbolic and in the second case to be elliptic.

The analysis of the higher terms of the Taylor series now reduces to the standard scheme of the theory of normal forms. The final answer (at least in the class of formal power series) has the form

$$x\partial_x - y\partial_y + xy\partial_z, \quad x\partial_y - y\partial_x + (x^2 + y^2)\partial_z$$

(in the hyperbolic and elliptic cases, respectively).

These formulae indicate, for example, that in the elliptic case the integral curves of the field of kernels of the 2-form, reduced to normal form, are screw curves, winding around the curve of degeneracy, and the pitch of the screw line decreases as the square of the distance from the axis (so that one twist of an integral curve close to the axis is very nearly a circle, Fig. 4).

Fig. 4.

It can be shown that the invariant thus constructed (the field of directions of the kernels) determines the symplectic type of the germ of the original four-dimensional submanifold. In fact, at no step of the construction has there been any loss of information: computations show that the field of kernels on the three-dimensional manifold determines a closed 2-form on it, uniquely up to a diffeomorphism, the restriction of the closed 2-form to the three-dimensional manifold of its degeneracy determines the germ of a closed 2-form on the four-dimensional manifold, uniquely up to a diffeomorphism, and finally, the germ of the closed 2-form on the four-dimensional manifold induced from the symplectic structure determines the symplectic type of the submanifold uniquely up to a symplectic diffeomorphism of the ambient space (by the theorem of Givental').

§3. Lagrangian manifolds, bundles, maps, and singularities

A. Lagrangian manifolds.

A *Lagrangian manifold* is the name for a submanifold of a symplectic manifold on which the symplectic structure vanishes and which has the greatest possible dimension (equal to half the dimension of the ambient manifold).

Example 1. The plane $p = 0$ in $\mathbf{R}^{2n}$ with Darboux coordinates is Lagrangian. All Lagrangian manifolds of the same dimension are locally symplectically diffeomorphic, that is, each of them is given locally by the formula $p = 0$ in suitable Darboux coordinates.

Example 2. The formula $p = ds$, where s is a function on V, defines a Lagrangian section on T^*V.

Example 3. The fibres of a cotangent bundle are Lagrangian.

Example 4. The manifold of all oriented normals to a smooth submanifold Γ (of arbitrary dimension) in the Euclidean space $\mathbf{R}^n$ is a Lagrangian submanifold of the symplectic space of lines in $\mathbf{R}^n$.

Example 5. The manifold of all monic polynomials of degree $2m$ that are divisible by x^m is Lagrangian.

B. Lagrangian bundles.

A *Lagrangian bundle* is a bundle whose fibres are Lagrangian.

Example 1. $\mathbf{R}^{2n} \to \mathbf{R}^n$, $(q, p) \mapsto q$.

Example 2. The cotangent bundle of any smooth manifold is Lagrangian.

Example 3. With each oriented line of a Euclidean space we associate its direction unit vector. The resulting map of the symplectic manifold of oriented lines of the Euclidean space to the sphere is a Lagrangian bundle (which coincides with the cotangent bundle of the sphere, except for the sign of the symplectic structure).

All Lagrangian bundles of the same dimension are locally equivalent (that is, they are given by the formula of Example 1 in suitable Darboux coordinates in a neighbourhood of each point of the bundle space).

C. **Lagrangian maps.**
Let L be an immersed Lagrangian submanifold in the space of a Lagrangian bundle. Its projection to the base of the bundle is said to be a *Lagrangian map* (Fig. 5). Thus, a Lagrangian map is a triple $L \to E \to B$, where the first arrow is the immersion of the Lagrangian submanifold and the second is a Lagrangian bundle.

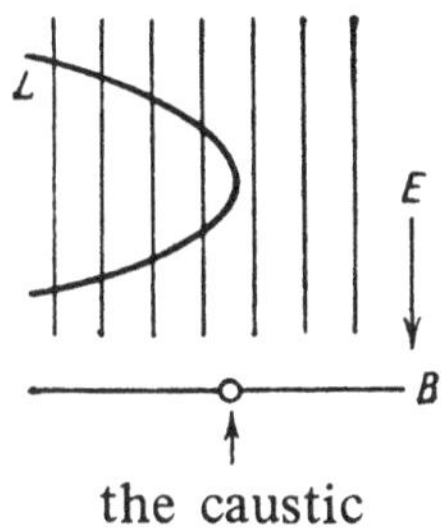

Fig. 5.

A *Lagrangian equivalence* of a pair of Lagrangian maps is defined as a symplectic diffeomorphism of the bundle spaces, sending fibres to fibres and one Lagrangian submanifold to the other. Thus, a Lagrangian equivalence is a (3 x 2)-commutative diagram whose horizontals are Lagrangian maps, while the verticals are diffeomorphisms, the middle one respecting the symplectic structure.

The set of critical values of a Lagrangian map is called its *caustic*. The caustics of equivalent maps are diffeomorphic.

Example 1 (the gradient map). $q \mapsto p = \partial S/\partial q$. The Lagrangian manifold is the graph of this map, and the bundle is the projection $(q, p) \mapsto p$.

Example 2 (the normal map). With each vector normal to a submanifold of Euclidean space we associate its end-point. This map is Lagrangian (L is represented in $T^*\mathbf{R}^n$ by the 1-forms $(\mathbf{n}, \cdot)$ at the ends of the normal vectors $\mathbf{n}$).

Its caustic is the envelope of the family of normals to the original manifold (for a hypersurface it is the locus of the centres of curvature).

Example 3 (the Gauss map). This is the map of a transversally oriented hypersurface in a Euclidean space to the sphere that associates a point of the surface with the direction unit vector of the normal at it.

The Gauss map is Lagrangian (the corresponding Lagrangian submanifold of the symplectic manifold of oriented lines in the Euclidean space is formed by the normals to the hypersurface).

These three examples served (in 1966) as motivation for setting up the theory of the singularities of Lagrangian maps, which is parallel to the general theory of singularities of Whitney, Thom, and Mather.

D. Lagrangian singularities.

Let $F(x, q)$ be the optical path length from a point x of a manifold, the source of light, to a point q of a manifold of observation. The phases of the waves on the observation manifold are determined by the Lagrangian manifold

$$(*) \qquad L = \{(p, q) \colon \exists x \colon \partial F/\partial x = 0, \quad p = \partial F/\partial q\}.$$

The family F of functions of x with parameter q is called the *generating family* of this manifold, and also of its Lagrangian map $(p, q) \mapsto q$.

Example (see [6]). The singularities with $\mu \leqslant n+1$

$$A_\mu \colon F = \pm x^{\mu+1} + q_1 x^{\mu-1} + \ldots + q_{\mu-1} x,$$
$$D_\mu \colon F = x_1^2 x_2 \pm x_2^{\mu-1} + q_1 x_2^{\mu-2} + \ldots + q_{\mu-2} x_2 + q_{\mu-1} x_1,$$
$$E_6 \colon F = x_1^3 \pm x_2^4 + q_1 x_1 x_2^2 + q_2 x_1 x_2 + q_3 x_2^2 + q_4 x_1 + q_5 x_2$$

exhaust, up to Lagrangian equivalence, the singularities of Lagrangian maps in general position of manifolds of dimension $n \leqslant 5$.

All the singularities A_μ, D_μ, and E_μ are stable and simple (do not have moduli). For large n generic singularities of Lagrangian maps have functional moduli (a classification of Lagrangian singularities for $n \leqslant 10$ is given in [7] and [8]).

Corollary 1. *A generic one-dimensional caustic has (apart from transversal self-intersections) only singularities of cusp type (A_3), two-dimensional ones have only swallow-tails (A_4) and singularities D_4, which Thom* [9] *has called the pyramid (D_4^-) and the purse (D_4^+) (Fig.* 6).

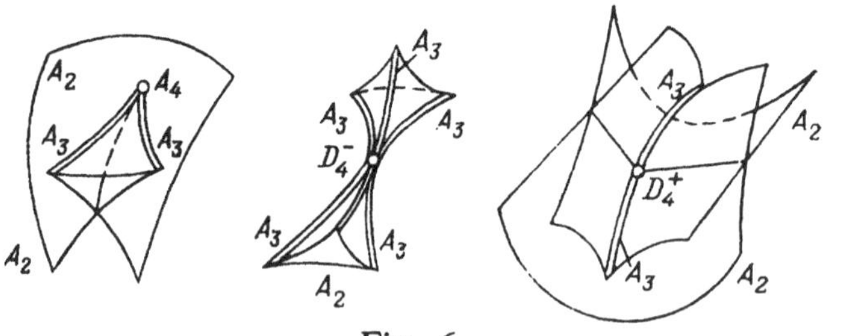

Fig. 6.

Corollary 2. *The Gauss map of a generic surface in* $\mathbf{R}^3$ *has as singularities only folds A_2 (on the curve of parabolic points) and pleats (or Whitney cusps) A_3 (at points where the asymptotic direction is tangent to this curve). See* [10], [11], [12].

Every Lagrangian manifold can be given locally by a generating family F of functions of x depending on the parameter q according to the formula (*). (For this formula to induce a smooth Lagrangian manifold it is necessary that the equation $\partial F/\partial x = 0$ be normally soluble with respect to x; the requirement of normal solubility is *included* in the definition of a generating family.)

Two families of functions are said to be *R^+-equivalent* if they can be mapped into one another by a smooth change of the independent variables, smoothly depending on the parameters, a smooth change of the parameters, and the addition of a smooth function of the parameters: in old-fashioned notation

$$F(x, q) \sim F(X(x, q), Q(q)) + \Phi(q).$$

Two families of functions (generally speaking, of different numbers of variables) are said to be *stably R^+-equivalent* if they become R^+-equivalent after the addition of non-degenerate quadratic forms in additional variables. For example, the function x^3 of x is stably equivalent to the function $x^3 - y^2$ of (x, y) but is not stably equivalent to the function x^3 of these two variables.

All generating families of Lagrangian equivalent maps are locally stably R^+-equivalent. Conversely, all stably R^+-equivalent generating families induce Lagrangian equivalent maps (*see* [8], [13]).

Thus, the classification of Lagrangian singularities (up to Lagrangian equivalence) reduces to that of families of functions (up to stable R^+-equivalence). For example, the Lagrangian stability of a singularity indicates the R^+-versality of the corresponding family, the known hierarchy of critical points of functions (see [7])

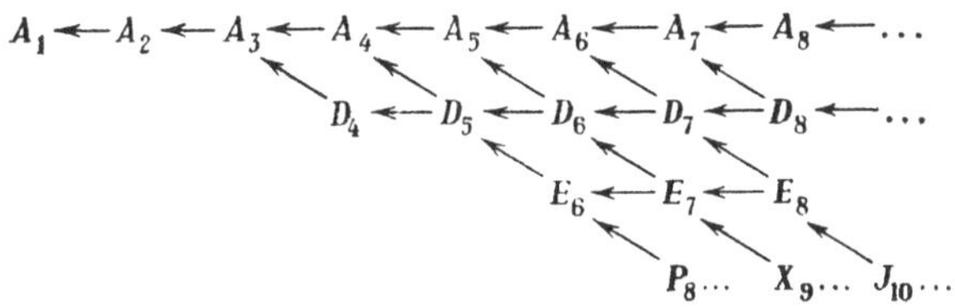

gives rise to a hierarchy of Lagrangian singularities, and so on.

Chapter II

APPLICATIONS OF THE THEORY OF LAGRANGIAN SINGULARITIES

Of the numerous applications of the classification of Lagrangian singularities we dwell here only on the following three: the estimation of integrals of rapidly oscillating functions (that is, the "stationary phase method"), the problem of the asymptotic number of integral points in a domain, and the theory of the cellular large-scale structure of the Universe.

§4. Oscillatory integrals

The caustics of Lagrangian maps can be described as manifolds associated with those points of the q-space at which the integral of the rapidly oscillating function

$$I(h, q) = \int e^{iF(x, q)/h} a(x, q)\, dx$$

decreases more slowly than at other points, as the wave length h tends to zero.

The real function F of variables x and parameters q is said to be the *phase* of the integral, and the function a its *amplitude*. The integral depends linearly on the amplitude. The amplitude is usually assumed to be finite. In speaking of the asymptotic behaviour of the integral, we have in mind that of the linear functional of the amplitude (supposed to be smooth) expressed by it. Thus, the asymptotic behaviour of the integral is that with respect to a "very bad" smooth amplitude, (giving a very slow decrease of the integral as $h \to 0$) and, consequently, also for any generic amplitude.

The simplest example of an oscillatory integral is the Fresnel integral $\int \cos (x^2/h)\, dx$. The "main contribution" to this integral is the neighbourhood of the point $x = 0$ (Fig. 7) of the phase $F = x^2$. The length of the first hump of $\cos (x^2/h)$ is of the order $h^{1/2}$, therefore, the whole integral decreases like $h^{1/2}$ (the contributions of subsequent humps and troughs cancel each other out). If the number of variables x_i is n and all the critical points of the phase F are Morse points ($\det \partial^2 F/\partial x_i\, \partial x_j \neq 0$), then the integral decreases like $h^{n/2}$. (To prove this it is enough to reduce the phase to a sum of squares and to use the Fresnel integral, Fig. 7.)

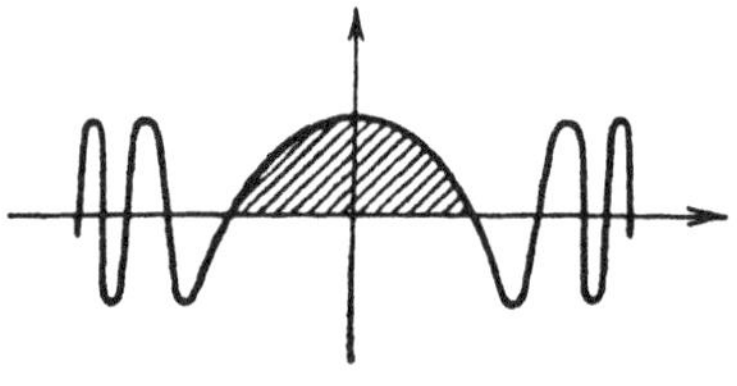

Fig. 7.

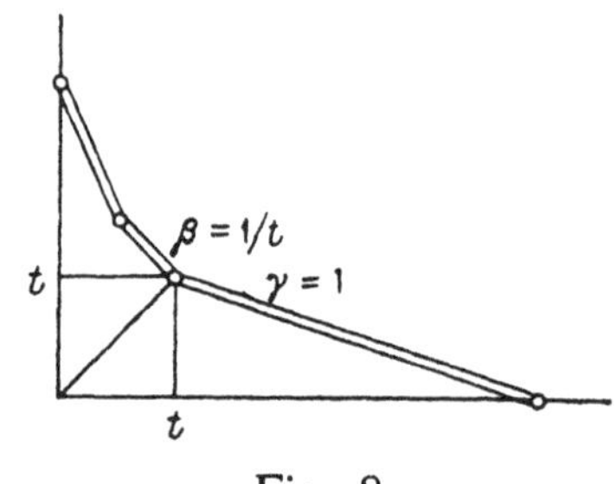

Fig. 8.

The coefficient of $h^{n/2}$ in the asymptotic expansion is proportional to $|\det (\partial^2 F/\partial x^2)|^{-1/2}$ (this quantity appears from the Jacobian of the change of variables reducing F to a sum of squares). Colin de Verdière [14] has established an estimate for an oscillatory integral by the sum of the contributions of the Morse points in the asymptotic expansion,

$$|I| \leqslant C h^{n/2} \sum |\det (\partial^2 F/\partial x^2)|^{-1/2},$$

which holds *uniformly* at non-caustic points as the caustic is approached. This estimate has been proved for phases depending, in general, on no more than 6 parameters (or, even more generally, determining only simple and parabolic Lagrangian singularities), but it has not been refuted in the general case.

If the point q belongs to the caustic, then the corresponding phase $F(\cdot, q)$ has, as a function of x, a degenerate (non-Morse) critical point. For such a value of q the integral decreases more slowly, and the index $\beta(q)$ of the principal term of the asymptotic expansion

$$I(h, q) \sim Ch^{\beta(q)} \log^{\gamma(q)} h + \ldots, \quad h \to 0,$$

is less than the usual value $n/2$.

In many cases the indices β and γ can be read off from the Newton boundary of the phase (Fig. 8). We consider the Taylor series of the phase at a critical point; transferring the critical point to the origin we find $F \sim \sum a_m x^m$ (the parameter q is fixed). The index m ranges over the non-negative orthant $\mathbf{Z}^n_+$ of the integral lattice $\mathbf{Z}^n$. The *support* of F is the set of values of m for which the coefficient a_m is different from zero. To each point of the support we attach a positive orthant $\mathbf{R}^n_+$. The boundary of the convex hull of the resulting "ladder" is called the *Newton boundary* of F.

We consider the point of intersection of the Newton boundary with the bisectrix of the coordinate orthant (Fig. 8). We denote the coordinates of this point by $(t, \ldots, t)$. The *exponent β of the given Newton boundary* is the number $1/t$. For example, the exponent β of the Newton boundary of the Morse function $x_1^2 + \ldots x_n^2$ is equal to $n/2$. The *exponent γ of the given Newton boundary* is the number of degrees of freedom of the support plane of the boundary at its point of intersection with the bisectrix. For the Morse function $\gamma = 0$.

The following theorem establishes a formula proposed by the author for the exponent β of the principal term of the asymptotic expansion. The Newton boundary is said to be *distant* if $t > 1$ for it.

Theorem (Varchenko [15]). 1) *The exponents β and γ of the principal term of the asymptotic expansion of an oscillatory integral with a phase having a given distant Newton boundary are equal to the exponents β and γ of the same diagram for typical phase functions with the given boundary.*

2) *For typical functions with a given not remote boundary the exponent β is not less than the exponent β of the boundary.*

3) *The exponent β for analytic phase functions depending on two parameters is equal to the exponent β of the boundary in a suitable coordinate system.*

The "non-typical" functions in this theory are those for which the Taylor coefficients corresponding to exponents lying on the Newton boundary satisfy some algebraic condition. If the boundary is distant, then $\beta(f) \leqslant \beta(\text{diagram})$ for *all* functions with the given boundary.

In other words, *if the boundary is remote, then the oscillatory integral always decreases no more quickly than the boundary prescribes and almost always at exactly this rate. If, however, the boundary is not remote, then it is asserted only that the integral almost always decreases no slower than the boundary prescribes.* In the latter case the decrease of the integral can turn out to become stable more quickly than the boundary prescribes. In that case the integral behaves as if the singularity of the phase were weaker than the boundary indicates: the fact is that the strong singularity indicated by the boundary can appear only in the complex domain and does not influence the asymptotic behaviour of the real oscillatory integral. The theorem asserts that such a phenomenon is impossible if the boundary is distant.

Varchenko has given an example of a near boundary in five-dimensional space for which this phenomenon occurs: for this boundary $\beta = 29/18$, while for the open set of phase functions with this boundary $\beta = 7/4 > 29/18$. Namely, the Newton boundary of the function of five variables $u^2 + x^9 + y^9 + z^9 + (u - (x^2 + x^4 + y^2 + z^2))\,v$ has this property. For functions of $n \leqslant 3$ variables such an example is impossible and the condition of remoteness in the first part of the theorem relating to β can be dropped (γ may be less than 1 even when $n = 3$, and for $f = xy$ even when $n = 2$).

We turn to an oscillatory integral with a phase $F(x, q)$ depending on a parameter q. If q does not belong to the caustic, then the integral decreases like $h^{n/2}$ as $h \to 0$, where n is the dimension of $\{x\}$. At non-singular points of the caustic the integral decreases more slowly than $h^{n/2-1/6}$.

For values of the parameter q belonging to the set of singularities of the caustic the integral usually decreases, as $h \to 0$, more slowly than the stronger singularity. In such cases the integral admits a uniform upper estimate (for parameters q close to q_0) of the size of the principal part of its asymptotic expansion for $q = q_0$ or at least of the size of $Ch^{\beta(q_0)-\varepsilon}$ for any $\varepsilon > 0$. For $n = \dim\{x\} = 1$ and a finite amplitude a such a uniform estimate (with $\varepsilon = 0$) was established by Vinogradov [16], and for $n = 2$ by Karpushkin [17]. For $n = 3$ Varchenko has constructed an example where β is not semicontinuous and for which, consequently, there is no uniform estimate: for $F = (qx_1^2 + x_1^4 + x_2^2 + x_3^2)^2 + x_1^{12} + x_2^{12} + x_3^{12}$ the exponent $\beta(q = 0) = 5/8$, $\beta(q < 0) = 7/12 < 5/8$. This phenomenon is inherent in generic families with $l \geqslant 73$ parameters. For $l \leqslant 7$ Karpushkin has established the uniform estimate above.

Pham [18] has conjectured that such a uniform estimate always holds if the definition of β is generalized by passing from oscillatory integrals on a real space to exponential integrals by the saddle-point method with respect to all possible (not necessarily real) chains. The hierarchy of singularities of such integrals has been studied by Vasil'ev [19].

Varchenko ([20], [21], [22]) has extracted from the asymptotic expansion of such integrals (for all amplitudes a and chains of integration) the mixed Hodge structure of the singularity, which had previously been determined in quite another way by Steenbrink [23]. Among the consequences of this result of Varchenko is the proof that the so-called intrinsic modality of semiquasihomogeneous functions coincides with the usual modality [24] and the proof that β is constant under deformations that do not change the multiplicity μ of the critical point [25]. For other applications, see [26].

Scherk, Steenbrink [27], and Pham [28] tried unsuccessfully to reprove Varchenko's theorem, while Saito [29] appears to have been successful.

In the mixed Hodge structure of a singularity there is hidden significant geometrical information still to be extracted: I mention only the estimate proved in [30] of the index of the gradient of a real function of $2m$ variables at a critical point, namely, $|\operatorname{ind}| \leqslant h_1^{m,m}$ in terms of the mean Hodge number. The semicontinuity of the (generalized) exponent β would follow from that of the spectrum of the singularity [31] (the collection of the so-called Steenbrink numbers [23]). The semicontinuity of the spectrum is corroborated by experimental material of Brieskorn [32] and Goryunov [33]: Goryunov, in particular, has shown that adjacencies of unimodal singularities to each other occur precisely when this does not contradict the semicontinuity of the spectrum (with the exception of the adjacency $Q_{11} \to J_{10}$, which is not a consequence of the semicontinuity of the inertia indices of the quadratic form of the singularity). Varchenko has used the semicontinuity of the inertia indices of the singularities $f(x)+y^a$ to prove the semicontinuity of the number of points of the spectrum on certain intervals of the real axis (see [34]); this result also supports the conjecture on the semicontinuity of the spectrum.

In October 1982 the conjecture on the semicontinuity of the spectrum was strongly reinforced by results in what would seem to be a quite unconnected classical problem in algebraic geometry: what is the greatest number ν of (Morse) singular points on a hypersurface of degree d in $\mathbf{C}P^n$? The number of singular points on a plane curve of degree d is maximal when the curve splits into d lines ($\nu = d(d-1)/2$). For surfaces of degree d the maximal number ν of singular points is known only for $d < 7$:

d	1	2	3	4	5	6	7
ν	0	1	4	16	31	64	$90 \leqslant \nu \leqslant 108$.

The unexpected connection between the problem on an estimate of the number of singular points and the semicontinuity of the spectrum was discovered in the following way. In 1981 Bruce [89] noticed that results in the local theory of singularities allowed him to make advances in the classical global problem on the estimate of ν. The upper estimate he obtained had for $n = 3$ the form $\nu \leqslant d^3/2 + O(d^2)$ (like, however, the estimates of his predecessors). The best of the known lower estimates increases like $d^3/3$. The problem of the semicontinuity of the spectrum turned out to be connected with shifting the coefficient 1/2 in the direction of 1/3.

In the spring of 1982 I talked on Bruce's work in a seminar on singularities and analyzed the values of ν given above. Following a general principle [30], I compared them with the number of integral points in the polyhedra connected with the problem and made the conjecture that the number of singular points ν does not exceed the number N of those integral points m in the cube $(0, d)^n$, for which the number $|m| = \sum m_k$ lies in the interval $(n-2)d/2+1 < |m| \leqslant nd/2$. For $n = 2$ the conjecture $\nu \leqslant N$ gives the exact estimate $\nu \leqslant d(d-1)/2$, while for $n = 3$ the estimate is of the form $\nu \leqslant 23d^3/48 + O(d^2)$, which is asymptotically better than Bruce's estimate even in the principal term. Chmutov has constructed an example with $\nu \geqslant 3d^3/8 + O(d^2)$.

While attempting to prove the conjecture that $\nu \leqslant N$, Givental' in October 1982 improved the smallest terms of Bruce's estimate. He pointed out the connection of the problem with the semicontinuity of the spectrum and to improve the estimate for ν he used inequalities of the Rayleigh - Fisher - Courant type, which in [31] were intended for the proof of the semicontinuity of the spectrum.

Immediately after the lecture by Givental', Varchenko skilfully applied to this situation the theory of mixed Hodge structures and a theorem of Steenbrink [90] on limits of Hodge structures. As a result, Varchenko obtained, on the one hand, a complete proof of the conjecture $\nu \leqslant N$, and on the other, a proof of the semicontinuity of the spectrum in the following cases: either a) the singularity is quasihomogeneous, and the deformation is lower, that is, it is obtained by adding monomials of lower quasihomogeneous degree; or b) the Newton diagram is distant. In the case b) the semicontinuity has been proved only for the left end of the spectrum, that is, for the "complex" exponent β; the condition of remoteness of the diagram is known to be satisfied if on writing the Taylor series in some coordinate system the point (1, ..., 1) does not lie strictly inside the Newton polyhedron. For functions of no more than three variables the semicontinuity of the "complex" β has been proved also without any restriction on remoteness, since in this case the near diagrams can easily be examined, and for them the behaviour of β is known.

§5. Integral points

Applications of the modern theory of oscillatory integrals to an asymptotic estimate of the number of integral points in a domain were apparently first published by Colin de Verdière [14].

Let G be a domain of volume V in Euclidean space $\mathbf{R}^n$, bounded by a smooth hypersurface, let λG be a homothetic domain, $N(\lambda)$ the number of integral points in G, and $R(\lambda) = \lambda^n V - N(\lambda)$ the deficiency of the asymptotic formula $N(\lambda) \sim \lambda^n V$ as $\lambda \to \infty$.

1°. *It is clear that* $|R(\lambda)| \leqslant C\lambda^{n-1}$. For the error is not greater than the number of integral points in a d-neighbourhood of the boundary of λG, where d is the diameter of the unit cube, and the area of the boundary increases like λ^{n-1} (Fig. 9).

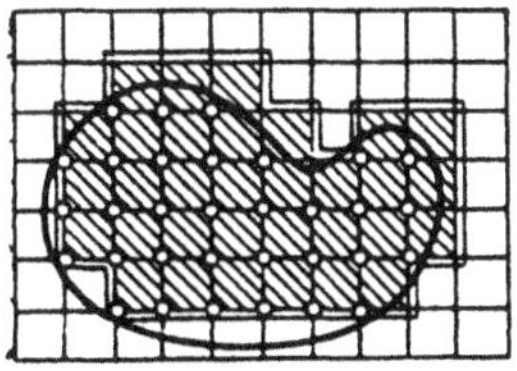

Fig. 9.

2°. On the other hand, the error in the order of magnitude is not less than the number of integral points on the boundary of the domain. For a sufficiently small change in λ changes $N(\lambda)$ by the number of boundary points. From this it follows that *for a ball with centre at the origin* $|R(\lambda)| \geqslant C\lambda^{n-2}$ *for some sufficiently large* λ.

For in the layer between balls of radii r and $r+a$ the number of integral points is of order r^{n-1}, while the number of spheres passing through them (that is, the number of integers between r^2 and $(r+a)^2$) is of the order r. By Dirichlet's box principle there is a sphere with cr^{n-2} points, and hence $|R(\lambda)| \geqslant C\lambda^{n-2}$ for sufficiently large λ.

3°. Using the theory of Lagrangian singularities, Colin de Verdière has established the estimate $|R(\lambda)| \leqslant C\lambda^{n-2+2/(n+1)}$ for a domain bounded by a smooth hypersurface in general position in $\mathbf{R}^n$ for $n \leqslant 7$. Whether or not this estimate is true for larger dimensions is not known. For $n = 2$ this estimate cannot be improved in the order of magnitude. For larger n it can probably be improved, but this has not been proved even if the curvature is positive.

4°. The remainder term R can be regarded as the sum of a large number of terms (of order λ^{n-1}) with "mathematical expectation" zero (they correspond to boxes cut by the boundary). If the terms were independent, then by the laws of probability theory their sum would be of order $\lambda^{(n-1)/2} \ll \lambda^{n-2}$.

The theory of Lagrangian singularities is justified by the following argument:

Theorem (Varchenko [35]). *The quadratic mean value* $|R(\lambda)|$ *over all the lattices obtained from the lattice of integral points by rotations and translations admits an upper estimate of the order* $\lambda^{(n-1)/2}$ *for any domain with a sufficiently smooth boundary in* $\mathbf{R}^n$.

The convex analytic case was studied earlier by Randol [36]. The crux of the proof is the same as in the proof that the canonical Maslov operator is unitary (the phase $f(x)-(q, x)$, regarded as a function of both variables, has only *Morse* critical points).

5°. The connection between the problem on integral points and the theory of oscillatory integrals consists in the following. We can regard an estimate of the remainder term R as one of the deficiency of the quadrature formula. In estimating the deficiency of the quadrature formula it is natural to apply the formula to a harmonic (and then decompose an arbitrary function into harmonics). For harmonics it is easy to compute the deficiency explicitly: it vanishes except for certain harmonics with large indices (proportional to λ). Therefore, the deficiency is basically determined by the Fourier coefficients of the characteristic function of the domain G with indices of order λ. But these coefficients can be expressed as oscillatory integrals over the boundary of the domain (the phase of the integral is the restriction of the linear function that serves as the phase of the harmonic to the boundary of the domain). Thus, the deficiency can be estimated by means of oscillatory integrals, for which the singularities of the phase reflect the flatness of the boundary of the domain. Thus, the singularities of the system of rays orthogonal to the boundary of the domain influence the asymptotic behaviour of the number of integral points in the stretched domain.

The remainder term $R(\lambda)$ can be written in the form

$$(*)\qquad R(\lambda)/\lambda^n = \int f\,dx - \sum f(z/\lambda)/\lambda^n, \quad z \in \mathbf{Z}^n,$$

which expresses the deficiency of the quadrature formula with step $1/\lambda$, applied to the characteristic function f of the domain G. Suppose, for simplicity, that G lies in the unit cube and continue f from it to $\mathbf{R}^n$ by periodicity. Suppose also that λ is integral (it is easy to get rid of both assumptions). For periodic functions in (*) we take the integral over the period cube and the sum at the λ^n points of this cube. (If we wish, we can take the integral over the torus $\mathbf{R}^n/\mathbf{Z}^n$ and the sum over the subgroup of the roots λ-th of unity of the torus.) The harmonics (the characters on the torus) are denoted by $e_k = e^{2\pi i(k,x)}$ (where k is a point of the dual lattice $\mathbf{Z}^{n*}$). For $f = e_k$ the deficiency of our quadrature formula is easy to compute:

$$\left(\sum/\lambda^n - \int\right) e_k = \begin{cases} 1 & \text{if } k \in \lambda \mathbf{Z}^{n*} \setminus 0, \\ 0 & \text{otherwise.} \end{cases}$$

If f is expanded in a Fourier series $\sum f_k e_k$, then $\left(\sum/\lambda^n - \int\right) f = \sum f_k$, $k \in \lambda\mathbf{Z}^{n*} \setminus 0$. The numbers of the harmonics in the sum are multiples of λ and, consequently, are large as $\lambda \to \infty$ ($|k| \geqslant \lambda$). But the

Fourier coefficient f_k is given by an integral over the period

$$f_k = \int f(x)\, e^{-2\pi i(x,k)} dx.$$

The exponential under the integral is rapidly oscillating for large $|k|$. If f is the characteristic function of G then the integral can be transformed into one over the boundary, whose phase is the function $-2\pi i\,(x, k)$. Degeneracies of the critical points of the restriction of this linear function to the boundary of the domain are responsible for the order of magnitude of the coefficients f_k, consequently, also of $R(\lambda)$.

It goes without saying that the Fourier series of the characteristic function converges badly, therefore, to justify the transition from estimates of the Fourier coefficients to estimates of R one has to smooth out f (the so-called van der Corput method). For particulars, see [14], [36], [37].

6°. Another inequality connected with integral points arose in [38] in the study of the statistics of the Newton diagrams of singularities.[1]

Theorem (Sevast'yanov, Konyagin). *The number of vertices of a convex integral polyhedron of volume V in $\mathbf{R}^n$ does not exceed $CV^{(n-1)/(n+1)}$. This quantity is an upper estimate for the number of faces of any fixed dimension.*

That the order of the estimate cannot be improved was proved in [38], where the estimate was also proved for $n = 2$.

Konyagin has also obtained a best possible estimate of the order of magnitude of the number of vertices of a Newton diagram bounding an area S by the quantity $CS^{1/3}(\log S)^{2/3}$, and two-sided estimates of the logarithm of the number of classes of convex polygons ($\asymp S^{1/3}$) and the number of Newton diagrams ($\asymp S^{1/3}(\log S)^{2/3}$).

7°. *Remark.* The interaction of the integral and the smooth structures, which arises in the problem of the integral points in a domain, is essential also in other questions of analysis. Let us consider, for example, the position of a smooth submanifold of Euclidean space in relation to integral hyperplanes.

The question how closely hyperplanes with not too large integral coefficients in their equations can approach a typical point of a typical submanifold is the subject of the so-called theory of Diophantine approximations to submanifolds ([39], [40]). This (unsolved) question is important in a number of applications, especially in the theory of resonances in non-linear oscillations (see [41], Ch. 4). The solution of this question depends on the distortion of the submanifold (flat places favour solidification of resonances). Nekhoroshev ([42], [43], [44]) has shown how the

[1] From the tables it is clear that for $\mu \leqslant 32$ the number ν of classes of singularities of functions with fixed multiplicity μ is a power of 2 when μ is a power of 2. The attempts to estimate the dependence of ν on μ mentioned above have led only to a very crude lower estimate for ν.

rate of evolution of the action variables in nearly integrable Hamiltonian systems can be estimated by the introduction of characteristics of the distortion of the level surfaces of the Hamiltonian function—the so-called *exponents of steepness.*

The values of these exponents for Hamiltonian functions in general position have been found so far only in the cases of 2 or 3 degrees of freedom (Landis [10]). The study of these exponents has led to a classification of tangential singularities (singularities of the position of the surface in projective space, in relation to tangent planes of various dimensions). We return to tangential singularities in §15.

§6. Metamorphoses of caustics

In a general one-parameter family of Lagrangian maps we can encounter (for special values of the parameters) singularities not in general position: when the parameter passes through such a value, the caustic experiences a metamorphosis.

Example. The motion of a medium of non-interacting particles is described by a one-parameter family of maps, $x \mapsto x + tv(x)$. If the velocity field V is potential, then these maps are Lagrangian.(1) The caustics of these maps are places of condensation of the particles. According to Zel'dovich [46], a model of this kind (taking gravitation into account) describes the way in which matter condenses from small perturbations of an originally homogeneous hot Universe.

The metamorphoses of caustics can be described as bifurcations of the sections of a large caustic in space-time by isochrones (see [47]). In the case of three-dimensional space the normal forms of the large caustics and the functions of time are as follows:

type	generating family	time		
A_3	$x^4+q_1x^2+q_2x$	q_3	or	$\pm q_1 \pm q_3^2 \pm q_4^2$
A_4	$x^5+q_1x^3+q_2x^2+q_3x$	q_4	or	$\pm q_1 \pm q_4^2$
A_5	$x^6+q_1x^4+q_2x^3+q_3x^2+q_4x$		$\pm$	q_1
D_4	$x_1^2x_2 \pm x_2^3+q_1x_2^2+q_2x_2+q_3x_1$	q_4	or	$\pm q_1+aq_2+q_3 \pm q_4^2$
D_5	$x_1^2x_2+x_2^4+q_1x_2^3+q_2x_2^2+q_3x_2+q_4x_1$		$\pm$	$q_1+q_4+aq_2$

Here $(q_1, \ldots, q_4)$ are the coordinates in space-time and a is a parameter.

(1)The Lagrangian property is preserved also when the particles are not free, but move in a potential field. This field may depend on time and may even be created by the particles themselves (for example, the Lagrangian property is preserved in the motion of a flow of particles in their own gravitational field). However, after the first metamorphosis (that is, just the formation of the caustic) the Lagrangian manifold ceases to be smooth (because of singularities of density, and consequently, also of the Hamiltonian, on the caustic). Apparently, the resulting singularities of the Lagrangian manifolds are rather weak and have no influence on the topology of the caustics and their metamorphoses (this has not been proved even for the one-dimensional problem).

The corresponding figures and a discussion of them can be found in [8] and [49] (first in [48]).

Here we consider only one metamorphosis: the birth of a caustic "from nothing" (of type A_3, with function of time $t = -q_1+q_3^2+q_4^2$). To get a picture of this metamorphosis of the caustic in three-dimensional space it is convenient to begin by analyzing the case of two-dimensional space (by putting $q_4 = 0$).

The large caustic in three-dimensional space-time (Fig. 10) has a semicubical edge of regression (the equation of this surface

$$\{q\colon \exists x\colon F_x = F_{xx} = 0\},\ F = x^4 + q_1x^2 + q_2x$$

in the space with the coordinates (q_1, q_2, q_3) has the form $8q_1^3+27q_2^2 = 0$). The isochrones t = const do not intersect the large caustic for $t < 0$. For $t = 0$ there appears a point caustic. It at once begins to grow (initially with infinite velocity). At the moment $t = \varepsilon$ the newly-born caustic has the form of a sickle of length of order $\varepsilon^{1/2}$ with two cusps and two points of inflection ("lips" in Thom's terminology).

In four-dimensional space-time the picture is analogous: the newly-born caustic has the form of a saucer (Fig. 11) of diameter of order $\varepsilon^{1/2}$,

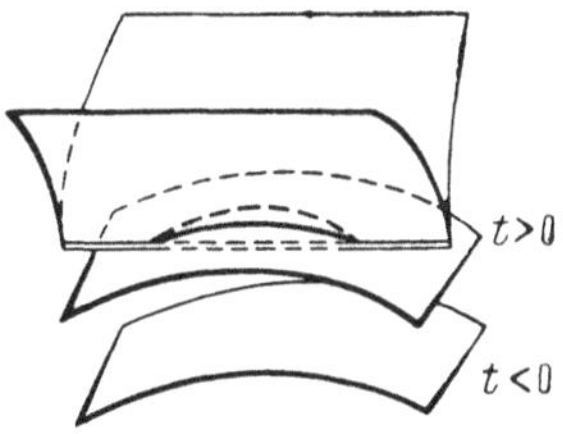

Fig. 10.

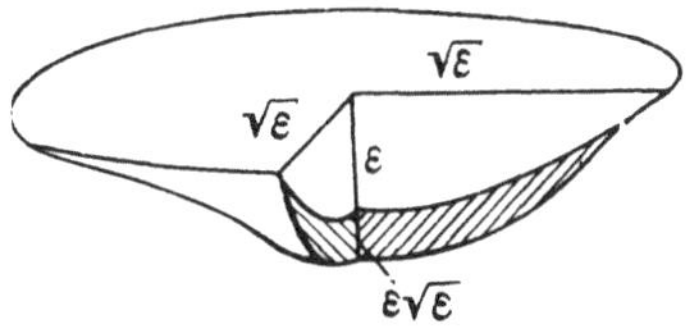

Fig. 11.

depth of order ε, and thickness of order $\varepsilon^{3/2}$. The edge of the saucer is an almost exactly elliptical edge of regression. Zel'dovich [50] calls such saucers pancakes; he interprets the metamorphosis of their birth (in conformity with the motion of dustlike material) as the birth of galaxies or their condensation.

The theory of singularities and of the metamorphoses of caustics predicts the pattern of future motion of the cellular structure: the material must condense chiefly on surfaces forming the walls of the cells, with greater density along certain distinguished curves (the singularities of the caustic) and an even greater concentration at special points. This inference is apparently confirmed by observing the distribution of densities in a galaxy.

Recently Shandarin [51] has proposed an analysis of the observational data from this point of view. As a measure of the cellularity of a structure he takes the parameter of flowing through ("percolation") defined as follows: we consider the union of the r-neighbourhoods of the given points and look for the smallest value of r for which the diameters of the connected components of the union become comparable with the size of the system.

For the observed distribution of the galaxies this radius turns out to be significantly smaller than for a random (Poisson) distribution of the same density or for a hierarchical distribution around random centres with the same correlation functions as for the observed distribution. For a complete corroboration of the cellularity of the observed structure one would have to compare also the maximal values of r for which the components of the complement of the union of the r-neighbourhoods are of a size comparable with that of the system.

CHAPTER III

CONTACT GEOMETRY

Just as symplectic geometry is the mathematical basis of classical mechanics, so its odd-dimensional analogue, contact geometry, lies at the base of geometrical optics. Although Hamilton's "optical-mechanical analogue" enables one to pass from symplectic geometry to contact geometry and vice versa, a direct description in terms of contact geometry is frequently preferable, if not for the ease of algebraic computations, then in any case for the geometrical intuition conveyed: it shows how geometrical facts are hidden behind the formulae of the Hamiltonian theory.

The relation between symplectic and contact geometry is similar to that between linear and projective spaces; to obtain from theorems of symplectic geometry their contact analogues we must everywhere replace functions by hypersurfaces, affine spaces by projective ones, and so on. In this chapter contact geometry is applied to the study of the singularities of fronts and their metamorphoses.

§7. Wave fronts

A caustic can be described as the trace of the edge of regression of a moving wave front. The theory of singularities of wave fronts is a particular case of the general theory of Legendrian singularities in contact geometry.

1°. Contact manifolds.

A *contact structure* on a manifold is defined as a maximally non-integrable field of hyperplanes: the form $d\alpha$ is *non-degenerate* in the plane $\alpha = 0$ if the field is given locally as the field of zeros of a form α (such a form α and the hyperplane $\alpha = 0$ are said to be be *contact*).

Example 1. $\mathbf{R}^{2n+1}$, $\alpha = dz - p_1\,dq_1 - \ldots - p_n\,dq_n$.

All the contact structures on a manifold of fixed dimension are locally contact-equivalent (and each contact form is given by the above formula in suitable local *Darboux coordinates*).

Example 2. *The manifold of* 1*-jets of functions* $J^1(M, \mathbf{R})$ *is contact.*

Example 3. A *contact element* on V is a hyperplane in the tangent space. All the contact elements on V form a bundle (the projectivization of the cotangent bundle). *The space of the projective cotangent bundle PT^*V is contact.*

The contact structures in Examples 2 and 3 are given by natural "integrability conditions" (the velocity of a contact element lies in a contact hyperplane if the velocity of the point of application belongs to the element).

For submanifolds of a contact manifold the theorem of Givental' (see §2) holds: *the outer local geometry is determined by the inner.*

2°. Legendrian submanifolds.
A submanifold of a contact manifold is said to be *Legendrian* if it is integral for the field of contact planes and has the greatest possible dimension (n in a $(2n+1)$-dimensional manifold).

Example 1. The fibres of the bundle $PT^*V \to V$ are Legendrian.

Example 2. The set of all 1-jets of a function on M is a Legendrian submanifold in the manifold of 1-jets $J^1(M, \mathbf{R})$.

Example 3. The set of all contact elements of V tangent to a fixed smooth submanifold (of any dimension) is Legendrian.

3°. Legendrian bundles.
A bundle is said to be *Legendrian* if its fibres are Legendrian.

Example 1. The projective cotangent bundle $PT^*V \to V$ (associating with a contact element on the manifold its point of application) is Legendrian.

Example 2. The bundle of the space of 1-jets of functions over the space of 0-jets, $J^1(M, \mathbf{R}) \to J^0(M, \mathbf{R})$ (the map of forgetting the derivatives, $(q, z, p) \mapsto (q, z)$), is Legendrian.

All Legendrian bundles of a fixed dimension are locally (in a neighbourhood of a point of the bundle space) Legendrian equivalent.

4°. Legendrian maps.
The projection of a Legendrian submanifold of a Legendrian bundle to the base is said to be a *Legendrian map.* Its image is called a *front.*

Example 1 (the Legendre transformation). We consider a smooth hypersurface Γ of the projective space P^n. The space of contact elements of a projective space is naturally isomorphic to the same space of elements for the dual space: $PT^*(P^n) \approx PT^*(P^{n*})$. For a point of the first space is a pair consisting of a point of P^n and a hyperplane containing it. The isomorphism associates with it the pair of the hyperplane and the point. *The isomorphism thus constructed respects the contact structures* (an instructive theorem of descriptive geometry, which lies at the base of the whole theory of the Legendre transformation).

We prove this theorem as follows. We consider an arbitrary smooth submanifold of the square of any manifold containing the diagonal. The following lemma is obvious.

Lemma. The tangent space of the submanifold at every point of the diagonal is invariant under the involution of the square that interchanges the factors.

Now we take as the original manifold the set of all flags consisting of a point of the projective space P^n and a hyperplane containing this point (that is, the space PT^*P^n). We say that one flag is *subordinate* to another if the point of the first lies in the hyperplane of the second. The graph of the relation "is subordinate to" is a submanifold of the direct product of the space of flags with itself that contains the diagonal. If the factors of the product are interchanged, this graph goes over to the graph of the relation "subordinates". According to the lemma, the tangent planes of both the graphs coincide at points of the diagonal. Hence, the intersections of the tangent planes of both graphs with the tangent plane to the product of the first factor with a point coincide, and these two intersections at the same time provide natural contact structures for the first factor, regarded on the one hand as $PT^*(P^n)$ and on the other as $PT^*(P^{n*})$. This proves the theorem.

A similar argument for complete flags (consisting of planes of all dimensions embedded in each other) lies at the base of the duality theory of curves in P^n (the dual curve is the edge of regression of the manifold of hyperplanes that touch the original curve).

Thus, the bundle $\pi\colon PT^*P^n \to P^{n*}$ that associates with a contact element the hyperplane containing it is a Legendrian map. The manifold of all contact elements tangent to a projective hypersurface Γ is a Legendrian submanifold of the space of this Legendrian bundle. Consequently, the restriction of π to this manifold is a Legendrian map. The front of this map is said to be the *hypersurface dual to* Γ. The affine version of the passage from Γ to the front is called the *Legendre transformation*. It is easy to see (with the help of the theorem proved above) that the passage to the dual hypersurface is an involution (the surface dual to the one constructed is the original one).

Instead of a hypersurface Γ we could have started with a smooth submanifold of any dimension: it also defines a Legendrian map. For example, the *front of a curve* is the set of hyperplanes tangent to it.

Example 2 (equidistants, or parallels). Along each normal to a smooth hypersurface Γ in Euclidean space $\mathbf{R}^n$ we lay an interval of length t. We obtain a Legendrian map $\Gamma \to \mathbf{R}^n$ whose front is said to be an *equidistant* or *parallel* of the given hypersurface (Fig. 12).

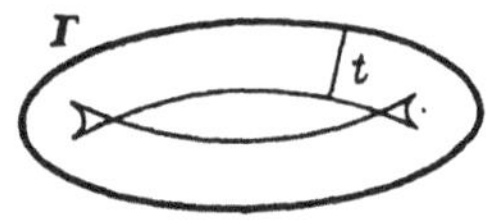

Fig. 12.

Every Legendrian map is locally equivalent to the map of Example 1 (and also to that of Example 2). Moreover, the front (in $J^1(M, \mathbf{R})$ or in PT^*V) determines its Legendrian map uniquely (with the possible exception of codimension infinity), because a surface defines its tangent plane. Therefore, the local theory of Legendrian maps is the theory of the singularities of Legendre transformations (and also the theory of the singularities of equidistants).

Remark. The word "front" comes from the term denoting the boundary of the propagation (of some disturbance or another). If the disturbance spreads with unit speed and initially is in the domain bounded by the initial front Γ (outside the ellipse in Fig. 12), then in time t the disturbance spreads into the domain bounded by the equidistant, at a distance t from the original front. Strictly speaking, the genuine front (the boundary of the domain affected by the disturbance) forms only part of the equidistant (the boundary of the lens-shaped figure in Fig. 12). The whole equidistant should be called a *wave front* (since in the propagation of a wave singularities could arise on the whole of the equidistant). As a rule, for brevity we omit the adjective "wave" in what follows.

§8. Singularities of fronts

The classification of Legendrian singularities reduces to that of families of hypersurfaces, just as the classification of Lagrangian ones reduces to that of families of functions.

We consider some Lagrangian submanifold of the space of a cotangent bundle. It can be "lifted" to a Legendrian submanifold of the space of 1-jets of functions on the base, at least locally, by setting $z = \int p\, dq$. The resulting Legendrian manifold projects to the original Lagrangian one under the natural map $J^1(M, \mathbf{R}) \to T^*M$ (the map of forgetting the values of the function).

Thus, a generating family $F(x, q)$ determines a Legendrian submanifold of the space of 1-jets of functions of q by the formula

$$\Lambda = \{(p, q;\ z)\colon \exists x\colon F_x = 0,\ p = F_q,\ z = F(x, q)\}.$$

The projection $(p, q;\ z) \mapsto (q;\ z)$ is a Legendrian bundle (the contact structure is $dz = p\, dq$). We obtain a Legendrian map of the manifold Λ onto its front in the space with the coordinates $(q;\ z)$.

The construction just described can give locally all Legendrian maps. The Legendrian equivalence of maps exactly corresponds in this bundle (over the $(q;\ z)$ space) to the stable equivalence of the families of hypersurfaces $z = F(x, q)$ in the x-space (with the usual understanding concerning stabilization, in which the hypersurface $\Phi(x) = 0$ is stably equivalent to the double of its complement), that is, branched along the double cover of the ambient space given by the equation $y^2 = \Phi(x)$).

Example 1. The generating family A_2

$$F(x, q) = x^3 + qx$$

determines the Legendrian curve in the three-dimensional space $(p, q;\ z)$:

$$q = -3x^2,\ p = x,\ z = -2x^3.$$

This curve is smooth, but its front (its image in the plane $(q;\ z)$) has a point of regression, or cusp.

Example 2. The generating family A_3 (see §3) gives a front diffeomorphic to the swallow-tail.

In a more general fashion, *the fronts given by the generating families A_μ, D_μ, or E_μ are locally diffeomorphic (in the complex domain) to the varieties of irregular orbits of the corresponding groups generated by reflections.* Since the connection between fronts and groups generated by reflections is essential for what follows, we describe it in more detail.

We consider the action of a finite group generated by reflections of the Euclidean space $\mathbf{R}^\mu$ on the complexification $\mathbf{C}^\mu$ of this space. The variety of orbits B is itself isomorphic to $\mathbf{C}^\mu$. The map $\mathbf{C}^\mu \to B$ that associates with a point its orbit is called the *Vieta map.*

Example 1. The group A_μ of permutations of the coordinates $(z_0, \ldots, z_\mu)$ acts irreducibly on the hyperplane $z_0 + \ldots + z_\mu = 0$ in $\mathbf{C}^{\mu+1}$. The variety of orbits is the space of polynomials

$$z^{\mu+1} + \lambda_1 z^{\mu-1} + \ldots + \lambda_\mu.$$

The Vieta map associates with the set of roots $(z_0, \ldots, z_\mu)$ the set of coefficients $(\lambda_1, \ldots, \lambda_\mu)$. It is given by the elementary symmetric polynomials $\sigma_2, \ldots, \sigma_{\mu+1}$ of z.

Example 2. The group D_μ acts on $\mathbf{C}^\mu$ by permuting the coordinate axes and changing the signs of an even number of coordinates. The Vieta map is given by the polynomials $\sigma_1(z_1^2, \ldots, z_\mu^2), \ldots, \sigma_{\mu-1}(z_1^2, \ldots, z_\mu^2);\ z_1 \ldots z_\mu$.

The orbit of a typical point under the action of a group generated by reflections consists of as many points as there are elements in the group. But some orbits consist of fewer points. Such orbits are said to be *irregular.* The variety of irregular orbits is the image of the mirrors under the Vieta map.

Example 1. The variety of irregular orbits of the group A_2 (generated by the reflections of the plane in three lines forming angles of 120°) is a semicubical parabola in the plane (Fig. 13).

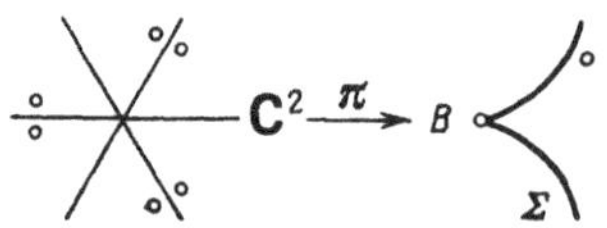

Fig. 13.

Example 2. The variety of irregular orbits of the group A_μ is that of polynomials $x^{\mu+1} + \lambda_1 x^{\mu-1} + \ldots + \lambda_\mu$ having at least one multiple root (the "generalized swallow-tail").

The Legendrian singularities A, D, and E are stable and simple (they do not have moduli). Generic Legendrian maps of manifolds of dimension n not exceeding 5 have only singularities that are Legendrian equivalent to A_μ, D_μ, or E_μ (with $\mu \leqslant n+1$). For large n the singularities A_μ, D_μ, E_μ are also stable and simple, but apart from them other singularities are encountered for generic Legendrian maps and these have moduli. For more details on the classification of Legendrian maps, see [52], [8].

From what has been said above it follows that a generic front in the plane has only ordinary (semicubical) cusp points (A_2) and points of transversal intersection (A_1A_1); these are all stable (they are preserved under a small perturbation of the front). Notice that our front is the projection of a smooth curve in three-dimensional space to the plane. The projection of a generic curve does not have cusps. The *Legendrian property* of the projected curve makes the projection more singular (the cusp point becomes unremovable).

In three-dimensional space a generic front has the following singularities: ribs or edges of regression of a semicubical cusp (A_2), swallow-tails (A_3), and transversal self-intersections (A_1A_1, A_1A_2, $A_1A_1A_1$); Fig. 14.

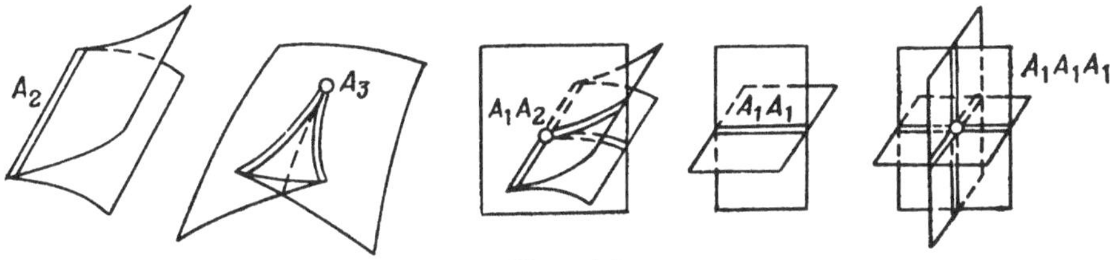

Fig. 14.

§9. Metamorphoses of fronts

In the propagation of a front its singularities may re-form. For example, in the propagation of a disturbance inside an ellipse the front (the equidistant, Fig. 12) is at first a smooth curve, but for large t four cusps appear on it.

It is convenient to study the metamorphoses of fronts by means of space-time, as for metamorphoses of caustics. The union of the momentary fronts in the isochrones (the planes t = const) of space-time is a hypersurface in space-time. It is not difficult to see that the hypersurface in space-time formed by the motion of a generic front is itself the front of a generic Legendrian map of a space of dimension greater by 1. For let $F_t(x, q)$ be the generating family of a Legendrian map depending on time. Then by adding the parameter t to the parameters q we can regard F as the generating family of a Legendrian map over space-time. We call the hypersurface in space-time formed by the moving front, the *large front*.

Example. The union of the equidistants of an ellipse placed in different planes t = const (where t is the distance to the ellipse) is the surface of the large front. It is situated in three-dimensional space-time and has four singularities diffeomorphic to the swallow-tail.

Thus, we know to what singularities on the large front the motion of a generic momentary front reduces: they are the singularities of generic fronts in a space whose dimension exceeds that of the original space by 1, in which the momentary front moves.

Example. If the original space is three-dimensional, then the singularities of a generic large front are diffeomorphic to the varieties of irregular orbits of the groups A_4 or D_4, or to the variety of irregular orbits of the group A_3 (the swallow-tail) multiplied by a line, or to the variety of irregular orbits of the group A_2 (the semicubical parabola) multiplied by a plane (of course, transversal self-intersections of branches of the kinds described are also possible).

The time function t has no critical points on space-time. Thus, the study of the metamorphoses of moving fronts reduces to that of functions without critical points on the space of the large front. We intend to reduce such functions to normal forms by diffeomorphisms preserving the large front. When the front is diffeomorphic to the variety of irregular orbits of a group generated by reflections, it is not difficult to do this by using the following two theorems.

Theorem 1 (see [47]). *The germ of a generic function that vanishes at the origin of the space of orbits of an irreducible group generated by reflections reduces, by a local diffeomorphism that preserves the variety of irregular orbits, to the germ of a linear function on the space of orbits, namely, that of the invariant of least (second) degree.*

Corollary. *The germ of a generic function of* $(\lambda_1, \ldots, \lambda_\mu)$ *at the origin reduces by a diffeomorphism that preserves the generalized swallow-tail*

$$\{\lambda \in \mathbf{C}^\mu \colon z^{\mu+1} + \lambda_1 z^{\mu-1} + \ldots + \lambda_\mu \text{ has a multiple root}\}$$

to the form λ_1 + const (*to be generic it is enough that* $\partial f/\partial\lambda_1|_0 \neq 0$).

In particular, a curve passing through the vertex of a semicubical parabola, but not tangent to it, can be straightened out by a diffeomorphism that preserves the parabola (can be mapped to any line passing through the origin that does not touch the parabola). The following theorem is proved with the help of the equivariant Morse lemma.

Theorem 2. *If the second differential at the origin of a function that is invariant under an irreducible compact group of linear transformations is non-degenerate, then the germ of the function at the origin can be transformed to its quadratic part by an equivariant diffeomorphism (commuting with the action of the group).*

This theorem is proved, like the ordinary Morse lemma, by the homotopy method (with integration over the group to ensure equivariance).

Though the corollary stated above easily follows from the easy Theorems 2 and 1, to have discovered it in this way would hardly have been possible. On the contrary, Theorems 1 and 2 were found after the corollary had been obtained, by more direct calculations, by using the explicit form of the vector fields tangent to the swallow-tail. We return to these vector fields in §§12 and 13.

For details of the proofs of both theorems, see [47]; the theorems are true not only in the complex and real analytic cases, but also in the smooth situation.

Theorem 1 allows us straightaway to normalize the function of time at the most singular points of the large front. However, the large front has fewer singular points. For example, the large front in three-dimensional space-time not only has point singularities of the swallow-tail type, but also has edges of regression. In a neighbourhood of a generic point of the edge of regression the isochrone is transversal to it. However, it is impossible in the large to get rid, by a small perturbation of the front, of individual points of the edge of regression, where the edge is tangent to the isochrone. In the neighbourhood of such points the function of time has still to be studied. We take for simplicity the holomorphic case. We consider the k-dimensional "rib" of the large front corresponding to a singularity X_μ (where X may be A, D, or E). We restrict the function of time to this "rib". If the point of the rib in question is not a critical point of the restriction, then in a neighbourhood of it the isochrones are transversal to the rib. In that case metamorphoses do not occur at the relevant point of space-time. In its neighbourhood the large front with its sections with the isochrones is diffeomorphic to the direct product of the momentary front with the time axis (this diffeomorphism preserves time, that is, it carries each isochrone into itself, but, in general, it does not preserve the points of space).

Generic functions on space-time, on restriction to the "rib" X_μ, have only distinguished isolated Morse critical points. Let us consider one of these. In a neighbourhood of such a point in space-time we take coordinates $(\tau_1, \ldots, \tau_k; \lambda_1, \ldots, \lambda_\mu)$ in such a way that the "rib" has the equation $\lambda = 0$, while the large front is defined by the condition "λ belongs to the variety of irregular orbits for X_μ". By the Morse lemma with parameters, the function of time locally reduces (in a suitable coordinate system τ) to the form $\tau_1^2 + \ldots + \tau_k^2 + \varphi(\lambda_1, \ldots, \lambda_\mu)$. For a generic function of time the germ of φ at the origin is generic. By Theorem 1, it reduces by a change of the coordinates λ that preserves the variety of irregular orbits to the linear form $\varphi = \lambda_1 + \text{const}$ (where λ_1 is the invariant of smallest degree).

We have thus obtained normal forms of the metamorphoses of fronts in spaces of dimension $\leqslant 5$ in general one-parameter families.

Example. The generic metamorphoses of each branch of the front on the plane reduce by diffeomorphisms of three-dimensional space-time that preserve time to the following normal forms (Fig. 15):

type	large front	function of time		
A_3	$\{\lambda:\ z^4+\lambda_1 z^2+\lambda_2 z+\lambda_3$ has a multiple root$\}$	$\pm\ \lambda_1$		
A_2	$\{\lambda:\ z^3+\lambda_1 z+\lambda_2$ has a multiple root $\}$	τ_1	or	$\pm\lambda_1\pm\tau_1^2$
A_1	$\lambda_1=0$	τ_1	or	$\pm\lambda_1\pm\tau_1^2\pm\tau_2^2$

In the table the moment of metamorphosis is transferred to $t = 0$. In Fig. 15 fragments of the front before, at, and after the moment of metamorphosis are shown (past and future may change places).

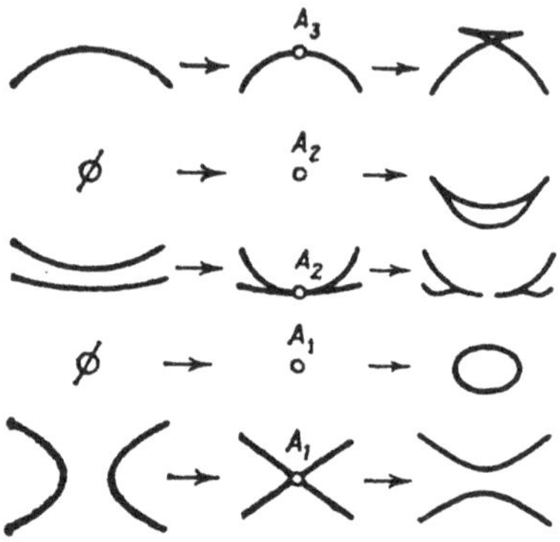

Fig. 15.

Of course, different branches of the front may freely move across each other, so that at particular moments of time one can have metamorphoses of the passage of one of the branches over a point of intersection of two of the others or over a cusp point, and also the tangency of two branches with the formation (or disappearance) of a pair of points of the intersection.

The metamorphoses of fronts in spaces of higher dimensions have been described in [47], [53], [8] (a description of the metamorphoses in three-dimensional space first appeared, it seems, in [54]).

Remark. Not all the metamorphoses of fronts described above are realized in the propagation of a front. In fact, the non-trivial metamorphoses A_1 and A_2 change the number of components of the corresponding Legendrian manifold and, consequently, are not realized as metamorphoses of equidistants. However, they occur as metamorphoses in one-parameter families of fronts in the following situation.

We consider the trace on the surface of the Earth of a front propagating in the air (for example, a shock wave). Although the topology of the Legendrian manifold corresponding to the front in three-dimensional space does not change as the front propagates, the Legendrian manifold determined by its trace on the surface of the Earth may itself experience metamorphoses in the course of time.

CHAPTER IV

THE CONVOLUTION OF INVARIANTS AND ITS GENERALIZATIONS

In this chapter we describe the algebraic technique that is needed for the reduction of various geometrical objects to normal forms. For example, the theory of the metamorphoses of wave fronts presented in §9 was based on the reduction to normal form of the time function by means of diffeomorphisms of space-time that preserve the swallow-tail. The Lie algebra of the group of such diffeomorphisms consists of vector fields tangent to the swallow-tail. The main results of this chapter express in a concentrated form the information about these fields that is provided by essentially elementary, but lengthy, computations with symmetric functions. Success is attained by the use of the theory of the invariants of groups generated by Euclidean reflections and connected with simple singularities. At the end of the chapter we state results of Varchenko and Givental' on period maps, which have allowed them to extend the theory of fields tangent to fronts to the case of non-simple quasihomogeneous singularities.

It turns out that the period map transfers the homology intersection form of a level set of a function to the base of a versal deformation, thus defining a field of 2-forms on the cotangent spaces of the base. In the general case this field replaces the Riemannian metric on the space of regular orbits of the group generated by Euclidean reflections.

§10. Vector fields tangent to a front

The front of a simple singularity maps biholomorphically onto the variety of irregular orbits of the corresponding group generated by reflections (§8). Therefore, we consider holomorphic vector fields tangent to the variety of irregular orbits.

Thus, we consider an arbitrary finite group generated by reflections in the Euclidean space $\mathbf{R}^\mu$. The group acts on the complexification $\mathbf{C}^\mu$ of this space. We denote by B the variety of orbits (it is also isomorphic to $\mathbf{C}^\mu$). We denote by $\pi:\mathbf{C}^\mu \to B$ the Vieta map that associates with a point its orbit. The image of the mirrors in $\mathbf{C}^\mu$ under the map π is precisely the variety of irregular orbits Σ in the space of orbits B (Fig. 13). We call the variety of irregular orbits the "generalized swallow-tail", having in mind the following example.

Example. In the case of A_μ the variety of orbits B consists of all complex polynomials of the form $z^{\mu+1} + \lambda_1 z^{\mu-1} + \dots + \lambda_\mu$. The map π associates with the set of roots of the polynomial its set of coefficients. The mirrors have the equations $z_i = z_j$ ($\mu = 2$ in Fig. 13), where z_i are the Cartesian coordinates in $\mathbf{C}^{\mu+1}$ ($\mathbf{C}^\mu$ is given by the equation $\sum z_i = 0$).

The variety of irregular orbits is

$$\Sigma=\{\lambda\in B:\ z^{\mu+1}+\lambda_1 z^{\mu-1}+\ldots+\lambda_\mu \text{ has a multiple root }\}.$$

For $\mu = 3$ this is the usual swallow-tail in $\mathbf{C}^3$.

We say that functions on the space of orbits are *invariant* (since for $\alpha: B\to\mathbf{C}$ the function $\pi^*\alpha: \mathbf{C}^\mu\to\mathbf{C}$ is invariant under our group and, conversely, every invariant function can be represented in the form $\pi^*\alpha$).

Definition. The *convolution of invariants* is the operation Φ on invariants that associates with a pair of invariant functions the scalar product of their Euclidean gradients in $\mathbf{C}^\mu$, so that

$$\pi^*\Phi(\alpha,\ \beta)=\langle\nabla\pi^*\alpha,\ \nabla\pi^*\beta\rangle,$$

where $\langle,\rangle$ and ∇ are the Euclidean scalar product and the gradient on $\mathbf{C}^\mu$

Example. We consider the action of the group A_2 on the plane. The convolution of the basic invariants λ_1 and λ_2 (in the notation of the preceding example) gives (after some manipulations with symmetric functions)

$$\Phi(\lambda_1,\ \lambda_1)=-2\lambda_1,\quad \Phi(\lambda_1,\ \lambda_2)=-3\lambda_2,\quad \Phi(\lambda_2,\ \lambda_2)=2\lambda_1^2/3.$$

Note that the operation of convolution of invariants is symmetric and bilinear, and satisfies the Leibniz identity, that is, for a fixed first argument it is differentiation of the second:

$$\Phi(\alpha,\ \beta\gamma)=\beta\Phi(\alpha,\ \gamma)+\gamma\Phi(\alpha,\ \beta).$$

Therefore, to each invariant $\alpha: B\to\mathbf{C}$ there corresponds a vector field v_α on the variety of orbits: the derivative of any invariant β along the field v_α is $\Phi(\alpha,\ \beta)$. We call v_α a *potential field with potential* α.

Theorem. *Every potential field v_α is tangent to the variety of irregular orbits.*

Proof. The Euclidean gradient of the invariant function $\pi^*\alpha$ is invariant under reflections and, consequently, it is tangent to the mirrors. The projection of this gradient to the variety of orbits is tangent to the projection of the mirrors, as required.

We choose any basis for the invariants (that is, a coordinate system $(\lambda_1, \ldots, \lambda_\mu)$ on B, not necessarily linear). From the function λ_i we construct μ vector fields on the variety of orbits touching the variety of irregular orbits $\{v_{\lambda_i}\}$. We shall call these μ fields the *basic fields.*

Example. For the basis (λ_1, λ_2) of the invariants of A_2 the basic fields have the form

$$v_{\lambda_1}=-2\lambda_1\partial/\partial\lambda_1-3\lambda_2\partial/\partial\lambda_2,\quad v_{\lambda_2}=-3\lambda_2\partial/\partial\lambda_1+2\lambda_1^2/3\partial/\partial\lambda_2,$$

as follows from the preceding example. The variety of irregular orbits in this case is the semicubical parabola $\Delta = 0$, where $\Delta = 4\lambda_1^3 + 27\lambda_2^2$. The first of the basic fields is the Eulerian and the second the Hamiltonian field with the Hamiltonian function Δ.

Theorem (see [55]). *At each point outside the variety of irregular orbits the vectors of the μ basic fields are linearly independent, while on the variety of irregular orbits the determinant of the components of the basic fields has a zero of the first order.*

Remark. At each point of the variety of irregular orbits the basic fields generate the whole tangent space to that stratum of the variety to which the given point belongs.

Knowing the basic fields, we can construct many other fields tangent to the variety of irregular orbits: it is sufficient to take any linear combination of the basic fields with functional coefficients.

One can show (see [56]) that we obtain in this way all vector fields, holomorphic at zero, that touch the variety of irregular orbits: every such field can be represented uniquely as a linear combination of the basic fields with coefficients that are holomorphic at zero.

Example. Every field that is holomorphic at zero and tangent to the semicubical parabola $\Delta = 0$ is (uniquely) representable in the form $\alpha_1 v_{\lambda_1} + \alpha_2 v_{\lambda_2}$ (in the notation of the preceding example) with coefficients α_1 and α_2 holomorphic at zero.

Remark. An infinitely differentiable vector field tangent to the real swallow-tail (the variety of real polynomials $z^4 + \lambda_1 z^2 + \lambda_2 z + \lambda_3$ with a multiple real root), is not always representable as a combination of the basic fields with smooth coefficients. In fact, the surface of the swallow-tail in $\mathbf{R}^3$ has a line of self-intersections, analytically continuable beyond the vertex of the tail as a curve not belonging to it. A smooth field touching the tail may not touch this continuation, but an analytic basic field is tangent to it.

In any case, a knowledge of the basic fields is very useful for the reduction of geometrical objects in the neighbourhood of a front to normal form, since the combinations of these fields with functional coefficients are known to be tangent to the front.

Explicit formulae for the components of the basic fields touching the variety of irregular orbits are given in [47], [57], and [58]. Since the matrix of components consists of the $\Phi(\lambda_i, \lambda_j)$, it is sufficient to compute the convolution of the basic invariants.

Example. The matrix $\Phi(\lambda_i, \lambda_j)$ for A_3 has (for the basis given above) the form

$$-\begin{bmatrix} 2\lambda_1 & 3\lambda_2 & 4\lambda_3 \\ 3\lambda_2 & 4\lambda_3 - \lambda_1^2 & -\lambda_1\lambda_2/2 \\ 4\lambda_3 & -\lambda_1\lambda_2/2 & 2\lambda_1\lambda_3 - 3\lambda_2^2/4 \end{bmatrix}$$

Consequently, a basis of the module of vector fields tangent to the swallow-tail in $\mathbf{C}^3$ is formed by the three fields

$$\begin{aligned} v_1 &= 2\lambda_1\partial_1 + 3\lambda_2\partial_2 + 4\lambda_3\partial_3, \\ v_2 &= 3\lambda_2\partial_1 + (4\lambda_3 - \lambda_1^2)\,\partial_2 - (\lambda_1\lambda_2/2)\,\partial_3, \\ v_3 &= 4\lambda_3\partial_1 - (\lambda_1\lambda_2/2)\,\partial_2 + (2\lambda_1\lambda_3 - 3\lambda_2^2/4)\,\partial_3. \end{aligned}$$

A general formula for the convolution of invariants of the singularities of the series A_μ was found by Fuks [47]. Let us denote by σ_i the i-th elementary-symmetric function of the roots of the polynomial $z^{\mu+1} + \lambda_1 z^{\mu-1} + \ldots + \lambda_\mu$:

$$\sigma_0 = 1,\ \sigma_1 = 0,\ \sigma_2 = \lambda_1,\ \sigma_3 = -\lambda_2,\ \ldots, \sigma_{\mu+1} = \pm\lambda_\mu,\ \sigma_{\mu+2} = \ldots = 0.$$

Then for $i \leqslant j$

$$\Phi(\sigma_{i+1}, \sigma_{j+1}) = \frac{i(\mu+1-j)}{\mu+1}\sigma_i\sigma_j - \sum(\beta-\alpha)\,\sigma_\alpha\sigma_\beta,$$

where the summation is over all α and β such that $\alpha+\beta = i+j$, $0 \leqslant \alpha < i$. Formulae for the convolution for all simple singularities are given in [57] and [58].

An effective device for computing fields tangent to a front was proposed by Zakalyukin in [53]. In the quasihomogeneous case his construction is as follows.

Let F be a versal deformation of a quasihomogeneous polynomial f,

$$F(x, \lambda) = f(x) + \lambda_1 g_1(x) + \ldots + \lambda_\mu g_\mu(x),$$

where the g_k are monomials generating over $\mathbf{C}$ the local algebra

$$Q_f = \mathbf{C}[[x]]/(\partial f/\partial x_i).$$

By the Weierstrass preparation theorem there are expansions

$$g_i(x)F(x,\lambda) = \sum h_{i,j}(x,\lambda)\partial F/\partial x_j + \sum v_{i,j}(\lambda)g_j(x).$$

The vector fields

$$v_i = \sum v_{i,j}(\lambda)\partial/\partial\lambda_j$$

are tangent to the front

$$\Sigma = \{\lambda \in \mathbf{C}^\mu\text{: the variety } V_\lambda = \{x : F(x, \lambda) = 0\} \text{ is singular}\}.$$

Every field that is holomorphic at zero and touches the front can be represented (in a unique way) as a combination of the fields v_i with coefficients holomorphic at zero.

The construction described is based on the following argument. We consider a family $(1 + \varepsilon g)F$ (where ε is small) near to F. This family is R-equivalent to a family induced from F. The corresponding change of the parameter λ is close to the identity. Its derivative with respect to ε at $\varepsilon = 0$ gives the required vector field. It is tangent to the front Σ, since the varieties V_λ, hence also the fronts of the constructed families, do not depend on ε, and a change of parameter takes fronts to fronts.

§11. The linearized convolution of invariants

The Lie algebra of vector fields touching a front (or variety of irregular orbits) is infinite-dimensional. However, we can manufacture from it a finite-dimensional Lie algebra if we replace each field by its linear part at zero. In most of the calculations this finite-dimensional algebra replaces the full algebra of fields tangent to the front. At the same time, it can simply be described in terms of multiplication in the local graded algebra of the corresponding singularity.

We denote by T the tangent space to the variety of irregular orbits at the origin and by T^* the dual space. The linearization of the convolution of invariants defines the operation of *linearized convolution* $\varphi: T^* \times T^* \to T^*$ by the formula $\varphi(da, db) = d(\Phi(a, b))$, where d is the differential of the invariant at zero of the space of orbits. It is easy to see that the value of φ does not depend on the invariants a and b, but only on their differentials at zero.

Example. We denote the differentials of the coordinates λ_i on B at the origin simply by λ_i. Then the operation of linearized convolution for the group A_3 is given (in the notation of the example of §10) by the triangular matrix

$$-\begin{bmatrix} 2\lambda_1 & 3\lambda_2 & 4\lambda_3 \\ 3\lambda_2 & 4\lambda_3 & 0 \\ 4\lambda_3 & 0 & 0 \end{bmatrix}$$

A similar triangular matrix gives the linearized convolution for A_μ (in the first row we have $2\lambda_1, \ldots, (\mu+1)\lambda_\mu$, while on each line parallel to the second diagonal all the elements are equal). This example shows that, in contrast to the full convolution, the linearized convolution is given by a simple formula. For groups generated by reflections connected with singularities the operation of linearized convolution admits a simple description in terms of the graded local algebra of the corresponding boundary singularity. These local algebras have the form (see [59], [55])

$$Q = \mathbf{C}[[x_1, x_2, x_3]]/(x_1\, \partial f/\partial x_1,\ \partial f/\partial x_2,\ \partial f/\partial x_3),$$

where f is given by the following table of simple boundary singularities of functions on $\mathbf{C}^3$ with boundary $x_1 = 0$:

type	normal form of f	type	normal form of f
A_μ	$x_1 + x_2^{\mu+1} + x_3^2$	E_6	$x_1 + x_2^3 + x_3^4$
B_μ	$x_1^\mu + x_2^2 + x_3^2$	E_7	$x_1 + x_2^3 + x_2 x_3^3$
C_μ	$x_1 x_2 + x_2^\mu + x_3^2$	E_8	$x_1 + x_2^3 + x_3^5$
D_μ	$x_1 + x_2^2 x_3 + x_3^{\mu-1}$	F_4	$x_1^2 + x_2^3 + x_3^2$

The connection between these functions and the corresponding groups of reflections consists in the following. We consider the versal deformation of the boundary singularity f on the space $\mathbf{C}^3$ with boundary $x_1 = 0$:

$$F(x, \lambda) = f(x) + \sum \lambda_i g_i(x), \quad 1 \leqslant i \leqslant \mu,$$

where the g_i are monomials whose images in Q generate Q as a linear space over $\mathbf{C}$.

Example. For the case F_4

$$F(x, \lambda) = f + \lambda_1 x_1 x_2 + \lambda_2 x_1 + \lambda_3 x_2 + \lambda_4.$$

Definition. The *bifurcation diagram* of a boundary singularity is the hypersurface in the base $\mathbf{C}^{\mu} = \{\lambda\}$ of the versal deformation F, formed by those values of the parameter λ for which the zero level surface

$$V_{\lambda} = \{x \in \mathbf{C}^{3}:\ F(x, \lambda) = 0\}$$

is either singular or is non-transversal to the plane $x_1 = 0$.

Theorem 1.(see [59]). *The bifurcation diagram of each of the boundary singularities A_{μ}, ..., F_4 goes over to the variety of irregular orbits of the corresponding group generated by reflections, under a suitable biholomorphic identification of the base of the versal deformation with the variety of orbits.*

There are many diffeomorphisms carrying the bifurcation diagram to the variety of irregular orbits. We call such diffeomorphisms (their germs at zero) *admissible identifications* of the base of the versal deformation with the variety of orbits.

Remark. For the only Weyl group G_2 not in this list, the local ring is constructed somewhat differently (see [59], [60]), while for the remaining groups of reflections that are not Weyl groups, namely, $I_2(p)$, H_3, and H_4, a direct construction of the algebra from the singularities is not known, although it is clear that one must obtain the algebras A_2, A_3, and D_4 with unusual, but known gradings.

Recently Lyashko has pointed out a problem of singularity theory that reduces to a bifurcation diagram biholomorphically equivalent to the variety of irregular orbits of H_3.

Lyashko considered the three-parameter family of smooth functions without singularities in the neighbourhood of the variety $x^2 = y^5$.

For H_4 a similar description has not yet been found; it appears that the variety of irregular orbits is diffeomorphic to the projection of the stratum in the base of the versal deformation of E_8 into the four-dimensional space of the coefficients of 1, x_2, x_3^3, $x_2x_3^3$.

The normal forms of the functions f listed above are quasihomogeneous with certain positive weights α_i. That means that $f = Df$, where $D = \sum \alpha_i x_i \partial/\partial x_i$ is the quasihomogeneous Euler operator. For example, for F_4 we have $\alpha_1 = 1/2$, $\alpha_2 = 1/3$, $\alpha_3 = 1/2$.

The operator D sends the ideal in the denominator of the definition of Q to itself and therefore acts on Q as a derivation of this algebra. We denote this derivation by D, as before, and call it the *Euler derivation.*

Theorem 2. *The linearized convolution of invariants $T^* \times T^* \to T^*$ is equivalent (as a bilinear operation) to the map $Q \times Q \to Q$ given by the formula $(p, q) \mapsto S(pq)$, where $S = D + (2/h)E$, and h is the Coxeter number of the group of reflections.*

Unfortunately, a simple uniform proof of this simple formula has not been found (the formula is proved in [55], [57] case by case) (see §12).

The identification of T^* with Q that realizes the equivalence of Theorem 2 is not natural. It is more natural to identify the *tangent* space T to the variety of orbits with the local algebra Q. For let us consider any admissible identification of the base of the versal deformation with the variety of orbits (the germ of a map, biholomorphic at zero, that sends the germ of the bifurcation diagram to the germ of the variety of irregular orbits). Its derivative at zero identifies the tangent space to the base of the versal deformation with T, which is already naturally identified with the local algebra (to the velocity of the deformation there corresponds its class in Q).

Thus, we obtain an admissible identification of T^* with Q^*. The next theorem describes the result of transferring the convolution to Q^*. To state it we need one further construction.

An element α of the space Q^* dual to the local algebra Q is said to be *admissible* if the linear function $\alpha: Q \to \mathbf{C}$ does not vanish over the whole annihilator of the maximal ideal in Q.

This annihilator is one-dimensional and is generated by the class of the Jacobian of the functions in the denominator of the definition of the algebra Q (see [8]). Therefore, the inadmissible elements form a hyperplane in Q^*, so that a generic element of Q is admissible.

To each element α of Q^* there corresponds a bilinear form $B_\alpha: Q \times Q \to \mathbf{C}$, given by the formula $B_\alpha(\beta, \gamma) = \alpha|(\beta\gamma)$. If α is admissible, then the form B_α is non-degenerate (see [8]). We denote by $N_\alpha: Q \to Q^*$ the operator of this form.

***Theorem* 3.** *Under any admissible identification the linearized convolution operation* φ *on* T^* *goes over to the operation* $\varphi_\alpha: Q^* \times Q^* \to Q^*$ *given by the relation* $\varphi_\alpha(a, b) = R^* N_\alpha (N_\alpha^{-1} a \cdot N_\alpha^{-1} b)$, *where* $R = E - D$ *and* α *is admissible.*

Every operation φ_α *with an admissible* α *can be obtained from the linearized convolution by a suitable choice of admissible identification. In particular, all the operations* φ_α *with an admissible* α *are equivalent.*

Theorems 2 and 3, which were stated in [55] as conjectures, were proved there only for the basic series of singularities. A general proof, applicable to all cases, was given by Givental' ([57], [58]). These theorems may be restated as propositions on the duality of the families of quadratic forms given by convolution and multiplication in the local algebra.

We fix an admissible identification of the tangent space T to the variety of orbits and the space of the local algebra Q. The linearized convolution of invariants φ then defines for each element q of Q the bilinear form

$$\psi_q: Q^* \times Q^* \to \mathbf{C}, \quad \psi_q(a, b) = \varphi(a, b) \mid q.$$

From the preceding formulae we obtain the following corollary.

Corollary. *The set of forms* $\{B_\alpha\}$ *on* Q *and* $\{\psi_\alpha\}$ *on* Q^* *are dual to each other in the sense that for each non-degenerate form* B_α *the dual form is one of the forms* ψ_q *and vice versa.*

The connection between the parameters α and q of mutually inverse forms is non-linear, but it can be expressed linearly in terms of the operation of inversion in the local algebra Q.

Formulae analogous to those described above are true also in the real case (see [55]). The real-dual forms B_α and ψ_q have the same indices.of inertia. But the signature of B_α is equal to the index of the corresponding singularity (see [55], [61]). Consequently, the indices of the boundary singularities are equal to the signatures of the corresponding forms ψ_q generated by the convolution of invariants. It is just this observation that led to the discovery of the above duality between the linearized convolution of invariants and multiplication in the local algebra of a singularity.

The theorems on the linearized convolution can be restated as theorems on the Lie algebras of linear vector fields. Just as the full convolution determines fields tangent to fronts, so the linearized convolution determines a linear family of linear vector fields on the tangent space T to the variety of orbits at zero; the fields are parametrized by the points of T^*. To an element α of T^* there corresponds a linear field w_α such that the derivative of the linear function β on T in the direction of the field w_α is $\varphi(\alpha, \beta)$. This is the linear part of the field v_α determined by the full convolution Φ (§10).

Theorem ([55]). *The fields w_α form a μ-dimensional Lie algebra. If $(\alpha_1, ..., \alpha_\mu)$ is a basis in T^* and the relevant group generated by reflections is irreducible, then the vectors of the basic fields w_{α_i} are linearly independent at every point of T, with the exception of the points of the hyperplane tangent to the variety of irregular orbits at zero.*

(This hyperplane is given by the equation $\alpha_\mu = 0$, where α_μ is the differential of the invariant of highest degree at zero.)

Notice that we have in this way connected with a group of reflections in μ-dimensional space (for example, with each Weyl group of rank μ) a somewhat unusual μ-dimensional Lie algebra. The corresponding group of linear transformations is that of the linear parts of the diffeomorphisms preserving the "generalized swallow-tail". It is clear that these linear transformations preserve the tangent plane to the swallow-tail at zero and also the whole set of tangent spaces to the various strata of the natural stratification of the swallow-tail (down to the direction corresponding to the annihilator of the maximal ideal). However, the Lie group in question is smaller than the group of all linear transformations preserving that set of tangent spaces: the transformations of the group preserve also some additional structure, some apparition, left over from the swallow-tail and its tangent spaces.

Theorems 2 and 3 allow us to describe the Lie algebra of the fields w_α in terms of the graded local algebra of the singularity.

For an element q of Q we denote by $M_q : Q \to Q$ the operator of multiplication by q. We define the linear operator W_q by the formula

$$W_q = M_q R, \quad R = E - D,$$

where $D : Q \to Q$ is a derivation.

Theorem ([55]). *For any derivation D of any commutative algebra Q the operators W_q form a Lie algebra.*

If D is the Euler derivation of the local algebra of the singularity, then the Lie algebra $\{W_q\}$ is isomorphic to the Lie algebra $\{w_\beta\}$ of the linearized vector fields tangent to the variety of irregular orbits. Moreover, the connection between the parameters β and q of the corresponding fields is linear.

After an admissible identification of Q^* with T^* this connection is given by the formula $\beta = N_\alpha q$ (where α depends on the choice of the admissible identification).

§12. Period maps and intersection forms

Recently Varchenko and Givental' [26] have indicated a new point of view on the convolution of invariants, connecting it with the period maps and the intersection forms of singularities. This has allowed them to extend the results in §11 on the linearized convolution to quasihomogeneous non-simple singularities, by menas of the asymptotic expansions of integrals that were obtained in the papers of Varchenko [21], [22] (see §4).

1°. We need some (simple) information about period maps for an arbitrary locally trivial fibre bundle. With such a bundle there are associated bundles of the homology and cohomology groups of the fibres with complex coefficients (the base is the same). These new bundles are not only locally trivial, but are canonically locally trivial (for, an integral cycle in one fibre can be shifted to a neighbouring fibre in a homologically unique way). A *period map* is a section of the cohomology bundle.

Example. Suppose that on the space of a (smooth) bundle there is given a differential form that is closed on the fibres. Then the map that associates with each point of the base the cohomology class of the form on the fibre over this point is a period map. It is called the *period map of the given form.*

We return to an arbitrary period map with a smooth base.

2°. Suppose that on the base there is also given a vector field. Then for any smooth period map there is defined its *derivative in the direction of the field*, and this derivative also is a period map. For, nearby fibres of the cohomology bundle can be identified canonically with each other by an integral local trivialization. After the identification a section becomes (locally) a map into one fibre and can be differentiated like an ordinary function.

3°. Let us consider, in particular, the case when the base is a complex manifold and the *complex dimensions of the base and the fibre of the cohomology bundle are the same.*

Definition. A period map is said to be *non-degenerate* if its derivatives in the directions of arbitrary **C**-linear independent vectors are at each point linearly independent. In other words, a period map is non-degenerate if the corresponding local map of the base to a fibre is a diffeomorphism.

4°. We consider some non-degenerate period map. Its differentiation along all the vectors applied at a point of the base induces an isomorphism of the tangent space at this point of the base with the space of the cohomology of the fibre over it.

Thus, *every non-degenerate period map induces an isomorphism of the tangent bundle of the base with the cohomology bundle.*

The dual isomorphism maps the homology bundle onto the cotangent bundle of the base. Hence, a non-degenerate period map transfers to the base the cohomology and homology structures in the fibres of the bundle.

5°. Suppose that the fibres of the original bundle are real oriented *even-dimensional* manifolds. We consider the homology and cohomology groups of the *middle dimension.* In this case a bilinear form is defined on the homology space of the fibre: the intersection index. This form is symmetric if the dimension of the fibre is a multiple of 4 and skew-symmetric otherwise. It is non-degenerate if the fibre is closed (compact and without boundary) but it may have a kernel otherwise. We suppose now that the intersection form is *non-degenerate.* Then on the space of cohomology groups of the fibres there is a dual (also non-degenerate) form.

In this situation a non-degenerate period map induces on the base of the bundle a symmetric (if the dimension of the fibre is a multiple of 4) or exterior differential form. It is obtained from the field of forms on the cohomology groups of the fibres dual to the intersection forms under the isomorphism of the cohomology bundle with the tangent bundle given by the period map.

In the situation we have described a non-degenerate period map induces on the base of the original bundle either the structure of a complex pseudo-Riemannian manifold *of zero curvature* or a *symplectic structure* (if the dimension of the fibre is of the form $4p+2$). For our differential 2-form is locally induced from a form with constant coefficients, and therefore, it is closed in the skew-symmetric case and induces a flat metric in the symmetric case.

6°. Varchenko and Givental' apply all these constructions to a versal deformation of a critical point of a function $f:(\mathbf{C}^n, 0) \to (\mathbf{C}, 0)$. We fix a deformation

$$F(x, \lambda) = f(x) + \lambda_1 g_1(x) + \ldots + \lambda_\mu g_\mu(x)$$

(where the images of the monomials g_i in the local algebra $Q = \mathbf{C}[\![x]\!]/(\partial f/\partial x)$ form a basis of it over $\mathbf{C}$). It is convenient to assume that $g_\mu \equiv 1$, so that λ_μ is the "constant term" of the deformation. We fix a sufficiently small ball B with centre at the origin in the x-space $\mathbf{C}^n$ and then a sufficiently small ball Λ (depending on B) in the λ-space $\mathbf{C}^\mu$. A point λ of the ball Λ is said to be a *bifurcation point* if the level variety $V_\lambda = \{x \in B: F(x, \lambda) = 0\}$ is singular. The bifurcation points form the *bifurcation variety* Σ. Over $\Lambda \setminus \Sigma$ the varieties V_λ form a locally trivial bundle. We now construct the period maps for this bundle. We note straightaway that the dimension of the base (μ) is equal to that of the cohomology space of the fibre. The real dimension of the fibre of V_λ is even (and divisible by 4 for an odd number of arguments n of the original function). Therefore, the preceding constructions are applicable provided only that the intersection form and the period map are non-degenerate.

7°. We consider a holomorphic differential $(n-1)$-form on $B \times \Lambda$. Such a form ω is closed on each fibre V_λ. Consequently, it determines a period map: to each point of the base $\lambda \in \Lambda \setminus \Sigma$ there corresponds the cohomology class $[\omega]_\lambda \in H^{n-1}(V_\lambda, \mathbf{C})$ of the restriction of ω to V_λ.

Definition. This map $\lambda \mapsto [\omega]_\lambda$ is called the *period map* of ω.

Similar definitions can be given for germs (if ω is the germ of a form at the origin, then its representative determines a period map for sufficiently small λ).

Example. Let $f = x_1^2 + \dots + x_n^2$ be the Morse function. The multiplicity μ is 1. The versal deformation is $F = f + \lambda$. The fibres V_λ are homotopy-equivalent to the "vanishing sphere" S_λ^{n-1}. The sphere S_λ, $\lambda < 0$, consists of the intersection of the fibre with the real space. The radius of this sphere $r = |\lambda|^{1/2}$ is small when $|\lambda|$ is small.

The integral of a generic $(n-1)$-form on the vanishing sphere decreases like $|\lambda|^{n/2}$ as $|\lambda| \to 0$. For this integral is equal to the flux of the vector field corresponding to the form across the surface of the sphere. The flux is equal to the integral of the divergence of the field over the ball of radius r. The integral of a function over a sphere of radius r has an expansion

$$c_n r^n + c_{n+2} r^{n+2} + \dots,$$

whose first term is proportional to the volume of the ball (and to the value of the function at zero).

From this it follows that the period map for a Morse function has for $|\lambda| \to 0$ the asymptotic expansion $[\omega]_\lambda \mid S_\lambda = c\lambda^{n/2} + \dots$ (for a generic form, $c \neq 0$).

8°. Turning to a versal deformation of an arbitrary holomorphic function, we consider the derivative of the period map of the form ω in the direction of the vector field $\partial_\mu = \partial/\partial\lambda_\mu$ (I recall that we have chosen the versal deformation in such a way that $g_\mu \equiv 1$, so that ∂_μ is a field transversal to the tangent plane to the bifurcation diagram).

Definition. The k-th derivative of the period map of ω in the direction of ∂_μ is called the k-th *adjoint period map* (of ω).

Example. For a Morse function the value of the k-th adjoint period map of a generic form on the vanishing cycle S_λ (in the notation of the preceding example) is asymptotically equal to $c\lambda^{(n/2)-k}$ if n is odd or if $k \leqslant n/2$. It has a higher order of zero if n is even and $k > n/2$ (in the latter case the principal term of the expansion disappears on differentiation).

9°. Returning to the versal deformation of an arbitrary holomorphic function, we consider the k-th adjoint period map of a generic holomorphic form. We choose integral bases for the cohomology groups of the fibres over a neighbourhood of a point of the base depending continuously on the point of the base (the "constant" basis of the canonical trivialization). We form the determinant of the matrix of the derivatives of the components of the period maps on this basis in the direction of the basis vectors $\partial/\partial\lambda_i$.

Example. For a Morse function the determinant is proportional to $\lambda^{(n/2)-k-1}$, except when n is even and $k \geqslant n/2$.

This determinant (not only for a Morse function, but also for any function) is defined up to its sign (distinct bases of the fibre differ by transformations with determinant ± 1). The square of the determinant is a holomorphic function on $\Lambda \setminus \Sigma$ (a consequence of the holomorphic dependence of the integral of a holomorphic form over a cycle on the parameter). It turns out that this holomorphic function is meromorphic over Λ. (To prove this it is sufficient to study the behaviour of the integrals on approach to a generic (non-singular) point of the bifurcation diagram, and there the Picard-Lefschetz formulae apply, since Morse functions are non-degenerate. We need not worry about more complicated singular points of the bifurcation diagram so far as this study is concerned, since they form a set of complex codimension 2 in Λ, therefore, the assertion concerning the meromorphism extends to them by general theorems of the theory of functions of several complex variables.)

10°. The calculation for Morse singularities has been given above. It shows that the order of zero of the determinant of the period map of a generic holomorphic form at a generic point of the bifurcation diagram is equal to $(n-2)/2$. A generic line in the direction $\partial/\partial\lambda_\mu$ intersects the bifurcation diagram in μ points. Therefore, the order of zero of the restriction of the determinant to a curve passing through the origin and transversal to the tangent plane to the bifurcation diagram (for example, to the λ_μ-axis), is equal to $\mu(n-2)/2$. Under each differentiation of the components of a matrix of order μ with respect to λ_μ the order of zero of the determinant decreases, generally speaking, by μ. After k differentiations the expected order of zero has gone down to $\mu(n-2k-2)/2$ (provided only that the coefficient of the principal term does not vanish).

Definition. The k-th adjoint period map of a form is said to be *infinitesimally non-degenerate* if the order of zero of the restriction of the determinant to the λ_μ-axis at the origin is $\mu(n-2k-2)/2$.

For example, for a Morse function of n variables the k-th adjoint period map of a generic form is infinitesimally non-degenerate for odd n and infinitesimally degenerate for even n and $k \geqslant n/2$.

11°. Now we can begin to state the results proved in the paper of Varchenko and Givental'.

***Theorem* 1.** *If the k-th adjoint period map of a form is infinitesimally non-degenerate, then it is non-degenerate.*

***Theorem* 2.** *The period map of a generic form is infinitesimally non-degenerate. If the intersection form is non-degenerate, then for any k the k-th adjoint period map of a generic form is infinitesimally non-degenerate.*

Here a "generic form" means a "form whose jet does not belong to any proper analytic subset of the space of jets of sufficiently high order".

Assertions close to these theorems were announced also in a preprint of Saito [62].

12°. The classification of period maps is based on the following notion of equivalence, which we define in the general situation of an arbitrary period map with a smooth base.

An *isotopy* of the base is a smooth path in the group of diffeomorphisms of the base beginning at the identity diffeomorphism. Every isotopy of the base carries over in the natural way to an isotopy of the cohomology bundle with integral (and therefore also with complex) coefficients: an integral cycle over a point of the base moving under the isotopy gets shifted continuously into a neighbouring fibre.

An isotopy of the base shifts each period map into a new period map (into the image of the first under the action of a finite transformation of the isotopy of the cohomology bundle).

Definition. Two period maps are said to be *equivalent* if one of them goes over into the other under a suitable isotopy of the base.

13°. We turn to our special case of bundles over $\Lambda \setminus \Sigma$ (Λ is the base of a versal deformation of f, and Σ is the bifurcation set). In this case we can define isotopies and equivalences by means of *local holomorphic diffeomorphisms* of Λ that preserve Σ.

***Theorem* 3.** *If the k-th adjoint period map of a holomorphic form is infinitesimally non-degenerate, then it is stable, that is, the k-th adjoint period map of any nearby holomorphic form is equivalent to it (in a neighbourhood of a suitable point close to the origin).*

***Theorem* 4.** *If f is a quasihomogeneous function, then all infinitesimally non-degenerate k-th adjoint period maps of holomorphic forms are equivalent to each other.*

14°. We return to 2-forms carried by non-degenerate period maps into the fibres of $T^*(\Lambda \setminus \Sigma)$ from the intersection form on the homology groups of the level sets of functions of n variables. We call this field of 2-forms the *intersection form of the period map.*

***Theorem* 5.** *The intersection form of an infinitesimally non-degenerate k-th adjoint period map is holomorphic in* $\Lambda \setminus \Sigma$, *while for* $n \leqslant 2k+2$ *it can be continued holomorphically to* $T^*\Lambda$.

***Theorem* 6.** *The intersection form mentioned is stable (that is, the intersection forms of the k-th adjoint period maps of nearby holomorphic forms go over to it under suitable local holomorphic diffeomorphisms of the pair* (Λ, Σ)*).*

***Theorem* 7.** *If f is a quasihomogeneous function, then the germs of the intersection forms of the infinitesimally non-degenerate k-th adjoint period maps of holomorphic forms go over into each other by biholomorphic diffeomorphisms of the pair* (Λ, Σ).

15°. Theorem 5 allows us to carry over to the case of non-simple singularities of functions of n variables the operation of convolution of invariants defined in §10. For let us fix an infinitesimally non-degenerate k-th adjoint period map with $2k+2 \geqslant n$ (the most important case is $n = 2k+1$). We associate with a pair of functions holomorphic at the origin of the base Λ of the versal deformation a new function whose value at each point of $\Lambda \setminus \Sigma$ is the value of the intersection form of the period map on the differentials of the given functions at these points. By Theorem 5 the resulting function $\Lambda \setminus \Sigma$ can be continued holomorphically to Σ. It depends bilinearly and symmetrically on the original functions for odd n (skew-symmetrically for even n). We obtain the usual convolution for $n = 2k+1$ if we begin with a simple singularity of functions of n variables (of type A_μ, D_μ, or E_μ).

16°. Now suppose in addition that the intersection form is non-degenerate. In that case the non-degenerate period map induces a 2-form not only on the cotangent, but also on the tangent space to the base at each point: it is induced from the form on the cohomology space by the dual intersection form. We call it the *inverse intersection form of the period map.*

***Theorem* 8.** *If* $n = 2k+2$, *then the inverse intersection form of an infinitesimally non-degenerate k-th adjoint period map of a holomorphic form can be continued from* $\Lambda \setminus \Sigma$ *to a holomorphic symplectic structure on the whole base* Λ *of the versal deformation.*

Moreover, the restriction of the symplectic structure to the strata of the bifurcation set Σ "remembers" the degeneracy types of the critical points over these strata (for example, the stratum $(\mu/2)A_1$ is Lagrangian, since the corresponding $\mu/2$ vanishing cycles vanish on the same level and therefore do not intersect).

17°. We consider now the case of an odd number of arguments of f. In this case the intersection form is symmetric.

Definition. A *principal period map* for a function of $n = 2k+1$ variables is an infinitesimally non-degenerate k-th adjoint map of a holomorphic form.

The choice of just this value of k is dictated by the behaviour of the Vieta map of a simple singularity on the mirrors. For a Morse function the period map has the asymptotic expansion $a\lambda^{(n/2)} + \ldots$ Therefore, the k-th adjoint map has the asymptotic expansion $b\lambda^{(n/2)-k} + \ldots$ If $n = 2k+1$, then the first term of the sequence is $b\lambda^{1/2}$. In other words, for this choice of k for a Morse function the k-th adjoint period map has for $\lambda = 0$ the simplest branching of order 2, like $\lambda^{1/2}$. From this it is easy to deduce that for any function of n variables and this choice of k the infinitesimally non-degenerate k-th adjoint period map has the same singularity as the map inverse to the Vieta map has on the variety of irregular orbits of the group of reflections.

18°. ***Theorem*** **9.** *If the intersection form is non-degenerate, then the isomorphism $T^*(\Lambda \setminus \Sigma) \to T_*(\Lambda \setminus \Sigma)$ given by any principal period map isomorphically takes the module of germs of differential 1-forms on Λ holomorphic at the origin to the module of germs of vector fields on Λ holomorphic at the origin and tangent to Σ.*

Theorem 9 generalizes the construction given in § 10 of fields tangent to a front from the convolution of the invariants of a group generated by Euclidean reflections. The Euclidean metric in the general case is replaced by the intersection form of a principal period map.

19°. We consider a principal period map of a simple singularity (A_μ, D_μ, or E_μ). For a simple singularity the sign of the intersection form of the fibre is determined. The monodromy group (the image of the representation of the fundamental group of the base $\Lambda \setminus \Sigma$ in the cohomology group of the fibre $H^{n-1}(V_\lambda)$) is in this case generated by Euclidean reflections.

Any period map associates with every point of the base the elements of the cohomology group of the fibre over the point, which are determined up to the action of the monodromy group. In the case of a simple singularity we obtain in this way a map of the manifold $\Lambda \setminus \Sigma$ into the variety of orbits of a group generated by reflections. The next theorem follows from the results of Looijenga (see [63]).

Theorem **10.** *A principal period map of a simple singularity can be continued to an isomorphism of the base Λ with the space of orbits of the corresponding group of reflections that takes Σ into the variety of irregular orbits. This isomorphism takes the intersection form of the period map into the form that determines the convolution of invariants of the group of reflections.*

Thus, the convolution determined by a principal period map of any function generalizes the convolution of the invariants of a group of reflections.

20°. We now pass to the analogue of the linearized convolution of invariants for non-simple singularities. The difficulty is that in contrast to the case of simple singularities, the convolution of two functions that vanish at the origin may give a function that does not vanish there. In this situation the linear part of the result of convolution is, generally speaking, not determined by the linear parts of the functions convoluted at the origin.

In the case of a *quasihomogeneous* (not necessarily simple) function of an odd number of variables with a *non-degenerate* intersection form this difficulty does not occur. For by Theorem 9 the holomorphic vector field ∇a corresponding to the differential of a function a to be convoluted is tangent to the bifurcation diagram Σ (this uses the non-degeneracy of the intersection form). By a theorem of Zakalyukin [53], a holomorphic field tangent to the bifurcation diagram of zeros of a *quasihomogeneous* function vanishes at the origin in Λ. We denote the value of the intersection form on da, db by $\langle \nabla a, \nabla b \rangle$. Then the functions

$$\langle \nabla(ab), \nabla c \rangle = a \langle \nabla b, \nabla c \rangle + b \langle \nabla a, \nabla c \rangle = a(\nabla b \mid c) + b(\nabla a \mid c),$$

where a, b, and c vanish at $0 \in \Lambda$, have at the origin a zero of the *second order*, since the fields ∇b and ∇a vanish at the origin. This proves that the linear part of $\langle \nabla u, \nabla v \rangle$ at the origin is independent of the terms of degree higher than 1 of the Taylor series of u and v at the origin.

21°. Thus, a principal period map of a quasihomogeneous function with a non-degenerate intersection form determines a linearized convolution operation $L: T_0^*\Lambda \times T_0^*\Lambda \to T_0^*\Lambda$.

The tangent space $T_0\Lambda$ to the base of the versal deformation can be identified canonically with the space of the local algebra Q (to the rate of deformation there corresponds its contiguity class in the local algebra). Consequently, $T_0^*\Lambda$ is identified with Q^*. Hence, the operation of "linearized convolution" L is identified with some operation $l: Q^* \times Q^* \to Q^*$. This operation depends on the holomorphic form whose period map determines L. Varchenko and Givental' have proved the following formula for this operation on Q^*.

Theorem 11. *The operation l constructed from any principal period map of a quasihomogeneous function coincides with one of the operations φ_α constructed in Theorem* 3 *of* §11 *with respect to the graded local algebra Q of this function. Each operation φ_α with an admissible α is an operation l for some principal period map.*

As far as I know, there remains in this theory the unsolved question of the connection between the signatures of the forms φ_α defined by the "linearized convolution" and the Levine-Eisenbud forms B_α [61] (see §10) for *real* non-simple functions f.

For non-quasihomogeneous singularities one may be able perhaps to define a linearized convolution that makes the base of the versal deformation smaller, but no such theory has yet been developed. Of interest also are questions what the image of a many-valued principal period map is and whether it is single-sheeted (in the quasihomogeneous case the image does not depend on the period map and in the simplest cases it is by its very nature the "upper half-plane").

Chapter V

LAGRANGIAN AND LEGENDRIAN TOPOLOGY

The classification of metamorphoses of caustics and fronts in generic one-parameter families allows us to study some questions of the global geometry and topology of Lagrangian and Legendrian maps.

In this chapter we define Lagrangian (respectively, Legendrian) cobordisms of Lagrangian (respectively, Legendrian) submanifolds immersed in the space of a cotangent bundle (respectively, in a space of contact elements or in a space of 1-jets of functions). We describe also the beginnings of a theory of Lagrangian and Legendrian characteristic classes—cohomology classes dual to the manifolds of singularities of various types of Lagrangian and Legendrian maps.

§13. Lagrangian and Legendrian cobordisms

Ths simplest example of a Legendrian cobordism is the relation between the traces on the surface of the Earth at different moments of time of a front propagated through the air. These traces are not necessarily homeomorphic. Their cobordancy imposes restrictions on the types of the singularities (for example, the parities of the numbers of cusps on cobordant fronts are the same).

The definition of a Lagrangian cobordism is based on the concept of a Lagrangian boundary. We consider a Lagrangian submanifold of the space of the cotangent bundle of a manifold with boundary. From a physical point of view, a Lagrangian submanifold describes the asymptotic behaviour of a wave state in a domain with boundary. The wave state in the domain determines the state on the boundary. Its asymptotic behaviour determines a Lagrangian submanifold of the space of the cotangent bundle over the boundary of the original manifold. It is called the Lagrangian boundary of the original Lagrangian manifold.

In other words, the embedding $\partial V \to V$ induces a projection of the boundary of the space of the cotangent bundle onto the space of the cotangent bundle of the boundary: $\partial(T^*V) \to T^*(\partial V)$ (to a 1-form at a point of the boundary there corresponds its restriction to the boundary). A Lagrangian (in general, immersed) submanifold of the space T^*V transversal to the boundary intersects the latter in a smooth (immersed) submanifold. The projection of the intersection into $T^*(\partial V)$ is an immersed Lagrangian submanifold of the space of the cotangent bundle of the boundary, the so-called *Lagrangian boundary* of the original Lagrangian manifold.

A Legendrian boundary is defined similarly. For example, we consider an (immersed) Legendrian submanifold of the spaces of 1-jets of functions on a manifold M with boundary ∂M. The boundary of the space of 1-jets $J^1(M, \mathbf{R})$ is fibred over the manifold of 1-jets of functions on the boundary (the map $\partial J^1(M, \mathbf{R}) \to J^1(\partial M, \mathbf{R})$ associates with the 1-jet of a function at a point of the boundary the 1-jet of the restriction of the function to the boundary). A Legendrian submanifold transversal to the boundary of the space of jets intersects this boundary in an (immersed) submanifold. The projection of the intersection into $J^1(\partial M, \mathbf{R})$ is called the *Legendrian boundary* of the original manifold. This is a Legendrian manifold (immersed in the space of 1-jets of functions on the boundary) of dimension 1 less than the original manifold. The "physical meaning" of a Legendrian boundary is simple: the front of a Legendrian boundary is the ordinary boundary of the original front.

Let L_0 and L_1 be two closed Lagrangian submanifolds (compact without boundary) of the space of the cotangent bundle T^*V. A Lagrangian (cylindrical) cobordism between them (Fig. 16) is a Lagrangian submanifold of the space $T^*(V \times [0, 1])$ of the cotangent bundle of the cylinder over V, whose Lagrangian boundary is the difference of $L_1 \times 1$ and $L_0 \times 0$ (for oriented cobordisms a change of orientation of a manifold changes its sign; for non-oriented ones the orientations do not matter). The manifolds L_1 and L_0 are said to be Lagrangian oriented (respectively, non-oriented) cylindrically *cobordant* if there is an oriented (respectively, non-oriented) cobordism between them.

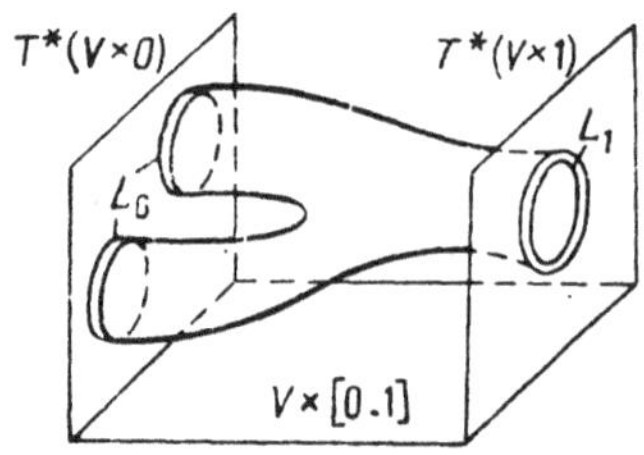

Fig. 16.

Classes of cylindrically Lagrangian cobordant manifolds form a semigroup, which turns out to be a group (for details, see [64], [65]). The resulting groups are called groups of classes of Lagrangian oriented (respectively, non-oriented) cylindrical cobordism over V (more briefly, *groups of Lagrangian cobordism over V*). The case $V = \mathbf{R}^n$ is especially important (the cobordism groups over $\mathbf{R}^n$ and the ring formed from all these play the role of the groups and rings of coefficients in bordism theory).

Legendrian cobordisms of submanifolds of a space of 1-jets of functions, of a space of non-oriented contact elements, or of a space of oriented contact elements are defined similarly.

In the Legendrian case instead of a cobordism of Legendrian manifolds we can speak directly of a cobordism of fronts (since a Legendrian manifold is defined uniquely by its front). It is only required that a cobordism of fronts is transversal to the boundary of the base of the Legendrian bundle (one can ask that in a neighbourhood of the boundary of the base the front-cobordism should be the direct product of the front-boundary and a half-interval).

Fig. 17.

The lists of the metamorphoses of Lagrangian and Legendrian singularities (§ §6 and 9) allow us to calculate the groups of Lagrangian and Legendrian cobordisms in small dimensions. For example, the group of Legendrian oriented cobordisms of Legendrian submanifolds of the space of 1-jets of functions of a single variable is free cyclic. A generator of this group is the cobordism class of the Legendrian manifold whose front has the form of a bow-tie (a figure of eight with cuspidal vertices, Fig. 17). We assign to a 1-jet its coordinates (x, y, p) $(y = f(x), p = \partial f/\partial x)$. In these coordinates the bow-tie is given by the equations

$$x = \cos t, \quad y = \sin 4t - 2 \sin 2t$$

(on the corresponding Legendrian manifold $p = 8 \sin 3t$).

The group L_n of classes of oriented Legendrian cobordism of submanifolds of the space of 1-jets of functions in $\mathbf{R}^n$ is isomorphic to the stable homotopy group

$$L_n \approx \lim \pi_{n+k} (T\lambda_k), \quad k \to \infty,$$

where T is the Thom space and λ_k is the tautological bundle over the Grassmannian $U(k)/SO(k)$ of oriented Lagrangian planes in $\mathbf{R}^{2k}$ (Eliashberg).

The graded algebra of cobordism classes $\oplus L_n$ is isomorphic modulo torsion to an exterior algebra with generators of degree $4k+1$. The torsion, as far as I am aware, has not been computed. Eliashberg has expressed in homotopy terms other groups of cobordism classes that were introduced in [64] and [65], for example, groups of classes of Lagrangian cobordism, groups of the cobordism classes of fronts on manifolds, and groups of classes of non-oriented cobordism. But since the corresponding homotopy calculations have not been carried out, these results have up to now been used more for the computation of homotopy invariants by means of metamorphoses of caustics and wave fronts than for the computation of the groups of Lagrangian and Legendrian cobordisms by means of homotopy invariants. Here we present only some of the results obtained in [64] and [65].

The group of classes of non-oriented cobordism of Legendrian curves in the space of 1-jets is trivial. Fig. 18 shows a series of metamorphoses of the bow-tie, which prove this. It reduces it to the empty set and is a non-oriented cobordism of the bow-tie with zero. This cobordism glues the bow-tie to a Möbius band.

Fig. 18.

The same series of metamorphoses together with the inverse series immerses the Klein bottle as a Legendrian manifold in $\mathbf{R}^5$ (and consequently, also as a Lagrangian manifold in $\mathbf{R}^4$).

The projective plane does not admit a Lagrangian immersion in $\mathbf{R}^4$ (nor, therefore, a Legendrian immersion in $\mathbf{R}^5$). Compact connected two-dimensional manifolds with *even* Euler characteristic admit Legendrian immersions in $\mathbf{R}^5$, but with *odd* characteristic do not even admit Lagrangian immersion in $\mathbf{R}^4$.

Remark. The question of Lagrangian *embeddings* of projections of two-dimensional Legendrian manifolds from $\mathbf{R}^5$ in $\mathbf{R}^4$ is unsolved. The corresponding one-dimensional problem is easily solved: the projection of a compact Legendrian one-dimensional manifold from the space of jets of functions on the line to the space of the cotangent bundle of the line must have self-intersections. In other words, the following theorem holds.

Theorem. *A closed curve in the plane that has a continuous non-vertical tangent everywhere (even at the cusp points) has a vertical chord with parallel tangents at its ends (Fig. 19).*

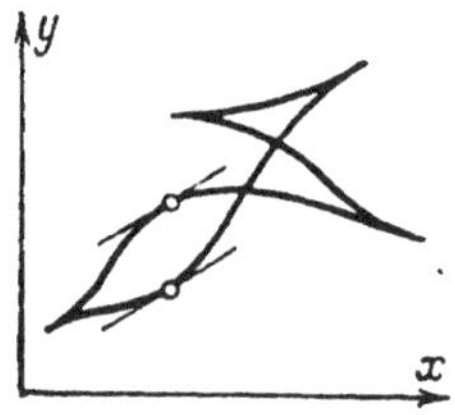

Fig. 19.

Proof. Our curve in the (x, y)-plane is the front of a Legendrian submanifold of the (x, y, p)-space of 1-jets of functions. The projection of each of its connected components onto the (x, p)-plane is a connected curve that bounds a domain of zero area $\left(\oint p\,dx = 0\right)$. Hence, this projection has a point of self-intersection, as required.

The old conjecture on *the impossibility of the Lagrangian embedding of a closed manifold in* $\mathbf{R}^4$ *that can be lifted to a Legendrian embedding in* $\mathbf{R}^5$ (that is, has a one-valued derivative function $\int p\,dx$), means geometrically that *a closed surface with everywhere non-vertical tangent plane in three-dimensional space has a vertical chord with parallel tangent planes at its ends.*

Here it is supposed that the only singularities of the surface are edges of regression, swallow-tails, and self-intersections (or at least that the surface has a tangent plane continuously depending on the point of the branch of the surface).

An oriented closed n-dimensional manifold with *non-zero* Euler characteristic cannot be embedded in $\mathbf{R}^{2n}$ as a Lagrangian submanifold (see [66]). For: 1°. The index of self-intersection of an embedded oriented manifold in a tubular neighbourhood of it is the same as in the ambient space (for an immersed submanifold this is not so). 2°. The index of self-intersection in Euclidean space is zero. 3°. The index of self-intersection in a tubular neighbourhood is equal to the Euler characteristic of the normal bundle. 4°. For a Lagrangian manifold in $\mathbf{R}^{2n}$ the normal bundle is isomorphic to the tangent bundle. From 1°.-4°. it is clear that the Euler characteristic of a Lagrangian manifold embedded in $\mathbf{R}^{2n}$ is zero.

In particular, a sphere cannot be embedded in $\mathbf{R}^4$ as a Lagrangian submanifold (it is easy to construct an immersion taking as the front, for example, the saucer of §9).

The assertion concerning chords with parallel tangent planes is proved, therefore, for all oriented two-dimensional fronts, except the torus. The torus admits a Lagrangian embedding in $\mathbf{R}^4$, but it would appear that there is no embedding with a single-valued derivative function.

The group of classes of a *Lagrangian* oriented cobordism of curves in the plane is isomorphic to $\mathbf{R}+\mathbf{Z}$. The cobordism class of a curve is determined by two numbers, which are cobordism invariants: the integral of the form $p\,dx$ and the Maslov index (which for a plane curve is twice the number of turns of the tangent vector of the curve and takes only even values). Thus, all curves with a fixed value of the Maslov index and bounding a domain of fixed area are Lagrangian oriented cobordant.

Example. We consider an integrable system with two degrees of freedom, for example, a particle in a central field in the (q_1, q_2)-plane. The level surfaces of the two first integrals are Lagrangian submanifolds in $\mathbf{R}^4$. We fix one of the coordinates, $q_2 =$ const, and consider the "trace" of the Lagrangian level manifold of the integrals on the (q_1, p_1)-plane (that is, we forget the value p_2). Suppose, for example, that the level manifold of the integrals is a torus in the phase space $\{(p, q)\}$, projected into the configuration plane $\{(q_1, q_2)\}$ as an annulus (Fig. 20). For large values of q_2 (A in Fig. 20) the trace on the (q_1, p_1)-plane is empty. When q_2 decreases and reaches a critical value, a point trace is born. For values less than the critical value ($q_2 = B$), the trace is cobordant to zero. Consequently, the trace that is born has the form not of an ellipse, but of a figure of eight.

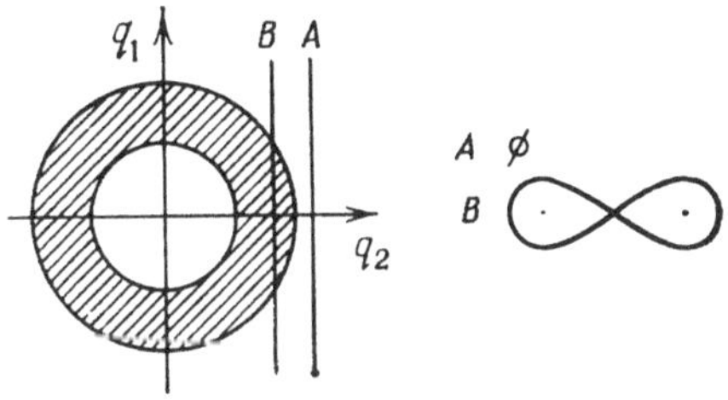

Fig. 20.

In [64], [65] scores of theories of Lagrangian and Legendrian cobordism are defined (taking account or not of the orientation of Lagrangian or Legendrian manifolds, cobordisms, bases, and contact elements), and results are given of the computation of the groups of cobordism classes on two-dimensional surfaces.

Example. The group of classes of Legendrian cobordism of oriented and armed (transversally oriented) fronts on the projective plane is free cyclic. As a generator we can take the class of the curve $x^2 = y^3$.

The only cobordism invariant of armed oriented fronts in $\mathbf{R}P^2$ is the index, which is equal to the number of cusps with signs taken into account (a point is regarded as positive if on going round the front in its given orientation the passage from branch to branch at a cusp is on the side of the normal arming the front). For example, the index of the bow-tie (Fig. 17) is 2, since at both cusps the passage is on the side of the normal (upwards in Fig. 17). The index is defined for fronts on any surface, oriented or not. On the projective plane the index can also take odd values. The index of a curve on the projective plane can also be computed as the number of points of inflection of the front with signs taken into account: a point of inflection is regarded as positive if after the point of inflection the front turns to the side away from the normal arming it. For example, on the bow-tie (Fig. 17) of the six points of inflection the four near to the cusps are positive. The indices of projectively dual curves are opposite to each other.

An expression of the Maslov index of curves on a Lagrangian manifold in terms of the points of inflection of the front in Euclidean space was found by Tvorogov ([67], [68]).

The index of an armed one-dimensional front, as introduced above, is connected with the Maslov index of curves on two-dimensional Lagrangian manifolds (the index of the intersection with the manifold of singularities of the Lagrangian projection) in the following way. An armed front on V determines a conical Lagrangian manifold in the space T^*V of the cotangent bundle of the manifold V: it consists of the 1-forms that vanish on a contact element touching the front and are positive on the armed normal. For a generic front this conical manifold is smoothly immersed in T^*V. A Riemannian metric on V determines an immersion of the front in this conical Lagrangian manifold (to a point of the front there corresponds a 1-form that is 1 on the positive unit normal). The index of a one-dimensional front, defined above as the number of inflection points with signs taken into account, is equal to the Maslov index of the curve constructed on the conical Lagrangian manifold (see [65]).

§14. Lagrangian and Legendrian characteristic classes

Lagrangian and Legendrian characteristic classes are the cohomology classes of closed Lagrangian and Legendrian manifolds (compact without boundary), dual to suitable manifolds of singularities. The corresponding characteristic numbers are invariants of Lagrangian (Legendrian) cobordism.

The simplest of these characteristic classes is the Maslov class: it is dual to the whole manifold of singularities of the projection of a generic Lagrangian manifold to the base (that is, the manifold A_2 in the notation of §3). This one-dimensional cohomology class assigns an integer to a closed oriented curve on a Lagrangian manifold (the number of singularities of A_2 with sign taken into account).

Vasil'ev [69] has constructed cohomology classes on generic Lagrangian submanifolds of T^*V corresponding to more complicated singularities. In the oriented case, for example, he has determined an integral four-dimensional class dual to A_5 (it is of order 2), five-dimensional integral classes dual to A_6 and E_6 (they are cohomologous to each other), and a six-dimensional integral class dual to P_8. Modulo torsion, $A_2A_6 = 3P_8$.

Similar results have been obtained for Legendrian singularities. For example, the number of A_6-singularities on a closed oriented Legendrian submanifold in the space of 1-jets of functions in $\mathbf{R}^5$ coincides (with signs taken into account) with the number of E_6-singularities and is the same for Legendrian oriented cobordant manifolds. Vasil'ev has also proved that the numbers of the singularities

$$A_4 \cup D_4,\ D_5,\ D_6,\ D_7,\ A_6 \cup E_6,\ E_7 \cup P_8,\ A_8 \cup D_8,\ D_8 \cup E_8,\ P_9$$

are even and that the number of A_6-singularities (counted with signs) is divisible by 3 on closed manifolds of an appropriate number of dimensions. (Of course, here and in everything that follows we are speaking of singularities of generic Lagrangian and Legendrian maps.)

The method of constructing Lagrangian and Legendrian characteristic classes consists in the following. We consider any class of the classification of the space of germs of functions at a critical point according to the types of singularity ($A_\mu^\pm$, $D_\mu^\pm$, . . .), that is, any type of Lagrangian or Legendrian singularity (the signs $\pm$ in the notation refer to real inequivalent, but complex equivalent singularities, for example, $A_3^\pm \cong \pm x^4$). Suppose that this class has codimension c (for example, codim $A_2 = 1$, more generally, codim $A_\mu^\pm$ = codim $D_\mu^\pm$ = codim $E_\mu^\pm = \mu - 1$). To determine the intersection index of the manifold of singularities of a given class with an oriented submanifold of dimension c we have to know the transversal orientation of our class, that is, the orientation of a fibre of the normal bundle. It must be given consistently at all points of the class, and it must be invariant under the group that specifies the equivalence of the singularity of the type in question. If such an orientation exists, then the relevant class of singularities is said to (invariantly) *coorientable.*

Maslov [70] remarked that the class A_2 is coorientable and showed explicitly its coorientation (it is given by the direction of the growth of the index of inertia of the second differential of the generating function).

Theorem (Vasil'ev). *The classes of stable R-equivalence $A_{4p+1}^\pm$, A_{2p} are coorientable, while $A_{4p-1}^\pm$, $D_p^\pm$ are not, for any p. Of the classes of codimension $\leqslant 7$ those that are coorientable are precisely $A_{4p+1}^\pm$, A_{2p}, $E_6^\pm$, E_7, E_8, $P_8^{1,2}$, P_9, while the remaining ones (that is, $A_{4p-1}^\pm$, $D_p^\pm$, X_9) are not.*

Here A, ..., X is the notation for the singularity classes according to the classification of [8].

Thus, let us consider any coorientable class. Taken by itself, the coorientation is insufficient to determine the intersection index: we need also some condition on the adjacency to the class of singularities in question on that part of its boundary that consists of singularities of codimension 1 greater.

Example. We consider the manifold of singularities A_2. This class of codimension 1 is transversally oriented. The boundary has codimension 2. We consider any generic boundary point. We draw through it a two-dimensional transversal to the boundary. The trace of A_2 in this plane is a curve with a singular point. With the exception of this point, the curve is transversally oriented. If we were to obtain topologically at this point one of the cases of Fig. 21 a) or b), then the intersection index of curves with A_2 could not be defined. For in that case the intersection index of A_2 with a small circle surrounding the boundary point would differ from zero, while the circle, of course, is homologous to zero and no cocycle on it can take a non-zero value.

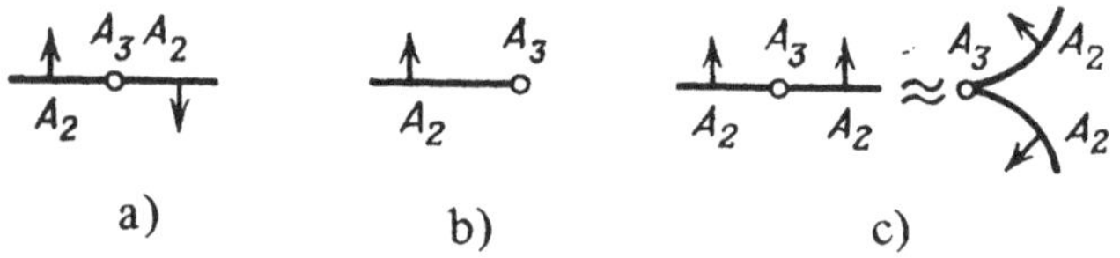

Fig. 21.

In fact, the coorientation of the class A_2 close to its boundary point A_3 is as shown in Fig. 21 c). This is the correct disposition of A_2 around A_3 and enables us to define the Maslov index of classes of homologous curves.

The condition of the "correct disposition" of a class close to a more complicated singularity is that it must be a *cocycle* of some complex. The cochains of this complex ω are formal sums of the cooriented classes. The grading is given by the codimension of the classes. The coboundary operator is determined by the adjacency of classes of the next codimension (taking coorientation into account). For example, Fig. 21 a) corresponds to $\delta A_2 = 2A_3$, Fig. 21 b) to $\delta A_2 = A_3$, and Fig. 21 c) to the true situation, $\delta A_2 = 0$.

Vasil'ev [69] has computed explicitly the first segment of the complex ω of cooriented classes of stable R-equivalent functions. He has obtained the following cohomology groups:

i	1	2	3	4	5	6	7
$H^i(\omega)$	$\mathbf{Z}$	0	0	$\mathbf{Z}_2$	$\mathbf{Z}$	$\mathbf{Z}$	?
generator	A_2	—	—	A_5	A_6	P_8	?
zeros	—	—	—	A_5	A_6-E_6	E_7+3P_8	P_9

These classes induce Lagrangian characteristic cohomology classes with integer coefficients on Lagrangian submanifolds of spaces of cotangent bundles T^*V. For example, the class A_2 induces the Maslov index of curves on Lagrangian manifolds. The intersection index with the transversally oriented manifold of A_6-singularities determines a five-dimensional cohomology class on a Lagrangian manifold in T^*V. The last row of the table contains some restrictions on the coexistence of singularities, which follow from the calculations of Vasil'ev. For example, the number of A_5-singularities for the Lagrangian projection of a four-dimensional closed oriented Lagrangian manifold, with the signs given by coorientation taken into account, is zero (that is, the class A_5 in the complex cannot be realized by singularities). For six-dimensional manifolds the sum of the number of E_7-singularities and three times the number of P_8-singularities (all with signs taken into account) is zero, and so on.

In the non-oriented case coorientations are unimportant, and the corresponding complex ν consists of all the (not necessarily coorientable) classes with coefficients in $\mathbf{Z}_2$. Their cohomology groups are as follows (see [69]):

i	1	2	3	4	5	6
$H^i(\nu)$	$\mathbf{Z}_2$	$\mathbf{Z}_2$	$\mathbf{Z}_2$	$\mathbf{Z}_2$	$\mathbf{Z}_2$	$\mathbf{Z}_2^2$
generator	A_2	A_3	A_4, D_4	A_5	A_6, D_6	A_7, E_7

These cohomology classes induce Lagrangian characteristic cohomology classes with coefficients in $\mathbf{Z}_2$ on Lagrangian manifolds in T^*V.

For the projection of a Lagrangian boundary the numbers of singularities of any type are even. For the projection of the Lagrangian boundary of an oriented compact manifold the number of singularities of each of the types A_2, A_6, E_6, E_7, P_8 (with signs taken into account) is zero.

A similar theory has been constructed for Legendrian singularities. In this case one has to begin with a coarser classification of the germs of functions at critical points than R-equivalence (Legendrian equivalence corresponds to the classification not of the functions themselves, but of their zero level hypersurfaces). For the definition of coorientedness one must require invariance under this group (which is, generally speaking, wider than the group of R-equivalences). The final results on Legendrian characteristic classes in small dimensions ($i \leqslant 6$) are as follows.

For Legendrian submanifolds of spaces of 1-jets of functions and spaces of armed contact elements one obtains the same complexes as for Lagrangian singularities (both in the oriented and in the non-oriented cases). Thus, all the results on the parity and equality of the numbers of singularities remain true for Legendrian manifolds and cobordisms.

For Legendrian submanifolds of the space of unarmed contact elements PT^*V the first few cohomology groups of the complexes ω (in the oriented case) and ν (in the unoriented case) are the following:

i	1	2	3	4	5	6
$H^i(\omega)$	0	0	0	$\mathbf{Z}_2$	$\mathbf{Z}$	$\mathbf{Z}^2$
generator	—	—	—	A_5	E_6	E_7, P_8
$H^i(\nu)$	$\mathbf{Z}_2$	$\mathbf{Z}_2$	$\mathbf{Z}_2^2$	$\mathbf{Z}_2$	$\mathbf{Z}_2^3$	$\mathbf{Z}_2^3$
generator	A_2	A_3	A_4, D_4	A_5	A_6, D_6, E_6	A_7, E_7, P_8

For a Legendrian boundary the number of singularities of any type is even, while for that of an oriented manifold the number of singularities of types E_6, E_7, and P_8 is zero (with signs taken into account).

The conditions listed above do not exhaust all the conditions for the coexistence of singularities of Lagrangian and Legendrian maps. For example, on a two-dimensional compact front the number of swallow-tails (A_3-singularities) is even, since a curve of self-intersection of a front begins and ends at swallow-tails.

Alekseev has proved that the number of points in which the front of a closed Legendrian submanifold of the space of 1-jets of functions on a two-dimensional manifold intersects its edge of regression is even: $|A_1A_2| = 0 \bmod 2$. In particular, if a closed surface in $\mathbf{R}^3$ has as singularities only edges of regression, swallow-tails, and transversal self-intersections and has no vertical tangent planes (even at singular points), then the number of points where the edge of regression pierces through the surface is even. His proof is based on the construction of a cobordism of the given front with a front without points of type A_1A_2 and on the fact that for cobordant fronts the parities of the numbers of points of type A_1A_2 are the same.

On a closed front in $\mathbf{R}^3$ the number of points of type A_1A_2 along each closed curve obtained from an edge of regression by joining vertices of swallow-tails is also even. This follows from the fact that the front determines in $\mathbf{R}^3$ a cycle homologous to zero modulo 2. The assertion remains true if we replace $\mathbf{R}^3$ by the space of jets $J^0(V^2, \mathbf{R})$.

Vasil'ev has generalized the theorem of Alekseev to the n-dimensional case, replacing the edge of regression (A_2) by a one-dimensional stratum of a front. We denote by $\{\Xi\}$ the difference reduced modulo 2 between the number of points where the stratum Ξ pierces the front and the intersection index of the closure of this stratum with the front in the $\mathbf{Z}_2$-homology of the space ambient to the front. Then

$$\{A_2\} = 0,\ \{A_3\} = 0,\ \{A_4\} = 0,\ \{A_5\} = \#E_6,\ \{A_6\} = \#E_7,\ \{A_7\} = 0;$$
$$\{D_4\} = 0,\ \{D_5\} = 0,\ \{D_6\} = \#E_7,\ \{D_7\} = 0;$$
$$\{E_6\} = \#E_7,\ \{P_8\} = 0,\ \{E_7\} = \#E_8 + \langle E_7, \text{dis}\rangle$$

(in the last formula dis denotes the disarming one-dimensional cocycle in the $\mathbf{Z}_2$-cohomology of the Legendrian manifold; this cocycle in induced from the first Stiefel-Whitney class of the tautological bundle by the identity embedding in PT^*V).

For each component of the closure of the one-dimensional stratum similar formulae are true with $\#E_i$ replaced by the number of points of E_i in the closure of the component. (In the expressions for $\{A_7\}$ and $\{E_7\}$ in this case one may perhaps have to add $\#P_9$.)

On closed fronts of suitable dimensions the numbers of A_5- and A_7-singularities are even, while the number of E_6-points coincides modulo 2 with the number of points of intersection of the strata A_5 and A_1.

The latter results were obtained by Vasil'ev [71] by means of a generalization of the complex of non-oriented classes ν, in which instead of singularities one considers "multi-singularities", that is, sets of certain Legendrian singularities. The differential is defined by the condition of scattering a more complicated multi-singularity into simpler ones by a generic one-parameter deformation of the front, on which all the singularities constituting the multi-singularity fall at one point. The number of multi-singularities of a given type on a compact front modulo 2 is an invariant of Legendrian cobordism if this type is a cocycle of the complex of multi-singularities. This number is even if this type is a coboundary.

The cohomology groups of a complex of multi-singularities can be computed by means of the following spectral sequence. We assign to each multi-singularity two numbers: p, the sum of the codimensions of the classes of the constituent singularities (in the space of germs of functions with critical point O and critical value 0), and q, the number of singularities constituting the multi-singularity. For example, $p(A_1A_1) = 0$, since codim $A_\mu = \mu - 1$; $q(A_1A_1) = 2$. The codimension in $\mathbf{R}^n$ of the manifold of multi-singularities of a given type on a front in $\mathbf{R}^n$ is $p+q$. For example, the codimension of the curve A_1A_1 of self-intersections of the swallow-tail in $\mathbf{R}^3$ is 2.

We form the group of (p, q)-cochains $E_1^{p,q}$ from formal sums of types of multi-singularities with fixed p and q (with coefficients in $\mathbf{Z}_2$). The beginning of the table of generators of the free $\mathbf{Z}_2$-modules $E_1^{p,q}$ has the form

$q = 3$	$A_1A_1A_1$	$A_1A_1A_2$	$A_1A_1A_3$, $A_1A_2A_2$
$q = 2$	A_1A_1	A_1A_2	A_1A_3, A_2A_2
$q = 1$	A_1	A_2	A_3
$p = 0$	$p = 0$	$p = 1$	$p = 2$

The whole complex of multi-singularities, as a $\mathbf{Z}_2$-module, is the direct sum of all the groups $E_1^{p,q}$. The differential, defined by the adjacencies of the classes, raises p (for example, $\delta(A_1A_1) = A_3$, since curves of self-intersection of fronts end at swallow-tails).

The filtration of the complex of multi-singularities by the subspaces $F^p = \bigoplus E_1^{i,q}$, $i \geqslant p$, generates the required spectral sequence.

Vasil'ev has shown that for $p \leqslant 5$ in this spectral sequence $E_\infty^{p,q} = E_3^{p,q}$ and he has calculated the first 7 cohomology groups:

i	1	2	3	4	5	6	7
$\dim H^i(S)$	1	1	1	1	1	2	3
$\dim H^i(P)$	1	1	1	2	2	5	7

Here S is the complex of multi-singularities of armed fronts and P the complex of multi-singularities of unarmed fronts (in the definition of equivalence it is permitted in the first case to multiply a function by a positive function, and in the second case by any non-zero function).

An obvious deficiency of these theories is the absence of examples to show that the Lagrangian and Legendrian characteristic classes thus constructed are non-trivial (different from zero). The existence of non-trivial classes follows from a comparison of the work of Eliashberg (see §10), the work of Fuks [72], and the computations of Vasil'ev [69], according to which the six-dimensional generator of the cohomology group of a Lagrangian Grassmannian induces a class dual to P_8, while the five-dimensional generator induces a class dual to A_6.

But even independently of whether or not the cohomology classes of the Lagrangian or Legendrian manifolds that arise are trivial, the complexes themselves constructed from the singularity and multi-singularity classes (and not only their cohomology groups) contain in concentrated form important information on the adjacencies to each other of the various strata of caustics and fronts of Lagrangian and Legendrian maps.

It seems certain that in the combinatorics of the adjacencies of strata there are hidden algebraic structures still not understood—I mention only that the number of classes of critical points of multiplicity $\mu = 2^a$ is (within the limits of the tables, that is, for $\mu \leqslant 16$) a power of 2, while there is an almost complete parallelism between the beginning of the hierarchy of degenerate critical points and the "admissible sequences" of Steenrod $(x_1 \geqslant 2x_2 \geqslant 4x_3 \geqslant ...)$:

$A_2\ A_3\ A_4\ A_5\ A_6\ A_7\ A_8\ \ldots$	1 2 3 4 5 6 7 ...
$D_4\ D_5\ D_6\ D_7\ D_8\ \ldots$	2,1 3,1 4,1 5,1 6,1 ...
$E_6\ E_7\ E_8\ \ldots$	? 4,2 5,2 ...
$P_8\ \ldots$	4,2,1 ...

(the number of terms of a sequence is equal to the corank of the singularity, and the sum is equal to the codimension of the orbit; the gap at E_6 is explained, possibly, by the relation $A_6 \sim E_6$ in the Vasil'ev complex).

Chapter VI

PROJECTIONS

Projections surround us on all sides. For example, we distinguish objects by the apparent contours of their projections on the retina of the eye, that is, by the sets of critical values of the corresponding projection maps.

The theory of the singularities of projections lies also at the base of the theory of the bifurcations of equilibrium positions of dynamical systems. In this case the projected manifold lies in the product of the parameter space and the phase space of the dynamical system and is formed by the equilibrium positions corresponding to all values of the parameters. The singularities of the projection of the equilibrium manifold to the parameter space are responsible for bifurcations of the equilibrium positions under a change of the parameters.

The difference between the theory of singularities of projections of submanifolds and the general theory of singularities of maps is in the smaller stock of singularities and, this is the main difference, the narrower equivalence relation: two projections of submanifolds from the spaces of bundles onto the base are regarded as equivalent if one goes into the other under a diffeomorphism of the space of the first bundle to the space of the second that takes a fibre of the first into a fibre of the second.

In this chapter a survey is given of work on the classification of the singularities of projections up to equivalence.

§15. Singularities of projections of surfaces to a plane

We consider a surface in three-dimensional space. By the classical theorem of Whitney, under its projection to a plane in the generic case the singularities are only folds (on curves) and pleats (Whitney cusps) (at individual points). However, if we project the surface not in a generic, but in a specially chosen direction, then we may get more complicated singularities. In other words, if we look at a generic surface from a suitable point of view, then we may see an unusual picture. It turns out that there are fourteen different local views of a generic surface; their classification is based on an analysis of vector fields tangent to generalized swallow-tails.

Let us make precise what "local views" of a surface we regard as different.

By the *projection* of a surface M in $\mathbf{R}P^3$ from a point O not on the surface we mean a diagram $M \to \mathbf{R}P^3 \setminus O \to \mathbf{R}P^2$, where the first arrow is inclusion and the second the projection that associates with each point the line joining it to the centre of projection O. The germ of the projection at a point is the germ of the diagram.

An *equivalence* of two (germs of) projections is a commutative 3×2 diagram whose horizontals are the projection germs and whose verticals are diffeomorphisms.

The plane $\mathbf{R}P^2$ to which we project is called the *base* of the projection. Thus, the equivalence in question is a diffeomorphism of the three-dimensional space containing the surface *fibred over the base.*

The composite maps $M \to \mathbf{R}P^2$ for equivalent projections are left-right equivalent (they go over into each other under diffeomorphisms of the source and target). The apparent contours and the sets of critical values of equivalent projections are diffeomorphic. Therefore, the classification of projection-germs up to equivalence leads to a classification of germs of apparent contours (and sets of critical values) up to diffeomorphism. However, the classification of projections up to our fibred equivalence is a priori finer than that of the singularities of apparent contours or even that of the composite maps up to left-right equivalence.

According to computations of Platonova [73], corrected by Shcherbak [74], [8], the hierarchy of projection-germs of a generic surface is as follows:

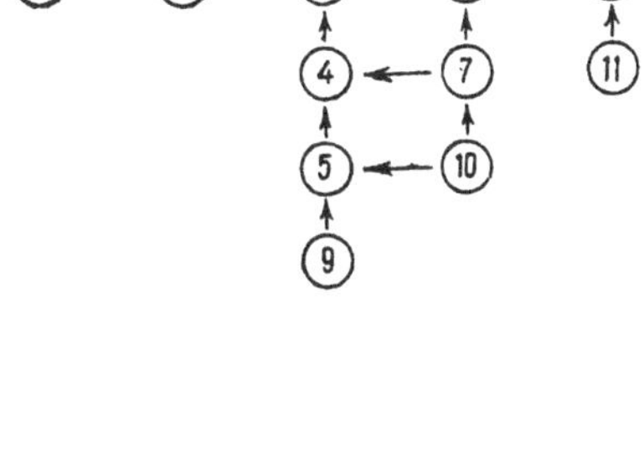

Here the numbers denote projection-germs equivalent to projection-germs at zero of the surface $z = f(x, y)$ by the pencil of lines parallel to the x-axis, where, in the non-singular case 1, $f = x$, while the remaining functions f are given by the following table:

type	$f(x, y)$	type	$f(x, y)$
2	x^2	7	$x^4+x^2y+xy^2$
3	x^3+xy	8	$x^5 \pm x^3y+xy$
4	$x^3 \pm xy^2$	9	$x^3 \pm xy^4$
5	x^3+xy^3	10	$x^4+x^2y+xy^3$
6	x^4+xy	11	x^5+xy

Theorem. *For any generic surface and for any centre of projection in three-dimensional space the projection germ is equivalent to one of these fourteen germs at zero.*

The singularities of the apparent contours and sets of critical values corresponding to the singularities listed are shown in Fig. 22.

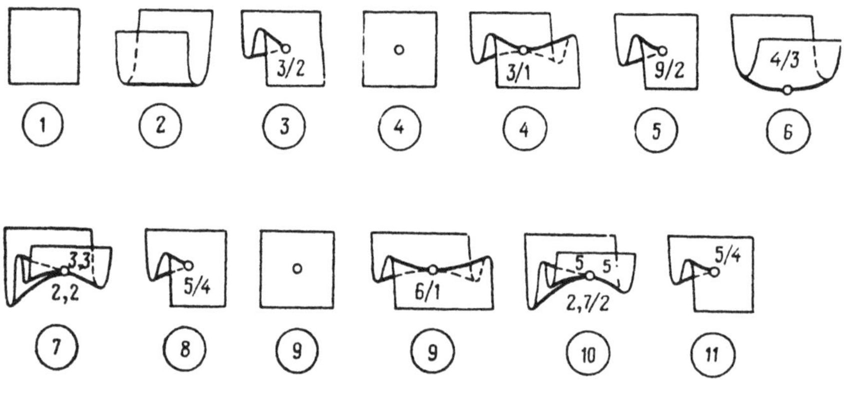

Fig. 22.

For projection from a generic centre the projection-germ at a generic point of the surface belongs to type 1, at points of some curve to type 2 (fold), and at particular points to type 3 (pleat or Whitney cusp). Namely, pleats occur for projection along an asymptotic ray of the surface. In other words, to see a pleat it is enough to look at a generic surface along one of its asymptotic directions.

To describe the remaining singularities, which occur only for specially chosen centres of projection, we require a projective classification of the points of a generic surface in three-dimensional space (Platonova [11],

Landis [10]; Fig. 23).

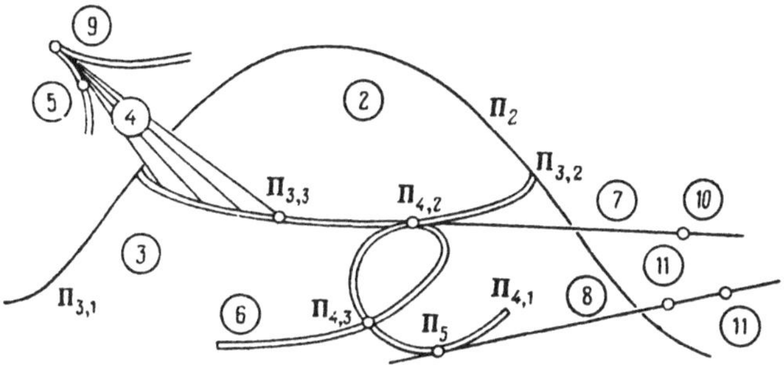

Fig. 23.

A smooth *curve of parabolic points*(1) $\Pi_{3,2}$ divides such a surface into a *domain of elliptic points* Π_2 (there are no real tangent lines of higher than the first order) and a *domain of hyperbolic points* $\Pi_{3,1}$ (there are two such tangents; they are called *asymptotic lines* and their directions at the point of tangency *asymptotic directions*).

In the domain of hyperbolic points there is also a distinguished smoothly immersed *curve of asymptotic inflections* $\Pi_{4,1}$ (at its points the order of tangency of an asymptotic line with the surface is greater than two). On this curve there lie particular *bi-inflection points* Π_5 (the order of tangency is four) and points of *self-intersection* $\Pi_{4,3}$ (both the asymptotic lines have third order tangency). At the points $\Pi_{4,2}$ the curve of asymptotic inflections is simply tangent to the curve of parabolic points (at such a point the unique asymptotic direction is tangent to the curve of parabolic points). Finally, on the curve of parabolic points there are also individual distinguished points $\Pi_{3,3}$ about which we shall say something below.

Therefore, the hierarchy of classes of points on a generic surface is as follows:

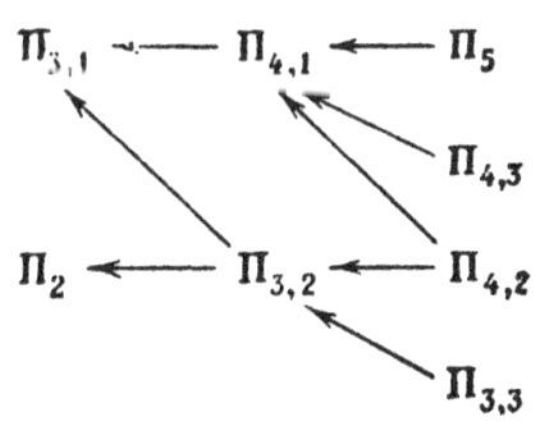

The projectively dual surface has an edge of regression at points corresponding to $\Pi_{3,2}$ and $\Pi_{3,3}$, and a swallow-tail at points corresponding to $\Pi_{4,2}$.

(1)The notation is such that the first index indicates the maximal multiplicity of the intersection of the surface with a real tangent line (the order of tangency is 1 less).

Platonova has found normal forms to which the p-jets of the surface at points of each class can be reduced by projective transformations:

class	normal form	restrictions	p	codim
Π_2	x^2+y^2	—	2	0
$\Pi_{3,1}$	$xy+ax^3+y^3$	$a \neq 0$	3	0
$\Pi_{3,2}$	$y^2+x^3+xy^3+ax^4$	—	4	1
$\Pi_{4,1}$	$xy+y^3+x^4+hx^3y$	—	4	1
$\Pi_{4,2}$	$y^2+x^2y+vx^4$	$v \neq 0$	4	2
$\Pi_{4,3}$	$xy+x^4+ax^3y+bxy^3+cy^4$	$c \neq 0$	4	2
Π_5	$xy+y^3 \pm x^3y+\varphi_5$	$\varphi_5(0, y) \neq 0$	5	2
$\Pi_{3,3}$	$y^2+x^3+ax^4+by^5 \pm xy^4+x^2\varphi_3$	—	5	2

Here φ_r is a homogeneous polynomial of degree r in x and y.

Theorem. *For a generic surface its p-jet at each point can be reduced by a projective transformation to one of the normal forms in the table.*

Remark. Kulikov has computed the number of points of the classes listed and the degrees of the curves formed by points of the various classes on an algebraic surface of degree d in $\mathbf{C}P^3$. His results are as follows [75]:

Π_5	$\Pi_{4,3}$	$\Pi_{4,2}$	$\Pi_{3,3}$
$5d(d-4)(7d-12)$	$5d(7d^2-28d+30)$	$2d(d-2)(11d-24)$	$10d(d-2)(7d-16)$
curve	$\Pi_{3,2}$	$\Pi_{4,1}$	
degree	$4d(d-2)$	$d(11d-24)$	

Corollary. *On a real surface of sufficiently high odd degree in* $\mathbf{R}P^3$ *there is at least one real point of asymptotic bi-inflection, at least one point of inflection of both asymptotic directions, and at least one curve of parabolic points.*

We return to the description of singularities of projections from non-generic centres (Fig. 23). The singularities **4** and **6** are realized for projections from points of certain surfaces (namely, the ruled surfaces generated by the asymptotic lines at points of the parabolic curve and of the curve of asymptotic inflections, respectively). Thus, to see the projection singularity **4** one has to look at a generic surface from a generic point of the asymptotic line at a generic parabolic point, while to see the singularity **6** one has to look from a generic point of an asymptotic tangent of the third order.

The singularities **5**, **7**, and **8** are realized for projections from points of certain curves. Namely, to see the singularity **5** one has to look at the surface from a generic point of the edge of regression of the surface generated by the asymptotic lines at parabolic points, because this ruled surface is developable and has an edge of regression.

The singularity **7** is realized by projection from a generic point of an asymptotic line touching the curve of parabolic points (that is, drawn through a point of type $\Pi_{4,2}$), while the singularity **8** is realized by projection from a generic point of an asymptotic tangent of the fourth order.

The singularities **9**, **10**, **11**, which are harder to observe, are realized only by projections from particular points. Namely, the singularity **9** is realized by projection from a cusp of the edge of regression of the developable surface described above (the corresponding asymptotic line passes through a point of type $\Pi_{3,3}$). The singularity **11** is realized by projection from two "focal" points on an asymptotic tangent of the fourth order, while the singularity **10** is realized by projection from a special point on an asymptotic line touching the curve of parabolic points (at a point of type $\Pi_{4,1}$).

The connection between the classifications of projections and of vector fields tangent to swallow-tails consists in the following. We regard the intersection of one of the lines of projection with the surface as a zero-dimensional submanifold of the line. For a generic line the intersection is transversal, but at places of singularities the points of intersection coincide with one another. Thus, each singularity of the projection of a surface in three-dimensional space to the plane defines a two-parameter deformation of the zero-dimensional submanifold of the line consisting of μ coincident points (where μ is the multiplicity of the intersection of the line of projection with the surface at the point of tangency in question): the two parameters indicate the number of the line in the projecting bundle.

This deformation, like any other, is equivalent to one induced from a versal deformation. Consequently, it determines a map of the base of the deformation (the germ of $\mathbf{R}P^2$ at the point corresponding to the relevant line) into the base of a versal deformation (that is, the space in which the generalized swallow-tail of the singularity A_μ lies). For example, if $\mu = 3$, then we obtain a surface passing through the vertex of the ordinary swallow-tail in $\mathbf{R}^3$.

Under the action of a diffeomorphism of the base of the versal deformation preserving the (generalized) swallow-tail, the surface thus constructed is carried to a new surface. This new surface determines a new two-parameter deformation of the μ-fold point, and consequently, a new projection, which is equivalent to the original one (in the sense of our equivalence relation, see [55]).

Thus, to reduce the projection to normal form it is sufficient to reduce to normal form the corresponding surface in the space of the generalized swallow-tail by a diffeomorphism preserving the swallow-tail. The latter reduction can be achieved at once with the help of the vector fields tangent to the swallow-tail and of the standard quasihomogeneous technique (see [55], [47]).

The projections of a surface from all possible points of three-dimensional space form a three-parameter family. In a generic three-parameter family of projections of surfaces to the plane there occur not only singularities from the list **1**–**11**, but also some others. According to computations of Goryunov, these are equivalent to the germs at zero of projections of the surfaces $z = x^4 \pm x^2y^2 + xy^2$, $x^5 + x^3y + x^2y$ and $x^6 + ax^4y + x^3y + xy$ along the x-axis.

The theorem of Platonova and Shcherbak shows that such germs cannot be realized as singularities of projections of generic surfaces by pencils of lines.

Let us consider, for example, the latter singularity, with $\mu = 6$. For a generic projection there occur pleat points, at which $\mu = 3$. Therefore, in generic three-parameter families of projections there are necessarily (for particular values of the parameters) points with $\mu = 6$. At the same time, a generic surface does not have tangent lines with tangency higher than the fourth order (that is, intersection multiplicity $\mu > 5$): this follows from the fact that the manifold of tangents is three-dimensional, and every increase in the order of tangency by 1 imposes one further condition on the tangent. Thus, in the projections of a generic surface from all points of three-dimensional space a singularity with $\mu = 6$ cannot occur.

This argument shows also that a singularity with $\mu = 6$ does not occur even for projections along the extremals of any variational problem (not necessarily straight lines) and even along the curves of any four-parameter family of curves in three-dimensional space.

For similar reasons the remaining two singularities of codimension 3 of Goryunov's list cannot be realized either (as Shcherbak has shown): they are also determined by properties of the line, but not by properties of a point on it. The singularity of the type $x^5 + x^3y + x^2y$ can arise only by projection from points of an asymptotic line having tangency of the fourth order with the surface and applied at a parabolic point. A generic surface does not have such tangents (from dimensional arguments).

The singularity of the type $x^4 \pm x^2y^2 + xy^2$ can arise only for projections from points of an asymptotic line having tangency of the third order with the surface and applied at a point where the parabolic curve fails to be smooth.

Projections by pencils of *parallel* lines were classified in [55] (10 types, excluding the four singularities **9**, **10**, **11**). This list coincides with that of singularities in two-parameter families of generic projections (Goryunov) and with the list of singularities of left-right codimension not greater than two (see [55], [12]).

If we look at a surface from one of the very special points (**9**, **10**, **11**), then by slightly turning the head we can see the surface as shown in the drawings of Figs. 24–27 composed by Shcherbak. The middle of each

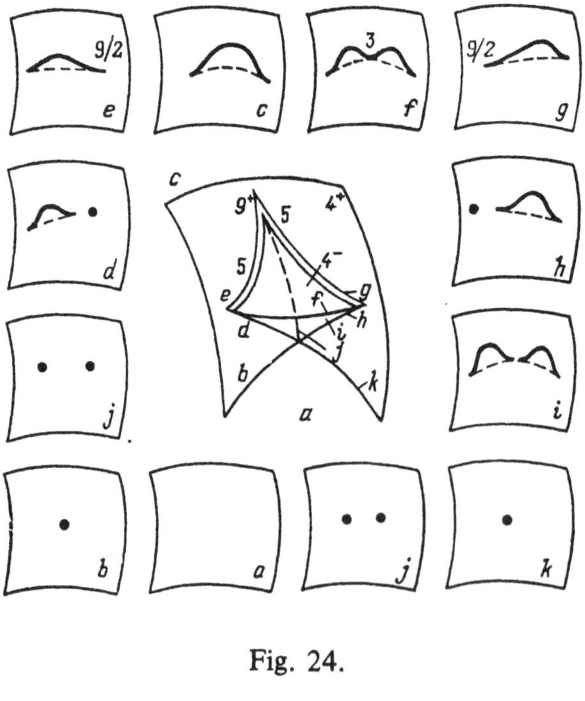

Fig. 24.

picture shows the bifurcation diagram formed by the centres of the non-generic projections, and the domains of its complement corresponding to the various pictures are marked.

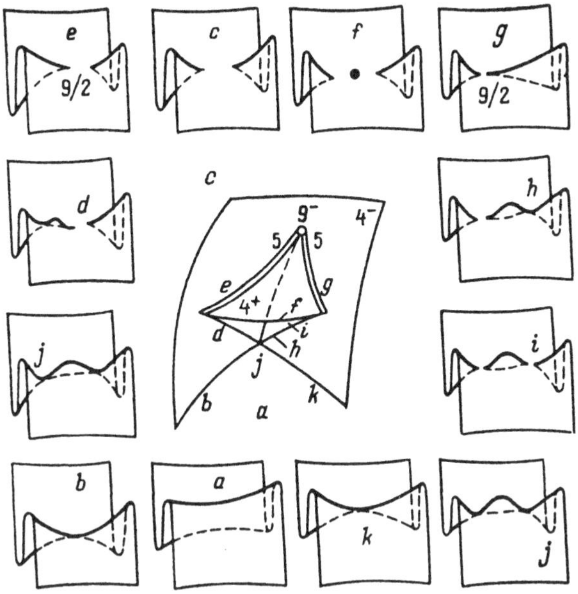

Fig. 25.

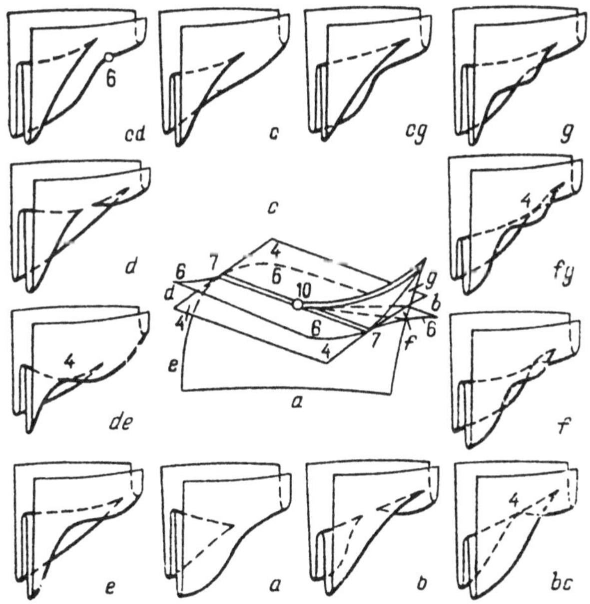

Fig. 26.

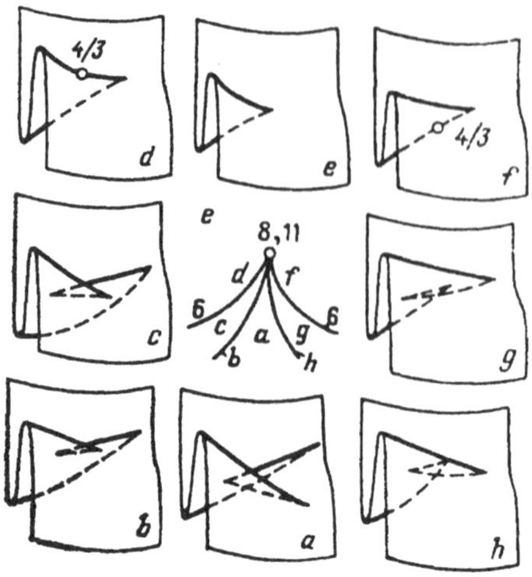

Fig. 27.

§ 16. Singularities of projections of complete intersections

The classification of the singularities of various objects shows that algebraically more natural answers are obtained when we study the so-called simple singularities, that is, singularities not having moduli. Thus, the classification of the simple critical points of functions and simple singularities of hypersurfaces (and also the classifications equivalent to these of simple Lagrangian and Legendrian singularities, or the singularities of caustics and wave-fronts) reduce to the list A_μ, D_μ, E_6, E_7, E_8 of Dynkin diagrams with simple bonds (the angles between the non-orthogonal simple roots are 120°) [7]. The classification of the simple critical points of functions on manifolds with boundary reduces to the same list complemented by the diagrams B_μ, C_μ, F_4 (admitting also the angle 135°) [59]. Therefore, it is natural in each classification problem of the theory of singularities first of all to look for the simple objects. Moreover, for each simple object one should at the same time establish its bifurcation diagram: these diagrams serve in their turn as finger-prints of the singularities. For example, it is just the fact that the topologies of the bifurcation diagrams are the same that allows us to identify the list B_μ, C_μ, F_4 in the list of boundary singularities [59].

A classification of the simple singularities of the projections $V \to E \to B$ of a submanifold V to the base of the bundle $E \to B$ up to fibred diffeomorphisms of the bundle space over the base has been made by Goryunov [76].

He considered not necessarily smooth submanifolds V, but assumed that V is a complete intersection (that is, given by the same number of equations in E as its codimension). These assumptions are natural from the point of view of the theories of the metamorphoses and cobordisms of projections, where degenerate projections appear for bifurcating values of the parameters on which the projection depends (that is, in one of the fibres of $V \to E \to B$ of the right fibration of the diagram $V' \to E' \to B' \to P$, where V' is a smooth submanifold of the space of the bundle $E' \to B'$).

A *suspension* over the projection $V \to E \to B$ is defined to be an embedding of E as a sub-bundle in the space of a large bundle with the same base. A *stable equivalence* is one of appropriate suspensions. A projection is said to be "*onto*" if the dimension of the projected space V is not less than that of the base B.

The projection of V onto B determines a family of submanifolds in the fibres of the bundle $E \to B$ (namely, the family of intersections of V with the fibres). The germ of the projection $V \to E \to B$ can be regarded as a deformation of the submanifold V_0 of a distinguished fibre (the codimension of V_0 in the fibre is equal to that of V in E).

***Theorem* 1** (see [76]). *A simple germ of a projection "onto" is stably fibre-equivalent to a deformation either of a hypersurface or of a curve in three-dimensional space or of a (multiple) point of the plane.*

Goryunov has found a list of normal forms of simple singularities in all three cases. Some of his lists can be identified with the lists of simple singularities in other classification problems. For example, he has proved the following result.

***Theorem* 2.** *A deformation of the germ of a curve in* $\mathbf{C}^3$ *that is not equivalent to a plane curve determines a simple projection-germ if and only if it is a versal deformation of a simple singularity of a curve in* $\mathbf{C}^3$.

A list of non-planar simple singularities of curves in $\mathbf{C}^3$ has been published by Giusti [77]. It consists of an infinite series of curves S_μ: $x^2 + y^2 + z^{\mu-3} = yz = 0$, $\mu \geqslant 5$, and also the 10 exceptional curves $f_1 = f_2 = 0$ of the following table:

type	f		type	f	
T_7	$x^2+y^3+z^3$,	yz	U_9	x^2+yz,	$xy+z^4$
T_8	$x^2+y^3+z^4$,	yz	W_8	x^2+z^3,	y^2+xz
T_9	$x^2+y^3+z^5$,	yz	W_9	x^2+yz^2,	y^2+xz
U_7	x^2+yz,	$xy+z^3$	Z_9	x^2+z^3,	y^2+z^3
U_8	x^2+yz+z^3,	xy	Z_{10}	x^2+yz^2,	y^2+z^3

I recall the definition of a versal deformation of a germ of a submanifold at a point (for motivation, see [8]).

A deformation of a germ $\{x\colon f(x) = 0\}$ is given by a germ $F(x, \lambda)$ such that $F(x, 0) = f(x)$. A deformation is *versal* if the images of the vectors of the initial velocities of the deformation $(\partial F/\partial \lambda_i\,|_{\lambda=0})$ in the quotient module $\mathcal{O}^m/\langle \partial f/\partial x_i,\ f_i e_j\rangle$ generate it as a space over $\mathbf{C}$ (here $\mathcal{O} = \mathbf{C}[\![x]\!]$ is the ring of Taylor series at the origin, $\mathcal{O}^m = \oplus \mathcal{O} e_i$ is the free $\mathcal{O}$-module with basis $\{e_i\}$, and the diamond brackets denote the submodule generated by the elements contained within them).

Example. For the curve S_5 ($f_1 = x^2+y^2+z^2$, $f_2 = yz$) we may take as a versal deformation $F_1 = f_1+\lambda_1$, $F_2 = f_2+\lambda_2+\lambda_3 x+\lambda_4 y+\lambda_5 z$. The projection of the surface $F_1 = F_2 = 0$ to the λ-space has a simple singularity at zero. Thus, Giusti's list presented above is part of Goryunov's list (and consists of the part that refers to projections of spaces of the corresponding dimensions).

From Theorem 2 and Giusti's list there follows the smoothness of the manifolds V for the simple projections $V \to E \to B$ described in the theorem that are stably inequivalent to projections of hypersurfaces (that is, such that V is not contained in the space of a subbundle of codimension 1 in E). For the other simple projections the complete intersection V happens not to be smooth.

Goryunov's list of simple projections is very extensive, and we limit ourselves to projections to the complex line.(1)

Theorem 3 ([76], [79]). 1) *For a simple projection-germ to the line the projected manifold (up to stable fibre-equivalence) is either a hypersurface, or a (singular) curve in three-dimensional space, or a (multiple) point of the plane or three-dimensional space.*

2) *A simple projection-germ of a hypersurface in the space of the bundle* $(x, u) \mapsto u$ *is reducible by a complex fibred diffeomorphism* $(x, u) \mapsto (h(x, u), v(u))$ *to the projection to the u-axis of the germ at zero of one of the hypersurfaces of the following list:*

type	hypersurface	type	hypersurface
A_μ, $\mu \geqslant 0$	$u+x_1^{\mu+1}+q=0$	E_8	$u+x_1^3+x_2^5+q=0$
D_μ, $\mu \geqslant 4$	$u+x_1^2x_2+x_2^{\mu-1}+q=0$	B_μ, $\mu \geqslant 2$	$x_1^2+u^\mu+q=0$
E_6	$u+x_1^3+x_2^4+q=0$	C_μ, $\mu \geqslant 3$	$x_1^\mu+ux_1+q=0$
E_7	$u+x_1^3+x_1x_2^3+q=0$	F_4	$x_1^3+u^2+q=0$,

where q denotes the sum of the squares of the unwritten variables ($q = x_2^2 + \ldots + x_n^2$ *for* A_μ, B_μ, C_μ, F_4; $q = x_3^2 + \ldots + x_n^2$ *for* D_μ, E_μ).

3) *The simple projections of germs of curves from three-dimensional space to the line form two series that reduce to the projections to the u-axis of the germs at zero of curves of the following list:*

type	curve
$C^{k,l}$, $2 \leqslant k \leqslant l$	$xy=0$, $x^k+y^l+u=0$
F_{2k+1}, $2 \leqslant k$	$x^2+y^3=0$, $y^k+u=0$
F_{2k+4}, $1 \leqslant k$	$x^2+y^3=0$, $xy^k+u=0$.

4) *The simple projections to the line of zero-dimensional complete intersections, lying in the plane* $\mathbf{C}^2$ *or the space* $\mathbf{C}^3$, *are equivalent to the projections* $(x, u) \mapsto u$ *or* $(x, y, u) \mapsto u$ *of multiple points, given by the*

(1)In [78] there are also tables of real normal forms (which are longer, since some complex singularities have several real forms).

equations of the following list:

type	point	restrictions
A_μ	$x=u^{\mu+1}=0$	$\mu\geqslant 0$
$X_{k,l}$	$x^k+u=x^l=0$	$2\leqslant k\leqslant l$
U_μ	$x^2+u^{\mu-2}=ux=0$	$4\leqslant\mu$
V_5	$x^2=u^2=0$	—
V_6	$x^3+u^2=ux=0$	—
Γ_{ab}	$xy=x^2+y^a=u+y^b$	$a=2,\ b=2,\ 3$ or $a=3,\ b=2,\ 3,\ 4.$

We list some consequences of this classification.

1°. Boundary singularities.

The simple singularities of projections onto a line are classified, up to complex stable equivalence, by the Weyl groups A_μ, ..., F_4, that is, the same as the boundary singularities(1) [59]. The connection with boundary singularities is explained by the fact that the list of hypersurfaces that are simple relative to our fibred equivalence over B coincides with the list of hypersurfaces that are simple relative to the group of diffeomorphisms preserving the hypersurface "boundary" $u = 0$ (or even with the list of "boundary singularities" of the functions specifying these hypersurfaces).

We remark incidentally that Matov [80], [81] has recently completed the classification of the unimodal boundary singularities begun in [59] and has classified the bimodal ones; for the fibred equivalence considered here there is as yet no such classification.

Recently Shcherbak has discovered an interesting duality for boundary singularities. With every boundary singularity there are associated two ordinary singularities: one of them is the singularity of the function on the ambient manifold (ignoring the boundary), and the other is the singularity of the restriction to the boundary. It turns out that these two functions can change places. For every boundary singularity there is a dual singularity for which the restriction of the function to the boundary becomes the function on the ambient manifold, while the function on the ambient manifold becomes the restriction to the boundary (of course, after the interchange the functions have to be stabilized by adding squares of new variables). The duality of which we are speaking was in essence noticed already by Lagrange, whose "multiplier rule" converts boundary extrema to ordinary ones.

If $f(x;\ y)$ ($x \in \mathbf{R}$, $y \in \mathbf{R}^{n-1}$) is a function on a manifold with boundary $x = 0$ (more precisely, the germ of the function at the critical point 0), then the Lagrange-dual function is given by the formula $f^*(z;\ x,\ y) = zx+f(x,\ y)$ (the equation of the boundary is $z = 0$). It is easy to see that the restriction of f^* to the boundary is stably equivalent to f, while the restriction of f to the boundary is stably equivalent to f^*; moreover, f^{**} is stably equivalent to f (as functions on a manifold with boundary).

The bifurcation diagrams of f and f^* are diffeomorphic, but the components forming them are interchanged (one corresponds to the degeneracy of the zero level manifold and the other to the non-transversality to the boundary).

Thus, the boundary singularities of functions are by their nature "products" of two ordinary singularities. For example, $C_\mu \sim (A_1,\ A^{\mu-1})$, $B_\mu \sim (A_{\mu-1},\ A^1)$, $F_4 \sim (A_2A^2)$ (the Milnor number is written on top for the restriction to the boundary and at the bottom for a singularity in the ambient space).

(1)The ordinary (not boundary) singularities A_μ, ..., E_8 correspond here to the projections of *smooth* hypersurfaces.

The "product" is not determined uniquely by its multiplier, and not every two critical points can be multiplied. For example, A_2 cannot be multipled by A^3 and, in general, A_{2k} cannot be multiplied by A^p with $p > 2k$. The products of A_p by A^p form a family of m-modal singularities, $m = [(p-1)/2]$:

$$F_p^p = x^2 + y^{p+1} + xa(y), \quad a(y) = a_1 y^{p-1} + \ldots + a_m y^{p-m}.$$

Shcherbak has found all "products" of simple singularities.

The Dynkin diagram of a *simple* product splits into the Dynkin diagrams of its factors by deleting the double bond (the bond is directed towards the diagram of the restriction of the function to the boundary). The modality of the product (in examples) is not less than the sum of the modality of the factors.

2°. Projections of curves.

The simple projections of curves form two series: a doubly-indexed series C and a simply-indexed series F, so denoted because they generalize the singularities C_μ and F_4 of the list of hypersurfaces.

Goryunov has extended to these cases the usual theorems in the theory of singularities on bifurcation diagrams, and has also constructed Dynkin diagrams for them (Fig. 28). Thus, for example, the "exceptional" Lie algebra F_4 actually turns out to be the ancestral member of the whole series F_μ of related objects.

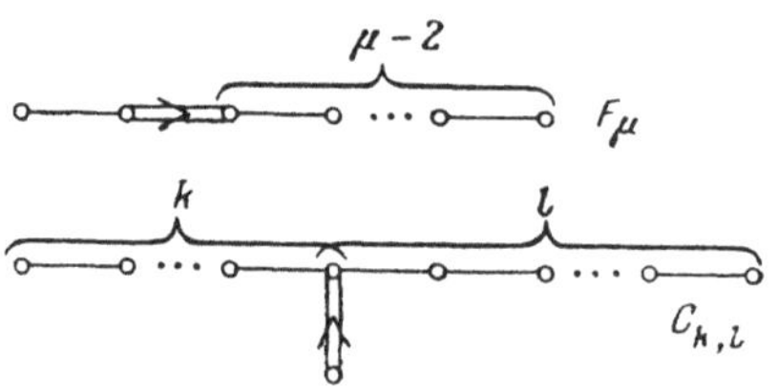

Fig. 28.

3°. The intersection form.

The intersection form for a projection to the line $V \to E \to \mathbf{C}$ is determined by the same construction as was used in [59] in the case of boundary singularities. We fix a (sufficiently small) ball of radius r with centre at the origin in the space E. We consider apart from the projected complete intersection V_0 (given in this ball by the equation $f = 0$) the neighbouring non-singular complete intersection V_ε (given by the equation $f = \varepsilon$, where ε is a point sufficiently close to the origin; the smallness of $|\varepsilon|$ is chosen to depend on r). For a generic ε the manifold V_ε intersects transversally the fibre of the bundle $E \to \mathbf{C}$ passing through the origin in a submanifold V_ε^0 of complex codimension 1 in V_ε. Now we form the two-sheeted covering $\tilde{V}_\varepsilon \to V_\varepsilon$, branched along V_ε^0. We denote by $\tilde{H}$ the group of integral homology of middle dimension of the manifold $\tilde{V}_\varepsilon$; interchange of the sheets of the covering acts on $\tilde{H}$ as an involution. We denote by H^- the anti-invariant part of $\tilde{H}$. The intersection form on $\tilde{H}$ induces a bilinear form on H^-. It is called the *intersection form of the projection.*

4°. The Dynkin diagram.
If the complex dimension of V is even, then the intersection form is symmetric, and the Dynkin diagram with short and long roots is defined in the usual way (see [59]). But if the dimension of V is odd (for example, if V is a curve, as will be in the case of interest to us of the singularities F_μ and $C_{k,l}$), then we have to take care of the difference between short and long roots in a special way. For boundary singularities of functions we introduce a stabilization operation (by adding the square of a superfluous variable), turning a function with a skew-symmetric intersection form into one with a symmetric form. In the present case of complete intersections of codimension greater than 1 there is no such stabilization (or rather none is known). Therefore, to distinguish short from long roots we proceed as follows.

Apart from the part H^- of the group $\widetilde{H}$ that is anti-invariant with respect to the involution, we consider the invariant part H^+. The subgroup H^- naturally projects into the quotient group $\widetilde{H}/H^+$. We say that an anti-invariant element $h \in H^-$ is *short* if its image in $\widetilde{H}/H^+$ is not divisible by 2, and *long* if it is.

The Dynkin diagram of a skew-symmetric form is a graph whose vertices represent basis vectors; two vertices are joined by the same number of edges as the value of the form on the given pair of basis vectors. The edge is so oriented that this value is positive. An edge joining a long element of the basis to a short one is furnished with the symbol $<$, open in the direction of the long element. When the graph is a tree, the orientations of the edges are not shown (because one can take them at will, by changing the signs of the basis elements). With these conventions the Dynkin diagrams of the boundary singularities $A_\mu, \ldots, F_4$ are the ordinary Dynkin diagrams of the corresponding Lie algebras.

Theorem (see [76]). *The intersection forms of the singularities F_μ and $C_{k,l}$ are given in suitable bases by the Dynkin diagrams of Fig.* 28.

5°. Versal deformations of projections.
In fact, in [76] more is proved: the bases mentioned can be chosen to be *distinguished.* We adapt to the case of projections of complete intersections the usual definition of a distinguished basis (see [59]). We consider a versal deformation of the projection of a complete intersection V, given in E by a system of equations $f(x, u) = 0$, to the base B of the bundle $(x, u) \mapsto u$. A deformation can be given by equations $F(x, u, \lambda) = 0$, where λ is the parameter of the deformation, ranging over a neighbourhood of the origin in some finite-dimensional space; here $F(x, u, 0) \equiv f(x, u)$. The versality of the deformation is defined as usual (see [59]). As a versal deformation we can take a deformation for which

$$F = f + \sum \lambda_i g_i(x, u),$$

where the g_i determine a $\mathbf{C}$-basis of the quotient-space.(1)

$$Q_f = \{\alpha\}/\{\sum f_i\beta_i + \sum \gamma_j \partial f/\partial x_j + \sum \delta_s \partial f/\partial u_s\};$$

here $\alpha, \beta, \gamma, \delta$ are functions of x and u that are holomorphic at 0 (α and each of the β_i are vector-valued functions with the same number of components as f, while each of the γ_j and δ_s is a scalar), and $\delta_s = \delta_s(u)$ does not depend on x.

Example. The curve $C_{2,2}$ in $\mathbf{C}^3$ is given by the equations $f_1 = f_2 = 0$, where $f_1 = xy$, $f_2 = x^2+y^2+u$. After some stereometric exercises (the "crossword" technique, see [8]) we find: $\dim_{\mathbf{C}} Q_f = 3$ and a three-parameter versal deformation is determined by the formulae

$$F_1 = f_1 + \lambda_1 + \lambda_2 x + \lambda_3 y, \quad F_2 = f_2.$$

6°. Vanishing cycles and semicycles.

From the map F that determines the versal deformation we construct the maps

$$F_\lambda\colon (x, u) \mapsto F(x, u, \lambda), \quad F_\lambda^0\colon (x, 0) \mapsto F(x, 0, \lambda).$$

The sets of critical values of these maps are hypersurfaces in the space of values of the map F. If a point ε of this space is not a critical value of the map F_λ then the manifold $V_{\lambda,\varepsilon}$, given in the space $\{(x, u)\}$ by the equation $F_\lambda = \varepsilon$ is non-singular, while if ε is a non-critical value also for F_λ^0, then this manifold $V_{\lambda,\varepsilon}$ transversally intersects the hyperplane $u = 0$ in a submanifold $V_{\lambda,\varepsilon}^0$.

From the pair $(V_{\lambda,\varepsilon}, V_{\lambda,\varepsilon}^0)$ we can construct an intersection form in the same way as we did in 3° for the pair $(V_\varepsilon, V_\varepsilon^0)$ corresponding to $\lambda = 0$. The points (λ, ε) and $(0, \varepsilon)$ can be joined by a path avoiding the hypersurfaces of critical pairs. Over this path the pairs thus constructed form a smooth bundle. The trivialization of this bundle (which is stationary in a neighbourhood of the boundary) allows us to coordinate all the constructions made earlier for $\lambda = 0$ with the analogous constructions for $\lambda \neq 0$. In particular, the path described determines an isomorphism of the two-sheeted coverings, the anti-invariant homology groups H^- for various (λ, ε), and the intersection forms on them.

We use it to define vanishing cycles and semicycles on complete intersections on a manifold with boundary.

We choose a generic point λ near to the origin and draw through a non-critical point ε a generic complex line. The complete inverse image of this line under the map F_λ is a smooth manifold whose dimension exceeds that of $V_{\lambda,\varepsilon}$ by 1; in it there is distinguished the smooth hypersurface $u = 0$ (the complete inverse image of the same line under the map F_λ^0).

(1)The first term of the denominator is generated by the infinitesimal changes of the system of equations $f = 0$ that do not change V, and the others by the infinitesimal diffeomorphisms of F fibred over B.

Suppose that this line intersects transversally the set of critical values of F_λ in a points and the set of critical values of F_λ^0 in a^0 points (all the points are in general position). We move ε to these critical points (by non-intersecting paths, as can always be done by the definition of distinguished bases). As ε approaches any of the a points of the first type on $V_{\lambda,\varepsilon}$, the Picard-Lefschetz cycle vanishes, and as it approaches any of the a^0 points of the second type, the "semicycle" [59] realizing the relative homology class in $H(V_{\lambda,\varepsilon}, V_{\lambda,\varepsilon}^0)$ vanishes. These $a+a^0$ cycles and semicycles generate the whole relative homology group, but they need not be independent. From each of the vanishing cycles (or semicycles) we construct in the usual way (see [59]) an anti-invariant cycle in $H_{\tilde{\lambda},\varepsilon}^-$. These cycles are said to be *distinguished* (a vanishing cycle generates a long distinguished cycle and a semicycle a short one). The Dynkin diagrams in Fig. 28 are in the basis of distinguished cycles (obtained by discarding the superfluous distinguished cycles from the $a+a^0$ cycles we have constructed).

Remark 1. The need to discard the superfluous cycles is clearly a defect of this construction—it is a manifestation of an analogous defect in the general theory of homology groups of complete intersections: the distinguished bases should probably be constructed not in the homology groups of the complete intersection itself, but for a whole flag of submanifolds lying inside each other with codimension 1 all at the same time.

We regard a non-singular complete intersection (a Milnor fibre) as a level set of a function on a complete intersection of dimension greater by 1. The critical points of this function determine vanishing cycles in the homology groups of middle dimension of the Milnor fibre and even distinguished systems of such cycles. However, these systems, generally speaking, are not bases (there are too many critical points). We consider, therefore, the free Abelian group generated by the vanishing cycles of the system, together with the homomorphism to the homology of the Milnor fibre that takes the generators to their homology classes. To study the kernel of this homomorphism we must look at the homology of the complete intersection of dimension greater by 1 on which the relevant Milnor fibre lies. Repeating the preceding construction, we obtain again a free Abelian group generated by vanishing cycles (of dimension greater by 1), and so on. As a result we obtain the "resolution", on each level of which there are an intersection form, distinguished systems of vanishing cycles, monodromy, a mixed Hodge structure, and so on. Unfortunately, neither the geometry nor the algebra of this situation has been studied in as much detail as they deserve.

Remark 2. For projections of complete intersections to the line we can also define another type of distinguished bases (which leads, in general, to other Dynkin diagrams). The projection to the line can itself be regarded as a function. The restriction of this function u to the manifold $V_{\lambda,0}$ (in the notation introduced above) for a generic λ is a Morse function with μ non-degenerate critical points and μ distinct critical values. The number of critical values is equal to that of spheres in the bouquet that is homotopy equivalent to the quotient space $V_{\lambda,0}/V_{\lambda,0}^0$, and is therefore equal to the rank of the group H^- of anti-invariant homology classes of the two-sheeted covering $\tilde{V}_{\lambda,0} \to V_{\lambda,0}$, branched along $V_{\lambda,0}^0$. (If the projection is simple, then $\mu = \nu + 1$, where ν is the dimension of the space of parameters of a miniversal deformation.) As u goes from the non-critical value ($u = 0$) to the μ critical values along μ distinguished paths on the complex u-axis, the level set of u on $V_{\lambda,0}$ degenerates. For the degeneracies the μ semicycles, and consequently, the μ homology classes of $V_{\lambda,0}/V_{\lambda,0}^0$, vanish. The corresponding anti-invariant cycles in the homology groups of the two-sheeted covering over $V_{\lambda,0}$ with branching along $V_{\lambda,0}^0$ then form a distinguished basis (in the new sense) of μ short cycles in H^-.

For the singularities $A_\mu, \ldots, E_8$ this construction leads to the usual Dynkin diagrams, but already for the B_μ we get unusual ones: in the symmetric case the diagram splits into points (that is, the distinguished basis in the new sense consists of roots ε_i in terms of which all the remaining roots are expressible in the form $\pm\varepsilon_i$, $\pm\varepsilon_i \pm \varepsilon_j$).

Further consequences of Goryunov's classification concern the structure of the bifurcation diagrams of projections. To describe them I recall what is known about bifurcation manifolds of singularities in general.

§17. The geometry of bifurcation diagrams

One of the simplest bifurcation diagrams is the semicubical parabola, formed in the (a, b)-parameter plane by polynomials x^3+ax+b with a multiple root. This curve appears as a bifurcation diagram in many problems of singularity theory. For example, we can regard it as the bifurcation diagram of the zeros of the function x^3: it is formed by those points of the base of a versal deformation of this function for which the zero level manifold of the deformed function is singular. This bifurcation diagram has a number of remarkable properties. For example, the complement to this (complex) curve in $\mathbf{C}^2$ is a $K(\pi, 1)$-space: all the homotopy groups of this complement, apart from the fundamental group, are trivial (the fundamental group is the Artin braid group with three threads).

Next we consider a generic holomorphic vector field on the plane of our semicubical parabola. As an example we can take the constant field $\partial/\partial b$ on the plane of polynomials of the third degree: the vector field at the cusp of the semicubical parabola is transversal to the tangent field to the curve there. Any generic vector field transversal to the tangent field in a neighbourhood of the cusp of the semicubical parabola is equivalent to this special field (can be transformed to it by a diffeomorphism preserving the semicubical parabola).

It turns out that both these properties (to be an Eilenberg-Maclane space and to admit straightening of a vector field by a diffeomorphism that preserves the diagram) are satisfied for a wide class of bifurcation diagrams, for example, for the surface of the swallow-tail, which is formed by the polynomials x^4+ax^2+bx+c in three-dimensional space (see Fig. 2), for bifurcation diagrams of functions on manifolds with boundary, and for bifurcation diagrams of simple projections, about which we shall say something later.

We consider to begin with functions on a manifold with boundary. Let $f:(\mathbf{C}^n, 0) \to (\mathbf{C}, 0)$ be the germ of a holomorphic function on the manifold $\mathbf{C}^n$ with fixed submanifold $\mathbf{C}^{n-1}$(the "boundary"), given by the equation $x_1 = 0$ (where the x_i are the coordinates in $\mathbf{C}^n$). Then the origin is said to be a *boundary singularity* for f if the restriction of f to the boundary has this point as a critical point.

The *multiplicity* μ *of the boundary singularity* is the dimension over $\mathbf{C}$ of the *boundary local algebra*

$$Q_{f|x_1} = \mathbf{C}[[x_1, \ldots, x_n]]/(x_1f_1, f_2, \ldots, f_n), \; f_i = \partial f/\partial x_i.$$

As a versal deformation we choose

$$F(x, \lambda) = f(x) + \lambda_1 g_1(x) + \dots + \lambda_\mu g_\mu(x),$$

where the g_i are monomials that generate after factorization a **C**-basis in the boundary local algebra. We find it convenient to assume that $g_\mu \equiv 1$. The space $\mathbf{C}^\mu$ with the coordinates $(\lambda_1, \dots, \lambda_\mu)$ is called the *base* of the versal deformation of the boundary singularity of f, and $\mathbf{C}^{\mu-1}$ with the coordinates $(\lambda_1, \dots, \lambda_{\mu-1})$ the *truncated base.*

Example. For the boundary singularity C_3

$$f = x_2^3 + x_1 x_2, \quad F = f + \lambda_1 x_2^2 + \lambda_2 x_2 + \lambda_3.$$

Definition. A number α is said to be a *critical value* of a function $F(\cdot, \lambda)$ if the manifold $F(\cdot, \lambda) = 0$ is either singular or non-transversal to the boundary.

Definition. The *bifurcation diagram of zeros* of a boundary singularity of f is the set of all points λ of the base of the versal deformation for which $F(\cdot, \lambda) = 0$ has zero as a critical value. The bifurcation diagram of zeros is denoted by Σ.

Example. The bifurcation diagram of zeros of the boundary singularity C_3 is formed by the points (a, b, c) for which the polynomial x^3+ax^2+bx+c has either a multiple or a zero root. This diagram, in the real case, is illustrated in Fig. 29. It consists of two irreducible components: one of these is diffeomorphic to a cylinder over a semicubical parabola (the surface of multiple roots), while the other, the plane $c = 0$, corresponds to the zero roots. The first component corresponds to manifolds $F(\cdot, \lambda) = 0$ that are not transversal to the boundary, and the second to singular manifolds (in this example $\lambda_1 = a$, $\lambda_2 = b$, $\lambda_3 = c$).

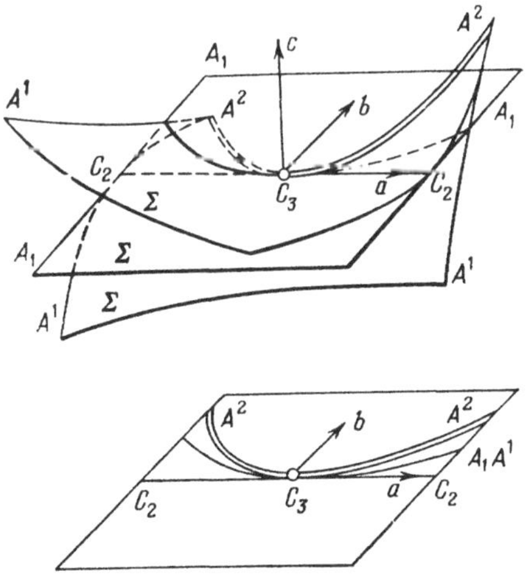

Fig. 29.

For simple boundary singularities the bifurcation diagrams of zeros are the varieties of irregular orbits of the corresponding groups generated by reflections (see §11). These algebraic varieties have a well-defined tangent plane at the origin ($\lambda_\mu = 0$ in our notation).

Theorem 1. *The germ of a holomorphic vector field in a neighbourhood of the origin of the bifurcation diagram of zeros of a simple boundary singularity that is transversal to the tangent plane at* 0 *goes into the germ of the constant field* $\partial/\partial\lambda_\mu$ *under a biholomorphic diffeomorphism that preserves the bifurcation diagram.*

This theorem, which was conjectured in [55], was proved by Lyashko [84] by means of Theorem 2 below; Zakalyukin [45] has extended it to non-simple singularities.

Definition. The *bifurcation diagram of functions* of a boundary singularity f is the set of all the points $\bar{\lambda}$ of the base of the truncated versal deformation for which $F(\cdot, \bar{\lambda}, \lambda_\mu)$ has fewer than μ distinct critical values.

Example. The bifurcation diagram of functions for the boundary singularity C_3 is formed by the line $b = 0$ and the two parabolas in the (a, b)-plane tangent to it at the origin (Fig. 29).

The following construction generalizes to the boundary case the construction of Lyashko [85] and Looijenga [86]. We associate with each point $\bar{\lambda}$ the value of λ_μ for which the sum of all the μ critical values of the function $F(\cdot, \bar{\lambda}, \lambda_\mu)$ is zero. We form the polynomial of degree μ whose roots are these critical values and obtain a map of the base of the truncated deformation into the space $\mathbf{C}^{\mu-1}$ of polynomials of degree μ with leading coefficient 1 for which the sum of the roots is zero.

Theorem 2 (see [84]). *If f is a simple boundary singularity, then the map constructed above is holomorphic and proper. Under this map the complement to the bifurcation diagram of the function covers the manifold of polynomials without multiple roots.*

Corollary. *The complement to the bifurcation diagram of fucntions for a simple boundary singularity is a* $K(\pi, 1)$*-space, where π is a subgroup of finite index in the Artin braid group with μ threads.*

Theorem 1 can be derived from Theorem 2 in the following way: the μ moments of passage of a phase curve of the vector field through points of the bifurcation diagram of zeros predetermine the straightened curve $\bar{\lambda} = \text{const}$, up to finitely many possibilities.

Theorem 1 is applicable also to non-boundary singularities. For example, it allows us to straighten a vector field in the neighbourhood of the ordinary swallow-tail. To straighten a field in the neighbourhood of the bifurcation diagram of zeros of a function φ we have to apply the theorem to the function $f(x) = x_0 + \varphi(x_1, \ldots, x_n)$ on the manifold with boundary $x_0 = 0$.

In that case the bifurcation diagram of zeros consists entirely of a single component: it coincides with the bifurcation diagram of zeros of the restriction of f to the boundary, that is, with the bifurcation diagram of zeros for the non-boundary singularity φ, since all the level manifolds of the functions $F(\cdot, \lambda)$ are non-singular.

Corollary. *The projection of the swallow-tail formed in $\mathbf{C}^3$ by the polynomials x^4+ax^2+bx+c with multiple roots to the plane along the c-axis is stable: every nearby projection is equivalent to the one in question in a neighbourhood of the origin.*

Of course, there is a similar corollary for all simple singularities of functions (ordinary or boundary) and also for all groups generated by reflections (among them the unitary ones).

Applications of Theorem 1 to the study of normal forms and metamorphoses of various geometrical objects in a neighbourhood of a bifurcation diagram (in particular, to the study of projections of smooth surfaces) were given in [55], where this theorem first appeared. Apart from Theorem 1, Theorem 3 below (in which the bifurcation diagram of zeros is replaced by a general hypersurface V in $\mathbf{C}^\mu$, in general, singular) is frequently of use in reductions to normal form.

Suppose that the restriction of the bundle $\pi\colon \mathbf{C}^\mu \to \mathbf{C}^{\mu-1}$ to the germ of the hypersurface $V \hookrightarrow \mathbf{C}^\mu$ is proper. Then V can be given by an equation $R(x) = 0$ that is polynomial along the fibres of π, whose degree k is equal to the multiplicity of the intersection of the fibre with V at zero. Let $D\colon \mathbf{C}^{\mu-1} \to \mathbf{C}$ be the discriminant of this polynomial. We call the variety of zeros of the discriminant the *bifurcation variety of the projection* $V \hookrightarrow \mathbf{C}^\mu \to \mathbf{C}^{\mu-1}$.

Definition. A vector field v *preserves* the variety given by the equation $\varphi = 0$ if the derivative of φ in the direction of v belongs to the ideal generated by φ.

Theorem 3 [84]. *Suppose that the dimension of the set of points of the base over which more than two of the points of the hypersurface coincide (so that the number of distinct points of the hypersurface over a point of the base does not exceed $k-2$) is strictly less than the dimension of the bifurcation variety. Then every germ of a holomorphic vector field on the base that preserves the bifurcation variety can be lifted to the germ of a field holomorphic in the bundle space that preserves the hypersurface concerned.*

Corollary. *A holomorphic field that preserves the bifurcation diagram of functions of a simple singularity lifts to a holomorphic vector field that preserves the bifurcation diagram of zeros.*

For example, a field tangent to the semicubical parabola, the projection of the edge of regression of the swallow-tail to the plane, lifts to a field tangent to the whole swallow-tail in three-dimensional space.

To lift diffeomorphisms that preserve bifurcation diagrams it remains to study the group of connected components of the groups of germs of biholomorphic diffeomorphisms that preserve the swallow-tail. Lyashko has computed these groups.

***Theorem* 4** [84]. *For any finite group W generated by reflections, the group π_0 of connected components of the group of germs of biholomorphic diffeomorphisms of the space of orbits at the origin that preserve the variety of irregular orbits coincides with the factor group of the group of automorphisms of W that preserve the set of all the reflections contained in W by the normal subgroup formed by the inner automorphisms. The group π_0 consists of two elements for $W = D_{2k}, B_2, F_4, G_2, I_2(2k)$ and is isomorphic to the group of permutations of three elements for D_4, while in the remaining cases it is trivial.*

We turn now to simple projections of complete intersections. Goryunov has extended Lyashko's theorems to this case. The list of simple projections (§16) contains that of boundary singularities. However, even for the boundary singularities Goryunov's results do not coincide with Lyashko's theorems: Goryunov obtains other $K(\pi, 1)$ spaces from the same bifurcation diagrams and straightens other vector fields.

Example. We consider again the variety of polynomials x^3+ax^2+bx+c having multiple roots (it is diffeomorphic to a cylinder over a semicubical parabola), which occurs in the bifurcation diagram of zeros for C_3, but we

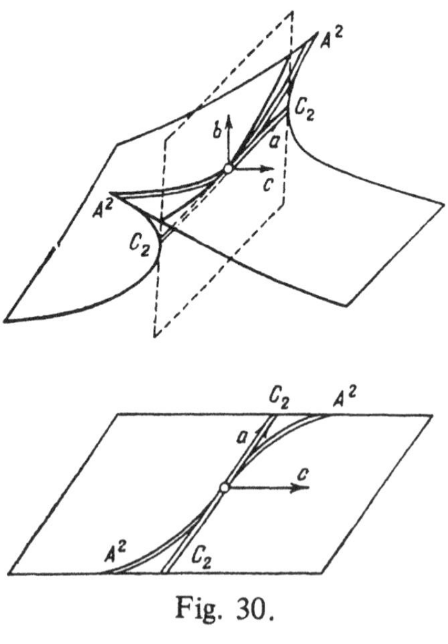

Fig. 30.

project it not along the c-axis, as in Fig. 29, but along the b-axis, as in Fig. 30. (The vertical plane, denoted in Fig. 30 by dots, does not for the moment come into the variety under consideration.) The *apparent contour* of our projection onto the (a, c)-plane consists of two smooth curves (A^2 and C_2), cubically tangent at zero. Applied to this case Goryunov's results are as follows.

***Theorem* 5.** *The complement to the apparent contour is a $K(\pi, 1)$-space, where π is a subgroup of finite index (equal to 8) in the Artin braid group with three threads.*

***Theorem* 6.** *The projection of the variety of polynomials x^3+ax^2+bx+c with multiple roots to the plane along the b-axis is stable in the sense that any nearby holomorphic projection (not necessarily linear) of the same variety is locally holomorphically equivalent to it.*

***Theorem* 7.** *The germ of the vector field $\partial/\partial b$ in the neighbourhood of the origin is stable with respect to the variety of polynomials with multiple roots in the sense that the germ of every nearby holomorphic field at a suitable nearby point can be transformed to the germ of the field $\partial/\partial b$ at zero by a biholomorphic diffeomorphism that preserves the variety of polynomials with multiple roots.*

***Theorem* 8.** *In a neighbourhood of each point of the edge of regression of the cylinder over the semicubical parabola $u^2 = v^3$ in the space with the coordinates (u, v, w) a generic holomorphic vector field is reducible to one of the two normal forms $\partial/\partial u$ (at a generic point) and $\partial/\partial v+w\partial/\partial u$ (at particular points) by a biholomorphic diffeomorphism that preserves the cylinder.*

To state Goryunov's theorem for all simple projections to the line, we consider the projection of the complete intersection V_0, given by the equation $f(x, u) = 0$, to the u-axis. A deformation of this projection is a family of projections to the u-axis of complete intersections V_λ, given by equations $F(x, u, \lambda) = 0$ (where $F(x, u, 0) \equiv f(x, u)$). A versal deformation is defined in the usual way (see §16, 5°). Below we suppose all the time that F induces a miniversal deformation of a simple germ of a holomorphic projection, that is, that the number ν of parameters λ_i is as small as possible.

Example. For a simple projection-germ of type C_3 the complete intersection V_0 is a plane curve $f = 0$, where $f = x^3+ux$. The miniversal deformation is a two-parameter family of curves $F(x, u, \lambda) = 0$, where $F = f+\lambda_1x^2+\lambda_2$. Connected with the deformation of the projection is the "total intersection"

$$W = \{(x, u, \lambda) \colon F(x, u, \lambda) = 0\}.$$

If the deformation is versal, then the germ of W at 0 is smooth.

Definition. The *bifurcation diagram of zeros of a projection* is the germ at 0 of the set of critical values of the map $(x, u, \lambda) \mapsto (u, \lambda)$ restricted to the germ of the manifold W.

Example. The bifurcation diagram of zeros of the projection C_3 consists of those (u, λ) for which the polynomial $x^3 + \lambda_1 x^2 + ux + \lambda_2$ has a multiple root (in Fig. 30, λ_1, u, λ_2 are denoted by a, b, c).

We consider now the natural projection of the bifurcation diagram of zeros of the projection to the space of parameters λ along the u-axis.

Definition. The *bifurcation diagram of the projection* is the union of the projection of the set of singular points of the bifurcation diagram of zeros and the set of critical values of the projection of the smooth part of the bifurcation diagram of zeros to the parameter space λ.

Example. The projection of the smooth part of the bifurcation diagram of zeros of C_3 to the λ-plane (the (a, c)-plane in Fig. 30) has as set of critical values a line, while the projection of the singularities of the diagram of zeros is a cubical parabola. Together these form the bifurcation diagram of the projection C_3.

Example. The projection of the swallow-tail to the plane along a direction transverse to the tangent plane does not have any critical values under the smooth part of the tail. The set of singularities of the tail consists of two components: the edge of regression and the curve of self-intersections. Their projections to the plane form the bifurcation diagram of the projection A_3 (here $f = u + x^4$, $F = f + \lambda_1 x^2 + \lambda_2 x$).

In the general case the bifurcation diagram of the projection consists of three hypersurfaces: one is the projection of the edge of regression of the bifurcation diagram of zeros, another is the projection of the self-intersection variety, and the third is the set of critical values of the projection of the regular part.

The bifurcation diagram of the projection can also be defined as the set of all values of the parameter λ of the miniversal deformation for which either the variety $V_\lambda = \{(x, u)\colon F(x, \lambda, u) = 0\}$ is singular or the restriction of u to this manifold is not a Morse function.

***Theorem* 9** (see [76]). *The germ of the complement to the bifurcation diagram of a simple projection to the line is a* $K(\pi, 1)$*-space, where* π *is a subgroup of finite index of the Artin braid group with* $\nu + 1$ *threads* $(\nu = \dim\{\lambda\})$.

***Theorem* 10** [76]. *In a neighbourhood of the origin the germ of the vector field* $\partial/\partial u$ *is stable relative to the bifurcation diagram of zeros of a simple projection in the sense that the germ of every nearby holomorphic field at a suitable nearby point can be transformed to the germ of the field* $\partial/\partial u$ *at zero by a biholomorphic diffeomorphism that preserves the bifurcation diagram of zeros.*

A similar theorem has been proved also for non-simple projections, provided only that they have a quasihomogeneous versal deformation.

Example. We consider in $\mathbf{C}^\mu$ the hypersurface Σ of polynomials of the form $x^\mu + a_1 x^{\mu-1} + \ldots + a_\mu$ having multiple roots. This hypersurface is diffeomorphic to the direct product of the swallow-tail in $\mathbf{C}^{\mu-1}$ and $\mathbf{C}$. The product of the singular point of the tail and $\mathbf{C}$ determines the singular curve of Σ. The theorems of Lyashko and Goryunov allow us to reduce to normal form germs of generic holomorphic vector fields in $\mathbf{C}^\mu$ at *all* points of the singular curve.

For at generic points on this curve the field vector is transversal to the tangent plane to Σ. In a neighbourhood of such a point the field can be reduced to the normal form $\partial/\partial a_\mu$ by a diffeomorphism preserving Σ (by Theorem 1). At particular points the field vector lies in the tangent plane, but at the rest the field is in general position. In a neighbourhood of such a point the field can be reduced to the normal form $\partial/\partial a_{\mu-1}$ by a diffeomorphism preserving Σ, by Theorem 10.

Remark. Goryunov has also proved similar theorems for some situations arising in the theory of simple singularities, for example, for simple projections of complete intersections with boundary to a line (such projections are stably equivalent to projections of hypersurfaces with boundary), and also for simple linear singularities of functions. *Linear singularities*, introduced by Siersma [87], are functions with non-isolated critical points that form a curve, and on a hypersurface transversal to this curve the singularity is Morse. Goryunov discovered that the simple linear singularities are classified by the Coxeter groups. The hierarchy of non-isolated singularities reflects the hierarchy of *series* of singularities (for example, the whole series D_μ corresponds to *one* non-isolated singularity D_∞). Thus, to one Coxeter class there corresponds a whole series of singularities (in the notation of [8]: $A_\mu \sim J_{\mu+1,\infty}$, $B_\mu \sim T_{2\mu+2,\infty}$, $C_\mu \sim Z_{\mu-2,\infty}$, $F_4 \sim W_{1,\infty}$).

In these situations Goryunov defines bifurcation diagrams and proves the $K(\pi, 1)$-property of their complements. It is interesting to note that the complement to the bifurcation diagram of a simple object is a $K(\pi, 1)$-space not for any classification theory, but only for a sufficiently "good" one.

Example. We consider the bifurcation diagram of zeros of a zero-dimensional complete intersection $x^2+u^2 = 0$, $xu = 0$ ($I_{2,2}$ in Giusti's list [77] of simple singularities of complete intersections). Knörrer [88] has shown that the complement to this diagram in $\mathbf{C}^4$ has a non-trivial group π_3, and consequently is not a $K(\pi, 1)$-space. However, Goryunov [79] has discovered that the complement to the bifurcation diagram of the *projection* of the same complete intersection to the u-axis is a $K(\pi, 1)$-space (the bifurcation diagram of this projection is obtained from the bifurcation diagram of the complete intersection $\{\lambda$: the variety $(u^2 + \lambda_1 u + \lambda_2 + \lambda_3 x + x^2 = 0,$ $ux - \lambda_4 = 0)$ is singular$\}$ by the addition of the plane $\lambda_4 = 0$, over the points of which u takes fewer than four values at points of the variety in question).

Thus, the classification of the singularities of simple projections of complete intersections is a "better" problem than simply the classification of complete intersections (this is perceived also by other signs, for example, by the status of the theory of distinguished bases).

However, for simple projections of complete intersections (even zero-dimensional ones in $\mathbf{C}^2$) the $K(\pi, 1)$-conjecture has not been proved in all cases (in the notation of §16 not for $X_{k,l}$, $3 \leqslant k < l$, V_5, and V_6).

Remark. The proof of the $K(\pi, 1)$-property in many cases follows the scheme of Lyashko and Looijenga (see Theorem 1 above). Unfortunately, so far no single theorem has been found that justifies an argument of this type in all cases. It is clear that the proof is based on the fact that the quasihomogeneous weights of the components of the map of Lyashko-Looijenga type are positive. But in spite of this fact, in each case one has to prove separately that outside the bifurcation diagram the map is open (the Jacobian does not degenerate).

To conclude this section we say a few words about some special bifurcation diagrams that occur often in various problems.

Definition. A *monomial curve of type* $(a_1, \ldots, a_n)$ is a curve with parametric equations $x_k = t^{a_k}$. We suppose that $a_1 < \ldots < a_n$. A monomial curve can be regarded as lying in $\mathbf{R}^n$, $\mathbf{C}^n$, $\mathbf{R}P^n$, or $\mathbf{C}P^n$. It is easy to verify the next theorem.

Theorem 11. *The germs of a monomial curve at zero and at infinity are each projectively equivalent to the dual of the other (the curve projectively dual to a given one is the set of all the osculating planes of the given curve).*

Definition. The *swallow-tail of type* $(a_0, \ldots, a_n)$ in $\mathbf{R}^n$ (in $\mathbf{C}^n$) is the closure of the set of all polynomials of the form $t^{a_n} + A_1 t^{a_{n-1}} + \ldots + A_n t^{a_0}$ with real (complex) coefficients having a non-zero multiple root (in the closure we add polynomials with a zero root of multiplicity greater than a_0).

Example. The ordinary swallow-tail is of type (0, 1, 2; 4).

Definition. The *reamer* of a smooth curve $\varphi: \mathbf{R} \to \mathbf{R}^n$ is the hypersurface formed by the closest planes of codimension 2 (the closure of the union of the planes spanned by the vectors $\varphi^{\cdot}, \varphi^{\cdot\cdot}, \ldots, \varphi^{(n-2)}$ at each point of the curve). The evolvents of curves in $\mathbf{C}^n$, $\mathbf{R}P^n$, and $\mathbf{C}P^n$ are defined similarly.

Example. The reamer of a curve in three-dimensional space is the union of its tangent lines. The reamer of the monomial curve (2, 3, 4) is diffeomorphic to the swallow-tail.

The next theorem follows from Theorem 11.

***Theorem* 12.** *The swallow-tail of type* $(a_0, \ldots, ; a_n)$ *is linearly diffeomorphic to the reamer of the monomial curve of type* $(a_n - a_{n-1}, \ldots, a_n - a_0)$.

Corollary. *The swallow-tail of type* $(a_0, \ldots; a_n)$ *is linearly diffeomorphic to the projection to the space with the coordinates* $(\lambda_{a_0}, \ldots, \lambda_{a_{n-1}})$ *of the variety of polynomials* $t^{a_n} + \lambda_{a_{n-1}} t^{a_{n-1}} + \ldots + \lambda_{a_0} t^{a_0}$, *having a root of such a multiplicity that the projection is a hypersurface.*

Here the multiplicity has to be $a_n - n + 2$.

Example. The ordinary swallow-tail is diffeomorphic to the variety

$$\{(B, C, D): \exists A: t^4 + At^3 + Bt^2 + Ct + D = (t + \alpha)^3(t + \beta)\}.$$

The reamer of a monomial curve can, therefore, be defined in three ways: as the union of the closest planes, as the variety of polynomials of a special type (having few terms) with multiple roots, and as the projection of a set of polynomials with roots of higher multiplicity to a coordinate subspace. Such reamers in three-dimensional space form a frequently encountered element of various bifurcation diagrams and deserve detailed study.

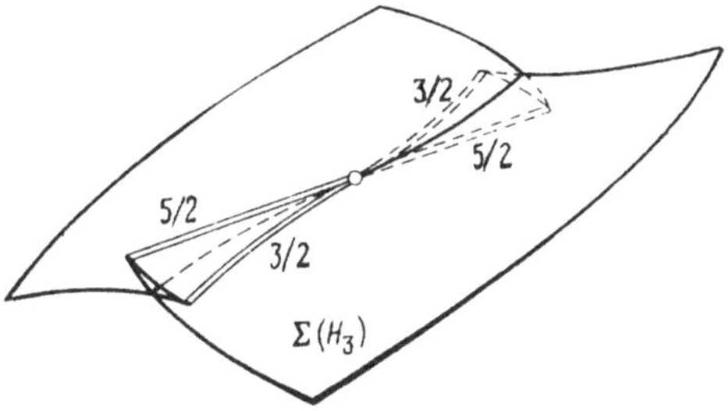

Fig. 31.

Example. Fig. 31 illustrates the reamer of the monomial curve (1, 3, 5). Lyashko has proved that this is a surface with two edges of regression (one of type 3/2 and the other of type 5/2) tangent to each other cubically and diffeomorphic to the variety of irregular orbits of the Coxeter group H_3 (that is, the group of symmetries of the icosahedron). Shcherbak has proved that the reamers of all the curves $x = t$, $t = t^3 + \ldots$, $z = t^5 + \ldots$ are locally diffeomorphic to the same surface. From this there follows, amongst other things, a proof of the conjecture of Givental' that the graph I drew of the many-valued function of time in the problem of going around an obstacle bounded by a plane curve with a cubic point of inflection is in the neighbourhood of this point diffeomorphic to the variety of irregular orbits of the group of symmetries of the icosahedron, drawn by Lyashko. This is because this graph, as is easy to see, is the reamer of a curve of the form described above.

For further details concerning the problem of going around an obstacle, see [107].

References

[1] J. Martinet, Sur les singularités des formes différentielles, Ann. Inst. Fourier **20**:1 (1970), 95-178. MR **44** # 3333.

[2] R. Roussarie, Modèles locaux de champs et de formes, Astérisque, No. 30 (1975). MR **55** # 13444.

[3] M. Golubitsky and D. Tishler, An example of moduli for singular symplectic forms, Invent. Math. **38** (1977), 219-225. MR **55** # 9131.

[4] A. Weinstein, Lagrangian submanifolds and Hamiltonian systems, Ann. of Math. **98** (1973), 377-410. MR **48** # 9761.

[5] R.B. Melrose, Equivalence of glancing hypersurfaces, Invent. Math. **37** (1976), 165-191. MR **55** # 9173.

[6] V.I. Arnol'd, Integrals of rapidly oscillating functions and singularities of projections of Lagrangian manifolds, Funktsional. Anal. i Prilozhen. **6**:3 (1972), 61-62. MR **50** # 8594.
= Functional Anal. Appl. **6** (1972), 222-225.

[7] ———, Normal forms for functions near degenerate critical points, the Weyl groups A_k, D_k, E_k and Lagrangian singularities, Funktsional. Anal. i Prilozhen. **6**:4 (1972), 3-25. MR **50** # 8595.
= Functional Anal. Appl. **6** (1972), 254-272.

[8] ———, A.N. Varchenko, and S.M. Gusein-Zade, *Osobennosti differentsiruemykh otobrazhenii* (Singularities of maps). I, Nauka, Moscow 1982. English translation to be published by Birkhäuser-Boston, Inc.

[9] R. Thom, Topological models in biology, Topology **8** (1969), 313-335. MR **39** # 6629.

[10] E.E. Landis, Tangential singularities, Funktsional. Anal. i Prilozhen. **15**:2 (1981), 36-49. MR **83f**:58011.
= Functional Anal. Appl. **15** (1981), 103-114.

[11] O.A. Platonova, Singularities of the mutual disposition of a surface and a line, Uspekhi Mat. Nauk **36**:1 (1981), 221-222. MR **82d**:14003.
= Russian Math. Surveys **36**:1 (1981), 248-249.

[12] T.F. Banchoff, T. Gaffney, and C. McCrory, Cusps of Gauss mappings, Research Notes in Mathematics **55**, Pitman, Boston 1982.

[13] L. Hörmander, Fourier integral operators. I, Acta. Math. **127** (1971), 79-183. MR **52** # 9299.

[14] Y. Colin de Verdière, Nombre de points entiers dans une famille homothétique de domains de $\mathbf{R}^n$, Ann. Sci. École Norm. Sup. (4) **10** (1977), 559-575. MR **58** # 563.

[15] A.N. Varchenko, Newton polyhedra and estimates of oscillatory integrals, Funktsional. Anal. i Prilozhen. **10**:3 (1976), 13-38. MR **54** # 10248.
= Functional Anal. Appl. **10** (1976), 175-196.

[16] I.M. Vinogradov, *Metod trigonometricheskikh summ v teorii chisel* (The method of trigonometric sums in the theory of numbers), Nauka, Moscow 1971. MR **53** # 13135.

[17] V.N. Karpushkin, Uniform estimates of oscillatory integrals in $\mathbf{R}^2$, Dokl. Akad. Nauk SSSR **254** (1980), 28-31. MR **83i**:32015.
= Soviet Math. Dokl. **22** (1980), 301-304.

[18] F. Pham, Caustics and microfunctions, Publ. Res. Inst. Math. Sci. **12** (1976/77), 357-366. MR **58** # 28656.

[19] V.A. Vasil'ev, Asymptotic behaviour of exponential integrals in the complex domain, Funktsional. Anal. i Prilozhen. **13**:4 (1979), 1-12. MR **81m**:41035.
= Functional Anal. Appl. **13** (1979), 239-247.

[20] A.N. Varchenko, Gauss-Manin connection of isolated singular point and Bernstein polynomial, Bull. Soc. Math. France (2) **104** (1980), 205-223. MR **83e**:32008.
[21] ———, The asymptotic behaviour of holomorphic forms determines a mixed Hodge structure, Dokl. Akad. Nauk SSSR **255** (1980), 1035-1038. MR **82g**:14010.
= Soviet Math. Dokl. **22** (1980), 772-775.
[22] ———, An asymptotic mixed Hodge structure in vanishing cohomology, Izv. Akad. Nauk SSSR Ser. Mat. **45** (1981), 540-591. MR **82m**:32007.
= Math. USSR-Izv. **18** (1982), 469-512.
[23] J.H.M. Steenbrink, Mixed Hodge structure on the vanishing cohomology, in: Real and complex singularities, Nordic Summer School, Oslo 1976, Sijthoff and Noordhoff, Alphen aan den Rijn 1977. MR **58** # 5670.
[24] A.N. Varchenko, A lower estimate for the codimension of the μ = const stratum in terms of a mixed Hodge structure, Vestnik. Moskov. Univ. Ser. Mat. Mekh. **1982**, no. 6, 28-31.
= Moscow Univ. Math. Bull. **37**:6 (1982), 30-33.
[25] ———, The complex index of a singularity does not change along a μ = const stratum, Funktsional. Anal. i Prilozhen. **16**:1 (1982), 1-12. MR **83j**:32005.
= Functional Anal. Appl. **16** (1982), 1-9.
[26] ——— and A.B. Givental', The period map and the intersection form, Funktsional. Anal. i Prilozhen. **16**:2 (1982), 7-20.
= Functional Anal. Appl. **16** (1982), 83-93.
[27] J. Scherk and J. Steenbrink, On the mixed Hodge structure on the cohomology of the Milnor fibre, Preprint, 29 pages.
[28] F. Pham, Structures de Hodge mixtes associés à un germ de fonctions à un point critique isolé, Exposé au Coll. d'Analyse et Topologie sur les espaces singuliers à Marseille-Luminy, juillet 1981.
[29] M. Saito, Gauss-Manin system and mixed Hodge structure, Proc. Japan Acad. Ser. A. **58**:1 (1982), 29-32. MR **83f**:14006.
[30] V.I. Arnol'd, The index of a singular point of a vector field, the Petrovskii-Oleinik inequalities and mixed Hodge structures, Funktsional. Anal. i Prilozhen. **12**:2 (1978), 1-14. MR **58** # 16685.
= Functional Anal. Appl. **12** (1978), 1-11.
[31] ———, On some problems in singularity theory, Proc. Indian Acad. Sci. Math. Sci. **90** (1981), 1-9. MR **83j**:58011.
[32] E. Brieskorn, Die Hierarchie der 1-modulären Singularitäten, Manuscripta Math. **27** (1979), 183-219. MR **80k**:14005.
[33] V.V. Goryunov, Adjacencies of spectra of certain singularities, Vest. Moskov. Univ. Ser. Mat. Mekh. **1981**, no. 4, 19-23. MR **83k**:58014.
= Moscow Univ. Math. Bull. **36**:4 (1981), 22-25.
[34] A.N. Varchenko, The spectrum and decompositions of a critical point of a function, Dokl. Akad. Nauk SSSR **270** (1983), 267-270.
[35] ———, On the number of integral points in a domain, Uspekhi Mat. Nauk **37**:3 (1982), 177-178. MR **83j**:10058.
= Russian Math. Surveys **37**:3 (1982), 223-224.
[36] B. Randol, On the asymptotic behaviour of the Fourier transform of the indicator function of a convex set, Trans. Amer. Math. Soc. **139** (1969), 279-285. MR **40** # 4678b.
[37] A.N. Varchenko, On the number of integral points in families of homothetic domains, Funktsional. Anal. i Prilozhen. **17**:2 (1983), 1-6.
= Functional Anal. Appl. **17** (1983).

[38] V.I. Arnol'd, The statistics of integral convex polygons, Funktsional. Anal. i Prilozhen. **14**:2 (1980), 1-3. MR **81g**:52011.
= Functional Anal. Appl. **14** (1980), 79-81.

[39] A.S. Pyartli, Diophantine approximations on submanifolds of Euclidean space, Funktsional. Anal. i Prilozhen. **3**:4 (1969), 59-62. MR **40** # 4209.
= Functional Anal. Appl. **3** (1969), 303-306.

[40] V.G. Sprindzhuk, *Problema Malera v metricheskoi teorii chisel,* Nauka i Tekhnika, Minsk 1967. MR **39** # 6832.
Translation: Mahler's problem in metric number theory, Amer. Math. Soc., Providence, RI 1969. MR **39** # 6833.

[41] V.I. Arnol'd, *Dopolnitel'nye glavy teorii obyknovennykh differentsial'nykh uravnenii,* Nauka, Moscow 1978. MR **80i**:34001.
Translation: Chapitres supplémentaires de la théorie des équations différentielles ordinaires, Mir, Moscow 1980. MR **83a**:34003.

[42] N.N. Nekhoroshev, An exponential estimate of the time of stability of nearly integrable Hamiltonian systems. I, Uspekhi Mat. Nauk **32**:6 (1977), 5-66. MR **58** # 18570.
= Russian Math. Surveys **32**:6 (1977), 1-65.

[43] ———, An exponential estimate of the time of stability of nearly integrable Hamiltonian systems. II, Trudy Sem. Petrovsk. **5** (1979), 5-50. MR **80m**:58017.

[44] ———, Stable lower estimates for smooth maps and for the gradients of smooth functions, Mat. Sb. **90** (1972), 432-478. MR **54** # 1283.
= Math. USSR-Sb. **19** (1973), 425-467.

[45] V.M. Zakalyukin, Legendrian maps in Hamiltonian systems, in: *Nekotorye problemy mekhaniki* (Some problems of mechanics), MAI, Moscow 1977, 11-16.

[46] Ya.B. Zel'dovich, The separation of homogeneous matter into parts under the action of gravitation, Astrofizika, **6** (1970), 319-335. PA 70-69358.
= Astrophysics **6**:2 (1970), 164-174.

[47] V.I. Arnold, Wave front evolution and equivariant Morse lemma, Comm. Pure Appl. Math. **29** (1976), 557-582 (in English). MR **55** # 9148.

[48] ———, *Teoriya katastrof* (Catastrophe theory), Priroda **1979**, no. 10, 54-63.

[49] ———, Bifurcations of singularities of potential flows in a collisionless medium and metamorphoses of caustics in three-dimensional space, Trudy Sem. Petrovsk. **8** (1982), 21-57.

[50] ———, S.F. Shandarin, and Ya.B. Zeldovich, The large scale structure of the Universe. I, Geophys. Astrophys. Fluid Dynamics **20** (1982), 111-130 (in English). See MR **83h**:83052.

[51] S.F. Shandarin, The theory of flow and the cellular-reticulated structure of the Universe, Preprint no. 137, Inst. Prikl. Mat., Moscow 1982.

[52] V.M. Zakalyukin, Lagrangian and Legendrian singularities, Funktsional. Anal. i Prilozhen. **10**:1 (1976), 26-36. MR **54** # 1294.
= Functional Anal. Appl. **10** (1976), 23-31.

[53] ———, Bifurcations of wave-fronts depending on one parameter, Funktsional. Anal. i Prilozhen. **10**:2 (1976), 69-70. MR **58** # 24332.
= Functional Anal. Appl. **10** (1976), 139-140.

[54] V.I. Arnold, Critical points of smooth functions, Proc. Internat. Congress Mathematicians Vancouver 1974, vol. 1, 19-39, Canadian Math. Congress, Montreat 1975. MR **55** # 4218.

[55] V.I. Arnol'd, Indices of singular points of 1-forms on a manifold with boundary, convolutions of invariants of groups generated by reflections, and singular projections of smooth surfaces, Uspekhi Mat. Nauk **34**:2 (1979), 3-38. MR **81e**:58041.
= Russian Math. Surveys **34**:2 (1979), 1-42.
[56] L. Solomon, Invariants of finite reflection groups, Nagoya Math. J. **22** (1963), 57-64. MR **27** # 4872.
[57] A.B. Givental', Convolution of invariants of groups generated by reflections and associated with simple singularities of functions, Funktsional. Anal. i Prilozhen. **14**:2 (1980), 4-14. MR **82f**:58019.
= Functional Anal. Appl. **14** (1980), 81-89.
[58] ———, The asymptotic behaviour of the intersection form of a quasihomogenous function, Funktsional. Anal. i Prilozhen. **16**:1 (1982), 63-65.
= Functional Anal. Appl. **16** (1982), 294-297.
[59] V.I. Arnol'd, Critical points of functions on a manifold with boundary, the simple Lie groups B_k, C_k, F_4, and singularities of evolutes, Uspekhi Mat. Nauk **33**:5 (1978), 91-105. MR **80j**:58008.
= Russian Math. Surveys **33**:5 (1978), 99-116.
[60] C.T.C. Wall, A note on symmetry of singularities, Bull. London Math. Soc. **12** (1980), 169-175. MR **81f**:32009.
[61] D. Eisenbud and H. Levine, An algebraic formula for the degree of a C^∞ map germ, Ann. of Math. (2) **106** (1977), 19-44. MR **57** # 7651.
[62] K. Saito, Primitive forms for universal unfolding of a function with an isolated critical point, Preprint Research Inst. Math. Sci., Kyoto 1981.
[63] E. Looijenga, Homogeneous spaces associated with certain semiuniversal deformations, Proc. Internat. Congress Mathematicians Helsinki 1978, 529-536, Acad. Sci. Fennica, Helsinki 1980. MR **81j**:14006.
[64] V.I. Arnol'd, Lagrangian and Legendrian cobordisms. I, Funktsional. Anal. i Prilozhen. **14**:3 (1980), 1-13. MR **83a**:57049a.
= Functional Anal. Appl. **14** (1980), 167-177.
[65] ———, Lagrangian and Legendrian cobordisms. II, Funktsional. Anal. i Prilozhen. **14**:4 (1980), 8-17. MR **83a**:57049b.
= Functional Anal. Appl. **14** (1980), 252-260.
[66] R.O. Wells, Compact real submanifolds of a complex manifold with nondegenerate holomorphic tangent bundles, Math. Ann. **179** (1969), 123-129. MR **38** #6104.
[67] V.B. Tvorogov, The conjugate Fourier integral operator in a problem concerning the expansion of cracks, Moscow 1979, depos. at VINITI no. 2872-79.
[68] ———, The sharp front and singularities of the solutions of a class of non-hyperbolic equations, Dokl. Akad. Nauk SSSR **244** (1979), 1327-1331.
= Soviet Math. Dokl. **20** (1979), 240-244.
[69] V.A. Vasil'ev, Characteristic classes of Lagrangian and Legendrian manifolds dual to singularities of caustics and wave fronts, Funktsional. Anal. i Prilozhen. **15**:3 (1981), 10-22. MR **83a** # 58018.
= Functional Anal. Appl. **15** (1981), 164-173.
[70] V.P. Maslov, *Teoriya vozmushchenii i asimptoticheskie metody* (The theory of perturbations and asymptotic methods), Izdat. Moskov. Gos. Univ., Moscow 1965.
[71] V.A. Vasil'ev, The self-intersections of wave-fronts and Legendrian (Lagrangian) characteristic numbers, Funktsional. Anal. i Prilozhen. **16**:2 (1982), 68-69. MR **83j**:58026.
= Functional Anal. Appl. **16** (1982), 131-133.

[72] D.B. Fuks, The Maslov-Arnol'd characteristic classes, Dokl. Akad. Nauk SSSR **178** (1968), 303-306. MR **37** # 934.
= Soviet Math. Dokl. **9** (1968), 96-99.
[73] O.A. Platonova, Singularities of systems of rays near an obstruction, depos. at VINITI, 11 Feb. 1981, 647-681.
[74] O.P. Shcherbak, Projectively dual space curves and Legendrian singularities, Trudy Tbilisi Inst. Mat. **13-14** (1982), 232-233; 280-336.
[75] V.S. Kulikov, The calculation of the singularities of the embedding of a generic algebraic surface in the projective space P^3, Funktsional. Anal. i Prilozhen. **17** (1983).
= Functional Anal. Appl. **17** (1983).
[76] V.V. Goryunov, The geometry of the bifurcation diagrams of simple projections onto a line, Funktsional. Anal. i Prilozhen. **15**:2 (1981), 1-8. MR **82i**:58017.
= Functional Anal. Appl. **15** (1981), 77-82.
[77] M. Giusti, Sur les singularités isolées d'intersections complètes quasihomogènes, Ann. Inst. Fourier **27**:3 (1977), 163-192. MR **57** # 702.
[78] V.V. Goryunov, Singularities of projections of surfaces, Dissertation, Moskov. Gos. Univ., Moscow 1981.
[79] ———, Projections of 0-dimensional complete intersections onto a line and the $K(\pi, 1)$-conjecture, Uspekhi Mat. Nauk **37**:3 (1982), 179-180.
= Russian Math. Surveys **37**:3 (1982), 206-208.
[80] V.I. Matov, Singularities of the maximum function on a manifold with boundary, Trudy Sem. Petrovsk. **6** (1981), 195-222. MR **82j**:58022.
[81] ———, Unimodal and bimodal germs of functions on a manifold with boundary, Trudy Sem. Petrovsk. **7** (1981), 174-189. MR **83j**:58024.
[82] J. McKay, Cartan matrices, finite groups of quaternions, and Kleinian singularities, Proc. Amer. Math. Soc. **81** (1981), 153-154.
[83] D. Drucker and D. Frohardt, Irreducible root systems and finite linear groups of degree two, Bull. London Math. Soc. **14** (1982), 142-148. MR **83j**:20046.
[84] O.V. Lyashko, The geometry of bifurcation diagrams, Uspekhi Mat. Nauk **34**:3 (1979), 205-206. MR **80k**:32010.
= Russian Math. Surveys **34**:3 (1979), 209-210.
[85] V.I. Arnol'd, Critical points of functions and the classification of caustics, Uspekhi Mat. Nauk **29**:3 (1974), 243-244.
[86] E. Looijenga, The complement of the bifurcation variety of a simple singularity, Invent. Math. **23** (1974), 105-116. MR **54** # 10661.
[87] D. Siersma, Singularities of functions on boundaries, corners, etc., Preprint 92, Univ. of Amsterdam 1976.
[88] H. Knörrer, Zum $K(\pi, 1)$-Problem für isolierte Singularitäten von vollständigen Durchschnitten, Comp. Math. **45** (1982), 333-340. MR **83h**:14003.
[89] J.W. Bruce, An upper bound for the number of singularities on a projective hypersurface, Bull. London Math. Soc. **13** (1981), 47-50. MR **82h**:14014.
[90] J.H.M. Steenbrink, Limits of Hodge structures, Invent. Math. **31** (1976), 229-257. MR **55** # 2894.
[91] O.A. Platonova, Singularities in the problem of the quickest way round an obstacle, Funktsional. Anal. i Prilozhen. **15**:2 (1981), 86-87. MR **82k**:58022.
= Functional Anal. Appl. **15** (1981), 147-148.
[92] V.I. Arnol'd, Ya.B. Zel'dovich, and S.F. Shandarin, Elements of the large-scale structure of the Universe, Uspekhi Mat. Nauk **36**:3 (1981), 244-245.
[93] ———, The covering up of a caustic by the edge of regression of a moving front, Uspekhi Mat. Nauk **36**:4 (1981), 233.

[94] V.I. Arnol'd, Lagrangian manifolds with singularities, asymptotic rays and the open swallow-tail, Funktsional. Anal. i Prilozhen. **15**:4 (1981), 1-14. MR **83c**:58011. = Functional Anal. Appl. **15** (1981), 235-246.

[95] A.B. Givental', Varieties of polynomials having a root of fixed comultiplicity and the generalized Newton equation, Funktsional. Anal. i Prilozhen. **16**:1 (1982), 13-18. MR **83j**:58020. = Functional Anal. Appl. **16** (1982), 10-14.

[96] ———, Lagrangian manifolds with singularities and irreducible sl_2-modules, Uspekhi Mat. Nauk (to appear).

[97] V.I. Arnol'd, Singularities of Legendre varieties, of evolvents and of fronts at an obstacle, Ergodic Theory and Dynamical Systems **2** (1982) (in English).

[98] S.M. Voronin, The analytic classification of pairs of involutions and its applications, Funktsional. Anal. i Prilozhen. **16**:2 (1982), 21-29. MR **83j**:58013. = Functional Anal. Appl. **16** (1982), 94-100.

[99] V.I. Arnol'd, *Matematicheskie metody klassicheskoi mekhaniki,* Nauka, Moscow 1974. MR **57** # 14032.
Translation: Les méthodes mathématiques de la mécanique classique, Mir, Moscow 1976. MR **57** # 14033.

[100] R.B. Melrose, Equivalence of glancing hypersurfaces. II, Math. Ann. **255** (1981), 159-198.

[101] E.E. Landis, Tangential singularities in contact geometry, Uspekhi Mat. Nauk **37**:4 (1982), 96.

[102] S.M. Voronin, The analytic classification of germs of conformal maps $(\mathbf{C}, 0) \to$ $\to (\mathbf{C}, 0)$, Funktsional. Anal. i Prilozhen. **15**:1 (1981), 1-17. MR **82h**:58008. = Functional Anal. Appl. **15** (1981), 1-13.

[103] J. Écalle, Théorie iterative: introduction à la théorie des invariants holomorphes, J. Math. Pures Appl. **54** (1975), 183-258. MR **58** # 17637.

[104] D. Hilbert, Über die vollen Invariantensysteme, Math. Ann. **42** (1893), 313-373.

[105] A.G. Gasparyan, The application of multi-dimensional matrices to the study of polynomials, Dokl. Akad. Nauk Arm. SSR **70**:3 (1980), 133-141. MR **82c**:15035.

[106] V.I. Matov, The topological classification of germs of the maximum and minimax functions of a family of functions in general position, Uspekhi Mat. Nauk **37**:4 (1982), 167-168. = Russian Math. Surveys **37**:4 (1982), 127-128.

[107] V.I. Arnol'd, Singularities in variational calculus, Itogi Nauki, Contemporary problems in Mathematics **22** (1983), 3-55.

[108] V.M. Zakalyukin, Bifurcations of fronts, caustics depending on a parameter, versality of maps, Itogi Nauki, Contemporary Problems in Mathematics **22** (1983), 53-93.

[109] O.V. Lyashko, The geometry of bifurcation diagrams, Itogi Nauki, Contemporary Problems in Mathematics **22** (1983), 94-129.

[110] A.N. Varchenko, The asymptotic behaviour of integrals and Hodge structures, Itogi Nauki, Contemporary Problems in Mathematics **22** (1983), 130-166.

[111] V.V. Goryunov, Singularities of projections of complete intersections, Itogi Nauki, Contemporary Problems in Mathematics **22** (1983), 167-206.

[112] A.G. Khovanskii, Newton polygons (resolution of singularities), Itogi Nauki, Contemporary Problems in Mathematics **22** (1983), 207-239.

[113] = [106]

[114] O.P. Shcherbak, Singularities of Frankel maps of submanifolds of projective space, Uspekhi Mat. Nauk **37**:4 (1982), 95-96.

[115] A.B. Givental', The maximal number of singular points on a projective hypersurface, Funktsional. Anal. i Prilozhen. **17**:3 (1983), 93-112.
= Functional Anal. Appl. **17** (1983).
[116] A.N. Varchenko, On semicontinuity of a spectrum and an upper estimate of the number of singular points of a projective hypersurface, Dokl. Akad. Nauk SSSR **270** (1983), 1309-1312.
= Soviet Math. Dokl.
[117] ———, On semicontinuity of the complex singularity index, Funktsional. Anal. i Prilozhen. **17**:4 (1983), 65-66.
= Functional. Anal. Appl. **17** (1983).
[118] V.D. Sedykh, Singularities of convex hulls, Sibirskii Mat. Zh. **24** (1983), 158-175.
= Siberian Math. J. **24** (1983).
[119] ———, Functional moduli of singularities of the convex hulls of manifolds of codimension 1 and 2, Mat. Sb. **119** (1982), 233-247.
= Math. USSR-Sb.
[120] V.V. Goryunov, Bifurcation diagrams of certain simple and quasihomogeneous singularities, Funktsional. Anal. i Prilozhen. **17**:2 (1983), 23-37.
= Funktsional Anal. Appl. **17** (1983).
[121] E.E. Landis, Tangential singularities in contact geometry, Uspekhi Mat. Nauk **37**:4 (1982), 96.
[122] O.V. Lyashko, Classification of critical points of functions on a manifold with singular boundary, Funktsional. Anal. i Prilozhen. **17**:3 (1983), 28-36.
= Functional Anal. Appl. **17** (1983).
[123] O.P. Shcherbak, The singularities of a family of involutes in the neighbourhood of a point of inflection of a curve, and the group H_3 generated by reflections, Funktsional. Anal. i Prilozhen. **17**:2 (1983), 70-72.
= Functional Anal. Appl. **17** (1983).

Translated by I.R. Porteous

Moscow State University

Received by the Editors 3 December 1982

V. I. ARNOLD

Singularities of Ray Systems

The simplest example of a ray system is the system of all normals to a given surface in Euclidean space. Hamilton (1824) turned the theory of ray system into a part of symplectic geometry; since Maslov's thesis (1965) ray systems are called Lagrangian submanifolds.

The normals to a surface foliate some neighbourhood of that surface; but away from that neighbourhood various normals start intersecting one another (Fig. 1). The resulting complicated and beautiful geometry

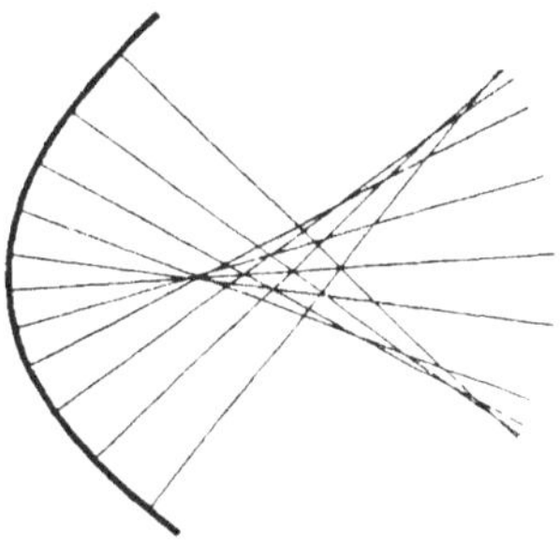

Fig. 1

was hidden up to 1972, when the relation between singularities of ray systems and Euclidean reflection groups was discovered.

This relation, for which there is no a priori reason, turned out to be a powerful method for the analysis of singularities. By 1978 it became clear that the Euclidean reflection groups also govern the singularities of Huygens evolvents.

Huygens (1654) discovered that the evolvent of a plane curve has a cusp singularity at each point of contact with the curve (Fig. 2). Plane curve evolvents and their higher-dimensional generalizations are the wave fronts on manifolds with boundary. The singularities of wave fronts, as well as those of ray systems, are classified by reflection groups.

Originally publ. in: Proc. of the Internat. Congress Math. (Warsaw, 1983). Vol. 1, 27-49, © 1984 Wydawnictwo Naukowe PWN (Polish Scientific Publishers PWN), Warszawa (Warsaw), Poland.

Reprint by Springer-Verlag Berlin Heidelberg 2018 with kind permission

While the ray and front systems on manifolds without boundary are related to the A, D and E series of the Weyl groups, the singularities of evolvents are described by the B, C, F series (those having Dynkin diagrams with double connections).

Fig. 2

The relation of the remaining reflection groups $(I_2(p),\ H_3, H_4)$ to singularity theory was unknown until recently. This situation has changed since the fall of 1982 when it was discovered that the group H_3 (the group of symmetries of the icosahedron) governs the singularities of evolvent systems at the inflection points of plane curves.

The icosahedron appears at an inflection point as mystically as it does in Kepler's law of planetary distances. I believe, however, that in our problem the appearance of the icosahedron is more relevant than in Kepler's case; I hope that the remaining group H_4 will appear naturally in the analysis of the more complicated singularities of ray systems and wave fronts.

The main theme of this paper is the application of the relation between singularities of ray systems and reflection groups. The results I shall discuss are now included in symplectic and contact geometries under the names of Lagrangian and Legéndrian singularity theories. But one may consider them as part of the calculus of variations, or of control theory, of PDE theory, or of classical mechanics, of optics, or of wave theory, of algebraic geometry, or of general singularity theory. Some of these results deal with objects so basic, that it seems strange that the classics have missed them. For example, the local classification of projections of surfaces in general position in the usual 3-space was discovered only in 1981.

The number of nonequivalent projection germs is 14: the point neighbourhoods on generic surfaces generate exactly 14 different patterns when the surfaces are seen from different points of 3-space.

The reason is perhaps the difficulty of the proofs: they depend on the relations (sometimes unexpected) to invariant theory, Lie algebras, reflexion groups, algebraic geometry, and Deligne mixed Hodge structures. Some of the results were stimulated by applications of singularity theory to perturbation analysis of Hamiltonian dynamical systems, and even to number theory, but most new concepts came from the problem of bypassing an obstacle in Euclidean 3-space.

In order to describe these new results I must recall some well-known notions.

1. Symplectic geometry

A *symplectic structure* on an even-dimensional smooth manifold is a closed nondegenerate differential 2-form on it.

Examples: 1. The oriented area element defines a symplectic structure on the plane. 2. The direct product of symplectic manifolds has a natural symplectic structure. 3. The phase space of classical mechanics (the total space of the cotangent bundle of a smooth manifold) has a natural "$dp \wedge dq$" symplectic structure. 4. One may equip the manifold of oriented lines in Euclidean space with the symplectic structure of the total space of the cotangent bundle of the sphere, since these two manifolds are diffeomorphic. 5. The *characteristic direction* at a point of a hypersurface in a symplectic manifold is the skew-ortho-complement to the tangent plane. The *characteristics* on a hypersurface are the integral lines of its field of characteristic directions. The manifold of characteristics inherits a symplectic structure from the original manifold. 6. In particular, the manifold of extremals of general variational problem, lying at the same level manifold of the Hamiltonian function, is equipped with a natural symplectic structure. 7. Consider the space of odd-degree binary forms. There exists a unique (up to constant multiple) nondegenerate SL_2-invariant bilinear skew form on this even-dimensional linear space. This form defines a natural symplectic structure on the space of binary forms. 8. The binary forms in x and y, with coefficient in front of x^{2k+1} equal to 1, form a hyperplane in the space of all forms. The manifold of characteristics of this hyperplane can be identified with the manifold of even-degree polynomials in x of the form $x^{2k}+\dots$ We have thus equipped this space of even-degree polynomials with a symplectic structure. 9. The one-parameter group of shifts along the x-axis preserves this symplectic structure. The Hamiltonian function of this group is a polynomial of degree 2, known already to Hilbert (1893). The manifold of characteristics of a level hypersurface of the Hamiltonian function can be identified with the manifold of polynomials $x^{2k-1}+\dots$ with sum of roots equal to 0. We thus get a natural symplectic structure on this space of polynomials.

THEOREM (G. Darboux, 1882). *All the symplectic structures on manifolds of a fixed dimension are locally diffeomorphic.*

Thus, every symplectic structure is locally reducible to the normal form $\sum dp_i \wedge dq_i$ by a suitable choice of local "*Darboux coordinates*" p_i, q_i.
Let us now consider submanifolds of a symplectic manifold. The restriction to the submanifold of the symplectic structure is a closed 2-form, but it is not necessarily nondegenerate. In Euclidean space there is not only the inner geometry of a submanifold, but also an extensive theory of exterior curvatures. In the symplectic case the situation is much simpler:

THEOREM (A. B. Givental, 1981). *The germ of a submanifold of a symplectic manifold is determined (up to a symplectic diffeomorphism) by the restriction of the symplectic form to the tangent spaces of the submanifold.*

An intermediate theorem, dealing with vectors nontangent to the submanifold, was proved by A. Weinstein (1973). Unlike the Weinstein theorem, the Givental theorem implies the classification of the germs of generic submanifolds in a symplectic space: one uses the classification of the degeneracies of symplectic structures obtained by J. Martinet (1970) and his followers.

Examples: 1. The germs of a generic 2-surface in a symplectic manifold are locally symplectomorphic (symplectically diffeomorphic) to those of the surface $p_2 = p_1^2$, $q_1 = 0$, $p_3 = q_3 = \dots = 0$ (we use the Darboux coordinates). 2. On a 4-submanifold one encounters stably the curves of elliptic and hyperbolic Martinet singularities with normal forms

$$p_2 = p_1 p_3 \pm q_1 q_2 + q_3^3/6, \quad p_3 = 0, \quad p_4 = q_4 = \dots = 0.$$

The ellipticity and hyperbolicity concern the character of the motions in a dynamical system related intrinsically to the submanifold. The relevant divergence-free vector field on a 3-dimensional manifold has a curve of singular points. The classification at singular curves turns out to be less pathological than that at singular points (the latter being almost as difficult as the whole of celestial mechanics).

I have thus described the first steps of the symplectic singularity theory of smooth submanifolds.

A *Lagrangian submanifold* of a symplectic manifold is a submanifold on which the restriction of the symplectic structure vanishes, and which has highest possible dimension (equal to half of the dimension of the symplectic manifold).

Examples: 1. The fibers of the cotangent bundle. 2. The manifold of lines normal to a smooth submanifold (of arbitrary dimension) in Euclidean space. 3. The set of all polynomials $x^{2m}+\dots$ divisible by x^m.

A *Lagrangian fibration* is a fibration whose fibers are Lagrangian submanifolds.

Examples: 1. The cotangent bundle. 2. The fibration sending an oriented line in Euclidean space to the corresponding unit vector at the origin.

All Lagrangian fibrations of a given dimension are locally symplectomorphic (in the neighbourhood of each point of the total space).

A *Lagrangian mapping* is a diagram $V \to E \to B$, where the first arrow is an immersion of a Lagrangian submanifold, and the second is a Lagrangian fibration (Fig. 3).

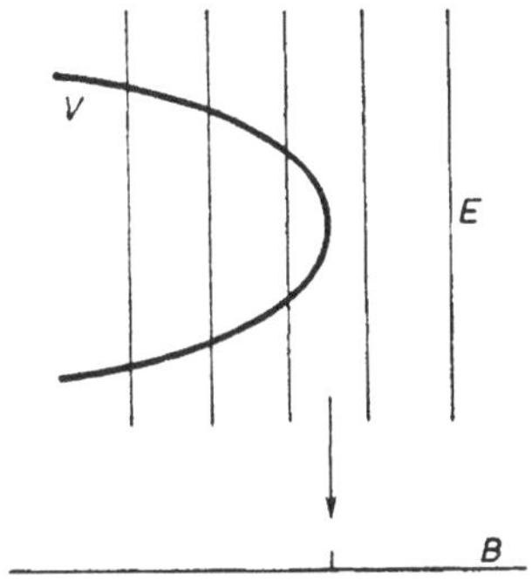

[Fig. 3

Examples: 1. The gradient mapping: $q \mapsto \partial S/\partial q$. 2. The normal mapping: associate to each vector normal to a submanifold in Euclidean space its end point. 3. The Gaussian mapping: associate to each point of a transversally oriented hypersurface in Euclidean space the unit vector at the origin in the direction of the normal at that point. (The corresponding Lagrangian manifold consists of the normals to that hypersurface.)

An *equivalence* between Lagrangian mappings is a fiber-preserving symplectomorphism between the total spaces of the fibrations, mapping the first Lagrangian submanifold onto the second one.

The set of critical values of a Lagrangian mapping is called its *caustic*. The caustics of equivalent mappings are diffeomorphic.

Example. The caustic of the normal mapping of a surface is the envelope of its normals, i.e., its focal surface (the surface of the curvature centers).

Every Lagrangian mapping is locally equivalent to a gradient one (to a normal one, to a Gaussian one). The singularities of generic gradient (normal, Gaussian) mappings are equivalent to those of generic Lagrangian mappings. These singularities are classified by the Euclidean reflection groups A, D, E.

Example. Consider a medium of dust-like particles moving inertially whose velocities form a potential field. After a time interval t a particle moves from x to $x+t\ \partial S/\partial x$. We obtain a one-parameter family of smooth mappings $\boldsymbol{R}^3 \to \boldsymbol{R}^3$.

These are Lagrangian mappings. Indeed, a potential field of velocities defines a Lagrangian section of the cotangent bundle. The phase flow of Newton's equation sends the initial Lagrangian manifold to new Lagrangian manifolds, which, however, need not be sections (for large t): their projections to the base space may have singularities (Fig. 4). The caustics

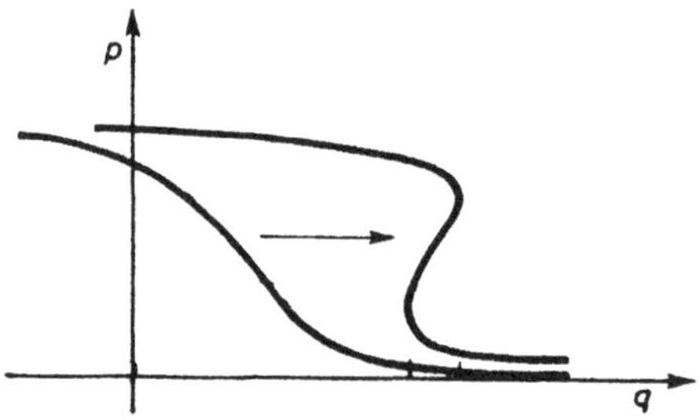

Fig. 4

of these mappings are the places where the density of particles becomes infinite. According to Ya. B. Zel'dovich (1970), a similar model (taking into account gravitation and expansion of the Universe) describes the generation of the large-scale nonuniformity of the distribution of matter in the Universe.

The theory of Lagrangian singularities implies that a new-born caustic has the shape of a saucer (at moment t after its birth the saucer's axes

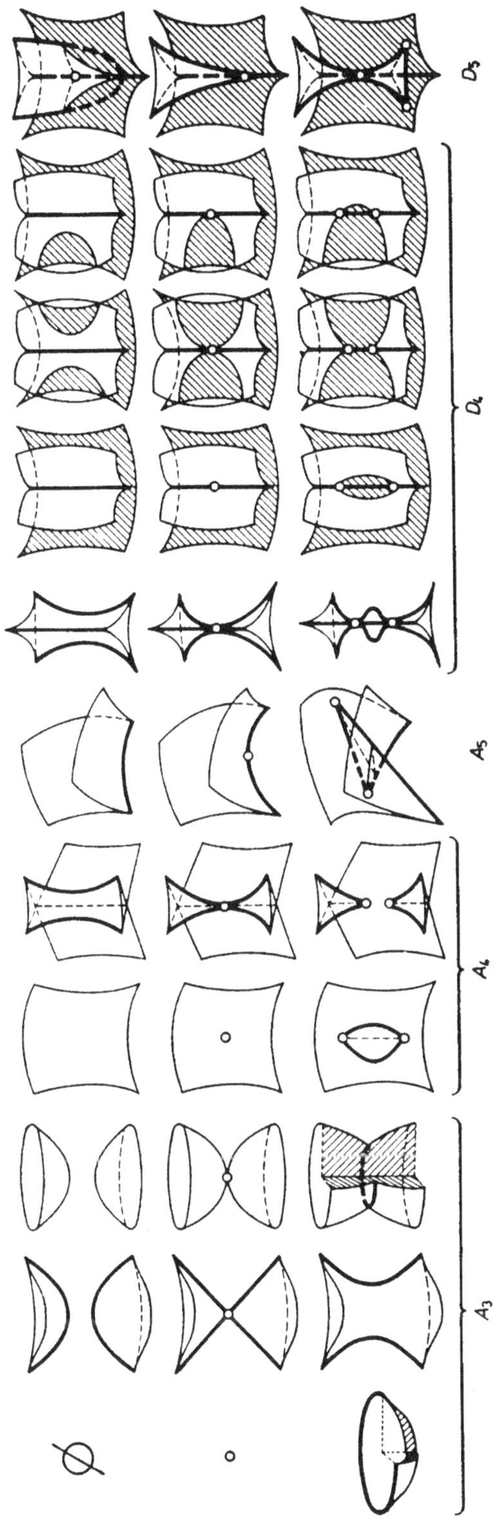

Fig. 5

are of order $t^{1/2}$, its depth of order t and thickness of order $t^{3/2}$). The saucer's birth corresponds to A_3. All metamorphoses of caustics in generic one-parameter families of Lagrangian mappings in 3-space are presented in Fig. 5 (1976).

THEOREM (1972). *The germs of generic Lagrangian mappings of manifolds of dimension* $\leqslant 5$ *are stable and simple (have no moduli) at every point. The simple stable germs of Lagrangian mappings are classified by the* A, D, E *Euclidean reflection groups, as explained below.*

2. Contact geometry

A *contact structure* on an odd-dimensional smooth manifold is a non-degenerate field of hyperplanes in the tangent spaces. The exact meaning of "nondegenerate" is irrelevant because of the "Darboux contact theorem": in the neighbourhood of a generic point, all generic fields of hyperplanes on a manifold of a fixed odd dimension are diffeomorphic.

Examples: 1. The *space of contact elements* of a smooth manifold consists of all its tangent hyperplanes. The velocity of an element belongs to the hyperplane defining the contact structure, if and only if the velocity of the contact point belongs to that element. 2. The space of 1-jets of functions $y = f(x)$ has a natural contact structure $dy = p\,dx$ ($p = \partial f/\partial x$ for the 1-jet of $y = f(x)$ at x).

The external geometry of a submanifold of a contact space is locally determined by the internal one, i.e., by the contact structure traces on the tangent spaces (the Givental contact theorem).

An integral submanifold of a contact manifold is said to be *Legendrian* if it has the highest possible dimension.

Examples: 1. The set of all contact elements tangent to a fixed submanifold (of arbitrary dimension). 2. In particular, the contact elements at a given point form a Legendrian manifold (the fibre of the bundle of contact elements). 3. The set of 1-jets of a function.

A fibration is said to be *Legendrian* if its fibers are Legendrian submanifolds.

Examples: 1. The projective cotangent fibration (a contact element is sent to its contact point). 2. The fibration of 1-jets of a function over its 0-jets (forgetting derivatives).

All Legendrian fibrations of a given dimension are locally contactomorphic (at every point of the total space).

The projection of a Legendrian submanifold to the base of a Legendrian fibration is called a *Legendrian mapping*. Its image is called a *front*.

Examples: 1. *The Legendre transformation*: A hypersurface in a projective space can be lifted to the space of its contact elements as a Legendrian submanifold. The manifold of contact elements of the projective space fibers over the dual projective space (associate to a contact element the hyperplane containing it). This fibration is Legendrian. The projection maps the lifted Legendrian manifold to the hypersurface which is projectively dual to the original hypersurface. Thus the projective dual of a smooth hypersurface is a Legendrian mapping front. 2. *The equidistant mapping*: Pick a point on every oriented normal to a hypersurface in Euclidean space, at distance t from the hypersurface (along the normal). We get a Legendrian mapping whose front is equidistant from the given hypersurface.

Legendrian equivalence, stability and simplicity are defined by analogy with the Lagrangian case.

Every Legendrian mapping is locally equivalent to a mapping defined by a Legendre transformation, and to an equidistant mapping. The local Legendrian singularity theory coincides with that of singularities of Legendre transformations (or equidistant mappings, or wave fronts).

THEOREM (1973). *The germs of generic Legendrian mappings of manifolds of dimension $\leqslant 5$ are stable and simple at every point. The simple stable germs of Legendrian mappings are classified by the A, D, E Euclidean reflection groups: the Legendrian mapping fronts are holomorphically equivalent to the varieties of singular orbits of the corresponding reflection groups.*

Example. The singularities of a generic wave-front in 3-space are (semicubical) cuspidal edges (A_2), and swallow-tails (A_3, Fig. 6: at these points the front is diffeomorphic to the surface in the space of polynomials x^4+ax^2+bx+c, consisting of the polynomials having multiple roots).

Remark. The necessity to complexify in the above theorem suggests that Euclidean reflection groups may have different real forms.

All Lagrangian singularities can be constructed from the Legendrian ones. For this, one considers Legendrian submanifolds of the space of 1-jets of functions. By forgetting the value of the functions one projects the jet space onto the phase space. The Legendrian manifolds' germs are projected isomorphically onto the Lagrangian ones. For instance, the caustic of a Lagrangian mapping is the projection of the cuspidal edge of the Legendre mapping front under a generic projection with 1-dimensional fibers.

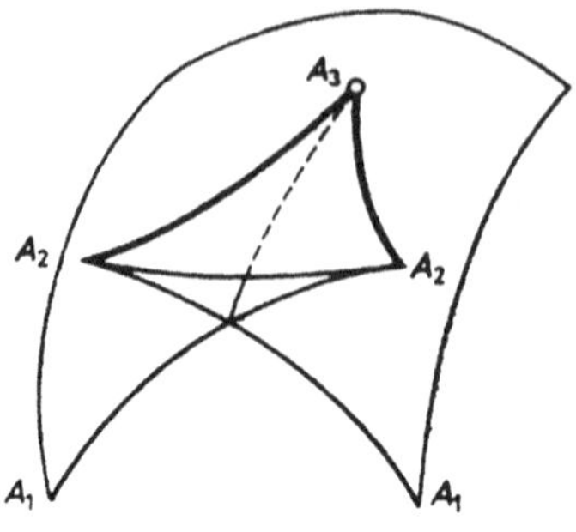

Fig. 6

THEOREM (O. V. Lyashko, 1979). *All holomorphic vector fields transversal to a front of a simple singularity can be mapped one onto another by front-preserving holomorphic diffeomorphisms germs.*

Example. A generic vector field in a neighbourhood of the "most singular point" of the swallow-tail $\{x^4+ax^2+bx+c = (x+a)^2\ldots\}$ is reducible to the normal form $\partial/\partial c$ (Fig. 7) by a swallow-tail-preserving diffeomorphism.

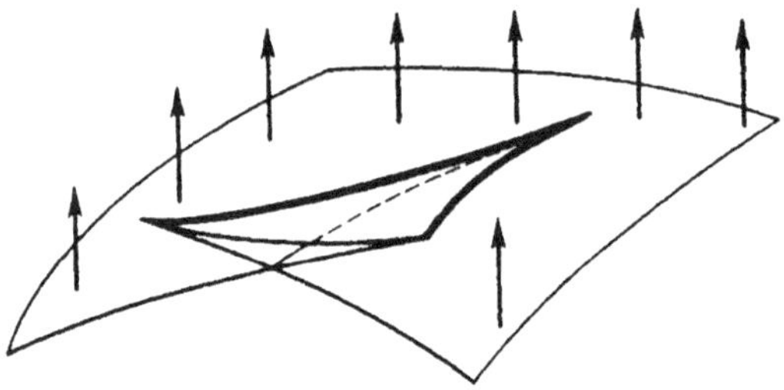

Fig. 7

The reduction to normal form of various geometric objects by wave-front or caustic-preserving diffeomorphisms is the main technical tool in the geometry of ray systems and wave fronts. For instance, the study of the metamorphoses of moving wave fronts is based on a result "dual" to the Lyashko theorem.

THEOREM (1976). *Generic holomorphic functions equal to 0 at the "most singular" point of a simple singularity front can be mapped one onto another by front-preserving holomorphic diffeomorphisms germs.*

Example. A generic function at the most singular point of a swallow-tail is reducible to the normal form a+const by a swallow-tail-preserving diffeomorphism.

The theorem above follows from the equivariant Morse lemma. We use it as follows. The momentary wave fronts form a "large front" in the space-time. Reduce the time function in the space-time to normal form by a large-front-preserving diffeomorphism. We obtain the normal form of the metamorphosis of the momentary front.

The infinitesimal diffeomorphisms preserving a front are the vector fields tangent to it. Their study leads to a "convolution operation" on the invariants of the reflection group. This operation associates to a pair of invariants (i.e., of functions on the orbit space) a new invariant – the scalar product of the gradients of the given functions (lifted from the orbit space to the Euclidean space).

The linearization of this operation is a bilinear symmetric operation on the space cotangent to the orbit space at 0.

THEOREM (1979). *The linearized convolution of the invariants is equivalent to the operation $(p, q) \mapsto S(p \cdot q)$ on the local algebra of the corresponding singularity, where $S = D + (2/h)E$, h is the Coxeter number, and D is the Euler quasihomogeneous derivation.*

For the exceptional groups this theorem was proved by A. B. Givental. In his joint work with A. N. Varchenko (1981) the theorem is extended to higher quasihomogeneous singularities. In this extension they substitute the Euclidean structure by the intersection form of a suitable non-degenerate period mapping. This period mapping comes from a family of holomorphic differential forms on the fibers of the Milnor fibration associated to a versal deformation of a function. A nondegenerate intersection form determines (according to the parity of the number of variables of the function) either a locally flat pseudo-Euclidean metric with a standard singularity at the Legendrian front, or a symplectic structure which is holomorphically extendable to the front.

Example. The set of odd-degree polynomials having highest coefficient equal to 1 and sum of the roots equal to 0 is thus equipped with a new symplectic structure. The variety of polynomials with maximal possible number of double roots is a Lagrangian subvariety.

3. Applications of Lagrangian and Legendrian singularities

The theory was first developed for the study of asymptotics of oscillatory integrals by the stationary phase method. I shall not discuss these (very important) applications here in detail, but shall rather mention: (1) Varchenko's (1976) proof of the formula describing the exponent of the main term of oscillatory integrals in terms of the Newton boundary of the phase function; (2) the example due to the same author of the nonsemicontinuity of this exponent; and (3) V. N. Karpushkin's (1981) proof of a uniform with respect to the parameters estimate from above of the double oscillatory integrals (for simple integrals such an estimate was obtained by I. M. Vinogradov, and for triple ones it was disproved by Varchenko's nonsemicontinuity example).

The uniform estimate also holds for all members of generic families of functions depending on a small number l of parameters (Duistermaat proved it in 1974 for $l \leqslant 6$; Colin de la Verdiere in 1977 for $l \leqslant 7$; Karpushkin in 1982 for $l \leqslant 9$); $l = 73$ is too large (the Varchenko example becomes possible).

The study of asymptotic expansions of oscillatory integrals in the complex domain has led Varchenko (1980–1981) to the construction of a mixed Hodge structure, which he calls the *asymptotic structure.* He has proved that its Hodge numbers coincide with the mixed Hodge numbers constructed algebraically by Steenbrink (1976). Among the corollaries of Varchenko's theory are: (1) the constancy of the Hodge structure invariants along the "μ = const" statum, and (2) the fact that the "inner modality" of quasihomogeneous functions coincides with their true modality. In real algebraic geometry the mixed structure gives some generalizations of the Petrovskii–Oleinik inequalities.

THEOREM (1978). *The local Poincaré index of a gradient vector field in $\boldsymbol{R}^{2n}$ is bounded from above by the middle Hodge number* $|\mathrm{ind}| \leqslant h_1^{n,n}$.

The singularity mixed Hodge structure associates to a finite multiplicity critical point of a function a finite set of rational numbers, the critical points *spectrum.* The spectrum's left end is the smallest exponent of the oscillatory integrals with a given phase function (along complex chains). The examples show the semicontinuity of this exponent, as well as of all the other spectrum points. For instance, the spectrum obtained through a deformation reducing the multiplicity by one, divides the initial spectrum (in the same way as the axes of an ellipsoid divide the axes of the initial ellipsoid).

The spectrum semicontinuity conjecture (1978) was recently confirmed by the works on an apparently unrelated to it algebraic geometry problem: *how large can the numbers of (Morse) singular points on a hypersurface of degree d in CP^n be?*

Bruce (1981) gives an estimate from above, which asymptotic (for the surfaces in CP^3) is $d^3/2 + \ldots$ (the best estimates from below are of order $3d^3/8$, S. Chmutov, 1983). Comparing the first exactly known answers (0, 1, 4, 16, 31, 64) with the mixed Hodge structure, I have formulated the following

CONJECTURE. *The number of singular points does not exceed the number of integer points m of the cube $(0, d)^n$, for which $(n-2)d/2+1 < \sum m_i \leqslant nd/2$.*

For surfaces in 3-space this implies an estimate from above $23d^3/48 + \ldots$ Trying to prove this cojecture A. B. Givental in October of 1982 improved the lower order terms in the Bruce estimate. His proof uses some Rayleigh–Fisher–Courant type inequalities and makes transparent the relation of the problem to the spectrum semicontinuity conjecture.

A. N. Varchenko immediately applied to this problem the Steenbrink (1976) theorem on the limits of the Hodge structures. Thus he proved both the conjectured estimate of the number of singular points and the spectrum semicontinuity (the last – for quasihomogenous function deformations, generated by adding lower weight monomials). The same way he proved the semicontinuity of the left end of the spectrum for all functions in 3 variables and for functions in n variables having "far away" Newton polyhedra.

I shall also mention the applications of Lagrangian singularities to the mechanical quadrature theory, i.e., to the problem of integer points in large domains. Let V be the volume of a smooth boundary domain G in the Euclidean $\mathbf{R}^n$, and $N(\lambda)$ the number of integer points inside λG, $R(\lambda) = \lambda^n V - N(\lambda)$. The Lagrangian singularity theory implies the following results:

THEOREM (Colin de Verdiere, 1977). *For $n \leqslant 7$ generically*

$$|R(\lambda)| \leqslant C\lambda^{n-2+2/(n+1)}.$$

THEOREM (Varchenko, 1981). *The average $|R(\lambda)|^2$ over all lattices obtained from the integer point lattice by rotations and shifts, does not exceed $C\lambda^{n-1}$.*

The convex analytic case was studied by Randol (1969). The exponent $(n-1)/2$ is what one might expect according to the law of large numbers (if the λ^{n-1} cells were divided by the boundary independently). The proof of the last theorem is inspired by the Duistermaat (1974) proof of the Maslov "canonical operator" unitarity.

The statistics of Newton diagrams of singularities has led to another inequality related to integer points.

THEOREM (K. A. Sevastianov, S. V. Konyagin, 1982). *The number of vertices of a volume V convex polyhedron in $\boldsymbol{R}^n$, whose vertices are integer points, does not exceed $CV^{(n-1)/(n+1)}$ (the same estimate holds also for the number of faces of arbitrary dimension).*

The influence of the boundary inflections on the remainder term of the asymptotic of the number of integer points is a particular case of the interrelations between the integer and smooth structures of $\boldsymbol{R}^n$, which are crucial for many branches of calculus.

For instance, the order of approximation of a typical point of a submanifold by the hyperplanes defined by the equations with not too large integer coefficients is essential for the resonance phenomena in the theory of nonlinear oscillations (the flattening of the fast frequencies' manifold enhances the sticking at resonances).

In his study of evolutions of action variables in Hamiltonian systems, N. N. Nehoroshev introduced "steepness exponents" of the unperturbed Hamiltonian function. The calculation of these exponents for a generic Hamiltonian function has inspired the theory of tangential singularities.

4. Tangential singularities

These are singularities of the arrangement of a projective surface with respect to its tangents of all dimensions.

Example. The tangential classification of points on a generic surface in 3-space (Fig. 8) was found by O. A. Platonova and E. E. Landis (1979). A line (p) of parabolic points divides the surface into the domain (e) of elliptic points and that of hyperbolic ones (h) containing the curve (f) of inflection points of the asymptotic lines with the biinflection points (b), the selfintersection points (c), and the points of tangency to the parabolic line (t).

This classification is useful both for Nehoroshev's exponent estimate and for the classification of projection degenerations.

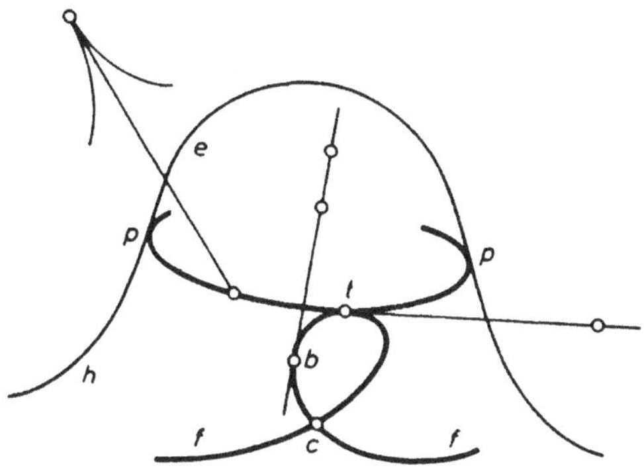

Fig. 8

THEOREM (O. A. Platonova, O. P. Shcherbak, 1981). *Project a generic surface from* $\boldsymbol{R}P^3$ *to a plane along the straight lines passing through a projection center (a point outside the surface).*

All the projections thus obtained are locally equivalent to the 14 *projections of surfaces* $z=f(x, y)$ *along the* x*-axis, where* f *is given by the list*

$$x, \quad x^2, \quad x^3+xy, \quad x^3 \pm xy^2, \quad x^3+xy^3, \quad x^4+xy, \quad x^4+x^2y+xy^2,$$
$$x^5 \pm x^3y+xy, \quad x^3 \pm xy^4, \quad x^4+x^2y+xy^3, \quad x^5+xy.$$

Here the projections are considered as the diagrams $V \to E \to B$ consisting of imbeddings and fibrations, and the equivalences are 3×2-diagrams, whose verticals are diffeomorphisms.

The singularities of a projection from a generic center are only Whitney folds and cusps (one sees a cusp along every asymptotic ray). Other singularities require special points of view. The finiteness of the list of normal forms of projections (and hence that of the list of visible contours) is not evident a priori, because there exists a continuum of nonequivalent singularities in generic 3-parameter families of projections of surfaces to the plane.

The hierarchy of tangencies may become more transparent in terms of the symplectic and contact geometries. Melrose (1976) remarked that the tangent ray geometry of a surface in Euclidean space depends on two hypersurfaces in the symplectic phase space: the first describes the metric and the second — the surface.

The same pair of hypersurfaces describes the hierarchy of asymptotic tangents. Thus we are able to transfer a large part of the geometry on the usual space surfaces to the general case of arbitrary hypersurface pairs in symplectic or contact spaces, using the geometrical intuition of the surface theory for the study of general variational problems with one-sided phase constraints.

Let Y and Z be two hypersurfaces in symplectic space X, intersecting transversally along a submanifold W. Projecting Y and Z onto their characteristics' manifolds U, V, we obtain a hexagonal diagram

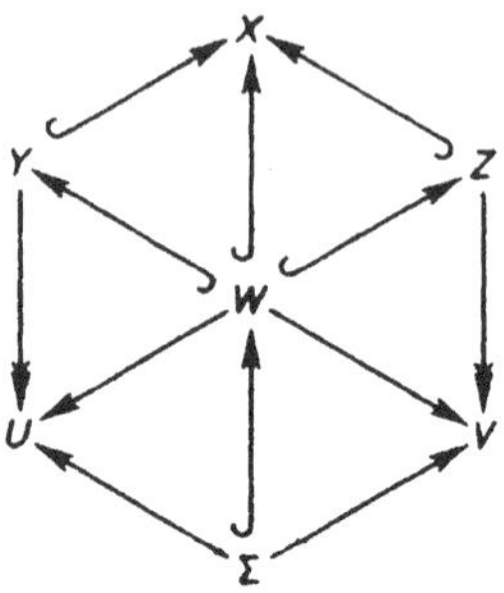

where Σ is the (common) manifold of critical points of the projections from W to U and to V.

Example. Let $X = \{q, p\}$ be the phase space of a Euclidean free particle (q is the particle position, p – its momentum); Y – the manifold of unit vectors ($p^2 = 1$); Z – the manifold of boundary vectors (q belongs to a hypersurface Γ). Then U is the ray space, V is Γ's tangent bundle space, W – the bundle space of the boundary (not necessarily tangent) unit vectors and Σ – the spherical tangent bundle space.

Singularities of both projections $W \to U$ and $W \to V$ at a nonasymptotic tangent unit vector are Whitney folds. Each projection defines an involution on W which is the identity on Σ.

Example. We have defined two involutions σ and τ on the manifold W of boundary unit vectors of a convex plane curve (Fig. 9). The product of involutions is the Birkhoff (1927) billiards transformation.

Melrose used the involution pairs to reduce the symplectic space hypersurface pairs to a local normal form by a C^∞-symplectomorphism (in the analytic case the series obtained are generically divergent, as is the case in the Ecalle (1975) and Voronin (1981) theories of dynamical systems at resonances).

At more complicate singularities (for instance, at asymptotic unit vectors) the symplectic space hypersurface pairs have moduli. However, one can reduce the pair formed by the first hypersurface and the intersection to simple normal form (at least formally), for the first two degeneracies of the fold. Thus we can study the singularities of the mapping which associates the ray to a boundary unit vector at the generic asymptotic and biasymptotic unit vectors.

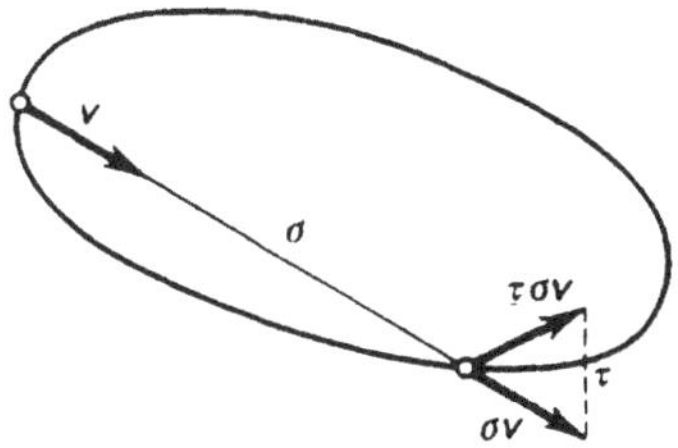

Fig. 9

The variety of critical values of this mapping is locally diffeomorphic to the product of the usual swallow-tail with a linear space. This variety lies in the symplectic space of straight lines in a standard manner:

THEOREM (1981). *All the generic symplectic structures at the point of the critical variety described above are locally reducible one into another by a critical-variety-preserving formal diffeomorphism.*

At a biasymptotic ray the variety of tangent rays is locally diffeomorphic to the product of a swallow-tail with a line. So the above theorem describes the symplectic geometry of the variety of tangent rays.

5. The obstacle problem

Consider an obstacle bounded by a smooth surface in Euclidean space. The *obstacle problem* requires a study of the singularities of the shortest path lenght from a point in the space to a fixed initial set, among paths avoiding the obstacle. This simple variational problem on a manifold with boundary is unsolved even for generic obstacles in 3-space.

The shortest path consists of segments of straight lines and of geodesics on the obstacle surfaces (Fig. 10). Hence let us consider the system of geodesics orthogonal to a fixed front. The system of all rays tangent to these geodesics is a Lagrangian variety in the symplectic space of all rays (as is every system of extremals of a variational problem). In the usual variational problems on manifolds without boundary the relevant Lagrangian variety is smooth (even in the presence of caustics). In the obstacle problem it may acquire singularities. The above theorem implies the following

COROLLARY (1981). *The Lagrangian variety in the generic obstacle problem has a semicubical cuspidal edge at the generic asymptotic rays and an "open swallow-tail" singularity at the bi-asymptotic rays.*

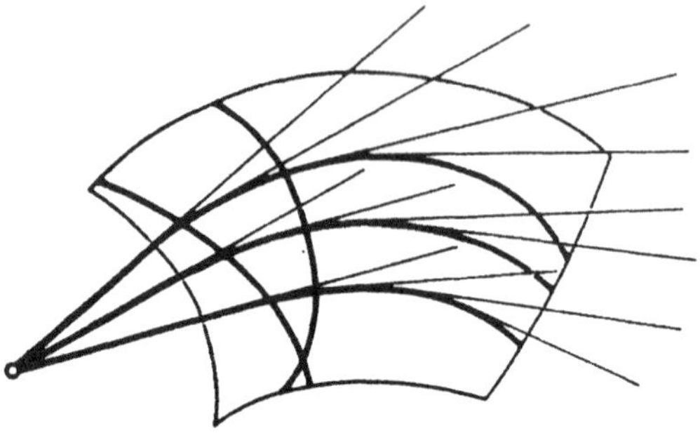

Fig. 10

The *open swallow-tail* is the surface in 4-space $\{x^5+Ax^3+Bx^2+Cx+D\}$ consisting of polynomials with at least triple roots. The differentiation of polynomials maps the open swallow-tail onto the usual one. The opening of the swallow-tail eliminates the selfinetersections but preserves the cuspidal edge (Fig. 11).

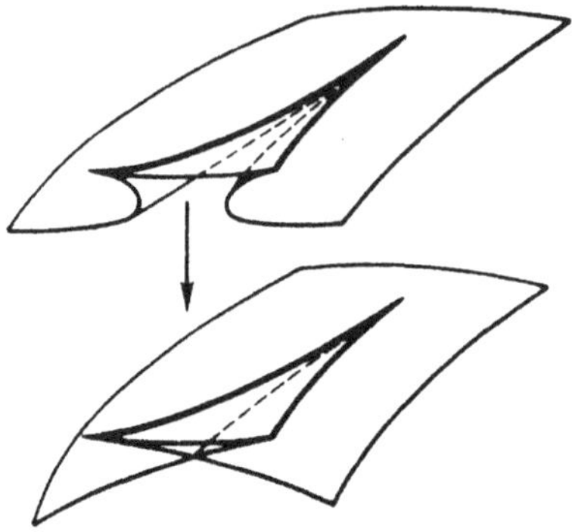

Fig. 11

THEOREM (1981). *The cuspidal edges of the wave fronts moving generically in 3-space form an open swallow-tail in the space-time (over the usual swallow-tail of the caustic).*

THEOREM (O. P. Shcherbak, 1982). *Consider a generic one-parameter family of space curves and suppose that for a given value of the parameter (of the time) the family curve has a biflatness point (of type 1, 2, 5). Then the projective dual curves form a surface in the space-time, which is locally diffeomorphic to the open swallow-tail.*

The open swallow-tail is a first representative of a large series of singularities. Consider the set of polynomials with a root of fixed comultiplicity k, $((x-a)^{n-k}(x^k+\dots))$ in the space of polynomials $x^n+\lambda_1 x^{n-2}+ +\dots+\lambda_{n-1}$. The differentiation of polynomials preserves the comultiplicities of the roots.

THEOREM (A. B. Givental, 1981). *The sequence of varieties of polynomials with roots of fixed comultiplicity stabilizes as the degree n increases, starting with $n = 2k+1$ (i.e., at the moment of the dissociation of the self-intersections).*

Example. The open swallow-tail is the first stable variety over the usual swallow-tail.

The following Givental theory of triads (1982) formalizes the appearance of the open swallow-tail in the obstacle problem.

DEFINITION. A *symplectic triad* (H, L, l) consists of a smooth hypersurt face H in a symplectic manifold, and of a Lagrangian manifold L tangento H (with first order tangency) along a Lagrangian manifold hypersurface l.

The *Lagrangian variety generated by the triad* is the image of l in the manifold of characteristics of H.

Example 1. In the obstacle problem with boundary $\Gamma \subset \boldsymbol{R}^n$ let us consider the distance, along the geodesics of Γ, to the initial front as a function $s\colon \Gamma\to\boldsymbol{R}$. The manifold L of all the extensions of the 1-forms ds from Γ to $\boldsymbol{R}^n$ forms a triad together with the hypersurface $H\colon p^2 = 1$.

This triad generates precisely the variety of rays tangent to the geodesics of our system of extremals on Γ.

Example 2. Consider the symplectic space of polynomials $\mathscr{F} = x^d + +\lambda_1 x^{d-1}+\dots+\lambda_d$ of an even degree $d = 2m$. The polynomials, divisible by x^m, form a Lagrangian submanifold L. Let h be the Hamiltonian function of the shifts along the x axis. (This polynomial in λ is

$$h = \sum(-1)^i \mathscr{F}^{(i)}\mathscr{F}^{(j)}, \quad i+j = d, \quad \mathscr{F}^{(i)} = d^i\mathscr{F}/dx^i.)$$

The hypersurface $h = 0$ is tangent to the Lagrangian manifold L along its hypersurface l of polynomials divisible by x^{m+1} and forms with them a triad.

The variety generated by this triad is the Lagrangian open swallow-tail of dimension $m-1$ (the set of polynomials $x^{d-1}+a_1x^{d+3}+\dots+a_{d-2}$ having a root of larger multiplicity than half the degree).

THEOREM (A. B. Givental, 1982). *The triad of Example 2 is stable. Germs of generic triads at all points are symplectically equivalent to those of Example 2.*

COROLLARY. *The variety of rays, tangent to the geodesics of the system of extremals in the generic obstacle problem is locally symplectically equivalent to the Lagrangian open swallow-tail.*

In contact geometry two sorts of Legendrian varieties are associated to the obstacle problem: the varieties of the contact elements of the fronts and the varieties of 1-jets of multi-valued time functions. The varieties of the first type ar egenerically the Legendrian open swallow-tails (they are diffeomorphically lifted Lagrangian swallow-tails). The varieties of the second type are the cylinders over the former.

Example. Consider the obstacle bounded by a plane curve with an ordinary inflection point. The fronts are the curve evolvents. They have two singularities: a usual cusp (of order 3/2) at the boundary curve of the obstacle and a 5/2-singularity at its inflectional tangent (Fig. 12). The

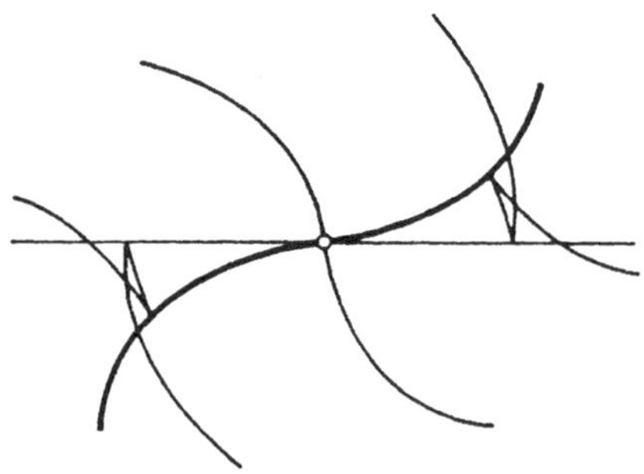

Fig. 12

Legendrian variety is nonsingular over the generic points of the obstacle curve, but over the inflectional tangent points the Legendrian variety has a cuspidal edge of order 3/2.

Let us consider the 3-space of the plane contact elements (fibered over the plane). All contact elements of all the evolvents of a generic curve form a surface in this 3-space. Let us consider the 3-space of polynomials x^3+ax^2+bx+c (fibered over the plane of their derivatives). All those polynomials having multiple roots form a surface in this 3-space.

THEOREM (1978). *The germ of the first surface at the tangent at an inflection point of a generic curve is diffeomorphic to the germ of the second surface at zero, by a fibre-preserving diffeomorphism.*

This surface (Fig. 13), together with the $c = 0$ surface representing the plane contact elements at the obstacle boundary points, forms the variety of singular orbits for the reflection group B_3. This remark has led to the boundary singularity theory (1978), of which I shall only mention the following.

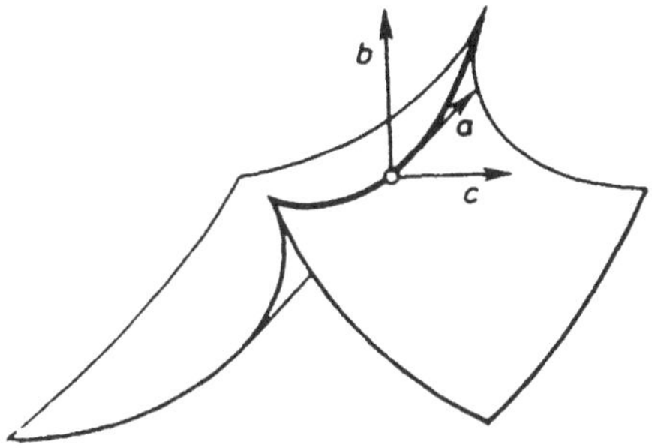

Fig. 13

Example (I. G. Shcherbak, 1982). Consider a generic curve on a generic surface in Euclidean 3-space. At some points the curve touches the surface curvature line. The boundary Lagrangian singularity theory implies that this situation is governed by the exceptional Weyl group F_4: the union of the focal sets of the surface and of the curve with all the surface normals at the points of the curve forms a variety which is locally diffeomorphic to the F_4 caustic.

The boundary Lagrangian singularity theory implies an amusing "Lagrange duality", which interchanges the singularity of a function on the ambient space with that of its restriction to the boundary: this duality is a modernized version of the "Lagrange multipliers rule" (I. G. Shcherbak, 1982).

Returning to an inflection point of a plane curve, consider the graph of the (multi-valued) time function for the obstacle problem. The level sets of this time function are the obstacle evolvents. Hence the graph has the shape drawn in Fig. 14 (1978); this surface has two cuspidal edges (of orders 3/2 and 5/2).

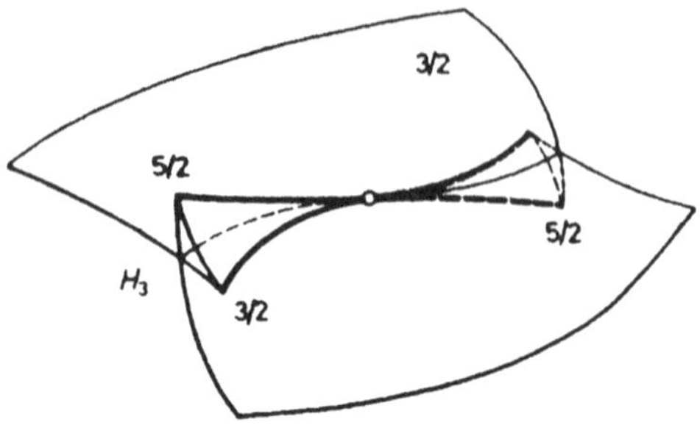

Fig. 14

When I showed this surface to A. B. Givental (1982), he recognized the singular orbit variety of the group H_3 of symmetries of the icosahedron drawn by O. V. Lyashko (1981). Givental's conjecture was rapidly confirmed:

THEOREM (O. P. Shcherbak, 1982). *The graph of the (multi-valued) time function in the generic plane obstacle problem is diffeomorphic to the variety of singular orbits at the inflection points of the obstacle boundary.*

THEOREM [1] (O. V. Lyashko, 1981). *The variety of singular orbits of H_3 is diffeomorphic of the space of polynomials $x^5+ax^4+bx^2+c$ having a multiple root.*

Lyashko's theorem describes the variety of singular orbits of H_3 as the union of the tangents to the curve (t, t^3, t^5) in 3-space, while Shcherbak deals with any curve $(t+\dots, t^3+\dots, t^5+\dots)$.

A generic front in the 3-space obstacle problem must have a singularity of the same type at the point of tangency of an asymptotic ray with the obstacle surface.

In this paper I have not even mentioned many important aspects of the Lagrangian and Legendrian singularity theory, especially the global ones, such as the theory of the coexistence of singularities (the Lagrangian and Legendrian cobordism theories reduced to homotopy problems by Ya. M. Eliashberg, the Lagrangian and Legendrian characteristic

[1] A similar description of $\Sigma(H_4)$ was found by O. P. Shcherbak in 1983: it is based on an inclusion of the H_4 graded local algebra defined by the invariants convolution, into the E_8 graded local algebra: $x^3+y^5+axy^3+by^3+cx+d$.

classes of V. A. Vasiliev, which are generalizations to higher singularities of V. P. Maslov's class, and so on).

I have not even mentioned the extensive classification of the simple projections (Goryunov, 1981), the theory in which, for instance, the exceptional root system F_4 is an ancestor of a whole family of descendants F_μ. One can find details of those theories and the extensive relevant bibliography in the surveys [1and [2].

In spite of the progress of the ray system geometry during the past three centuries, from Huygens up to now, the drawing of pictures very similar to those one finds in Huygens' works is still one of the main sources of new discoveries in this difficult domain where even the 3-dimensional problem is still unsolved and where numerous useful but unexpected interrelations with other branches of mathematics (such as relation of the obstacle problem to the group H_3 of symmetries of the icosahedron) still remain mysterious.

References

[1] Arnold V. I., Singularities of Rays Systems, *Uspekhi Mat. Nauk* **38** (2) (1983).
[2] Arnold V. I., Singularities in the Calculus of Variations, *Itogi Nauki, Sovremennye problemy mat.* **22**, VINITI, Moscow, 1983.
[3] Arnold V. I., *Catastrophes Theory*, Springer, 1983.

SINGULARITIES IN VARIATIONAL CALCULUS

V.I. ARNOLD

(The Petrovskiĭ seminar on Differential equations and mathematical problems of physics; meeting on April 18, 1984.[1])

1°. H_4 singularities. In 1972–1983 among the singularities of solutions of generic variational problems, there were successively found the varieties of irregular orbits of the Coxeter groups $A_\mu, \ldots, H_3$, so by 1984 only the group H_4 remained "jobless" (see [1] and [2]).

The group H_4 consists of the symmetries of a certain regular polyhedron in $\mathbb{R}^4$. Its 120 vertices lie on $S^3 \approx \mathrm{SU}(2)$ and form the binary group of the icosahedron (the binary group doubly covers the rotation group of the icosahedron for the covering $S^3 \to \mathrm{SO}(3)$). A variational interpretation of the variety of irregular orbits of the group H_4 has just been obtained by O.P. Shcherbak.

Consider an obstacle bounded by a smooth surface in the Euclidean space $\mathbb{R}^3$. The extremals joining a point outside of the obstacle to all points bypassing the obstacle form a pencil (a one-parameter family) of geodesics on the surface of the obstacle. The *time function* is by definition the distance to a fixed initial manifold (e.g., a point) along a stationary (not necessarily minimal) path consisting of segments of geodesics and their tangents. This distance is to be regarded as a (multi-valued) function of the destination point in the space (a solution of the Hamilton–Jacobi equation).

Theorem 1 (O.P. Shcherbak). *For a generic obstacle, the graph of the time function is locally diffeomorphic to the variety Σ of irregular orbits of the group H_4 at the focal (for the pencil) point of the asymptotic tangent to the geodesic of the pencil at a parabolic point of the surface.*

O.P. Shcherbak gave an explicit parametrization of Σ:

$$(a,\ b^2/2 + ac,\ c^2/2 + ab^3,\ b^5 + c^3/3 + ab^3c).$$

Theorem 2 (O.P. Shcherbak). *Consider the values of λ for which the curve $x^5 + y^3 + \lambda_1 x^3 y + \lambda_2 x^3 + \lambda_3 y + \lambda_4 = 0$ is singular. One of the irreducible components of this three dimensional hypersurface $\{\lambda\}$ is diffeomorphic to the variety of irregular orbits of the group H_4.*

The conjecture on a relation between H_4 and the described above subspace of the base of the versal unfolding of E_8 was proposed by the author in 1979 (see [3], page 26),[2] because the embedding of the local algebra D_4 into the local algebra E_8

[1]Translation of Uspekhi Mat. Nauk, **39** (1984), no. 5, p. 256.

[2]Page 27 of the English translation. (Editors' remark)

Originally publ. in: Usp. Mat. Nauk 39(5), 256, © 1984 Russ. Acad. Sciences. All rights reserved
Translated by Boris Khesin and Andrey Khesin
Springer-Verlag Berlin Heidelberg 2018 with kind permission

corresponding to this subspace induces on the former exactly the same grading as that determined by the operation of convolution of the H_4 invariants introduced in [3].

2°. Poisson structures on the plane and other powers of volume forms. A *Poisson structure* on the plane is given by a smooth bivector field $f\,\partial_x \wedge \partial_y$ and, hence, by the 2-form $f^{-1}\,dx \wedge dy$.

Theorem 3. *The hierarchy of singularities of Poisson structures on the plane starts with the types A, D, E (according to the singularities of the curves $f = 0$). For a fixed quasi-homogeneous curve, the dimension of the variety of the equivalence classes of structures is one less than the number of irreducible components of the curve (i.e., equal to the number $h^{1,1}$ of the mixed Hodge–Deligne–Steenbrink structure).*

The case A_0 provides an evident counterexample to A. Weinstein's conjecture on semi-simplicity of a structure whose equivalence class cannot be deformed by perturbations that do not contain a linear part. Theorem 3 can be carried over to any (even complex) powers of volume forms in $\mathbb{C}^n$ or $\mathbb{R}^n$.

Theorem 4. *Let f be a nondegenerate quasi-homogeneous polynomial. A form $f^\beta h\,dx_1 \wedge \cdots \wedge dx_n$, where $h(0) \neq 0$, can be reduced by a local diffeomorphism to a form where $h = \pm 1 + \varphi$, here φ is a quasi-homogeneous polynomial of such a weight that the weight of the form $f^\beta \varphi\,dx_1 \wedge \cdots \wedge dx_n$ is equal to 0 (e.g., $\varphi \equiv 0$ for $\beta > 0$).*

References

[1] V.I. Arnold. Singularities of ray systems. *Uspekhi Mat. Nauk*, 38 (1983), no. 2, 77–147. English translation: *Russian Math. Surv.*, 38 (1983), no. 2, 87–176.

[2] V.I. Arnold. Singularities in variational calculus. In: Itogi Nauki i Tekhniki. Sovremennye Problemy Matematiki (Principal Results of Science and Technology. Current Problems in Mathematics). Volume 22. VINITI, Moscow, 1983, 3–55. English translation: *J. Soviet Math.*, 27 (1984), no. 3, 2679–2713.

[3] V.I. Arnold. Indices of singular points of 1-forms on a manifold with boundary, convolution of invariants of reflection groups, and singular projections of smooth surfaces. *Uspekhi Mat. Nauk*, 34 (1979), no. 2, 3–38. English translation: *Russian Math. Surv.*, 34 (1979), no. 2, 1–42.

SINGULARITIES IN VARIATIONAL CALCULUS

V. I. Arnol'd

UDC 515.164.14+517.974

A survey is given of theories of singularities of systems of rays and wave fronts, that is, singularities of systems of extremals of variational problems and solutions of the Hamilton–Jacobi equations near caustics. The problem of passing about an obstacle bounded by a smooth surface of general position is studied in detail. Theorems are proved on the normal forms of Lagrangian manifolds with singularities formed by rays of the system of extremals of a variational problem in the symplectic space of all oriented lines which tear off from the surface of the obstacle as well as theorems on Legendre manifolds with singularities formed by contact elements of a wave front and 1-jets of a solution of the Hamilton–Jacobi equation.

We begin with an example: we consider the shortest path from a point of the plane to a given curve, for example, from a point of an elliptic region to its boundary (Fig. 1). The extremals of this variational problem are normals to the ellipse. The minimum value of the functional (the path length u) as a function of the initial point satisfies the Hamilton–Jacobi equation $(\nabla u)^2 = 1$ at points of smoothness. However, this function has singularities (on the segment joining foci of the ellipse). In the same way the system of extremals has singularities (on the astroid enveloping the family of normals to the ellipse). The enveloping system of extremals is called a caustic. The caustic of our problem has four cusp points. These singularities of the caustic are stable: any curve close to the ellipse (with derivatives) has near the astroid a caustic with four cusp points.

The level lines of a solution of the Hamilton–Jacobi equation are called fronts. In our example the fronts are equidistant curves of the ellipse. The equidistant curves close to the the ellipse are smooth curves. On the more distant equidistant curves there are singularities (the four cusp points of Fig. 2). These singularities are stable: the equidistant curves of any curve close to the ellipse have similar singularities.

Of course, the level line of the shortest distance to the ellipse is only part of the equidistant curve. However, many properties of singularities of solutions of variational problems are more easily understood if not only minima but also other critical points of functionals are considered (in the present case the equidistant curves are first studied globally and then the required parts are distinguished).

For example, we consider the distance to a curve as a function of a finite point. It is convenient to consider this function multivalued by taking account not only of the shortest distance but also the lengths of the remaining normals to the ellipse drawn from the given point.

The graph of the multivalued function of the distance from a point of the domain to the ellipse has in a neighborhood of a focus of the ellipse a characteristic singularity diffeomorphic to a "swallow tail" (Fig. 3). The swallow tail is a surface in the space of polynomials $x^4 + ax^2 + bx + c$ formed by polynomials with a multiple root. This singularity is stable: under small displacement of the ellipse the singularity on the graph diffeomorphic to the swallow tail is preserved.

The graph of the time function (motion with speed 1 to the boundary of the region) lies in space–time and intersects isochrones along equidistant curves. Hence, as the distance to the curve (say, to an ellipse) varies the equidistant curves experience the same metamorphosis (at the time of passing through the focus) as that experienced by sections of the swallow-tail surface in space–time by level surfaces of the time function.

It is proved in the theory of singularities that a function of general position in the space containing the swallow tail $(f(a, b, c)\partial f/\partial a|_0 \neq 0)$ can be reduced to the normal form

Translated from Itogi Nauki i Tekhniki, Seriya Sovremennye Problemy Matematiki, Vol. 22, pp. 3-55, 1983.

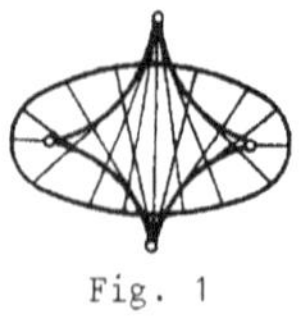

Fig. 1

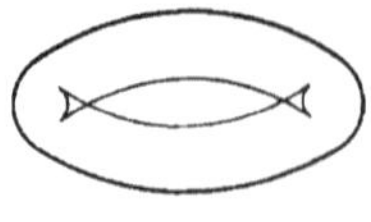

Fig. 2

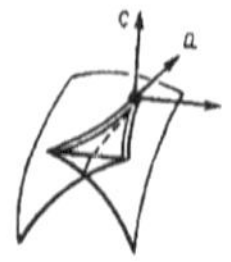

Fig. 3

Fig. 1. A system of extremals and its caustic.

Fig. 2. Singularities of an equidistant curve of an ellipse.

Fig. 3. Swallow tail.

$f = \pm a$ + const preserving the swallow tail by a local diffeomorphism of a neighborhood of the point $a = b = c = 0$ [30]. Hence, the metamorphoses of the equidistant curves are the same as the metamorphoses of sections of the swallow tail by planes a = const; the latter metamorphoses can be easily studied.

If we wish to restrict attention to minima, then it suffices to simply remove part of the swallow tail (namely, the pyramid abutting the cusp edges).

The schemes of applying the theory of singularities to the investigation of solutions of variational problems is the same in the general case as in the above example.

1. A list of standard singularities (generalizing cusp points and swallow tails) is set forth. These singularities are usually connected with the geometry of groups generated by reflections, and they can be studied in detail once and for all by suitable algebraic means (using Lie algebras, invariant theory, root systems, Dynkin diagrams, etc.).

2. It is proved that the standard singularities are stable and that only they are encountered in situations of general position. For example, the caustic of the circle consists of a single point — the center of the circle, but under small perturbation of general position this degenerate singularity at the center is converted into a curve close to the center whose only singularities are cusp points and points of transversal self-intersection.

3. After this the study of singularities and metamorphoses of various objects connected with a variational problem of general position reduces to the study of the corresponding properties of standard singularities. For example, the study of the metamorphoses of fronts reduces to the investigation of sections of generalized swallow tails by hypersurfaces.

In the present survey we formulate the basic results on the realization of this program obtained beginning in 1972 when the connection of singularities of solutions of variational problems with the theory of groups generated by reflections was first recognized [1]. Originally only the groups A, D, E (without double connections in diagrams) were used. After it was discovered in 1978 [5] that groups with double connections (B, C, F) are responsible for boundary singularities (for example, for the singularities of the distance to a manifold with boundary) only the groups of reflections $I_2(n)$, H_3, and H_4 remained unused.

In the autumn of 1982 it was discovered that the group H_3 (the symmetry group of the icosahedron) controls the singularities in the problem of passing about an obstacle in the plane bounded by a curve of general position with a point of inflection, i.e., the singularities of the family of Huygens evolvents of a plane curve near its point of inflection.

Thus far it is not known which variational problem leads to the group H_4, but there is little doubt that this group as well is encountered in the description of the singularities of systems of extremals and fronts of more involved variational problems.

The language of symplectic and contact geometry is the language in which it is most convenient here to formulate the results. The required general definitions and theorems of symplectic geometry are collected in Sec. 1 and those of contact geometry in Sec. 2. In Sec. 3 the problem of passing about an obstacle is discussed. Section 4 is devoted to applications of symplectic and contact geometries to the theory of asymptotic rays (i.e., to the hierarchy of extremals of the variational problem of passing about an obstacle). In Sec. 5 the algebraic machinery is prepared (generalized open swallow tails are studied) by means of which the main results on normal forms of systems of rays and fronts in the problem of passing about an obstacle are proved in Sec. 6.

1. Symplectic and Contact Geometry

A closed, nondegenerate differential 2-form on a manifold is called a symplectic structure.

Examples. 1. The element of area $dp\wedge dq$ gives a symplectic structure on the (p, q) plane. 2. The direct product of symplectic manifolds inherits from the factors a natural symplectic structure. 3. In particular, $\mathbf{R}^{2n}$ has the symplectic structure of the direct sum of n planes, $\Sigma dp_i\wedge dq_i$. 4. Phase space (the space of the cotangent bundle) $M = T^*V$ of any smooth manifold V has a natural symplectic structure which in the usual coordinates of phase space can be written in the form $\omega=\Sigma dp_i\wedge dq_i$. It is important to note that neither the Poincaré integral invariant $\omega = d\alpha$ nor the action form $\alpha = \Sigma p_i dq_i$ depends on the choice of coordinates on V (in other words, diffeomorphisms of V take α into α and ω into ω). The coordinate-free definition of the "action form" α in terms of the tangent fibering $\pi:TM \to M$ and the cotangent fibering $\rho:T^*V \to V$ is

$$\alpha(\xi)=\pi(\xi)\rho_*(\xi),\text{ where } \xi\in TM.$$

All symplectic structures on a fixed manifold are locally diffeomorphic (the Darboux theorem [6]).

We consider a hypersurface in a symplectic space. The skew-orthogonal complement to a tangent plane is called a characteristic direction on the hypersurface. The integral curves of the field of characteristic directions are called characteristics. Any hypersurface in a symplectic space is fibered into characteristics (at least locally).

Example. In the phase space $\mathbf{R}^{2n}$ over Euclidean space $\mathbf{R}^n$ we consider the hypersurface given the Euclidean metric $p^2 = 1$. The characteristics of this hypersurface are defined by the extremals of least action for a free particle. Each characteristic corresponds to an oriented line. The manifold of characteristics is the manifold of all oriented lines in $\mathbf{R}^n$. This manifold can be identified with the space of the cotangent bundle of the sphere S^{n-1} (to each line there corresponds its unit vector). We thus equip the manifold of oriented lines with a symplectic structure (borrowing it from T^*S^{n-1}).

The manifold of characteristics of a hypersurface in a symplectic space (for example, the manifold of extremals of any one-dimensional, regular variational problem) inherits a natural symplectic structure from the space itself (its value on the projections of two tangent vectors of phase space applied at the same point is equal to the original structure on these same vectors).

Example. The symplectic structure of the space of oriented lines in Euclidean space which is inherited from phase space differs from the symplectic structure taken from the space of the cotangent bundle of the sphere only in sign.

A one-parameter group of diffeomorphisms of a symplectic manifold preserving the symplectic structure has as velocity field a Hamiltonian field (the 1-form corresponding to this field under the isomorphism of the tangent and cotangent bundles defined by the symplectic 2-form is closed). If this closed 1-form is the differential of a function, then this function is called the Hamiltonian function of the group.

The phase curves of the original velocity field are characteristics of the level surfaces of the Hamiltonian function.

Example. We consider the linear space V of all binary forms (homogeneous polynomials in two variables x, y) of degree d. The dimension of this space is equal to d + 1. If d is odd on this space there is (a unique up to a factor) linear skew-symmetric form invariant under the action of SL_2 (the group of linear transformations of the x, y plane with determinant 1). On V it defines a symplectic structure "with constant coefficients." It is convenient to describe the explicit form of this structure ω by means of "Darboux coordinates" in which $\omega=\Sigma dp_i\wedge dq_i$; the binary form $\varphi(x, y)$ for which

$$\varphi(x, 1)=q_0x^d/d!+\ldots+q_nx^{n+1}/(n+1)!-p_nx^n/n!+\ldots+(-1)^np_0,\quad d=2n+1$$

has coordinates $p_0,\ldots,q_n$ (the signs in front of the p_i alternate).

The hyperplane q_1 = const can be identified with the space of polynomials of degree d in x with fixed leading coefficient. The function q_1 is the Hamiltonian function of translations along the axis of values of the polynomial. Therefore, a characteristic is determined by the derivative of the polynomial with respect to x. Thus, the manifold of characteristics of this hyperplane can be identified with the manifold of polynomials f of even degree $b = d - 1 = 2n$ with fixed leading coefficient. In this space we also obtain a symplectic structure. In Darboux coordinates $(p_1, \ldots, q_n)$:

$$f(x) = x^b/b! + q_1 x^{b-1}/(b-1)! + \ldots\ldots + q_n x^n/n! - p_n x^{n-1}/(n-1)! + \ldots + (-1)^{n-1} p_1.$$

Translations of f along the x axis preserve this symplectic structure. It is not hard to find the Hamiltonian function of the translations:

$$h = p_1 + q_1 p_2 + \ldots + q_{n-1} p_n + q_n^2/2.$$

(this polynomial was found already by Hilbert [35], cf. [17, 18]).

We consider the manifold of characteristics of the hyperplane $h = 0$. This manifold can be identified with the manifold of polynomials of degree $a = b - 1$ with fixed leading coefficient and root sum 0 (indeed, the translation annihilating the root sum is uniquely determined, and the value of p_1 is determined by the remaining coefficients and the condition $h = 0$). We obtain a symplectic structure in the space of polynomials g(x) of odd degree with fixed leading coefficient and root sum 0. In Darboux coordinates $q_2, \ldots, p_m$:

$$g(x) = x^a/a! + q_2 x^{a-2}/(a-2)! + \ldots + q_m x^m/m! - p_m x^{m-1}/(m-1)! + \ldots + (-1)^m p_2, \quad a = 2m+1.$$

This example will serve us below as a source of standard singularities.

A submanifold of a symplectic manifold is called a Lagrangian manifold if on it the symplectic 2-form is identically equal to zero and it has maximum possible dimension (equal to half the dimension of the symplectic manifold).

Examples. 1. The plane q = const in standard phase space is Lagrangian. 2. The fibers of the cotangent bundle are Lagrangian. 3. The set of all 1-forms equal to zero on the tangent spaces of a submanifold (of any dimension) forms a Lagrangian submanifold in the space of the cotangent bundle of the enveloping manifold. 4. The differentials of a solution of the Hamilton–Jacobi equation form a Lagrangian submanifold lying in the hypersurface of phase space defined by the Hamilton–Jacobi equation.

The last example makes it possible to consider Lagrangian submanifolds of the space of the cotangent bundle, which are not necessarily sections, as "multivalued solutions" of the Hamilton–Jacobi equation. For example, the differentials of all branches of the multivalued function of the distance from a point of the plane to an ellipse define a smooth (even over the caustic) Lagrangian manifold in phase space.

When carried over to this manifold the system of extremals and fronts becomes regular: all singularities arise only due to singularities of the projection of the Lagrangian manifold from phase space to configuration space.

A bundle with symplectic structure is called Lagrangian if its fibers are Lagrangian.

Example. The cotangent bundle is Lagrangian.

A Lagrangian mapping is a triple $L \to E \to B$ consisting of an immersion of a Lagrangian manifold and a Lagrangian fibering $E \to B$.

Singularities of the composite mapping $L \to B$ are called Lagrangian singularities.

The set of critical values of Lagrangian singularities is called a caustic of the Lagrangian mapping.

A symplectic mapping of fiber spaces taking the first Lagrangian submanifold into the second and fibers of the first bundle into the second is called an equivalence of Lagrangian mappings.

The caustics of equivalent mappings are diffeomorphic, but the classification of Lagrangian singularities up to Lagrangian equivalence is, generally speaking, finer than classification according to types of caustics.

A Lagrangian singularity is called stable if it is preserved under small perturbation, i.e., if any Lagrangian mapping nearby (together with derivatives) has at a suitable nearby point a singularity Lagrangian-equivalent to the given singularity.

A Lagrangian singularity is simple if the germs of all sufficiently close Lagrangian singularities are exhausted up to Lagrangian equivalence by a finite list.

THEOREM 1 [1]. Singularities of Lagrangian mappings of general position of manifolds of dimension less than 6 are simple and stable at each point.

THEOREM 2 [1]. The simple and stable singularities of Lagrangian mappings are exhausted by the list A_k, D_k, E_6, E_7, E_8 of singularities of the projection $(p, q) \mapsto q$ of Lagrangian manifolds given by the formulas

$$\{p, q: \exists x: F_x = 0, \ p = F_q\}, \qquad (*)$$

where the "generating family" F is as follows:

$$A_\mu: F = \pm x^{\mu+1} + q_1 x^{\mu-1} + \ldots + q_{\mu-1} x,$$

$$D_\mu: F = x_1^2 x_2 \pm x_2^{\mu-1} + q_1 x_2^{\mu-2} + \ldots + q_{\mu-2} x_2 + q_{\mu-1} x,$$

$$E_6: F = x_1^3 \pm x_2^4 + q_1 x_1 x_2^2 + q_2 x_1 x_2 + q_3 x_1 + q_4 x_2^2 + q_5 x_2,$$

$$E_7: F = x_1^3 + x_1 x_2^3 + q_1 x_2^4 + q_2 x_2^3 + q_3 x_2^2 + q_4 x_2 + q_5 x_1 x_2 + q_6 x_1,$$

$$E_8: F = x_1^3 + x_2^5 + q_1 x_1 x_2^3 + q_2 x_1 x_2^2 + q_3 x_1 x_2 + q_4 x_1 + q_5 x_2^3 + q_6 x_2^2 + q_7 x_2$$

(variables q_i with $i \geqslant \mu$ on which F actually does not depend may be included in the arguments of the function F).

COROLLARY [1]. The caustics of general Lagrangian mappings in three-dimensional space are locally diffeomorphic to a swallow tail (Fig. 3), a pyramid (Fig. 4), or a purse (Fig. 5), and these singularities are stable; all others decompose into combinations of these under a small general Lagrangian perturbation (of course, transversal sections of several branches with singularities of the type described are also stable).

The preceding formulas provide, in particular, normal forms of caustics of general position in such variational problems as the problem of the shortest distance to a point along a surface, which was considered already by Jacobi, the problem of the shortest distance to a hypersurface of Euclidean space, etc. as long as the dimension is less than 6.

Regarding the general case, the following is known.

THEOREM 3 [36, 12]. For Lagrangian singularity is given locally by a generating family of functions according to formula (*) in which the equation $F_x = 0$, $F_q = p$ is normally solvable for x [i.e., $\operatorname{rk}(F_{xx}, F_{xq}) = \dim\{x\}$].

THEOREM 4 [36, 12]. For Lagrangian equivalence of singularities stable, fibered R^+ equivalence of some (and then all) of their generating families is necessary and sufficient.

Two families of functions F_i of variables x and parameters q are called fibered R^+ equivalent if one goes over into the other under a smooth change of parameters, a smooth change of variables depending smoothly on the parameters, and addition of a smooth function of the parameters. A superstructure over a family of functions is obtained by addition of a nondegenerate quadratic form of supplementary variables. A stable equivalence is an equivalence of superstructures (with this definition it is possible to compare singularities of families of functions of a different number of variables).

The variational meaning of a generating family F is very simple: it is necessary to consider F as the value of a functional along an extremal leading from a point x of the initial manifold to a final point q. In considering problems with initial data on a k-dimensional manifold of general position Lagrangian singularities of corank k of general position are encountered (the corank of a Lagrangian singularity is the dimension of the intersection of the tangent space of the immersed Lagrangian manifold with a fiber of the Lagrangian fibration).

For example, the singularities of the distance to a curve of general position in a space of any dimension are exhausted by the list A_μ, while those of the distance to a two-dimensional surface of general position in a six-dimensional space are exhausted by the list A_μ, D_μ, E_μ, $\mu \leqslant 6$.

A classification of Lagrangian singularities of spaces of dimension $\leqslant 10$ is presented in [12].

Fig. 4

Fig. 5

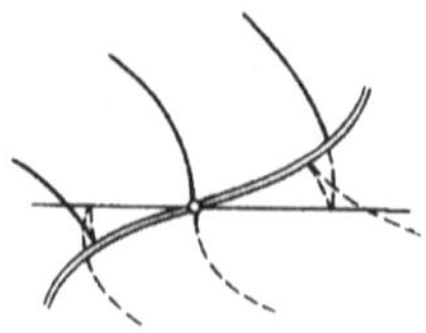
Fig. 6

Fig. 4. Pyramid (caustic D_4^-).

Fig. 5. Purse (caustic D_4^+).

Fig. 6. Family of evolvents of a plane curve.

2. Contact Geometry

A contact structure on an odd-dimensional manifold is a nondegenerate field of hyperplanes in the tangent spaces. The hyperplanes of the field are called contact hyperplanes. It is possible not to indicate explicitly exactly what the condition of nondegeneracy in this definition consists in, since by the "contact theorem of Darboux" all fields of general position in a fixed space are locally diffeomorphic.

Locally it is convenient to give a field of hyperplanes in $\mathbf{R}^{2n+1}$ as the field of zeros of a 1-form α. In this notation the condition of nondegeneracy can be written $\alpha \wedge (d\alpha)^n \neq 0$.

Examples. 1. The space $\mathbf{R}^{2n+1}$ with coordinates $q_1, \ldots, p_n$, z and the form $\alpha = dz - pdq$.

2. The space of 1-jets of functions $J^1(V^n, \mathbf{R})$ on an n-dimensional manifold V has a natural contact structure which is given locally by the formula of Example 1 (q_i are the coordinates on V, z the coordinate in $\mathbf{R}$, and p_i are the partial derivatives with respect to q_i).

This structure is natural in the sense that it does not depend on the choice of coordinates (it goes over into itself under diffeomorphisms of V^n and $\mathbf{R}$).

Indeed, we consider the 1-graph of a function $f: V^n \to \mathbf{R}$ (a section of the bundle J^1 over V^n assigning to a point of V^n the 1-jet of the function f at this point). The tangent spaces to all 1-graphs at a given point of the space of 1-jets belong to a contact hyperplane. A contact hyperplane is determined by this condition and hence, just as the 1-graphs, it is defined in a coordinate-free manner.

3. A contact element on M^m is a hyperplane in the tangent space. All contact elements on M form a manifold of dimension $2m - 1$. The manifold of contact elements has a natural contact structure: the velocity of displacement of an element belongs to a contact hyperplane if the velocity of the point of application belongs to the element. This structure is natural in the sense that it goes over into itself under all diffeomorphisms of the original manifold M onto itself.

We consider a hypersurface in a contact space transversal to the contact planes. A contact plane $\alpha = 0$ is equipped with the structure of a linear symplectic space defined (up to a factor) by the two-form $d\alpha$. The skew-orthogonal complement to the trace of the tangent space of a hypersurface in a contact plane is called a characteristic direction on the hypersurface. The integral curves of the field of characteristic directions on a hypersurface in a contact space are called its characteristics.

The manifold of all characteristics of a hypersurface inherits a contact structure from the original contact manifold (this is easily proved by means of Stokes' formula).

Example. In the space of 1-jets of functions on a Riemannian manifold we consider the hypersurface $p^2 = 1$ defining the Hamiltonian–Jacobi equation of the propagation of perturbations with speed 1 (the square is Riemannian). A characteristic of this hypersurface is a geodesic of the Riemannian manifold parametrized by length measured from some point.

In exactly the same way an extremal of any regular variational problem $\delta \int L(q, \dot{q}, t)dt = 0$ parametrized by the value of the S functional is a characteristic of the Hamilton–Jacobi hypersurface $p_0 + H(p, q, t) = 0$ in the space of 1-jets of functions $S(q, t)$ (where $dS = pdq + p_0 dt$, and H is the Hamiltonian function).

Thus, the manifold of all extremals parametrized by the value of the functional acquires a natural contact structure.

A Legendre submanifold of a contact manifold is a maximal integral submanifold of a field of contact planes. The dimension of Legendre submanifolds in a space of dimension $2n + 1$ is equal to n.

Examples. 1. The manifold of all contact elements passing through a given point is a Legendre submanifold of the manifold of contact elements. 2. The manifold of all contact elements tangent to a given hypersurface is a Legendre manifold. 3. The manifold of all contact elements tangent to a given submanifold of any dimension is a Legendre manifold. 4. The 1-graph of a function is a Legendre submanifold of the space of 1-jets. 5. The manifold of all 1-jets of functions with a fixed 0-jet is a Legendre manifold.

A Legendre bundle is a bundle whose fibers are Legendre manifolds.

Examples. 1. The manifold of all contact elements of a space is Legendre-fibered over this space (this is the projectivization of the cotangent bundle). 2. The manifold of transversally oriented contact elements is also Legendre-fibered over the space (this is the spherization of the cotangent bundle). 3. The manifold of contact elements of projective space is fibered both over this space as well as over the dual space (an element is assigned the hyperplane containing it). Both these fibrations are Legendre fibrations. 4. The manifold of 1-jets of functions is Legendre-fibered over the manifold of 0-jets.

A Legendre mapping is a triple $L \to E \to B$ formed by the immersion of the Legendre manifold in the space of a Legendre bundle. The image of the composite mapping $L \to B$ is called a front.

The dimension of a Legendre submanifold L is one less than the dimension of the base B; therefore, a front is, generally speaking, a hypersurface in the base.

Examples. 1. We consider a submanifold (of any dimension) in a Riemannian manifold. We assign to each unit vector of each normal the result of its translation along its geodesic a distance t. The final points obtained on the manifold form the front of a Legendre mapping. Instead, the manifold of unit vectors of normals can be identified with the Legendre manifold of transversally oriented contact elements, and the geodesic flow (moving each unit vector by t along the geodesic) preserves the contact structure.

2. We consider a smooth hypersurface in projective space. The dual hypersurface is the hypersurface in the dual space formed by tangent hyperplanes of the original hypersurface.

A dual hypersurface is a front of a Legendre mapping. Indeed, the manifold of contact elements of the original hypersurface is a Legendre submanifold of the space of contact elements; therefore, its projection into the dual projective space is a Legendre mapping.

3. We consider a "multivalued solution" of the Hamilton–Jacobi equation of a variational problem as a Legendre manifold of the space of 1-jets of functions. Its image in the space of 0-jets is the usual graph of a multivalued function (the value of the functional on extremals joining the initial manifold with the final point is considered a function of the final point). Thus, the graph of this multivalued solution is the front of a Legendre mapping.

The theory of singularities of Legendre mappings is the same as the theory of singularities of equidistant curves or fronts or Legendre transformations (the singularities of general position in each of these classes are the same).

An equivalence of Legendre mappings is a diffeomorphism between the spaces of Legendre bundles taking the contact structure of the first into the contact structure of the second, the first immersed Legendre manifold into the second, and fibers of the first bundle into fibers of the second.

Fronts of equivalent mappings are diffeomorphic.

A Legendre singularity at a point is called stable if it is preserved under small perturbation, i.e., if any Legendre mapping which is nearby (together with derivatives) has at a suitable nearby point a singularity Legendre-equivalent to the given singularity.

A Legendre singularity is simple if the germs of all sufficiently close Legendre singularities are exhausted, up to Legendre equivalence, by a finite list.

THEOREM 1 [3, 12]. The singularities of Legendre mappings of general position of manifolds of dimension less than 6 are simple and stable at each point.

THEOREM 2 [3, 12]. The simple and stable singularities of Legendre mappings are exhausted by the list A_μ, D_μ, E_μ of singularities of the projection $(p, q, f) \to (q, f)$ of Legendre manifolds given by the formula

$$\{p, q, f: \exists x: F(x, q) = f,\ F_x = 0,\ F_q = p\}, \tag{**}$$

where the "generating family" F is the same as in Theorem 2 of Sec. 1.

COROLLARY. A front of general position in three-dimensional space is locally diffeomorphic to a swallow tail or a cylinder over a semicubical parabola (or to the transversal intersections of several branches with such singularities).

The foregoing formulas provide, in particular, a description of the singularities of wave fronts of dimension less than 6.

The connection of these singularities with groups of reflections consists in the following.

THEOREM 3 [1, 2, 3]. The fronts A_μ, D_μ, E_μ are biholomorphically equivalent to the manifolds of nonregular orbits of the corresponding groups generated by reflections.

The caustics A, D, E can also be described in these terms.

THEOREM 4 [7, 23, 20]. A holomorphic vector field of general position in a neighborhood of a manifold of nonregular orbits of an irreducible group generated by reflections can be rectified (reduced to the standard form $\partial/\partial f$ in the notation of Theorem 2) by a biholomorphic diffeomorphism preserving the manifold of nonregular orbits.

Thus, the fibering of the manifold of orbits into curves transversal to the manifold of nonregular orbits is uniquely determined up to a diffeomorphism preserving the manifold of nonregular orbits.

We consider the mapping of the manifold of nonregular orbits onto the base of this bundle.

THEOREM 5 [1, 4]. A caustic of a Lagrangian singularity A_μ, D_μ or E_μ is diffeomorphic to the projection of a cusp edge of the manifold of nonregular orbits of the corresponding group generated by reflections onto the base of the bundle defined above.

These theorems make it possible to study in detail the geometry of caustics and fronts and also their metamorphoses in spaces of dimensions not too large. Regarding the general case, the following is known.

THEOREM 6 [12]. Any Legendre singularity is given locally by a generating family of hypersurfaces $F = f$ by formula (**) in which the equations $F = f$, $F_x = 0$, $F_q = p$ are normally solvable for x [i.e., $\operatorname{rk}(F_{xx}, F_{qx}, F_x) = \dim\{x\}$].

THEOREM 7 [12]. For Legendre equivalence of singularities stable, fibered V-equivalence of some (and then all) of their generating families of hyperplanes is necessary and sufficient.

Fibered stable V-equivalence of families of hypersurfaces $F(x, q) = f$ with parameters q and f is defined as fibered V-equivalence of multiple superstructures. A superstructure over a hypersurface (or a family) is defined as a two-sheeted covering of the enveloping space with ramification along the hypersurface. In the language of functions F an r-fold superstructure signifies the addition to F of a nondegenerate quadratic form of r new arguments. V-equivalence of functions is this equivalence of manifolds of zero level. Two families of hypersurfaces $G_i(X, Q) = 0$ depending on the parameters Q are fibered V-equivalent if one of the functions G_i goes over into the other under a smooth change of parameter, a smooth change of variable not depending on the parameter and multiplication by a nonvanishing smooth function of the variables and parameters.

A classification of Legendre singularities of mappings of spaces of dimension $\leqslant 10$ is given in [12].

3. Problem of Passing about an Obstacle

In Euclidean space we consider an obstacle bounded by a smooth surface of general position. The problem of passing about the obstacle consists in investigating the singularities

Fig. 7

Fig. 8

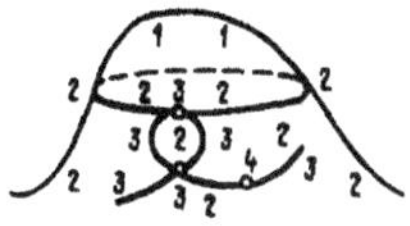

Fig. 9

Fig. 7. Extremals and a front near an obstacle.

Fig. 8. An open swallow tail.

Fig. 9. Classification of points of a surface according to orders of tangents.

of the distance in passing about the obstacle from an arbitrary point of the enveloping space to a fixed initial point of general position.

For an obstacle bounded by a plane curve the level lines of the distance are composed of the evolvents of the curve (Fig. 6). At a point of convexity of the obstacle the level lines of the distance therefore have a semicubic singularity (a cusp). Computations show that an inflection point of the obstacle the evolvent has a singularity of order 5/3. The Legendre manifold corresponding to such a front itself has a singularity (namely, a semicubic cusp).

Fronts of variational problems on a manifold without boundary (for example, equidistant curves) can also have singularities, but these are only singularities of projection: the original Legendre submanifold is nonsingular. For variational problems on a manifold with boundary (and generally for problems with one-sided constraints) the Legendre manifold itself formed by contact elements tangent to the front has singularities.

Lagrangian manifolds with singularities arise in the same way. The extremals connecting a fixed point of the manifold with all its points form a Lagrangian submanifold in the space of all extremals. This Lagrangian manifold is smooth if the original manifold has no boundary but may have singularities otherwise.

We consider, for example, the shortest paths from a fixed point of three-dimensional space to all possible points in passing about an obstacle (Fig. 7). Such a path begins from a segment of the line from the initial point to the point of tangency with the surface of the obstacle. Next there follows a segment of a geodesic tangent to the rectilinear segment. Thus, on the surface of the obstacle there arises a pencil (one-parameter family) of geodesics. Further portions of the shortest paths consist of segments of tangent breaking off from the geodesics of the pencil, etc.

At some points on the surface of the obstacle the geodesic pencils will have an asymptotic direction. Such points, generally speaking, form a smooth curve (a on Fig. 7). Rays breaking off at points of this curve form a cusp edge of the Lagrangian manifold of the broken-off rays. At individual points of the curve the geodesic pencil is tangent to the curve. The singularity of the manifold of broken-off rays at this point will, generally speaking, be an "open swallow tail."

An open swallow tail is the manifold of polynomials $x^5 + Ax^3 + Bx^2 + Cx + D$ having a root of multiplicity $\geqslant 3$. Differentiation of the polynomials maps the open tail into an ordinary tail diffeomorphically almost everywhere; only points of the line of self-intersection of the ordinary swallow tail have two preimages on the open swallow tail. Thus, opening of the tail annihilates self-intersection while preserving the cusp edge (Fig. 8).

Legendre and Lagrangian manifolds describing solutions of variational problems with one-sided constraints always have singularities of this type. Although the problem of passing about an obstacle has not been completely solved even in ordinary three-dimensional space, we shall indicate the normal forms of the Lagrangian and Legendre manifolds with singularities arising here, including the normal forms of the corresponding families of rays in the manifolds of all extremals. These normal forms are provided by the theory of invariants of binary forms; its connection with the questions considered of variational calculus was previously not obvious and was found as a result of an entire series of works [25, 26, 8-10, 13] in an attempt to understand the reasons for the unexpected coincidence of bifurcation diagrams, cancellations of many terms in lengthy calculations, and the universatility of open swallow tails encountered in various areas of mathematics.

On the other hand the graph of the time function in the problem of passing about an obstacle is locally diffeomorphic to the manifold of nonregular orbits of the symmetry group of the icosahedron (the group H_3 is Coxeter's classification of groups generated by reflections). This singular manifold appears in the problem of passing about an obstacle in three-dimensional space as a singularity of the front (at a point on the surface of the obstacle at which the ray is asymptotic).

Along an asymptotic ray both the fronts and the manifold of rays breaking off from the surface of the obstacle have singularities. We shall therefore begin from the geometry of asymptotic tangents of a surface.

4. Asymptotic Rays in Symplectic and Contact Geometry

A plane curve of general position has no tangents of higher than second order of tangency. A surface of general position in three-dimensional Euclidean space has no tangents of higher than fourth order. There are tangents of higher than first order (asymptotic tangents) in the entire region of hyperbolic points of the surface, there are tangents of third order on a line in it, and of fourth order at individual points of this line (Fig. 9).

The hierarcy of asymptotic tangents becomes especially clear if it is reformulated in terms of symplectic (or contact) geometry. Here we immediately extend the theory of asymptotic tangents to a much broader circle of problems, for example, to the case of submanifolds of a Riemannian manifold or to general variational problems with constraints.

We consider a hypersurface ∂M in a Riemannian manifold without boundary M. We shall call points of the phase space T^*M vectors. The Riemannian metric defines in phase space a hypersurface composed of all vectors of length 1. The surface ∂M also defines in phase space a hypersurface, namely, the hypersurface composed of all vectors applied at ∂M.

Melrose [37] noted that a considerable part of the exterior geometry of ∂M in M, for example, the billiard problem in a domain with boundary ∂M, can be developed in terms of the symplectic geometry of a constructed pair of hypersurfaces of a symplectic manifold: Neither the fact that this manifold is the space of the cotangent bundle nor the origin of the hyperplanes plays a role. Thus, we can use the geometric intuition accumulated in the study of submanifolds of Euclidean space and billiard problems in the much more general situation of arbitrary variational problems with constraints.

4.1. Geometry of a Pair of Hypersurfaces of a Symplectic Manifold. Thus, in an arbitrary symplectic manifold we consider two smooth hypersurfaces of general position, and we call one of them the "surface of unit vectors" and the other the surface of "vectors applied on the boundary" (briefly, boundary vectors). We assume that these two surfaces intersect transversally along a submanifold of "boundary unit vectors" (of codimension 2 in the original manifold).

A hypersurface in a symplectic manifold is locally fibered into characteristics (the integral curves of the field of skew-orthogonal complements of tangent planes). The characteristics of the surface of "unit vectors" are called "rays" (if the surface of "unit vectors" is transversally oriented, then the "rays" are also naturally oriented).

The manifold of all "rays" is even-dimensional and inherits a natural symplectic structure (cf. Sec. 1).

The manifold of all "unit vectors" is fibered over the manifold of "rays" and contains the submanifold of "boundary unit vectors" as a hypersurface. The imbedding and fibration determine a canonical projection of the manifold of "boundary unit vectors" into the manifold of "rays" (having the same dimension): to each "unit vector" we assign the "ray" issuing from it (Fig. 10).

The singularities of the canonical projection constructed on the basis of a pair of hypersurfaces of general position are of standard Whitney type; they are locally equivalent to the singularities A_k of the projection of the hypersurface $x^{k+1} + \lambda_1 x^{k-1} + \ldots + \lambda_k = 0$ from the bundle space $(x, y) \mapsto \lambda$ onto its base.

Example. $k = 1$ corresponds to an ordinary fold: the parabola $x^2 + \lambda_1 = 0$ is projected onto the λ_1 axis.

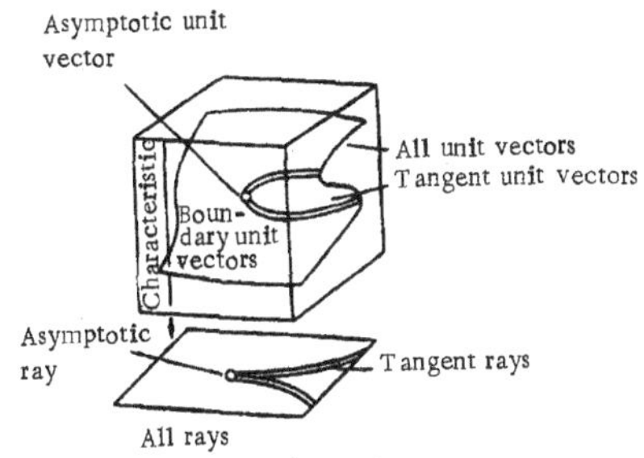

Fig. 10

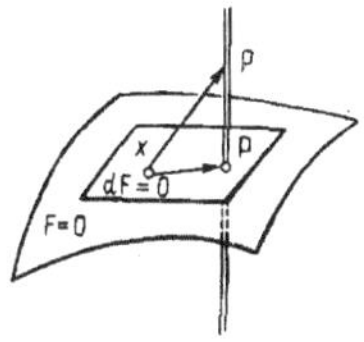

Fig. 11

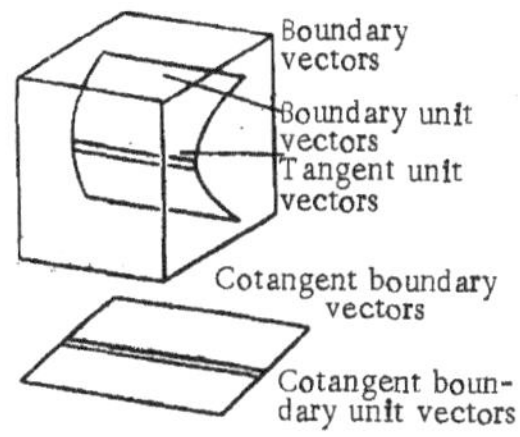

Fig. 12

Fig. 10. Canonical projection of the manifold of boundary unit vectors into the manifold of rays.

Fig. 11. A characteristic of the surface of tangent vectors.

Fig. 12. The second canonical projection.

Definition. "A unit vector applied at the boundary" is said to be "tangent" if it is a singular point of the canonical projection, "asymptotic" if it is a singular point more complex than a fold ($k > 1$), "biasymptotic" if $k > 2$, etc. (Fig. 10).

Example. For a surface $F(x) = 0$ in three-dimensional space and the surface of unit vectors $\{(x, p):p^2 = 1\}$ all the concepts introduced above are the objects of the same name of ordinary differential geometry.

Together with the canonical projection introduced above, it is also possible to consider a second canonical projection which is defined by the same construction but with the roles of both hypersurfaces interchanged.

Returning to the example considered above, we consider the momentum p a vector of Euclidean space. A characteristic of the surface of boundary vectors consists of vectors applied at a single point of the boundary whose end points run through a line perpendicular to the tangent plane of the boundary at this point (Fig. 11).

Having this example in mind, in the general case we call the manifold of characteristics of a hypersurface of "boundary vectors" the manifold of "(co)tangent vectors of the boundary." The second canonical projection maps the manifold of "boundary unit vectors" into the manifold of "boundary (co)tangent vectors." In the example the projection assigns to a boundary vector the point P of the intersection of the line passing through it with the tangent plane normal to it. From this it is evident that in this example (and generally in the Riemannian case) the second canonical projection has a unique fold singularity (A_1) on the manifold of tangent vectors (Fig. 12).

The meaning of the second canonical projection becomes clearer if the momentum p is understood as a cotangent vector of the enveloping space, i.e., as a 1-form on its tangent space at the point x. With this interpretation the characteristic is the line $p + RdF$. This line is uniquely determined by the restriction of the form p to the tangent plane of the boundary, $dF = 0$. Thus, the second canonical projection assigns to the 1-form p its restriction to the tangent space.

4.2. Normal Forms. If the original surface in Euclidean space is convex, then both the first and second canonical projection of the manifold of boundary unit vectors has on the manifold of tangent unit vectors a singularity of fold type (A_1). We suppose that the two hypersurfaces in symplectic space generate two projections of "fold" type. A mapping of fold type defines an involution of a neighborhood of the set of critical points which is stationary on the set of critical points (a permutation of sheets, i.e., a permutation of both preimages of a single point under a mapping). The first and second canonical projections have singularities of fold type and thus define on the manifold of "boundary unit vectors" two involutions which are stationary on the manifold of "tangent unit vectors"; we call them the first and second Melrose involutions.

Example. For a plane convex curve the first involution assigns to a unit vector applied on the curve a unit vector in the same direction at the second point of intersection of the

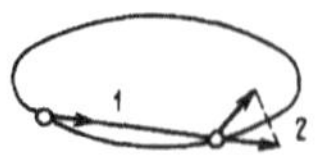

Fig. 13. Birkhoff's billiard transformation.

line defined by this unit vector with the curve (Fig. 13), while the second involution assigns the unit vector applied at this same point but symmetric relative to the tangent to the curve. The product of both involutions is precisely the Birkhoff billard transformation.

Melrose [37] proved that in the case where the singularities of both canonical projections are folds the original pair of hypersurfaces can be reduced by a symplectic diffeomorphism in a neighborhood of the point under consideration to the standard normal form

$$q_0=0; \quad q_0=p_0^2+p_1$$

in Darboux coordinates ($\omega=dp_0\wedge dq_0+\cdots$). Thus, all objects derived from the hypersurfaces, including the pair of involutions and the product, are also reduced to normal forms. It is interesting to note that the reduction to normal form of Melrose is possible in the C^∞-situation or at the level of formal series but not in the analytic situation. The pairs of involutions encountered in other problems (see [16]) also have a simple formal classification without moduli and have functional moduli in the analytic situation. In the problem of a pair of hypersurfaces of a symplectic manifold, as far as I know, functional moduli have not been announced (the same pertains to analytic classification of pairs of canonical involutions and their products). If we require reduction of the product of a pair of involutions to normal form in a neighborhood not of a point but of the entire stationary manifold, then the C^∞-classification already becomes unmanageable even when the stationary manifold is a circle, i.e., the case of mappings of a circular annulus. This is a well-known difficulty of the billard problem which is common to all nonintegrable problems of dynamics; cf. [33].

For singularities of the canonical projection more complex than a fold (for example, for creases) it is not possible to write out a single normal form of a pair of hypersurfaces (see [38]). However, it is possible to indicate normal forms of a pair consisting of the first hypersurface and the trace of the second surface on it in the case of singularities A_2 and A_3 (in particular, for ordinary asymptotic and biasymptotic rays on a surface in three-dimensional space).

THEOREM 1 [9]. In a neighborhood of a point where the canonical projection has the singularity A_k the submanifold of codimension 1 on the hypersurface $q_0 = 0$ can be reduced to the normal form $q_0 = 0 = F$ where

$$F=p_0^2+p_1 \text{ for } k=1,$$
$$F=p_0^3+p_1p_0+q_1 \text{ for } k=2,$$
$$F=p_0^4+p_1p_0^2+q_2p_0+p_2 \text{ for } k=3$$

in Darboux coordinates ($\omega=dp_0\wedge dq_0+\ldots+dp_n\wedge dq_n$) at least at the level of formal power series.

Thus, all objects depending only on the first hypersurface and the trace of the second surface on it, for example, the first canonical projection, are reduced to normal form. The critical values of the first canonical projection form a manifold of "tangent rays" which, generally speaking, has singularities (the semicubical cusp edge in the case A_2, the swallow tail multiplied by a linear space in the case A_3). This manifold of "tangent rays" lies in the symplectic space of all "rays" and has codimension 1 in it.

The reduction of a pair (a hypersurface and a submanifold of codimension 1 in it) to symplectic normal form brings with it the reduction to normal form of the manifold of "tangent rays" by means of a symplectic transformation.

Thus, the foregoing results provide normal forms of symplectic structure in neighborhoods of points of the products of semicubical parabolas or swallow tails by a linear space. Namely, we have the following result.

THEOREM 2 [9]. A symplectic structure of general position formally reduces to the form

$$\omega = dA\wedge dB + dC\wedge dD + \ldots$$

in a neighborhood of a point O on the hypersurface $A^3 = B^2$ (of general position, that is, nondegenerate on the space $A = B = 0$) and to the form

$$\omega = dA\wedge dD + dC\wedge dB + dE\wedge dF + \ldots$$

in a neighborhood of a point O on the hypersurface (of general position, that is nondegenerate on the space $B = C = 0$) defined by the condition: the polynomial $x^4 + Ax^2 + Bx + C$ has a multiple root.

These results can also be interpreted as reduction to normal forms of the products of semicubic parabolas or swallow tails by linear space by means of transformations preserving the symplectic structure of the enveloping space.

For $k > 3$ such a classification is not possible: moduli sppear. C^∞ and analytic versions of this classification are so far not known even for $k = 3$.

A contact version of the classification presented was constructed by E. E. Landis: the normal forms [29, 38] are

$$\alpha = dZ - BdA - \ldots$$

in a neighborhood of a cylinder over the semicubical parabola $A^3 = B^2$, $\alpha = dZ - DdA - CdB - EdF - \ldots$ in a neighborhood of the swallow tail.

4.3. Geodesics. The set of critical values of a canonical projection is a hypersurface in a symplectic manifold. The characteristics of this hypersurface are called geodesics.

Example. We consider a pair of hypersurfaces of phase space corresponding to a surface of Euclidean space. The critical values of the first canonical projection are tangent rays. In this case the "geodesics" are curves in the space of rays formed by the tangents of one geodesic line on the surface.

The critical values of the second canonical projection are (co)tangent unit vectors of the surface. In this case the "geodesics" are the trajectories of an ordinary geodesic flow of the surface.

As in this example, so in the general case we determine two families of "geodesic" lines: the first is constructed on the basis of the first projection and the second on the basis of the second projection.

There is still a third family: the family of characteristics on the manifold of critical points common to both projections (this manifold has codimension three; a characteristic direction on an odd-dimensional submanifold of a symplectic manifold is defined as the direction skew-orthogonal to all remaining tangential directions).

We call the characteristics of the manifold of critical points upper "geodesics." The canonical projections map upper "geodesics" into lower "geodesics" (where the latter are defined, i.e., at points of folds; see [9]).

From the normal forms of the first projection presented above it is possible to obtain normal forms for the first family of "geodesics" and for the upper "geodesics."

THEOREM 3 [9]. In the coordinates of Theorem 1 the equations of the upper "geodesics" have the form

$$q_0 = p_0 = 0,\ q_1 = \xi_1 \quad \text{for } k = 1,$$
$$q_0 = 0,\ p_0 = \xi,\ p_1 = -3\xi^2,\ q_1 = -2\xi^3 \quad \text{for } k = 2$$

(the unwritten coordinates are constant, ξ is the parameter along the "geodesics"). For $k = 3$ the description of the "geodesics" is as follows. We form the polynomial in p_0

$$\Phi = p_0^5/5 + p_1p_0^3/3 + q_2p_0^2/2 + p_2p_0 + q_1/2 = q_1/2 + \int F\,dp_0.$$

A "geodesic" in the space of rays with coordinates $(p_1, \ldots, q_n)$ (a "geodesic" of the first family) is given by the conditions: Φ has a root ξ of multiplicity $\geqslant 3$; the coordinates $(p_1, p_3, q_3, \ldots, p_n, q_n)$ are constant.

An upper geodesic is obtained by adding the conditions $q_0 = 0$, $p_0 = \xi$.

We consider, in particular, the case $k = 3$, $n = 2$ (corresponding to the study of a bi-asymptotic ray on a surface in ordinary three-dimensional Euclidean space). From the foregoing formulas we derive the following result.

THEOREM 4 [9]. A Lagrangian manifold of general position lying on the direct product of a swallow tail by a line in the symplectic space $\mathbf{R}^4$

$$\{(p_1, p_2, q_1, q_2) : \Phi' \text{ has a root of multiplicity } \geqslant 2\},$$

can be reduced by a symplectic diffeomorphism to the form

$$\{(p_1, p_2, q_1, q_2) : \Phi \text{ has a root of multiplicity } \geqslant 3\}.$$

Definition. An open swallow tail (Fig. 8) is a two-dimensional submanifold in four-dimensional space of polynomials $f = x^5 + Ax^3 + Bx^2 + Cx + D$ defined by the condition "f has a root of multiplicity $\geqslant 3$."

From Theorems 1-4 we obtain (at the level of formal series)

COROLLARY 5 [9]. A Lagrangian submanifold of general position on a three-dimensional manifold of tangent rays in a neighborhood of a "biasymptotic ray" is diffeomorphic to an open swallow tail.

In contrast to the theorems themselves, this corollary can also be proved in C^∞ and in the analytic case (see Sec. 6).

We shall apply Corollary 5 to a surface in three-dimensional Euclidean space. We consider a pencil of geodesics on the surface (a one-parameter family). All lines tangent to the geodesics of the pencil form a two-dimensional submanifold (with singularities) in the four-dimensional manifold of all lines.

COROLLARY 6 [9]. In a neighborhood of a biasymptotic tangent this manifold is diffeomorphic to an open swallow tail (if the surface and the pencil are of general position).

The extremals of the problem of passing about an obstacle in three-dimensional space are curves formed from segments of geodesics and segments of their tangents. The system of shortest paths connecting all points of space with a fixed initial point contains, generally speaking, a collection of tangents to the geodesics of the pencil and among them, generally speaking, are found individual biasymptotic tangents.

Therefore, Corollaries 5 and 6 describe one of the singularities of the system of extremals in the problem of passing about an obstacle of general position: this singularity is a Lagrangian open swallow tail.

The appearance of an open swallow tail as a system of extremals of the problem of passing about an obstacle is by no means obvious, and it is not easy to guess that with a bi-asymptotic tangent there is connected a polynomial of fifth degree (or at least to describe it geometrically).

The system of extremals of an ordinary variational problem is a Lagrangian submanifold of the symplectic manifold of all extremals. We see that for a problem with a one-sided constraint the system of extremals itself becomes, generally speaking, a Lagrangian manifold with singularities. These singularities are of special type: they are singularities of Lagrangian submanifolds of the manifold of "tangent rays" (which itself is not smooth; see Fig. 10).

The definition of Lagrangian manifolds with singularities occurring in the problem of passing about an obstacle is not the first that comes to mind. For example, it is possible to consider Lagrangian ideals (closed under Poisson bracket) or manifolds given by generating families with violation of the conditions of transversality (see Sec. 1). These definitions lead to hierarchy of Lagrangian manifolds with singularities different from that which arises in the problem of passing about an obstacle. I think that precisely the latter defition (a Lagrangian submanifold of the manifold of "tangent rays") leads to the most substantial theory of Lagrangian manifolds with singularities and that the semicubic parabola on the plane and the open Lagrangian swallow tail in four-dimensional space are the first examples in this (future) theory.

4.4. Contact Case. The results described above have contact analogues. The contact normal forms are obtained from the symplectic forms by contactization, i.e., passage from the symplectic phase space T^*V to the contact manifold of 1-jets of functions $J^1(V, \mathbf{R})$. The dimension of the latter is one greater; it is obtained from T^*V by the addition as new coordinate of the quantity of the generating function.

We denote the Darboux coordinates in J^1 by $(p_0,\ldots,q_n;\ z)$ (the contact structure is given by the form $\alpha = dz - pdq$). We assume that 1) the pair of hypersurfaces in contact space intersect transversally, 2) the intersection is transversal to the contact planes, 3) the characteristics of each of the hypersurfaces are tangent to their intersection at the point in question with first-order tangency.

THEOREM 7 (see [37]). The pair can be reduced by a contact C^∞-diffeomorphism to the local normal form of Melrose ($q_0 = 0$; $q_0 = p_0^2 + p_1$).

If the intersection is not transversal to the contact planes or if the order of tangency is higher than first, then the C^∞-classification has functional moduli [37, 38].

The theorem of Melrose is not applicable to the simplest (three-dimensional) contact space, since in three-dimensional contact space conditions 1), 2), 3) are not jointly satisfied. The line of intersection of a pair of surfaces of general position in three-dimensional contact space is tangent to the characteristics at individual points with first-order tangency, but it is not transversal to the contact plane at these points. In a neighborhood of such a point the pair can be reduced by a formal or C^∞(?)-contact diffeomorphism to the normal form

$$z=q_0;\ p_0^2=q_0.$$

In the analytic case the series leading to the normal form, generally speaking, diverge, since on the line of intersection there arises a pair of involutions having functional moduli (see [16]).

THEOREM 8 (see [22]). A submanifold of codimension 1 on a hypersurface in a contact space can be reduced by a formal contact diffeomorphism to one of the normal forms of Theorem 1 under the conditions of nondegeneracy indicated above.

We denote by Σ_k the set of those points of the submanifold where the order of tangency of the characteristics with the submanifold is greater than or equal to k (this is the same k as in Theorem 1).

The conditions of nondegeneracy are as follows:

k = 1: the submanifold is transversal to the contact planes;

k = 2: $\dim \operatorname{Ker} d\alpha|_{\{\alpha=0\}\cap T_x\Sigma_1} = 1$ (*)

k = 3: (*) and Σ_2 is transversal to the contact planes.

Generally speaking, these conditions are violated on a submanifold of codimension 1 in Σ_1.

Example. We consider a hypersurface in Euclidean space E. On the basis of this surface we construct the mapping exp of the manifold of all nonzero vectors applied at points of the surface into the manifold of all unit vectors of the enveloping space by assigning to the vector $\vec{xy}$ applied at the point x on the surface a unit vector of the same direction applied at the point y.

The mapping exp can be considered as the first canonical mapping for a pair of hypersurfaces of the contact space $J^1(E, \mathbf{R})$: as the first hypersurface we take the Hamilton–Jacobi equation of perturbations propagating with speed 1, $(\nabla z)^2 = 1$, and as the second we take a cylinder over the hypersurface given in E.

The preceding formulas provide normal forms of the singularities of the mapping exp.

THEOREM 9 (see [26, 22]). The singularities of the mapping exp for asymptotic vectors of a surface of general position in three-dimensional space are given by Table 1.

TABLE 1

Point	$\Pi_{3,1}$	$\Pi_{4,1}$	Π_5	$\Pi_{4,3}$	$\Pi_{3,2}\Pi_{3,3}$	$\Pi_{4,2}$
Singularities	A_2, A_2	A_2, A_3	A_2, A_4	A_3, A_3	A_2	A_3

(the types $\Pi_{i,j}$ of points on the surface are defined in [11], Sec. 15; the mapping A_k is equivalent to projection of the surface $x^{k+1} + \lambda_1 x^{k-1} + \ldots + \lambda_k = 0$ along the x axis onto the space with coordinates $\lambda_1, \ldots, \lambda_k$).

We call a unit vector of Euclidean space singular (relative to the given surface of general position) if the line containing it is tangent to the surface. The singular unit vectors form a four-dimensional hypersurface in the five-dimensional space of all unit vectors. From Theorem 9 we obtain the following:

COROLLARY. The hypersurface of singular unit vectors has singularities on the unit vectors of the asymptotic tangents: in a neighborhood of the unit vectors of an ordinary tangent it is diffeomorphic to the product of a curve with a semicubic cusp with three-dimensional space, in a neighborhood of unit vectors of a biasymptotic tangent it is diffeomorphic to the product of a swallow tail with the plane, and in a neighborhood of unit vectors of a tri-asymptotic tangent it is diffeomorphic to the product of a three-dimensional swallow tail with the line.

4.5. Applications to the Problem of Passing about an Obstacle. There are two Legendre manifolds connected with the variational problem of fastest arrival from a given subset at a variable point of space: the manifold of 1-jets of the (multivalued) time function and the manifold of contact elements of the front. The foregoing results make it possible to describe the singularities of these manifolds in the problem of passing about an obstacle of general position in spaces of 2 and 3 dimensions (the general case is investigated in Sec. 6).

In order to carry out this description, we must briefly repeat all the constructions connected with canonical projections in the contact case. For clarity we shall use the terminology of the problem of passing about an obstacle of general position in a Riemannian space, although the entire theory carries over easily to the general case of a pair of hypersurfaces of a contact manifold.

We begin from the manifold $J^1(M, \mathbf{R})$ of 1-jets of functions on a Riemannian manifold without boundary M equipped with a distinguished hypersurface ∂M. We denote the point of application of a jet by $x \in M$, the value of a jet by $t \in \mathbf{R}$, and the first differential of a jet at x by the letter p.

In the manifold of 1-jets on M we consider the two hypersurfaces

$$SJ^1(M, \mathbf{R}),\ \partial J^1(M, \mathbf{R})$$

(the first is given by the Hamilton–Jacobi equation $p^2 = 1$ of the problem of propagation of perturbations with speed 1, and the second is given by the equation $x \in \partial M$).

The characteristics of the first hypersurface are the trajectories of a geodesic flow in the spherical cotangent bundle ST*M equipped with the natural parameter t. The manifold of characteristics can be identified with the space of the spherical cotangent bundle over M. Indeed, we fix a time τ. We assign to each point of the parametrized geodesic flow the point corresponding to this value of the parameter $t = \tau$. We have defined the bundle $SJ^1(M, \mathbf{R}) \to ST^*M$. The fibers are characteristics, and we have thus identified the manifold of characteristics on $SJ^1(M, \mathbf{R})$ with the space ST*M. This identification depends on the choice of the time τ.

A characteristic of the second hypersurface consists of the 1-jets at a fixed point of ∂M of all extensions of a fixed function from ∂M to M. The manifold of characteristics can naturally be identified with the manifold $J^1(\partial M, \mathbf{R})$ of 1-jets of functions on ∂M.

The contact structures of manifolds of characteristics coincide with the standard contact structures in ST*M and $J^1(\partial M, \mathbf{R})$ (defined in Sec. 2).

The intersection of both our hypersurfaces is a "cylinder over the boundary unit vectors"

$$\Pi = \{(x, p, t) : p^2 = 1,\ x \in \partial M,\ t \text{ arbitrary}\}.$$

The first canonical projection $I:Ц\to ST^*M$ carries a boundary unit vector at time t into a unit vector of the same geodesic corresponding to the selected value of the parameter τ. The second projection $II:Ц\to J^1(\partial M, \mathbf{R})$ assigns to the 1-jet of a function on M at a point of ∂M the 1-jet of its restriction to ∂M.

This mapping has a singularity of fold type on the hypersurface Σ in Ц composed of unit vectors tangent to ∂M,

$$\Sigma=\{(x, p, t) : p^2=p_0{}^2=1,\ x\in\partial M,\ t \text{ arbitrary}\},$$

where p_0 is the restriction of the 1-form p to ∂M.

The set of critical values of the second canonical projection is the hypersurface $SJ^1(\partial M, \mathbf{R})$ given by the Hamilton–Jacobi equation $p_0^2 = 1$ on ∂M. We obtain the imbedding $SJ^1(\partial M, \mathbf{R}) \to Ц$ (the natural diffeomorphism on Σ) and the mapping $\exp:SJ^1(\partial M, \mathbf{R}) \to ST^*M$ which essentially coincides with the mapping denoted the same way on the preceding page.

The objects constructed form the hexagonal part of the commutative diagram (here $\hookrightarrow$ denotes an imbedding and $\twoheadrightarrow$ a fibering).

The system of geodesics on ∂M normal to a fixed hypersurface in ∂M and parametrized by the distance to this hypersurface defines a solution of the Hamilton–Jacobi equation $p_0^2 = 1$ on ∂M and hence a Legendre submanifold l imbedded in $SJ^1(\partial M, \mathbf{R})$. From this smooth Legendre manifold we shall construct two Legendre manifolds with singularities: the 1-graph of the time function and the manifold of contact elements of a front in the problem of passing about an obstacle.

To this end we consider the image of the manifold l under the mapping exp and the union Γ of the characteristics of the first hypersurface $SJ^1(M, \mathbf{R})$ passing through the point l.

<u>THEOREM 10.</u> $\exp l$ and Γ are Legendre manifolds with singularities in ST^*M and $J^1(M, \mathbf{R})$, respectively. Namely, the first consists of all contact elements of the front at time τ, while the second consists of the 1-jets of the (multivalued) time function in the problem of passing about an obstacle bounded by the submanifold ∂M under the condition that the value of the time function on ∂M is determined by the Legendre submanifold l in $SJ^1(\partial M, \mathbf{R})$.

It is convenient to describe the 1-graph Γ of the time function by means of the smooth space γ of the fibering $\gamma \twoheadrightarrow l$ with one-dimensional fibers which is induced by the fibering $SJ^1(M, \mathbf{R}) \twoheadrightarrow ST^*M$ with the base mapping $\exp|_l : l \to ST^*M$. Namely, Γ is the image of the manifold γ under the natural mapping of the spaces of the fiberings $\gamma \to SJ^1(M, \mathbf{R})$.

Theorems 10, 8, and 1 and Corollary 5 make it possible by a contact formal transformation to reduce the entire left side of the diagram below to normal form in the situation where k of Theorem 1 does not exceed 3 (for example, near the asymptotic and biasymptotic rays in the problem of passing about an obstacle in three-dimensional space). Namely, the following objects can be reduced to contact normal forms:

$$SJ^1(M, \mathbf{R}),\ Ц,\ SJ^1(\partial M, \mathbf{R}),\ ST^*M,\ l,\ \exp l,\ I,\ \exp,\ \exp|_l,\ \gamma,\ \Gamma.$$

If we exclude from this list Ц, $SJ^1(\partial M, \mathbf{R})$ and exp, then all the remaining objects can be reduced to normal form by a contact transformation even in C^∞ and the analytic sense and also for large values of k (i.e., in the multidimensional problem of passing about an obstacle); see Sec. 6. We shall not write out here the normal forms themselves, but rather indicate only the following:

<u>COROLLARY 11.</u> The manifolds of contact elements of the front and 1-jets of the time function in the problem of passing about an obstacle of general position in three-dimensional

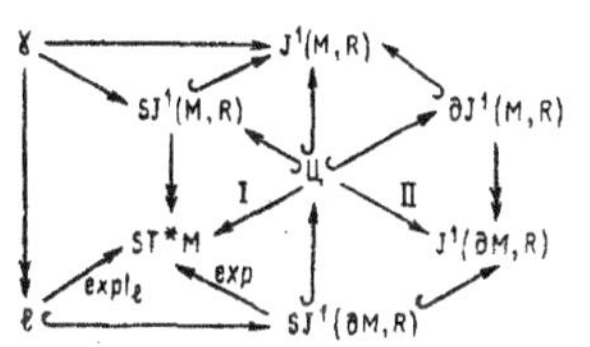

Fig. 13. Birkhoff's billiard transformation.

space have (in a neighborhood of an element of a ray breaking off along a biasymptotic direction) singularities diffeomorphic to the singularities at zero of the submanifolds of spaces of polynomials

$$\{x^5+A_1x^4+\ldots+A_5\} \text{ and } \{x^7+a_1x^6+\ldots+a_7\},$$

consisting of polynomials having a root of multiplicity not less than 3 with a sum of roots zero and a root of multiplicity not less than 5, respectively:

$$\exp l\simeq\{(x-\xi)^3(x^2+3\xi x+\eta)\},\quad \Gamma\simeq\{(x-\xi)^5(x^2+\eta x+\zeta)\}.$$

The time function on Γ has the form $t = \pm a_1 + \text{const}$, and each ray (characteristic) consists of all translations of a single polynomial.

In the problem of passing about an obstacle bounded by a curve on the plane in a neighborhood of elements of a tangent of inflection $\exp l\simeq\{(x-\xi)^2(x+2\xi)\}\subset\{x^3+A_1x^2+A_2x+A_3\}$, $\Gamma\simeq\{(x-\xi)^4(x+\eta)\}$. The expression for contact structures of the spaces of 1-jets and surfaces $SJ^1(M, \mathbf{R})$ in the coordinates A_i and a_i can be found in Sec. 6.

Example. It is evident from Corollary 11 that the manifold of contact elements of the front is diffeomorphic to an open swallow tail (in the spatial problem) and a semicubic parabola (in the plane problem), while the 1-graphs of the time function are diffeomorphic to cylinders over these manifolds.

By considering Fig. 6, it can be seen that the projection of the 1-graph of the time function in the plane problem onto the plane is nothing other than the projection of the bifurcation diagram B_3 considered by Goryunov [11, Sec. 17]

$$\{x^3+ax^2+bx+c : \exists \text{ a multiple root}\}\to\{a, c\}.$$

Figure 6 also makes it possible to visualize the 0-graph: in order to obtain it, it is only necessary to place each of the evolvents in its own plane (in the space-time isochrone). The surface obtained (Fig. 14) has two cusp edges (singularities of type 3/2 and 5/2) tangent to one another with second-order tangency and a line of self-intersection. It is diffeomorphic to the manifold of nonregular orbits of the group H_3 of symmetries of the icosahedron (see [11, Sec. 17]).

The time function in a neighborhood of the graph can be reduced to normal form by a transformation preserving the graph (according to the equivariant Morse lemma of [30] applied to the group H_3). The normal form of the graph is the closure of the set

$$\{(A, B, C) : x^5+Ax^4+Bx^2+C \text{ has multiple roots}\}.$$

The normal form of the time function is $t = A + \text{const}$.

5. Open Swallow Tail

In this section we study special algebraic manifolds: manifolds of polynomials of fixed degree having a root of fixed multiplicity. They must be studied, since they occur as normal forms of various singularities.

Example 1. The manifold of polynomials $x^5 + Ax^3 + Bx^2 + Cx + D$ having a triple root is called an open swallow tail. The manifold of rays breaking off from the surface of an obstacle is diffeomorphic to it (see Sec. 4).

Example 2. We consider a wave front propagating in three-dimensional space. The cusp edge of the front sweeps out a caustic. We suppose that on the caustic there is a swallow tail (this singularity is encountered on caustics of general position). The propagating front is represented by a hypersurface (a "large front") in space-time. In the present case this hypersurface has as singularity a three-dimensional swallow tail (diffeomorphic to the manifold of polynomials $x^5 + Ax^3 + Bx^2 + Cx + D$ with multiple roots). The cusp edges of the instantaneous fronts sweep out on the large front a two-dimensional (singular) manifold. To it belong those points (A, B, C, D) for which the polynomial has a root of multiplicity 3 or higher. This is the open swallow tail.

Differentiation of the polynomial lowers by one both its degree and the multiplicity of the root. Under differentiation the open swallow tail in the four-dimensional space of polynomials of degree 5 goes over into an ordinary swallow tail in the space of polynomials of degree 4 (Fig. 8).

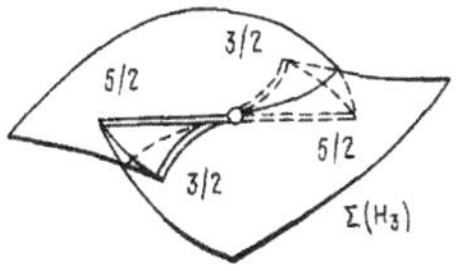

Fig. 14

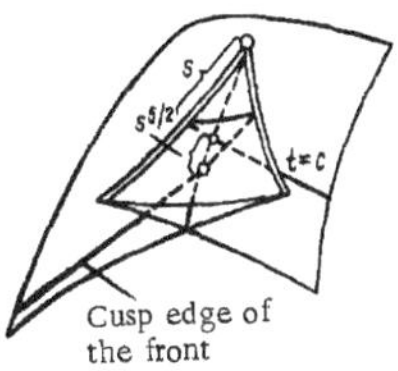

Fig. 15

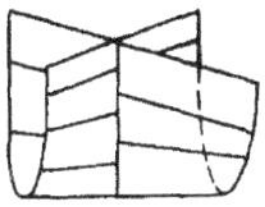

Fig. 16

Fig. 14. Graph of the time function.

Fig. 15. Sweeping-out a caustic by cusp edges of the fronts.

Fig. 16. Sweeping-out a Whitney umbrella.

The physical meaning of this mapping consists in the following. Space–time is fibered into world lines of events "occurring at a given place." The base of this fibering is three-dimensional. The large front in space–time can be reduced to the canonical form

$$\{(A, B, C, D) : x^5+Ax^3+Bx^2+Cx+D \text{ has a multiple root}\}$$

by a diffeomorphism taking world lines into lines parallel to the D axis (by a theorem of Lyashko [23] it is even possible to choose this diffeomorphism so that on each world line D differs from time by a constant).

Thus, differentiation of polynomials corresponds to projection of space–time onto space along world lines: a large front lies in the space of polynomials of degree 5, while a caustic lies in a space of polynomials of degree 4 (if our normal forms, i.e., suitable coordinates in a neighborhood of the singularity, are used).

5.1. Sweeping-Out of Caustics by Cusp Edges of the Wave Fronts. The investigation of the sweeping-out of a caustic by cusp edges of instantaneous fronts proceeds as follows: we first study the sweeping-out of the open tail on the large front and then its projection onto the surface of the caustic. To study the sweeping-out of the open tail on the large front it is necessary to reduce the time function to normal form. The time function of general position on the large front can be reduced to the form $t = \pm A + \text{const}$ by a diffeomorphism preserving the front (see [30]). However, this diffeomorphism already takes rectified world lines into some kind of curves. It can therefore not be used to study the sweeping-out of the caustic. The problem thus arises of reducing the time function to normal form by a diffeomorphism preserving the fibering of space–time into world lines.

If we were to require of the reducing diffeomorphism that it preserve also the entire large front, then we would obtain too small a group of diffeomorphism, and it would not suffice to reduce the families of cusp edges of instantaneous fronts to normal form. We shall therefore reduce to normal form only the restriction of the time function to the manifold formed by the cusp edges of the instantaneous fronts in space–time.

We shall formulate the final results.

Definition 1. A function of (A, B, C, D) equal to zero at the origin and having there a nonzero derivative with respect to A we call a time function.

Definition 2. We call two time functions equivalent if their restrictions to an open swallow tail go over into one another under a diffeomorphism of (A, B, C, D)-space preserving the open tail and taking the family of all lines parallel to the D axis into itself and also by a diffeomorphism of the axis of values of the function.

THEOREM 1 (see [10]). A time function is formally equivalent to the function $A \pm D$. In other words, the family of instantaneous cusp edges of the fronts can be reduced to the normal form $A \pm D = \text{const}$ at least at the level of formal series.

Remark 1. In the class of holomorphic functions this reduction is clearly impossible: generally speaking, the series diverge. Indeed, the cusp edges of a moving front pass through a point on the line of self-intersection of the caustic twice (Fig. 15). (It follows from Theorem 1 that both passage times are separated by an interval of order $s^{5/2}$, where s is the distance from the vertex of the caustic.)

We consider the preimage of the line of self-intersection of the swallow tail on the surface of the open tail. On this line there occur two involutions with a common fixed point over the vertex of the caustic (one involution represents two points of the large front with a common projection onto the caustic, while the other represents two points on a single isochrone).

The pair of involutions can be reduced to simple normal form in the class of formal power series, but in the holomorphic case the reducing series, as a rule, diverge: the pairs of involutions even have functional moduli (Voronin [15, 16], Ecalle [34]).

Remark 2. Several more theorems on the normal forms of sweep-outs of various surfaces by cusp edges of moving wave fronts and caustics of general position are proved in [10, 8, 13].

Example 3. In a neighborhood of a metamorphosis of "lip" or "nose-to-nose" type a caustic (or front) moving along the plane can be reduced by a formal diffeomorphism of space–time fibered over the time axis to the normal form

$$t=\lambda+\lambda^{3/2}\pm\tau^2$$

(λ, τ are coordinates on the plane). If the time is not transformed, then it is possible to reduce the family to the form

$$t=\varphi(\lambda)+\lambda^{3/2}\pm\tau^2,\quad \varphi'(0)\neq 0$$

but only by a formal rather than a genuine diffeomorphism (in contrast to the first normal form, reduction to which in the holomorphic case is realized, generally speaking, by divergent series).

Example 4. We consider the sweeping-out of the Whitney umbrella (the image of the mapping $\varphi(p, q) = (p, pq, q^2)$) by cusp edges of instantaneous caustics or fronts. Here the normal form of the time function has the form $t = q + a(q^2)$, and reduction to this form is realized by a genuine diffeomorphism, while the diffeomorphism is not admitted on the time axis (otherwise the normal form is $t = q$). In particular, the general family of smooth curves sweeping out the umbrella is converted into a family of lines by a diffeomorphism preserving the umbrella (Fig. 16).

5.2. Bicaustics. In [10] the manifold swept out by the cusp edges of instantaneous caustics moving in space (not in space–time) is called a bicaustic.

Definition. The projection of the set of singularities of the original caustic (not accounting for self-intersection) onto the base of the fibering is called a bicaustic (of a caustic situated in the space of a fibering with one-dimensional fibers).

In [10] a classification is presented of the singularities of bicaustics of general position on the plane and in three-dimensional space. Here we briefly reproduce the basic conclusions. On the plane a bicaustic of general position in a neighborhood of each of its points is diffeomorphic to one of the following curves in a neighborhood of zero:

$$u=0,\quad uv=0,\quad u^3=v^2,\quad u^3=au^2v^2+v^6\ (4a^3+27\neq 0).$$

These singularities are stable (only the parameter a in the normal form can change under perturbation).

The general bicaustic in three-dimensional space has only the following singularities (aside from self-intersections): 1) a cusp edge (composed of points of type A_4); 2) a swallow tail (a point of type A_5); 3) a Whitney umbrella (one of the points of type A_3); 4) lines D_4 with singular points D_5; 5) intersections of lines A_4 and D_4 with lines of self-intersection.

In a neighborhood of each regular point of the surface of singularities (A_3) of a general caustic in four-dimensional space projection of general position from four-dimensional space into three-dimensional space restricted to the surface A_3 is diffeomorphic either to an imbedding of the plane in three-dimensional space or to the mapping of the plane onto a Whitney umbrella ($(u, v) \mapsto (u, uv, v^2)$) in a neighborhood of the origin. Stability of the Whitney umbrella is well known.

Thus, a bicaustic of general position in three-dimensional space has as singularities (aside from transversal self-intersections) only singularities diffeomorphic to a Whitney

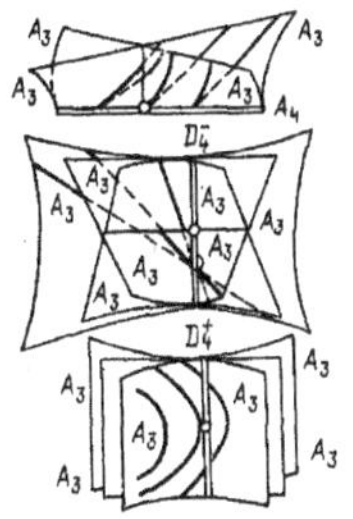

Fig. 17

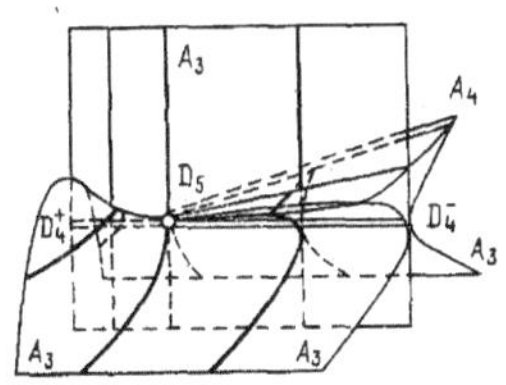

Fig. 18

Fig. 19

Fig. 17. Singularities of bicaustics and their sweep-outs by cusp edges of caustics.

Fig. 18. The singularity D_5 on a bicaustic.

Fig. 19. Diffeomorphic projection of the cusp edge of a swallow tail onto a plane semicubic parabola.

umbrella and singularities on lines swept out by point singularities of instantaneous caustics (A_4 or D_4).

The general caustic in three-dimensional space in a neighborhood of each point of the projection of the line A_4 (excepting individual points A_5, D_5) is diffeomorphic to one of the three surfaces

$$u^2=v^3,\quad u^2=v^3w^2,\quad u^2w=v^3w$$

in a neighborhood of the origin. These singularities are stable.

Thus, the projection of the line A_4 is an edge of a semicubic cusp of the bicaustic with the exception of individual points near which both parts of the bicaustic separated by the cusp edge intersect (Fig. 17).

The proof is based on the fact that the image of a surface of general position with a semicubic cusp edge under the mapping of folding three-dimensional space is locally diffeomorphic to the "folded Whitney umbrella" $u^2 = v^3w^2$.

The general bicaustic in three-dimensional space in a neighborhood of each point of the projection of the curve D_4^+ (the purse) is diffeomorphic to the plane, while D_4^- (the pyramid) is diffeomorphic to the surface $w(w - v^2)(w - Av^2) = 0$ where for common points of the curve $A = \alpha + u$, $\alpha \neq 0$ while for individual points $A = u$ or $A = \alpha \pm u^2$, $\alpha \neq 0, 1$ (Fig. 17). All these singularities are also stable (the value of the parameter α can change under perturbation).

In a neighborhood of points A_5 the general bicaustic in three-dimensional space is diffeomorphic to the swallow tail. The question of the normal form of a bicaustic in a neighborhood of the point D_5 (Fig. 18) has not been completely solved.

Of course, on a general bicaustic it is also possible to find transversal intersections of the branches enumerated.

However, we return to the open swallow tail: Theorem 1 can be considered as the investigation of the sweep-out of a bicaustic by cusp edges of instantaneous caustics in a neighborhood of the point A_5.

The proof of Theorem 1 is based on the study of vector fields tangent to the open tail.

THEOREM 2. Each vector field in three-dimensional space holomorphic at the point 0 and tangent to a swallow tail is the projection of some vector field holomorphic in a neighborhood of the point 0 of four-dimensional space and tangent to an open swallow tail.

In [10] this theorem was proved by direct but surprising computations in which many terms cancelled by "happy coincidence." By analyzing the result, A. B. Givental' arrived at the general theory presented below of open swallow tails in spaces of polynomials of any degree.

5.3. Multidimensional Open Swallow Tails. We begin from the tower formed by polynomials of all degrees with leading coefficient one (which are complex, to be specific, but all constructions can be carried out in the real case as well). The n-th story of the tower is the space

$$\mathbf{C}^n = \{x^n + \lambda_1 x^{n-1} + \ldots + \lambda_n\}.$$

The mapping of differentiation and division by the exponent of the degree projects each story onto the preceding story (the projection is a fibering with one-dimensional fibers).

Definition. The comultiplicity of a root of a polynomial is the difference between the degree and the multiplicity.

Differentiation of a polynomial does not change the comultiplicity of a root (as long as it remains a root). Polynomials having at least one root of comultiplicity $\leqslant k$ form an algebraic submanifold in each story of the tower. We obtain (for each fixed k) a chain of manifolds of polynomials of different degrees having at least one root of comultiplicity $\leqslant k$. Projection from each story onto the preceding story maps an upper manifold of the chain onto a lower manifold.

THEOREM 3 (see [18]). The chain described stabilizes at the $(2k + 1)$-st story (the first story where a root of comultiplicity $\leqslant k$ is unique). More precisely, each manifold of a higher story (i.e., a manifold of polynomials of degree $n > 2k + 1$ having a root of comultiplicity $\leqslant k$) lies on a smooth algebraic hypersurface which projects diffeomorphically onto the preceding story. Thus, all manifolds of polynomials of degree $n \geqslant 2k + 1$ with a root of comultiplicity $\leqslant k$ are diffeomorphic to one another (as manifolds with singularities and as affine algebraic manifolds).

Remark. An analogous theorem holds for a tower composed of polynomials with a sum of roots equal to zero. This is easily deduced from Theorem 3 by noting that differentiation commutes with translation.

Example. We consider the projection of a swallow tail formed by polynomials $x^4 + Ax^2 + Bx + C$ having a multiple root onto the plane with coordinates A, B (Fig. 19). The cusp edge projects diffeomorphically into a semicubic parabola, because it lies on the surface $C = -A^2/4$. Here $k = 1$, and the chain stabilizes at the story of cubic polynomials.

The hypersurface indicated in Theorem 3 can be indicated explicitly. For the first story over a stable story ($n = 2k + 2$) this was done by Hilbert [35]. The following result is easily verified.

THEOREM 4. If

$$x^n + \lambda_1 x^{n-1} + \ldots + \lambda_n = (x-a)^{k+2}(x^k + a_1 x^{k-1} + \ldots + a_k),$$

then

$$2n!\lambda_n = \sum (-1)^{i+1} i!(n-i)!\lambda_i\lambda_{n-i}, \quad 1 \leqslant i < n.$$

Thus, the hypersurface on which the second stable manifold lies is a paraboloid. The hypersurfaces of the next stories are also paraboloids. In order to indicate their equations explicitly, it is convenient to introduce the operators $G^{(s)}$ by the formula $G^{(s)}[F] = (d^sF/dx^s)/(s-r)!$ where r is a fixed number and the factorials of nonnegative numbers are defined as the products of the preceding nonnegative numbers cut off at some sufficiently distant site, for example, at $1 - r$: for $s < r$ we set

$$(s-r)! = (s-r)\ldots(1-r).$$

THEOREM 5. If $F = (x-a)^{r+1}(x^{\mu-1} + \ldots)$, where $r > \mu$, then

$$\sum (j-i)G^i[F]G^j[F] = 0$$

(the summation goes over $i + j = r + \mu$, $0 \leqslant i \leqslant \mu$).

Theorem 5 provides an explicit formula for the paraboloid lying at the $(r + \mu)$-th story and containing the manifold of polynomials with a root of comultiplicity $\leqslant k = \mu - 1$.

From Theorems 4 and 5 we obtain Theorem 3 and also the following corollary.

COROLLARY. All n coefficients of the polynomial

$$(x-a)^r(x^k+a_1x^{k-1}+\ldots+a_k)=x^n+\lambda_1x^{n-1}+\ldots+\lambda_n$$

can be expressed in terms of $\lambda_1,\ldots,\lambda_{2k+1}$ in polynomial fashion (in the form of universal, i.e., depending only on n and k, polynomials with rational coefficients).

The proofs of Theorems 4 and 5 can be obtained by direct computations (this involves rather clever identities for the combining numbers). A. B. Givental' noted that Theorems 3 and 4 follow from the following two identities (in which $F^{(i)} = d^iF/dx^i$).

THEOREM 6 (see [17, 18]). If n is even, then for any polynomial F of degree n the polynomial $R_0[F]=\sum_{i=0}^{n}(-1)^iF^{(i)}F^{(n-i)}$ is constant.

If F has a zero of comultiplicity <n/2, then this constant is zero.

If the degree n of the polynomial F is odd, then the polynomial $R_1[F] = \sum_{i=0}^{(n-1)/2}(n-2i)F^{(i)} \times F^{(n-1)}$ is linear, and $(d/dx)R_1[F] = R_0[dF/dx]$. If F has a root of comultiplicity $<(n-1)/2$, then the linear function $R_1[F]$ is zero.

The identities of Theorem 4 are obtained from Theorem 6 by the substitution $x = 0$.

Proof of Theorem 6 (according to [18]). We consider the "generalized Newton equation of order 2m for a free particle"

$$d^{2m}\xi/dt^{2m}=0.$$

The function

$$H=1/2\sum_{i=1}^{m-1}(-1)^{i-1}\xi^{(i)}\xi^{(2m-i)}\left(=\int\xi^{(2m)}\dot{\xi}dt\right)$$

is a first integral. In the phase space of the Newton equation we introduce a symplectic structure by means of the Darboux coordinates

$$q_i=\xi^{(i-1)},\quad p_i=(-1)^{i+1}\xi^{(2m-i)},\quad i=1,\ldots,m$$

[or, equivalently, by means of the variational principle $\delta\int L dt = 0$, where $L = (\xi^{(m)})^2/2$]. The Newton equation can then be written as a Hamiltonian system with Hamiltonian function H. Solutions of the Newton equation are polynomials of degree $\leqslant 2m-1$. Theorem 4 and the first part of Theorem 6 constitute a conservation law of H for the equation of a free particle.

The Euler homogeneous and quasihomogeneous fields

$$E_0=\sum\xi^{(i)}\partial/\partial\xi^{(i)},\quad E_1=\sum i\xi^{(i)}\partial/\partial\xi^{(i)}$$

take phase curves of the Hamiltonian system of a free particle into phase curves. The field $V_1 = (2m-1)/2E_0 - E_1$ is Hamiltonian (since the symplectic structure is homogeneous of degree 2 and quasihomogeneous of degree $2m-1$).

The Hamiltonian of the field V_1 is equal to

$$H_1=1/2\sum(-1)^{i-1}(2m+1-2i)\,\xi^{(i-1)}\xi^{(2m-i)}\quad(1\leqslant i\leqslant m).$$

From the homogeneity and quasihomogeneity of H we easily find the Poisson bracket $\{H, H_1\} = H$. This proves the second part of Theorem 6 (and Theorem 3).

Example. We consider manifolds of polynomials $x^n + \lambda_2x^{n-2} + \ldots + \lambda_n$ with a root of comultiplicity less than or equal to $k = 2$. In the manifold of polynomials of fourth degree this is an ordinary swallow tail, while in the manifold of polynomials of fifth degree it is an open swallow tail. The projection of the open tail onto the ordinary tail is not a diffeomorphism, since the points of the line of self-intersection each have 2 preimages. According to Theorem 3, the open swallow tail in the space of polynomials of degree $2k + 1 = 5$ is already stable: all manifolds of the chain lying above it are diffeomorphic to it.

As in this example, stabilization always occurs precisely at the story where all self-intersections disconnect. We therefore shall call the first stable manifold a "multidimensional open swallow tail." (The one-dimensional open swallow tail is the semicubic parabola, the two-dimensional open swallow tail lies in the space of polynomials of degree 5, etc.).

5.4. Vector Fields Tangent to a Stable Manifold of Polynomials with a Root of Comultiplicity $\geqslant k$. THEOREM 7 (see [18]). Each germ of a holomorphic vector field at zero of the space of polynomials $x^n + \lambda_2 x^{n-2} + \ldots + \lambda_n$ which is tangent to a multidimensional swallow tail (the manifold of polynomials with multiple roots) can be lifted to the germ at zero of a holomorphic vector field tangent to the stable manifold lying over it.

In other words, in the space of polynomials of any degree $N \geqslant (2n - 3, n)$ there exists a holomorphic germ of a vector field which is tangent to the manifold of polynomials with a root of comultiplicity $\leqslant n - 2$ and projects into a field on the space of polynomials of degree n under $(N - n)$-fold differentiation.

Remark 1. Similar theorems hold for spaces of polynomials without restrictions on the leading coefficient or on the sum of the roots. They also hold for polynomial vector fields in the whole space.

Remark 2. Fields tangent to (multidimensional) swallow tails are well known (see [30]). From the proof of Theorem 7 of [8] it is possible to derive explicit formulas for the lifted fields.

Remark 3. The restriction of stability $N \geqslant 2n - 3$ in Theorem 7 is essential: for $n = 5$, $N = 6$ the theorem is not true.

Remark 4. Any polynomial vector field in a space of polynomials of degree n can be represented as a sum of a polynomial field lifted from the space of polynomials of degree $n - 1$ and a field tangent to the manifold of polynomials with multiple roots. Therefore, Theorem 7 admits the following converse: each polynomial field tangent to a manifold of polynomials of degree n with a root of comultiplicity $\leqslant k$ can be represented as the sum of lifted fields tangent to the preceding swallow tails (i.e., manifolds of polynomials with multiple roots of degrees $n, n - 1, \ldots, k + 2$).

Below we shall need one further result concerning the open swallow tail.

5.5. Lagrangian Property of Open Swallow Tails. THEOREM 8 (see [18]). The open k-dimensional swallow tail

$$\{x^n + a_{n-2}x^{n-2} + \ldots + a_0 = (x - \xi)^{k+1}(x^k + \ldots)\}$$

is a Lagrangian submanifold (with singularities) in the space of polynomials of degree $n = 2k + 1$ with leading coefficient 1 and sum of roots 0 equipped with its natural symplectic structure (see Sec. 1).

Proof. We recall the symplectic structure in question. Our space is the space of characteristics of the hypersurface $h = 0$ in the symplectic space of polynomials of the next $[(n + 1)$-st$]$ degree where h is the Hamiltonian of translations along the x axis.

The open k-dimensional swallow tail is the projection onto the space of characteristics of the manifold of polynomials of degree $n + 1$ divisible by x^{k+2}. On this k-dimensional smooth submanifold in $(2k + 2)$-dimensional symplectic space the symplectic form is equal to zero (the submanifold is one of the coordinate planes in the Darboux coordinates of Sec. 1). Therefore, its projection into the 2k-dimensional space of characteristics is a Lagrangian manifold (generally speaking, with singularities).

It is not hard to compute an explicit expression for the symplectic structure of Theorem 8.

Example. We consider the ordinary open tail $(k = 2)$

$$\{x^5 + Ax^3 + Bx^2 + Cx + D = (x - \xi)^3(x^2 + 3\xi x + \eta)\}.$$

We start from the standard symplectic structure in the space of all polynomials of degree $2k + 3 = 7$

$$q_0 x_7/7! + q_1 x^6/6! + q_2 x^5/5! + q_3 x_4/4! - p_3 x^3/3! + p_2 x^2/2 - p_1 x + p_0,$$

given by the Darboux coordinates of Sec. 1. The hyperplane $q_0 = 1$ has lines parallel to the p_0 axis as characteristics. The space of characteristics can be identified with the manifold of derivatives, i.e., polynomials of degree $2k + 2 = 6$

$$x^6/6!+q_1x^5/5!+q_2x^4/4!+q_3x^3/3!-p_3x^2/2+p_2x-p_1$$

(the symplectic structure is $\Sigma dp_i\wedge dq_i$). The Hamiltonian of translations has the form

$$H=p_1+p_2q_1+p_3q_2+q_3^2/2.$$

On the surface $H = 0$ it is possible to take for coordinates $(q_1, q_2, q_3, p_3, p_2)$. A characteristic of the surface $H = 0$ is a polynomial with $H = 0$ considered up to translation along the x axis. The space of characteristics can therefore be identified with the space of polynomials of degree $2k + 2 = 6$ with $H = 0$ which have zero sum of the roots ($q_1 = 0$). On this space the coordinates are (p_2, q_2, p_3, q_3).

Therefore, the space indicated projects diffeomorphically onto the space of polynomials of degree $2k + 1 = 5$ under differentiation of the polynomials. We have thus identified the space of characteristics with the space of polynomials of degree $2k + 1 = 5$ with fixed leading and zero succeeding coefficient. We shall express the symplectic structure of the space of characteristics in terms of the remaining coefficients of the polynomial. The expressions for these coefficients in terms of our Darboux coordinates are $A = 20q_2$, $B = 60q_3$, $C = -120p_3$, $D = 120p_2$.

Thus, Theorem 8 asserts that the open tail is Lagrangian with symplectic structure $3dA\wedge dD-dB\wedge dC$ (in the general case:

$$\sum(-1)^i i!j!da_i\wedge da_j,\quad i+j=n-2).$$

5.6. Open Swallow Tail and Metamorphoses of Curves Projectively Dual to Smooth Curves. Let $\varphi: \mathbf{R}\to\mathbf{R}^3$ be a smooth mapping. We call it a curve of type (a, b, c) at 0 if the first derivative nonzero at 0 has index a, the first not collinear with it has index b, and the first not coplanar with them has index c.

Curves of type $(a_1,\ldots,a_n)$ in $\mathbf{R}^n$, $\mathbf{RP}^n$, $\mathbf{C}^n$, $\mathbf{CP}^n$ are defined similarly. A curve of finite type has an osculating hyperplane at each point.

Definition. The dual to $\varphi: \mathbf{R}\to\mathbf{RP}^n$ curve $\varphi^*: \mathbf{R}\to\mathbf{RP}^{n*}$ is the manifold of osculating hypersurfaces of the original curve. The next result is easily proved (see [11, Secs. 7, 17] and [27, 28]).

THEOREM 9. $\varphi^{**}=\varphi$; the dual to the type (a, b, c) is the type $(c - b, c - a, c)$ (in the n-dimensional case: $a_n - a_{n-1},\ldots,a_n - a_1, a_n$).

Points of smoothness are distinguished by the condition $a_1 = 1$. The origin of the hierarchy of points of smooth curves in $\mathbf{R}^3$ is as follows:

$$\begin{array}{l}(1,2,3)\text{ ordinary point} \leftarrow (1,2,4)\text{ flattening} \leftarrow (1,2,5)\text{ biflattening} \leftarrow \ldots \leftarrow\\ \qquad\qquad\qquad\qquad\quad (1,3,4)\text{ inflection} \leftarrow \ldots\end{array}$$

The codimension of the type is given by the formula $c = \Sigma(a_i - i)$ (the type of codimension c can be realized in nonremovable manner by a small perturbation of a family of curves depending on $c - 1$ parameters).

The manifold of all tangent hypersurfaces (hypersurfaces containing a tangent line) of a curve φ forms in the dual space a developed surface with a cusp edge φ^*. We call it the front of the Legendre singularity of φ (more briefly, the front of the curve φ).

Example. Suppose that φ has the simplest degeneracy of codimension 1 [a flattening (1,2,4)]. Then φ^* has the singularity (2,3,4) of the curve $x = t^2$, $y = t^3$, $z = t^4$. The tangent lines of this curve sweep out an ordinary swallow tail – the front of the curve φ.

If φ is a curve of general position in $\mathbf{R}^3$, then on it there may be individual points of flattening, but there are no points of other types. Hence, the front of a curve of general position is a surface with a cusp edge and swallow tails at individual points.

In a general one-parameter family of curves in $\mathbf{R}^3$ for individual values of the parameter there may also be points of biflattening. We consider the family of dual curves in $\mathbf{R}^{3*}$. We place each of them in its own three-dimensional space. We thus consider the family of dual curves as a two-dimensional surface in four-dimensional space $\mathbf{R}^{3*} \times \mathbf{R}$ ($\mathbf{R}$ is the axis of the parameter).

THEOREM 10 (see [9, 27, 28]). The surface constructed as locally diffeomorphic to an open swallow tail, and its partition into dual curves, is diffeomorphic to the partition of the space of polynomials $x^5 + Ax^3 + Bx^2 + Cx + D$ with triple roots into lines A = const.

For specific values of the parameter points of inflection of a curve of the original family may occur. In this case the normal form of the surface swept out by the dual curves in four-dimensional space, according to Shcherbak [27, 28], is as follows:

$$\{x^4 + Ax^3 + Bx^2 + Cx + D \text{ has a multiple root}\},$$

and the partition into dual curves is B = const. Shcherbak [28] proves also the following theorems regarding the fronts of the curves and their rearrangements:

1. The front of a curve dual to a curve of general position (i.e., the developed surface whose cusp edge is the original curve of general position in $\mathbf{R}^3$ or $\mathbf{RP}^3$) has at points of flattening of type (1,2,4) of the original curve a "folded Whitney umbrella" singularity (diffeomorphic to the germ of the surface $x^2y^3 = z^2$ at zero; see Fig. 17).

2. The front of a curve of general position in $\mathbf{RP}^n$ has as singularities only multidimensional swallow tails (locally diffeomorphic to bifurcation diagrams A_μ, $\mu \leqslant n$). The union of the front of the curve with the hyperplane dual to the point considered of the original curve is locally diffeomorphic to the bifurcation diagram B_μ.

3. In typical one-parameter families of curves in $\mathbf{RP}^n$ for individual values of the parameter there occur points of types $(1,\ldots,n-1, n+2)$ and $(1,\ldots,n-2, n, n+1)$. The normal forms of the corresponding metamorphoses of the fronts are as follows:

$$\{x^{n+2} + \lambda_1 x^n + \ldots + \lambda_{n+1} \text{ with a multiple root}\}, \text{ fronts: } \lambda_1 = \text{const};$$
$$\{x^{n+1} + \lambda_1 x^n + \ldots + \lambda_{n+1} \text{ with a multiple root}\}, \text{ fronts: } \lambda_2 = \text{const};$$

The theory of fronts of curves can be considered part of the theory of Legendre singularities. Fronts of manifolds of any dimension can be defined in more general fashion. Any submanifold M of projective space defines a Legendre mapping. Namely, the set L of all contact elements of projective space $\mathbf{RP}^n$ tangent to the submanifold M forms a Legendre manifold in $PT^*\mathbf{RP}^n$. The fibering assigning to each contact element the hyperplane containing it ($PT^*\mathbf{RP}^n \to \mathbf{RP}^{n*}$) is a Legendre fibering. The mapping $L \to \mathbf{RP}^{n*}$ is called a frontal mapping for M. It is a Legendre mapping. The image of L in $\mathbf{RP}^{n*}$ is called the front of the original submanifold in $\mathbf{RP}^n$. The front is a hypersurface in the dual space regardless of the dimension of the original manifold in projective space.

It is well known that swallow tails of all dimensions are developed surfaces. The multidimensional generalization of this fact is the following simple result.

THEOREM 11 (see [28]). Any stable germ of a Legendre mapping of corank m is Legendre-equivalent to the germ of a frontal mapping of an m-dimensional submanifold of projective space. Frontal mappings of curves of general position have only the singularities A_μ. Metamorphoses of Legendre mappings of corank m encountered in families of general position with a finite number of parameters are also Legendre-equivalent to metamorphoses of frontal mappings of m-dimensional manifolds.

6. Symplectic and Contract Triads

The problem of passing about an obstacle led us in Sec. 4 to open swallow tails: these are singularities of the Lagrangian manifold of rays breaking off from the surface in the symplectic space of all rays (or singularities of Legendre manifolds of two types: manifolds of contact elements of fronts and 1-jets of time functions).

In Sec. 5 we described the singularities of open swallow tails in terms of the geometry of spaces of polynomials and binary forms. We now combine the results of Secs. 4 and 5 in order to explain why manifolds of polynomials with roots of high multiplicity describe singularities of Lagrange and Legendre manifolds in multidimensional variational problems with one-sided constraints.

A. Symplectic Triads. The following definition due (just as Theorems 1-4 following it) to Givental' [19] is obtained from the constructions of Melrose [37] (see Sec. 4) by dropping the structures that are unnecessary for our purposes.

Definition 1. A symplectic triad is a collection (H, L, l) consisting of a hypersurface H in a symplectic manifold and a Lagrangian manifold L tangent to H along a submanifold l of codimension 1 in L with first-order tangency.

Example 1. We consider a region Γ on a hypersurface in a Riemannian manifold M. We define a pencil of geodesics on Γ of the time function $s:\Gamma \to \mathbf{R}$ (equal to the distance from some initial front on Γ). The time function satisfies the Hamilton–Jacobi equation $(\nabla s)^2 = 1$ (we assume it to be smooth in the region Γ considered).

There is a symplectic triad in phase space T^*M connected with this situation. The hypersurface H consists of all unit vectors ($p^2 = 1$). The manifold L is formed by all extensions of the algebraic 1-forms $p = ds|_q$ from the tangent spaces to the hypersurface Γ to the tangent spaces to the entire enveloping manifold M. The hypersurface H cuts out on L a submanifold l of codimension 1: it consists of extensions equal to zero on the normal to Γ. (If we identify 1-forms on M with vectors by means of the Riemannian metric, then l is composed of vectors tangent to Γ of the field grad s, while L consists of vectors projecting into these under orthogonal projection onto the tangent plane.)

THEOREM 1. The triple (H, L, l) constructed on the basis of a pencil of geodesics is a triad.

Proof. That the tangency of H and L along l has first order follows from the nondegenerate convexity of the Riemannian metric (in less regular variational problems this condition may be violated!). The Lagrangian property of L follows from the fact that $i^*pdq = d\pi^*s$ ($i:L \to T^*M$ is an imbedding, $\pi:L \to \Gamma$ is the natural projection, pdq is the canonical 1-form on T^*M). Thus, (H, L, l) is a triad.

We now consider an arbitrary symplectic triad.

Definition 2. The image of the manifold l in the space of characteristics of the hypersurface H is called the Lagrangian manifold with singularities generated by the triad (H, L, l). It is clear that this manifold is actually Lagrangian where it is nonsingular.

THEOREM 2. The manifold of all rays (geodesics of the enveloping space) tangent to the geodesics of the pencil considered on the hypersurface is the Lagrangian manifold with singularities generated by the triad of Theorem 1.

Indeed, the space of characteristics is the space of oriented geodesics of the enveloping space, while the initial conditions from l are unit vectors tangent to the geodesics of the pencil on the hypersurface.

Symplectically equivalent triads generate symplectically diffeomorphic Lagrangian manifolds with singularities.

It follows from Theorems 1 and 2 that to obtain the normal form of the manifold of rays breaking off from geodesics of the pencil it suffices to symplectically reduce the triad to normal form. The following example provides normal forms of triads.

Example 2. We consider the symplectic space V of homogeneous polynomials of degree $d = 2k + 2$ in x with leading coefficient 1 and its natural symplectic structure (see Sec. 1). A polynomial with Darboux coordinates (p, q) is given by the formula

$$\varphi(x, \quad 1) = x^d/d! + q_1x^{d-1}/(d-1)! + \ldots + q_{k+1}x^{k+1}/(k+1)! - \ p_{k+1}x^k/k! + \ldots + (-1)^{k+1}p_1$$

(the signs in front of the p_i alternate). We denote by h the Hamiltonian of translations

$$h = p_1 + q_1p_2 + \ldots + q_kp_{k+1} + q^2_{k+1}/2.$$

Let H be the hypersurface $h = 0$, and let L be the Lagrangian plane $p = 0$. The hypersurface H is quadratically tangent to the Lagrangian plane L along the subspace of codimension 1 in L given by the equations $p = 0$, $q_{k+1} = 0$. The next result is easily verified.

THEOREM 3. The Lagrangian manifold with singularities generated by the triad of Example 2 is symplectically diffeomorphic to a k-dimensional open swallow tail, i.e., to the manifold of polynomials with a root of multiplicity $>k$ in the symplectic space of polynomials of degree $2k + 1$ (with fixed leading coefficient and with sum of roots zero).

The main result on triads is the following.

THEOREM 4. The germ of the triad of Example 2 at zero is stable. A germ of a triad of general position at a point of quadratic tangency of the hypersurface with a Lagrangian manifold is symplectically equivalent to the germ of the triad of Example 2 or the superstructure over it.

[The superstructure over a triad (H, L, l) in $\mathbf{R}^{2a}$ with Darboux coordinates (p, q) is the triad in $\mathbf{R}^{2a} + \mathbf{R}^{2a'}$ with Darboux coordinates $(p, p'; q, q')$, where $H_{new} = H + \mathbf{R}^{2a'}$, $L_{new} = L + \mathbf{R}^{a'}$, and $\mathbf{R}^{a'}$ is the Lagrangian plane $p' = 0$ in $\mathbf{R}^{2a'}$.]

COROLLARY 5. The germ of a Lagrangian manifold with singularities generated by a triad of general position at each point is diffeomorphic to the product of an open swallow tail of suitable dimension with Euclidean space.

COROLLARY 6. The germ of the manifold of rays breaking off from the surface of an obstacle is diffeomorphic to the product of a swallow tail of suitable dimension with Euclidean space (for a pencil of geodesics of general position on the surface of an obstacle of general position).

Remark 1. The condition of general position in Theorem 4 is as follows. We consider the field of characteristic directions of the hypersurface H. At points of l the direction of the field is tangent to L. It is required that the field of tangent directions to L arising on the hypersurface l in L be general (reduce to the normal form $\partial/\partial q_1 + q_1\partial/\partial q_2 + \ldots + q_{a-1}\partial/\partial q_a$ by a diffeomorphism of L reducing l to the form $q_a = 0$).

Remark 2. Theorem 4 (and its corollaries) are true both in the smooth and analytic or holomorphic versions.

Proof of Theorem 4. 1°. We locally rectify the hypersurface H by choosing Darboux coordinates in which its equation has the form $q_1 = 0$ near the point 0.

2°. We choose a linear Lagrangian space in these coordinates which is transversal to H and L at 0. The parallel spaces form a Lagrangian fibering, and L is a section of it (near zero).

3°. We choose new Darboux coordinates (linear with respect to the old coordinates) in which fibers of the fibering have the form $q = \text{const}$. The equation of L then has the form $p = \partial S/\partial q$. H remains a hyperplane in the new coordinates.

4°. We go over to the Darboux coordinates $P = p - \partial S/\partial q$, $Q = q$. L has now become the zero section of the fibering $(P, Q) \mapsto Q$. We identify this Lagrangian fibering (locally) with the cotangent bundle T^*L [a point (P, Q) is interpreted as the 1-form PdQ at Q].

5°. The intersection of the hypersurface H with each fiber of the bundle T^*L is an affine hyperplane in this fiber.

6°. Giving an affine hyperplane in the space is the same as giving in the linear space itself a line passing through 0 and a linear 1-form on this line. Indeed, the hyperplane $\Pi = \{\xi : \xi|x = c\}$ defines the line $\{x : \xi|x = \eta|x \ \forall\xi, \eta\in\Pi\}$ and the 1-form $\xi|x$ on it, and the pair consisting of a line and a 1-form on it defines a hyperplane.

7°. The hypersurface H can thus be described in the intrinsic terms of L as a pair consisting of a field of directions on L and a field of 1-forms on lines of the field. In other words, H is converted into a pair consisting of a field of directions on L and a family of differential 1-forms on its integral curves.

The hypersurface l in L is defined in these terms as the set of points where the forms of the family vanish. The direction of the field at points of l is characteristic for H.

8°. The imbedding of the hypersurface l in L and the fibering of L into integral curves define a projection with one-dimensional fibers. For a triad of general position the singularities of this projection are Whitney singularities (see Sec. 4 and Remark 1 above).

9°. Hence, in L it is possible to choose coordinates (x, λ) so that the integral curves are the lines $\lambda = \text{const}$, while the hypersurface l is given by the equation $F = 0$ where $F = x_a + \lambda_1 x^{a-2} + \ldots + \lambda_{a-1}$.

10°. The family of differential 1-forms on the integral curves can be written in these coordinates in the form $f(x, \lambda)dx$. On the hypersurface l the function f vanishes. The fact that H is quadratically tangent to L along l implies that the order of the zero is precisely second order, i.e., $f(x, \lambda) = \varphi(x, \lambda)F^2(x, \lambda)$, where $\varphi(0, 0) \neq 0$.

11°. LEMMA. The family of differential 1-forms $\varphi F^2 dx = \varphi(x, \lambda)(x^a + \lambda_1 x^{a-2} + \ldots + \lambda_{a-1})^2 dx$, where $\varphi(0, 0) \neq 0$, fibered over λ by the change of variables $[x = x(X, \Lambda), \lambda = \lambda(\Lambda)]$ can be reduced to the form $(X^a+\Lambda_1 X^{a-2}+\ldots+\Lambda_{a-1})^2 dX$.

12°. Proof. This is essentially a versality theorem for the forms $x^a|dx|^{1/2}$. (Apparently, an analogous versality theorem is true in a much more general situation, for example, for forms $x^a|dx|^\alpha$. Unfortunately, a general proof rests on unproven generalizations of the Weierstrass–Malgrange preparation theorem.) In the case $\alpha = 1/k$ (for example, 1/2) the theorem reduces to an ordinary versality theorem for the function $\int x^{ak}dx$ (the situation is that the condition on the derivative of the function of being k-th degree is invariant under diffeomorphisms).

13°. We rewrite the constructions on 6° and 7° in explicit coordinate form. Let q_i be the coordinates on L, and let p_i be the corresponding momenta.

To the hypersurface $H\subset T^*L$ with equation

$$u_0+u_1p_1+\ldots+u_ap_a=0,\quad u_i=u_i(q),$$

in 6° and 7° we assigned a field of directions of a vector field on L

$$\dot q_1=u_1,\ldots,\dot q_a=u_a;\quad \dot q_i=0\quad (i>a)$$

and a family of differential 1-forms on its integral curves equal to

$$-u_0/u_1dq_1=\ldots=-u_0/u_adq_a.$$

In particular, if $u_1 = 1$, then the family of 1-forms has the form $-u_0dq_1$. Conversely, knowing the components of the field and the form, we can form from them the equations of the hypersurface H.

14°. We choose coordinates (X, Λ) on L according to the lemma of 11°. In these coordinates the field of directions has the form $\partial/\partial X$, while the family of 1-forms has the form $F^2(X, \Lambda)dX$. We denote the momenta corresponding to the coordinates by p_X and p_Λ. The equation of the hypersurface H can be written in the form

$$F^2(X,\Lambda)+p_X=0\ \text{(see 13°)}.$$

This formula already determines a certain normal form of the triad and therefore proves the first half of Theorem 4. It remains to transform the triad to the form indicated in the theorem.

15°. LEMMA. The triad

$$(H: p_x+F^2(X,\Lambda)=0,\ L: p=0,\ l=H\cap L)$$

is symplectically equivalent to the triad

$$(H: p_1+q_1p_2+\ldots+q_{a-1}p_a+q_a^2/2=0;\ L: p=0;\ l: p=q_a=0).$$

16°. To prove the lemma we shall rectify l (at the expense of distorting the field). It is convenient to first make the leading coefficient of F equal to $1/a!$ and to reduce the family of 1-forms to the form $-F^2(X, \Lambda)dX/2$ (this is achieved by dilations). After this we choose as new coordinates on L the derivatives of the polynomial F:

$$q_a=F,\ q_{a-1}=\partial F/\partial X,\ldots,\ q_1=\partial^{a-1}F/\partial X^{a-1}\ (=X)$$

(we retain the coordinates q_i on which neither the field of directions nor the 1-form depend).

17°. In the new coordinates the field of directions and the 1-form can easily be written out. Indeed, the derivatives of the new coordinates with respect to X are given by the formulas

$$\dot q_1=1,\quad \dot q_2=q_1,\ldots,\dot q_a=q_{a-1}.$$

On the integral curves of this field of directions $F^2dX/2 = (q_a^2/2)dq_1$. Hence, by the formulas of 13° the equation of the hypersurface H has the form $q_a^2/2 + p_1 + q_1p_2 + \ldots + q_{a-1}\cdot p_a = 0$. This proves the equivalence of the triads and Theorem 4.

Remark. The appearance of F^2 in the normal form of 14° apparently indicates the connection of the manifold of polynomials with a root of high multiplicity with the manifold diffeomorphic to it (according to [14]) of polynomials with a large number of multiple roots (cf. 16° in Sec. 12 of [11]). Each polynomial of even degree can be uniquely represented, first of all, as a sum of a square and a polynomial of degree two less and, secondly, as a sum of a polynomial with a zero root of half the multiplicity and a polynomial of half the degree. The normal form 17° is connected with the second decomposition, but the symplectic structure connected with the first decomposition may also be useful. The formulas presented in [14] unfortunately do not clarify the explicit form of the symplectic structure of the space of polynomials in which the manifold of polynomials with a large number of multiple roots is Lagrangian (they also do not explain why the manifold of polynomials $x^5 + Ax^3 + Bx^2 + Cx + D$ with multiple critical values is diffeomorphic to a swallow tail).

B. Contact Triads. We shall now describe the contact version of Theorem 4 (see [32]). Its purpose is the investigation of the Legendre manifolds with singularities arising in the problem of passing about an obstacle (the manifolds of contact elements of fronts and 1-jets of the time function; see Sec. 4).

Definition. A contact triad (H, L, l) consists of

1. a hypersurface H in a contact manifold transversal to the contact planes;
2. a Legendre submanifold L in the same contact manifold;
3. a smooth hypersurface l in L such that the hypersurface H is tangent to L along l with first-order tangency.

We shall investigate the germ of a triad at the point 0 from l.

Definition. We call the image of the germ of (H, L, l) at 0 under the projection of H onto its manifold of characteristics the Legendre manifold generated by the triad l at 0.

Example 1. We consider the contact manifold $J^1(M, \mathbf{R})$ of 1-jets of functions on a Riemannian manifold without boundary M equipped with a hypersurface ∂M (it may be supposed that M is Euclidean space, and the case of a plane is already significant).

The Hamilton–Jacobi equation for the propagation of perturbations with speed 1 along M, $(\nabla u)^2 = 1$, defines in J^1 a hypersurface H. We consider a pencil of geodesics on ∂M orthogonal to a fixed hypersurface and parametrized by the distance s to it. We consider all possible extensions of the function s from ∂M to M. The manifold of 1-jets of all extensions at all points of ∂M is a Legendre submanifold L in $J^1(M, \mathbf{R})$. It intersects H along a manifold of 1-jets of extensions whose derivative along the normal to ∂M is equal to zero. We denote this intersection by l.

THEOREM 5. The triple (H, L, l) constructed on the basis of the submanifold ∂M of the Riemannian manifold M and the pencil of parametrized geodesics on it is a contact triad.

Indeed, the order of tangency of H with L along l is first order due to the nondegenerate convexity of the metric.

Remark. In the notation of the diagram of 5°, Sec. 4, the construction of L from l is symmetric to the construction of γ from l (the symmetry permutes I and II; l on this diagram is not placed on the vertical axis only to economize space).

THEOREM 6. The Legendre manifold with singularities generated by the triad of Theorem 5 is the manifold of contact elements of a front in the problem of passing about an obstacle (lying in the manifold of oriented contact elements on M).

Proof. See Theorem 10 of Sec. 4.

To construct normal forms of contact triads we define natural contact structures in odd-dimensional spaces of polynomials. We start again from the space of binary forms of odd degree and its natural (SL_2-invariant) symplectic structure (see Sec. 1).

The projectivization of this even-dimensional space is an odd-dimensional projective space of all hypersurfaces (zero-dimensional submanifolds) of fixed odd degree on a projective line.

In the space of nonzero binary forms the skew-orthogonal complement to the radius vector gives a GL_2-invariant field of hyperplanes. This field determines a field of hyperplanes in the projective space of zero-dimensional submanifolds of a projective line of fixed odd degree. It is natural (invariant under the action of the group of projective transformations of the line on the space of zero-dimensional submanifolds of fixed degree).

We write out the explicit formula for this contact structure in terms of the coordinates (p, q) in the affine part of our projective space of dimension $d = 2a - 1$ formed by zero-dimensional submanifolds not containing the "infinitely distant" point $x = \infty$. This manifold is given by the condition $f(x) = 0$ where

$$f(x) = x^d/d! + q_2 x^{d-1}(d-1)! + \ldots + q_a x^a/a! - p_a x^{a-1}(a-1)! + \ldots + (-1)^a p_1 \qquad (*)$$

(the signs in front of the p_i alternate). It is easy to prove the following theorem.

THEOREM 7. The natural contact structure of the projective space of submanifolds is given in the coordinates (p, q) introduced by the 1-form $\alpha = p'dq' - d'dp' - dp_1$ (where $p' = p_2, \ldots, p_a$; $q' = q_2, \ldots, q_a$).

Definition. The neutral hypersurface of a one-parameter group of contact diffeomorphisms is the hypersurface formed by points where the velocity vector lies in the contact plane [$\alpha(v) = 0$ if v is the velocity vector and $\alpha = 0$ is the equation of the contact planes].

We assume that the neutral hypersurface is transversal to the contact surfaces. It is easy to see that the orbits of the group on the neutral hypersurface are the characteristics of the latter.

Translations of polynomials along the x axis act on the space of polynomials (*) while preserving its contact structure.

THEOREM 8. The equation of the neutral surface has the form $k = 0$ where

$$k = p_2 + p_3 q_2 + \ldots + p_a q_{a-1} + q_a^2/2.$$

Indeed, the field of the velocities of translations along the x axis has the components $\dot{q}_i = q_{i-1}$, $\dot{p} = -p_{i+1}$ (where $p_{a+1} = q_a$, $q_1 = 1$).

We shall now present a normal form of a contact triad. It is formed in the contact space of polynomials (*) by the neutral hypersurface $k = 0$ and the Legendre submanifold of polynomials having zero as a root of multiplicity $\geqslant a$.

THEOREM 9. The triple $(H: k=0;\ L: p=0;\ l: p=q_a=0)$ is a contact triad in the space R^{2a-1} with contact structure $\alpha = p'dq' - q'dp' - dp_1$.

The proof follows immediately from the formula for k.

Definition. The contact triad of Theorem 9 and the Legendre manifold with singularities generated by it are called the standard contact triad in R^{2a-1} and the standard Legendre manifold with singularities.

The dimension of this standard manifold is equal to $m = a - 2$. We shall denote the m-dimensional standard Legendre manifold by Σ^m.

From Theorem 9 we obtain:

COROLLARY 10. The standard Legendre manifold Σ^{a-2} consists precisely of those polynomials of the form (*) for which

$$q_2 = 0,\quad p_2 = -(q_3 p_4 + \ldots + p_a q_{a-1} + q_a^2/2),$$

and which have a root of multiplicity greater than a. The contact structure in this space is given by the 1-form

$$\alpha = p''dq'' - q''dp'' - dp_1, \text{ где } q'' = (q_3, \ldots, q_a),\ p'' = (p_3, \ldots, p_a).$$

Combining this result with the results of Sec. 5 regarding manifolds of polynomials with roots of high multiplicity, we obtain:

COROLLARY 11. The standard Legendre manifold Σ^m is diffeomorphic to an open swallow tail of dimension m, i.e., to the manifold of all polynomials of the form

$$x^{2m+1} + \lambda_1 x^{2m-1} + \ldots + \lambda_{2m} = (x - \xi)^{m+1}(x^{m-1} + \ldots).$$

Example. For $a = 3$ Corollary 10 gives the polynomials

$$x^5/120 + q_3 x^3/6 - p_3 x^2/2 - q_3^2 x/2 - p_1,\quad \alpha = p_3 dq_3 - q_3 dp_3 - dp_1.$$

Polynomials of this form with a root of multiplicity greater than 3 form the Legendre curve $\Sigma^1 = \{f = (x-\xi)^4(x+4\xi)/120\}$, i.e., $q_3 = -\xi^2/2$, $p_3 = -\xi^2/3$, $p = -\xi^5/30$. This curve is diffeomorphic to the semicubic parabola. Its singularities are responsible for the singularities of order 5/2 which evolvents of a plane curve have at points of a tangent of inflection.

Remark. The manifold Σ^{a-2} is quasihomogeneous. If we ascribe to the coordinates the weights

$$\deg q_i = i-1,\ \deg p_i = 2a-i\ (\deg k = 2a-2,\ \deg \alpha = 2a-1),$$

then Σ^{a-2} is the image of a space with coordinates of weights $(1,\ldots,a-2)$ under a quasihomogeneous mapping.

Example. For $a = 4$ the coordinates $(q_3, q_4, p_4, p_3, p_1)$ have weights (2, 3, 4, 5, 7). The Legendre manifold Σ^2 is the image of the plane (ξ, η), where $\deg \xi = 1$, $\deg \eta = 2$:

$$\Sigma^2 = \{f = (x-\xi)^6(x^2 + 5\xi x + \eta)/7!\}.$$

According to Sec. 5, this manifold is diffeomorphic to the manifold of polynomials of degree 6 with leading coefficient 1, sum of roots zero, and a root of multiplicity not less than 4.

Definition. The superstructure over the standard triad in $\mathbf{R}^{2a-1}$ is defined as the triad in $\mathbf{R}^{2b-1}$ where the contact structure has the form $\alpha = p'dq' - q'dp' - dp_1$ $[p' = (p_2,\ldots,p_b)$, $q' = (q_2,\ldots,q_b)]$. The triad (H, L, l) is defined by the equations

$$H: p_2 + p_3q_2 + \ldots + p_aq_{a-1} + q_a^2/2 = 0;\ L: p = 0;\ l: p = 0,\ q_a = 0\ (b \geqslant a).$$

THEOREM 12. The germs at 0 of standard contact triads and the superstructures over them are stable (as germs of contact triads considered up to contact equivalence). The germs of contact triads of general position at all points are contact-equivalent to the germs of standard-triad superstructures.

COROLLARY 13. The germs of the Legendre manifolds formed by contact elements tangent to a front in the problem of passing about an obstacle of general position (see Sec. 4) are contact-stable and contact-equivalent to germs of standard (possibly, superstructure) Legendre manifolds with singularities and are hence diffeomorphic to germs of open swallow tails.

Theorem 12 can be obtained from the symplectic Theorem 4 by contactization. For this we choose the 1-form α giving the contact structure so that the kernel of the form $d\alpha$ at points of the hypersurface contained in the triad is tangent to this hypersurface. The projection of the contact space (along integral curves of the field of the kernel) onto the even-dimensional space of the integral curves defines a symplectic structure on the latter (the image of the form $d\alpha$).

It is easy to verify that the projection of a contact triad is a contact triad. The original contact triad can be obtained from this symplectic contactization (by the addition of the generating function of the Lagrangian manifold as a new coordinate). Applying this construction to the standard symplectic triads of Theorem 4, we obtain the standard contact triads of Theorem 12.

We can now reduce the singularities of the 1-graph of the time function [the manifold of 1-jets of the time function lying in $J^1(M, \mathbf{R})$] to normal form.

We recall that the 1-graph of the time function (Γ in the notation of 5°, Sec. 4) is formed by all characteristics of the hypersurface H [$SJ^1(M, R)$ in the notation of 5°, Sec. 4] passing through points of the manifold l of the triad (H, L, l) of Theorem 5 (see Theorem 10 in Sec. 4).

We consider the problem of passing about an obstacle bounded by a surface of general position ∂M in a Riemannian manifold M. On ∂M we consider a pencil of geodesics of general position parametrized by the distance s to a fixed hypersurface on ∂M. Such a pencil defines a time function (which coincides with s on ∂M).

THEOREM 14. The germ of the manifold Γ of 1-jets of the time function in the situation described is contact-diffeomorphic to the germ of the manifold of polynomials (*) having a root of multiplicity greater than a. The 1-jets of the time function at points of a single ray (i.e., a single geodesic on M) go over under this diffeomorphism into polynomials obtained from one another by translation along the x axis.

Proof. By Theorem 5 the triple (H, L, l) constructed on the basis of s is a contact triad. For an obstacle and pencil of general position this triad (as is easy to show) is of general position. Hence, it is equivalent to the triad of Theorem 9 (by Theorem 12; superstructures are unnecessary, since we consider germs of standard triads not only at zero). For the triad of Theorem 9 the manifold l consists of polynomials with a zero root of multiplicity greater than a. The characteristics of the surface H of the triad of Theorem 9 are the orbits of the action of translations along the x axis (by Theorem 8). The orbits of points of l sweep out the manifold of all polynomials (*) with a root of multiplicity greater than a, as was required to prove.

Finally, we write out explicit formulas for the manifolds of contact elements of a front and the 1-graphs of time functions realized as submanifolds of spaces of polynomials of odd degree

$$x^n+a_1x^{n-1}+\ldots+a_n,\ n=2k+1.$$

The natural (SL_2-invariant) contact structure of this space of polynomials is given by (as is easily calculated) the equation

$$\Sigma i!j!(-1)^i a_i da_j=0,\ (\text{where } i+j=n,\ a_0=1).$$

From Theorems 14 and 12 we obtain:

COROLLARY 15. The 1-graph (manifold of 1-jets) of a time function of general position in the problem of passing about an obstacle in a space of dimension m is locally contact-equivalent to the Legendre manifold with singularities formed by polynomials of the form $(x-\xi)^{m+2}(x^{m-1}+\ldots)$ in the space of polynomials of degree $n = 2m - 1$ with its natural contact structure. The restriction of the time function to this manifold has the form $\pm a_1$ + const, and points of a single ray correspond to polynomials obtained from one another by translation along the x axis.

COROLLARY 16. The manifold of contact elements of a front of general position in the problem of passing about an obstacle in a space of dimension m is locally contact-equivalent to the Legendre manifold with singularities formed by the polynomials

$$x^n+\lambda_{n-1}x^{n-1}+\ldots+\lambda_0=(x-\xi)^m(x^{m-1}+m)\xi x^{m-2}+\ldots)$$

in the space of polynomials of degree $n = 2m - 1$ equipped with the antinatural contact structure

$$d\lambda_{n-1}=\Sigma i!j!(-1)^i(i+2)\lambda_i d\lambda_j,\ \text{where } 0\leqslant i\leqslant n-2,\ i+j=n-2.$$

The question of constructing the singularities of the fronts themselves and of the graphs of the time function in the multidimensional problem of passing about an obstacle remains open even in the topological formulation.

In particular, it is not known if the shortest time of passing about an obstacle of general position considered as a function of the final point is topologically equivalent to the Morse function (this is so for the functions of maximum and minimax

$$f(y)=\max_x F(x,y),\ f(z)=\min_y \max_x F(x,y,z)$$

of families of smooth functions of general position on a contact manifold [31, 24]).

It is also not known if in the problem of passing about an obstacle of general position the principle of brittleness of the good (Sec. 30 in [6]) is satisfied according to which the directions in the "bad" direction occupy no less than half of the tangent space at singular points of the boundary of a "good" set. It is clear only that "good" is unattainable.

LITERATURE CITED

1. V. I. Arnol'd, "Normal forms of functions near degenerate critical points, the Weyl groups A_k, D_k, E_k, and Lagrangian singularities," Funkts. Anal. Prilozhen., 6, No. 4, 3-25 (1972).
2. V. I. Arnol'd, "Remarks on the method of stationary phase and Coxeter numbers," Usp. Mat. Nauk, 28, No. 5, 17-44 (1973).
3. V. I. Arnol'd, "Contact manifolds, Legendre mappings, and singularities of wave fronts," Usp. Mat. Nauk, 29, No. 4, 153-154 (1974).
4. V. I. Arnol'd, "Critical points of smooth functions and their normal forms," Usp. Mat. Nauk, 30, No. 5, 3-65 (1975).

5. V. I. Arnol'd, "Critical points of functions on a manifold with boundary, the simple Lie groups B_k, C_k, F_4, and singularities of evolutes," Usp. Mat. Nauk, 33, No. 5, 91-105 (1978).
6. V. I. Arnol'd, Supplementary Chapters in the Theory of Ordinary Differential Equations [in Russian], Nauka, Moscow (1978).
7. V. I. Arnol'd, "Indices of singular points of 1-forms on a manifold with boundary, displacement of invariants of groups generated by reflections, and singular projections of smooth surfaces," Usp. Mat. Nauk, 34, No. 2, 3-38 (1979).
8. V. I. Arnol'd, "Sweeping out a caustic by a cusp edge of a moving front," Usp. Mat. Nauk, 36, No. 4, 233 (1981).
9. V. I. Arnol'd, "Lagrangian manifolds with singularities, asymptotic rays, and the open swallow tail," Funkts. Anal. Prilozhen., 15, No. 4, 1-14 (1981).
10. V. I. Arnol'd, "Potential flows of collisionless particles, Lagrangian singularities, and metamorphoses of caustics," Tr. Sem. im. I. G. Petrovskogo, 8, 5-53 (1982).
11. V. I. Arnol'd, "Singularities of systems of rays," Usp. Mat. Nauk, 36, No. 2, 77-147 (1983).
12. V. I. Arnol'd, A. N. Varchenko, and S. M. Gusein-Zade, Singularities of Differentiable Mappings [in Russian], Nauka, Moscow (1982).
13. V. I. Arnol'd, Ya. B. Zel'dovich, and S. F. Shandarin, "Elements of a large-scale structure of the universe," Usp. Mat. Nauk, 36, No. 3, 244-245 (1981).
14. A. N. Varchenko and A. B. Givental', "Mapping of periods and the form of intersections," Funkts. Anal., 16, No. 2, 7-20 (1982).
15. S. M. Voronin, "Analytic classification of germs of conformal mappings $(C, 0) \to (C, 0)$ with an identical linear part," Funkts. Anal., 15, No. 1, 1-17 (1981).
16. S. M. Voronin, "Analytic classification of pairs of involutions and its applications," Funkts. Anal., 16, No. 2, 21-29 (1982).
17. A. G. Gasparyan, "Application of multidimensional matrices to the investigation of polynomials," Dokl. Akad. Nauk Arm. SSR, 70, No. 3, 133-142 (1980).
18. A. B. Givental', "Manifolds of polynomials having a root of fixed comultiplicity and the generalized Newton equation," Funkts. Anal., 16, No. 1, 13-18 (1982).
19. A. B. Givental', "Lagrangian manifolds with singularities and irreducible sl_2-modules," Usp. Mat. Nauk, 38 (1983) (in the press).
20. V. M. Zakalyukin, "Legendre mappings in Hamiltonian systems," in: Some Problems of Mechanics [in Russian], Moscow Aviation Inst. (1977), pp. 11-16.
21. E. E. Landis, "Tangential singularities," Funkts. Anal., 15, No. 2, 36-49 (1981).
22. E. E. Landis, "Tangential singularities in contact geometry," Usp. Mat. Nauk, 37, No. 4, 96 (1982).
23. O. V. Lyashko, "The geometry of bifurcation diagrams," Usp. Mat. Nauk, 34, No. 4, 205-206 (1979).
24. V. I. Matov, "Topological classification of germs of the maximum and minimax functions of families of functions of general position," Usp. Mat. Nauk, 37, No. 4, 129-130 (1982).
25. O. A. Platonova, "Singularities in the problem of fastest passage about an obstacle," Funkts. Anal. Prilozhen., 15, No. 2, 86-87 (1981).
26. O. A. Platonova, "Singularities of the system of rays near an obstacle," Preprint, Moscow, Dep. VINITI February 11, 1981, No. 647-81.
27. O. P. Shcherbak, "Projectively dual spatial curves and Legendre singularities," Tr. Tbilis. Univ., Nos. 13-14, 232-233, 280-336 (1982).
28. O. P. Shcherbak, "Singularities of frontal mappings of submanifolds of projective space," Usp. Mat. Nauk, 37, No. 4, 95-96 (1982).
29. K. G. Yakobi, Lectures on Dynamics [in Russian], ONTI, Moscow (1936).
30. V. I. Arnold, "Wave front evolution and equivariant Morse lemma," Commun. Pure Appl. Math., 29, No. 6, 557-582 (1976).
31. V. I. Arnold, "On some problems in singularity theory," Proc. Indian Acad. Sci. Math., 90, No. 1, 1-9 (1981).
32. V. I. Arnold, "Singularities of Legendre varieties, of evolvents, and of fronts at an obstacle," Ergodic Theory Dynam. Sys., 2 (1982).
33. G. D. Birkhoff, "On the periodic motions of dynamical systems," Acta Math., 48 (1927).
34. J. Ecalle, "Théorie iterative. Introduction à la théorie des invariants holomorphes," J. Math. Pures Appl., 54, 183-258 (1975).
35. D. Hilbert, "Über die vollen Invariantensysteme," Math. Ann., 42, 313-373 (1893).
36. L. Hörmander, "Fourier integral operators. I," Acta Math., 127, 71-183 (1971).
37. R. Melrose, "Equivalence of glancing hypersurfaces," Invent. Math., 37, 165-191 (1976).
38. R. Melrose, "Equivalence of glancing hypersurfaces. II," Math. Ann., 255, No. 2, 159-198 (1981).

Singularities, bifurcations, and catastrophes

V.I. Arnol'd

M. V. Lomonosov. Moscow State University

Usp. Fiz. Nauk **141**, 569–590 (December 1983)

The theories of smooth-mapping singularities and dynamical-system bifurcations are reviewed. Mention is made of the applications to optics (caustic and wave-front metamorphoses) and to theories of short-wave asymptotics, the origin of large-scale structure in the universe, and loss of equilibrium and self-oscillation stability ("catastrophe theory").

PACS numbers: 02.30. – f, 03.40.Kf

CONTENTS

INTRODUCTION

The theory of the singularities of smooth mappings is a young branch of mathematics, a far-reaching generalization of studies of functional behavior at maximum and minimum points. Functions are replaced in this theory by mappings—sets comprising an arbitrary number of functions of any number of variables.

For a typical function of one variable, the critical points are maxima and minima, and are nondegenerate: the second derivative of the function does not vanish there. In the neighborhood of a nondegenerate maximum or minimum a function of a single variable can be put into the form $y = \pm x^2 + c$ by a suitable *smooth* change of the independent variable. By the word "smooth" we will mean "differentiable as many times as required." Deformations that have little effect on a function or on its derivatives up to the required order will be called small disturbances, or "wobbles."

Similarly, a typical smooth function of n variables can be reduced to the form

$$y = \pm x_1^2 \pm \cdots \pm x_n^2 + c$$

by an appropriate smooth change of independent variables. An "atypical" function, say $y = x^3$, can be converted to a typical function by introducing an arbitrarily small disturbance: $y = x^3 - \varepsilon x$, for example.

The term "typical" function here refers to a function which does not belong to a certain meager, or thin, set in the space of all functions. In functional space a meager set of atypical functions is singled out by an algebraic relation among the Taylor coefficients of a function that does *not* reduce to an identity. Accordingly the meager set will form a hypersurface in functional space—a submanifold of codimension one, analogous to a surface in ordinary 3-space, or to a curve in a plane.

The set complementary to this meager set of atypical functions is called the set of functions "in general position," or *generic* functions. Any function, then, can be put into general position by an arbitrarily small disturbance.

In other situations as well we can single out a thin set of "atypical" objects and a thick set of generic objects. For instance, a plane curve in general position will not contact any straight line at more than two points; a surface will not do so at more than four points. The Gibbs phase rule states that in an n-parameter generic system no more than $n + 1$ phases can occur simultaneously. A generic vector field in a plane can have node-, saddle-, or focus-type singular points, but not a center-type singularity.

By limiting the discussion to generic objects one can construct very simple models of phenomena, which nevertheless do exhibit the effects being studied. In mathematical language, such models are usually described as *normal forms*, like those given above for the critical points of functions.

This article offers the nonspecialist a concise review of several applications of modern singularity theory. Among these applications are the analysis of typical wave-front and caustic singularities, solutions of the eikonal or Hamilton-Jacobi equations, and their metamorphosis into single-parameter generic families. From the wave or quantum point of view we shall consider singularities in the asymptotic behavior of multidimensional oscillating or saddle-point integrals as several stationary-phase points or several saddle points merge. By classifying the singularities one can investigate the asymptotic properties of n-dimensional generaliza-

Originally publ. in: Usp. Fiz. Nauk 141(4), 569-590, © 1983 Russ. Acad. Sciences
English transl.: Soviet Phys. Usp. 26(12), 1025-1037, © 1983 American Institute of Physics.

tions of the Airy functions and kindred integrals.

The asymptotic analysis of these integrals involves the geometry of multidimensional "kaleidoscopes"—crystallographic groups generated by reflections. Some particularly simple wave-front and caustic singularities and integral asymptotics are described by simple Lie groups ($A_k \approx SU_{k+1}$, $D_k \approx O_{2k}$), which turn up in these problems in an altogether unexpected way.

To interpret the relationship between the stationary-phase integral asymptotics and the root systems of simple Lie groups, we turn to the topology of complex level manifolds of the phase function (Riemann surfaces and their n-dimensional generalizations). This topology is described by the theories of vanishing cycles and monodromy, which study how cycles on the Riemann surfaces and integrals over them behave as the complex phase approaches a critical value and circles around it.

Caustic bifurcation theory serves as the mathematical foundation of the "pancake" theory which Yakov B. Zel'dovich has invoked as a model for the development of large-scale structure in the universe.

The paper concludes by describing some applications of singularity theory to the theory of bifurcations of dynamical systems, including the theory of the loss of self-oscillation and equilibrium stability, with "strange attractors" being formed.

"Catastrophe theory," the term which the French mathematician René Thom introduced to designate the theories of singularities and bifurcations and their applications, is concerned with situations of this last type, wherein a smooth change in the parameters of a system is capable of triggering a sudden, abrupt change in its state or its regime of motion.

Although the "general position" arguments on which this theory relies have consciously and persistently been used by many physicists (such as A. A. Andronov and L. D. Landau as well as Ya. B. Zel'dovich), a systematic mathematical theory of singularities dates only from 1955, when the American mathematician Hassler Whitney[1] classified the singularities of mappings of generic surfaces into the plane.

This new mathematical theory interacts with its applications in much the same way as is true of analysis: in simple practical work the methods of differential and integral calculus can be replaced by elementary Huygens-type procedures, but in more complicated contexts only an intensive use of analysis will serve the purpose.

We begin by considering a problem in singularity theory which cannot be handled without some serious mathematical apparatus: the study of the singularities and metamorphoses of wave fronts and caustics.

1. WAVE-FRONT SINGULARITIES

Let us take a very simple example. Suppose a disturbance (light, sound, perhaps an epidemic) is propagating along a plane at unit velocity and at starting time exists on the outside of an ellipse (Fig. 1). To construct the wave front at elapsed time t, we have to lay off a segment of length t along each normal interior to the ellipse. So long as t remains small, the resultant wave front will form a smooth curve, but after a certain time interval the front will acquire a pair of cusps. These cusps will be stable: if we smoothly wiggle the original front (the ellipse), the singularities will not disappear but will merely shift slightly.

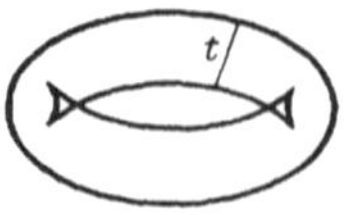

FIG. 1. How a wave front develops singularities.

The generic front in the neighborhood of a cusp may by a smooth change of coordinates be reduced to the standard "normal form" $x^2 = y^3$. That front will have no singularities in the plane other than such standard cusps and crossing points of the two branches. Any more complicated singularities will resolve into singularities of these types when the initial front is given a small general wobble.

To illustrate, the front of a disturbance propagating into the interior of a circle will collapse to a point; but if we replace the circle by a slightly different generic curve, then instead of collapsing to a point the front will collapse to a curve which remains close to the point but has an even number of cusps (Fig. 2).

Now one of the fundamental results of singularity theory is that in generic situations only certain standard, universal transforms will be encountered; they can be thoroughly studied once and for all, and have to be recognized in various guises. The simplest of these universal images is a semicubical-type cusp ($x^2 = y^3$). Next in complexity comes the *swallowtail*, a singular surface in 3-space (Fig. 3).

Swallowtails develop in many contexts of singularity theory. Let us return, for example, to the propagation of a wave front along a plane. We can describe the process in terms of spacetime (three-dimensional, in our case). As it travels over the plane, the front will form an image surface in three-dimensional spacetime. Sections of that surface by isochrones $t =$ const will represent instantaneous fronts.

A pair of cusps formed on an instantaneous wave front will then be imaged as a swallowtail singularity on our spacetime surface. This singularity will be stable: if we wobble the original ellipse, replacing it by a slightly different smooth curve, the swallowtail will merely shift a bit in spacetime; it will not disappear.

The same surface in spacetime can also be described as the graph of a multivalued function—the distance from the original front along the normals, or as the graph expressing the solution of the equation $(\nabla u)^2 = 1$, that is, the eikonal equation in optics or the Hamilton-Jacobi equation in mechanics. Thus the generic solutions of eikonal or Hamilton-

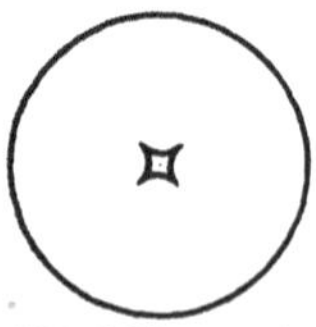

FIG. 2. The cusped wave front from a near-circular curve.

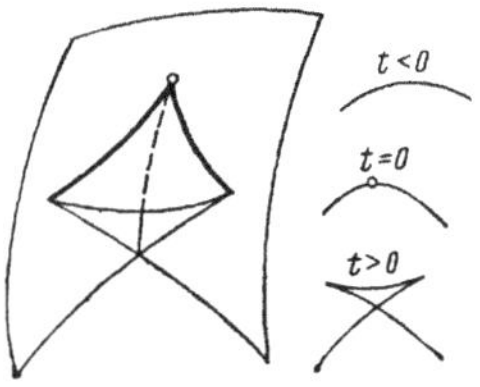

FIG. 3. A swallowtail, and three sections across it.

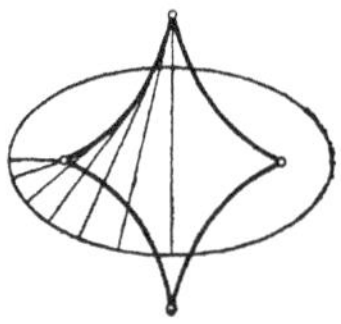

FIG. 4. An astroid, the caustic of the set of normals to an ellipse.

Jacobi equations in two independent variables will have singularities that cannot be eliminated by a small disturbance of the initial conditions: near a singular point the graph of the solution will have a swallowtail singularity or a cusp-type edge.

A swallowtail is a typical singularity on generic wave fronts in 3-space. One can obtain such a front, for example, by laying off a segment of length t on each inward normal to a triaxial ellipsoid (or a smooth surface differing slightly from an ellipsoid).

Generic wave-front singularities are equivalent to singularities in the Legendre transforms of smooth generic functions. Thus the theory of wave-front singularities, also called the theory of Legendre singularities, concurrently supplements the classification of the singularities of thermodynamic potentials (assuming that one of them is a smooth, although not a convex, function).

In mathematics, swallowtails occur chiefly as surfaces in $x^4 + ax^2 + bx + c$ polynomial space that are formed by multiple-root polynomials. As sketched in Fig. 3, one can easily visualize such a surface by cutting it with planes $a = \text{const}$.

There is another mathematical description of a swallowtail that is often useful: the surface formed in 3-space by all tangents to the curve $A = t^2$, $B = t^3$, $C = t^4$ at the point (A, B, C).

2. CAUSTIC SINGULARITIES

Disturbances that are being propagated can be described not only by their wave fronts but also by a system of rays. The rays will not intersect next to a smooth front, but a certain distance away they will produce focal points—intersections of infinitely close rays. Taken together, the focal points will form a *caustic* of the ray system. In the case of a disturbance advancing at constant velocity through Euclidean space, the focal points will be the centers of curvature of the fronts. The system of normals to an ellipse, for instance, forms an astroid caustic (Fig. 4), a semicubical curve with four cusps.

These singularities again are stable: if the ellipse is replaced by a slightly different smooth curve, the cusps will not disappear but will merely shift slightly. The caustics of generic ray systems in a plane have no singularities other than semicubical-type cusps (and self-crossing points).

The caustic comprising the set of normals to a circle consists of a single point, the center. On stretching the circle slightly into some generic curve, say an ellipse, the central caustic point will spread into a curve (for an ellipse, into a tiny astroid).

In 3-space as well, generic caustics have only standard singularities: cusp (semicubical type) edges, swallowtails, and two new types of point singularities called *pyramid* and *pucker* points (Fig. 5). At a pyramid singularity itself, three (smooth) cusp edges are mutually tangent. In the neighborhood of a pucker point, the caustic comprises two surfaces with similar singularities, intersecting along a pair of curves; the cusp edges of the two surfaces continue each other, forming a single smooth curve.

All these singularities are stable. For example, the tangency of the three caustic cusp edges at the vertex of a pyramid singularity cannot be eliminated by wobbling the initial ray system. Every more complicated type of caustic singularity in 3-space will break down into singularities of the types mentioned above if the generic position is disturbed slightly.

3. SHORT-WAVE ASYMPTOTIC SINGULARITIES

Caustic singularities can also be described as singularities in the asymptotic behavior of integrals of rapidly oscillating functions (such as the Fresnel and Airy integrals):

$$I_\lambda(h) = \int e^{iF(x,\lambda)/h} a(x,\lambda)\,dx, \quad h \to 0;$$

here the small parameter h represents a wavelength, the *phase*[1] F is a real function, and the smooth function a, the *amplitude*, differs from zero only within some restricted region.

According to the stationary-phase principle, the main contribution to the asymptotic behavior as $h \to 0$ will come from the stationary points of the phase F (at which $\partial F/\partial x = 0$). For generic λ, these stationary points will be nonde-

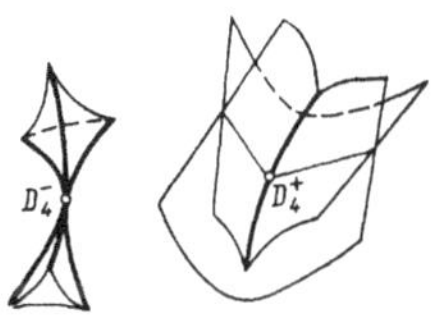

FIG. 5. Pyramid- and pucker-type caustic singularities.

[1]Think of $F(x, \lambda)$ as the optical path length from a light source at point x to an observer at point λ.

generate: $\det \partial^2 F/\partial x_i \partial x_j \neq 0$. A change of variable will reduce the integral in the neighborhood of a stationary point to a Fresnel-type integral, and it will therefore fall off like $h^{n/2}$ as $h \to 0$, where n is the dimensionality of the x-space.

A typical phase function, regarded as a function of the point x, will have no degenerate critical points. If, however, we consider not an individual phase function of point x but a whole family, depending on one or more parameters λ, then for certain exceptional values of those parameters we may indeed encounter degenerate critical points.

For example, the family of x-functions $x^3 + \lambda x$ with a zero value for the parameter λ has a degenerate critical point: $x = 0$. Any single-parameter family of functions close to this one will, for some near-zero value of that parameter, have a degenerate critical point of the same kind.

The values of the parameter λ for which the phase has a degenerate critical point are called the *caustic* values. As $h \to 0$ the integral falls off for the caustic values more slowly than usual, like $h^{(n/2)-\beta}$. The exponents β defined in this way measure the degree to which light, for example, is concentrated toward the caustic, and they depend on the character of the caustic singularity An ordinary caustic point will have $\beta = 1/6$ (reflecting the asymptotic behavior of the Airy integral); for ordinary points along a caustic cusp edge, $\beta = 1/4$; for swallowtails, $\beta = 3/10$; and for pyramid and pucker points,[2] $\beta = 1/3$.

4. FRONT AND CAUSTIC METAMORPHOSES

As a wave front advances through 3-space, its singularities (cusp edges) will slide along a caustic. When the front passes through a *singular* point of the caustic (a swallowtail vertex, a pyramid or pucker point), it will undergo a metamorphosis. The metamorphoses of moving generic fronts are stable, and take the standard forms illustrated in Fig. 6. All more complicated metamorphoses can be dispensed with, because if the initial front is wobbled, such metamorphoses will break down into the small number of standard ones.[3,4]

Caustics can also metamorphose if they depend on parameters, say upon time, An example of time-dependent caustics is furnished by Zel'dovich's theory,[5] wherein density singularities would develop in a dustlike medium. Although this theory is designed to interpret the formation of large-scale irregularities in the distribution of matter in an expanding universe from small fluctuations in the primordial velocity field of the gravitating particles, the essential features of the phenomenon can best be understood by considering a very simple model of noninteracting particles that move by inertia through Euclidean 3-space.

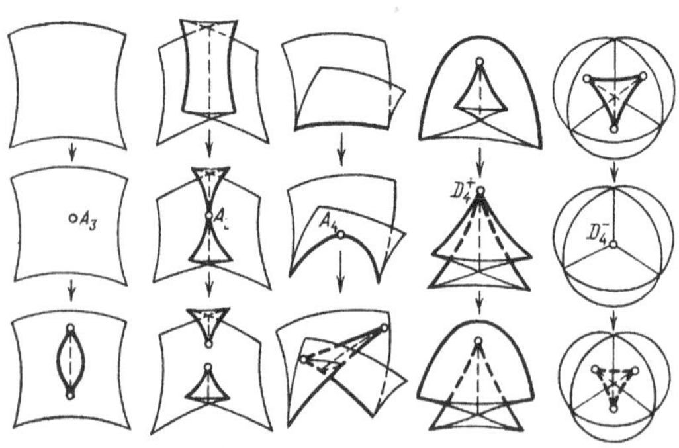

FIG. 6. The metamorphoses of generic wave fronts.

In time t a particle will travel from point $\mathbf{q}$ to point $\mathbf{q} + t\mathbf{v}(\mathbf{q})$, where $\mathbf{v}$ designates the initial velocity field. Assume that the field $\mathbf{v}$ can be derived from a potential: $\mathbf{v} = \partial S/\partial \mathbf{q}$ (such an assumption is physically prompted by the fact that in an expanding universe irrotational perturbations will grow faster than vortex perturbations). The graph of the velocity field $\mathbf{p} = \mathbf{v}(\mathbf{q})$ will form a three-dimensional submanifold in six-dimensional phase space $\{(\mathbf{p}, \mathbf{q})\}$ (Fig. 7).

In view of the potential property of the field, the action integral $\int \mathbf{p}\, d\mathbf{q}$ will depend only on the ends of the integration path and will not change if the path itself is deformed, provided the path remains on the manifold referred to. Submanifolds of phase space that possess this property are said to be Lagrangian. Thus an initial potential velocity field will be imaged by a Lagrangian manifold in phase space.

Transformations of the phase flow of any Hamiltonian differential equation will carry Lagrangian phase-space submanifolds into Lagrangian submanifolds. This will be true, in particular, for the phase flow of the equation of a free particle. Accordingly, the velocity field of a medium will remain irrotational if it was so at initial time, and if the forces have that property as well.

The initial velocity field will be represented by a phase-space submanifold nondegenerately projected onto configuration space (a nonvertical tangent to the graph in Fig. 7). If the Hamiltonian phase flow undergoes transformations, an initially Lagrangian manifold will remain Lagrangian but, generally speaking, will (after a long enough time t) lose its property of nonverticality: the projection onto configuration space will cease to be one–one. The points where verticality occurs are called the *singularities* of the projection of the Lagrangian manifold onto configuration space. The projections of these singularities onto configuration space are called the *caustics*.

Caustics will be formed if fluctuations in the original velocity field make the particles begin to overtake one another. At time t the caustic will consist of the points where particles issuing from infinitely close points of space come into collision. The particle distribution along the caustic will have infinite density.

The term "caustic," borrowed from optics, is employed here because in a mathematical sense we are dealing with the same phenomenon in both cases—with the critical values of

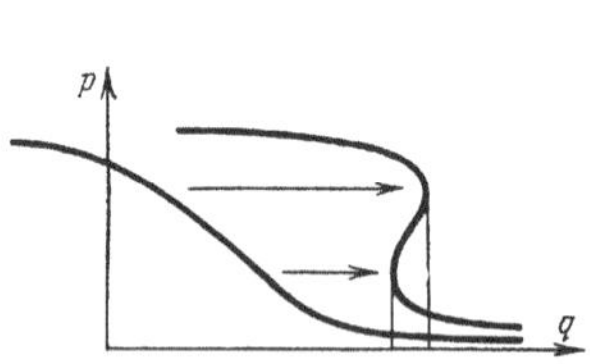

FIG. 7. The development of Lagrangian mapping singularities as a Lagrangian manifold is transported by phase flow.

a Lagrangian mapping, that is, a mapping of the projection of a Lagrangian phase-space submanifold onto configuration space. In the optical context, a Lagrangian manifold is defined by the condition $p = \partial u/\partial q$, where the solution u of the eikonal equation $(\nabla u)^2 = 1$ may be multivalued. In other words, the Lagrangian manifold corresponding to a given ray system (say the rays emerging from a point source, a line, or a surface) will consist of the "front delay vectors" at all points of the medium.

Once the caustic has formed, the particle motion will become multistream: several streams (initially three, as a rule) will pass through a single point of space. The Lagrangian-manifold concept is the mathematical equivalent of a multistream irrotational flow in physics.

From the mathematical standpoint, then, the particle motion specifies a single-parameter family of Lagrangian mappings of 3-spaces: the value of the parameter t stipulates a map taking the point q into the point $Q = q + tv(q)$. The critical points of this mapping are determined by the condition that det $\partial Q/\partial q = 0$. The critical values (that is, the images of the critical points) will form a caustic. At any definite instant the caustic will form a surface in configuration space.

As the parameter t changes, the caustic will move, and it may metamorphose. The first metamorphosis will be the creation of the caustic itself. Zel'dovich calls the newborn caustic a "pancake" (*blin*). Indeed, a short time ε after it is born the caustic will be shaped like an elliptical dish roughly $\varepsilon^{3/2}$ thick, ε deep, and with axes of the order of $\sqrt{\varepsilon}$ (Fig. 8). In a scattering medium, optical caustics could become visible; V. M. Zakalyukin suggests that an observer might perceive them as flying saucers.

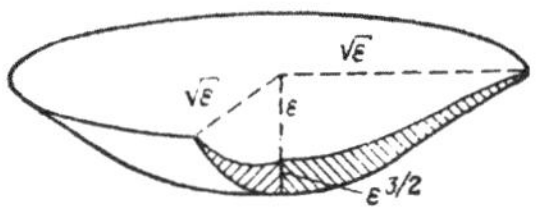

FIG. 8. A newborn caustic—a Zel'dovich pancake.

After the pancakes have formed, the caustics may experience further metamorphoses (Fig. 9). To sort them out one has to apply fairly elaborate mathematical apparatus from the theory of Lagrangian and Legendre singularities[4,6,7] (the singularities of Legendre transformations).

5. THE *A, D, E* CLASSIFICATION

The classification of the singularities that can occur in all the problems we have mentioned (and in many others) turns out to be related in a remarkable way to the classification of objects which would seem to have nothing to do with such problems: regular polyhedra in Euclidean space, for instance, and simple Lie groups (thus, swallowtails on a caustic correspond to the SU_5 group; pucker points, to the orthogonal-matrix groups O_8).

Although the reasons why such disparate theories should be interrelated are not yet fully understood, these relationships have proved to be an exceptionally powerful tool for studying singularities. Only by devising this technique has it been feasible to work out the classification of front and caustic singularities and metamorphoses outlined above.

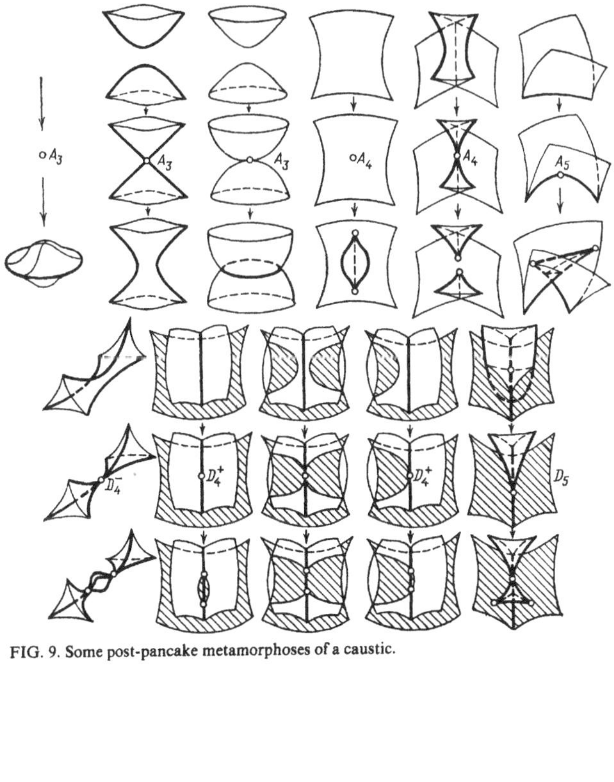

FIG. 9. Some post-pancake metamorphoses of a caustic.

An unexpected benefit of all these classifications has been the finding that a short, universal list of simple objects repeats itself in various guises (Lie groups, regular polyhedra, caustics, wave fronts, Riemann surfaces, ...). The list comprises two infinite sequences (A_k, D_k) and three exceptional objects (E_6, E_7, E_8), as depicted in Fig. 10. In the theory of regular polyhedra the A_k represent the regular polygons; the D_k, the dihedra; E_6, the tetrahedron; E_7, the octahedron; E_8, the icosahedron. Corresponding to each regular polyhedron is a particular wave-front singularity, caustic singularity, oscillating Airy-type integral, and so on.

In Fig. 10 all these objects are represented graphically, joined by lines in *Dynkin diagrams*. These diagrams may be interpreted in entirely different ways, depending on the character of the objects under study. But whatever approach is adopted, the diagrams are always present and play an important role. For example, all the possible decays of wave-front singularities may be exactly described by decompositions of corresponding Dynkin diagrams.

In order to explain how complicated singularities pertaining to wave fronts, caustics, and stationary-phase and saddle-point integrals are composed of elementary, very simple singularities, we need certain information from the geometry of complex manifolds. In fact, most of the roads from caustics, fronts, and oscillating integrals to Lie groups, Dynkin diagrams, and regular polyhedra take us through the theory of the critical points of holomorphic functions—that is, through the geometry of Riemann surfaces and their n-dimensional generalizations. In applications to optics, the holomorphic function that comes into play is the phase—the optical path length. But in other circumstances, such as in thermodynamics, the holomorphic function may have a completely different physical meaning.

We therefore consider at the outset a holomorphic function (a polynomial, say) of several complex variables: $f(z_1,\ldots,z_n)$. It turns out that the asymptotic behavior of the integrals associated with such a function (for example, integrals of the form $\int e^{if/h}\,dz$) and the geometry of the associated singularities (fronts, caustics, ...) will be fundamentally influenced by the topology of the complex level manifold $f(z_1,\ldots,z_n)=c$ [or, in the case of a function of only two variables, by the topology of the Riemann surface $f(z_1,z_2)=c$].

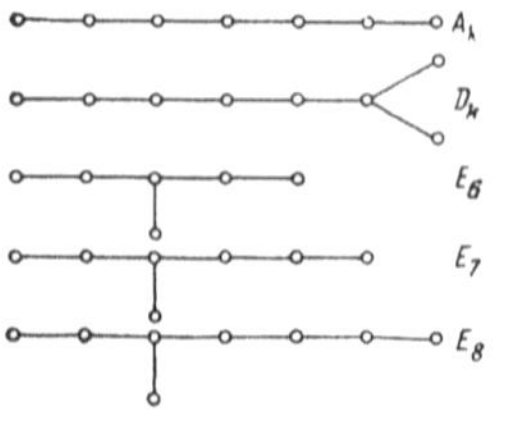

FIG. 10. Dynkin diagrams.

6. MONODROMY

The level manifold (a line, a surface) of a real function will undergo topological rearrangements when the function takes on critical values (called Morse rearrangements, or Morsifications). For example, the hyperbola $x^2-y^2=c$ will interchange branches as c passes through the critical value zero. In the complex case, the plane of function values is not divided by critical values, so a complex level manifold will be altered only at the instant of passage through a critical value (the manifold will become singular); after passage the level manifold will instantly revert to its former state.

On the other hand, in the complex case instead of passing *through* a critical value, a circuit can be taken *around* it. As the function values undergo a continuous change, so also will the level manifold. After the value has returned to the original point in the function-value plane, the level manifold, continuously changing, will similarly return to its original position. But along the way it will be capable of "flipping over," so that each individual point of the level manifold will in general find itself returning to a different point of the same manifold (like the reversal that occurs when a rectangle is spliced into a Möbius strip).

Such a mapping of the nonsingular level manifold of a function into itself is called a *monodromy*. Along with the topology of the level manifold, a monodromy conveys much information on the singularities that may arise as a critical value is approached (on the asymptotics of integrals, for instance). With present-day mathematical techniques one can study all this topology in detail and extract from it any information needed on the asymptotics (this approach is known as the Hodge theory of mixed structures, an n-dimensional generalization of the Legendre theory of elliptic integrals of the first, second, and third kinds).

Some idea of the geometry of the complex objects that may develop can be gained from the following example (which in this theory plays the role of $1/z$ in the elementary theory of residues).

Let $f=z_1^2+z_2^2$. The level manifold $z_1^2+z_2^2=c$ will be a two-sheet Riemann surface, $z_2=(c-z_1^2)^{1/2}$, with two branch points A, B. (Fig. 11). After a complete circuit of the noncritical value c around the critical value $c=0$, the pair of branch points will make a half revolution, and the Riemann surface $f=c$ will revert to its former position.

Topologically, the "complex circle" $z_1^2+z_2^2=c\,(c\neq 0)$ represents a cylinder (the profiles joining the branch points on the two sheets are drawn in Fig. 12 as circles going once around the cylinder). The progressive change in the Riemann surface as c varies is depicted in Fig. 11. This process may occur in such a way that outside a sufficiently large (compared with $\sqrt{c}$) sphere the surface points remain almost motionless. As a result the path Δ_0 on the original surface, gradually deforming, will ultimately become the path Δ_1. Figure 12 shows that the monodromy will twist our level

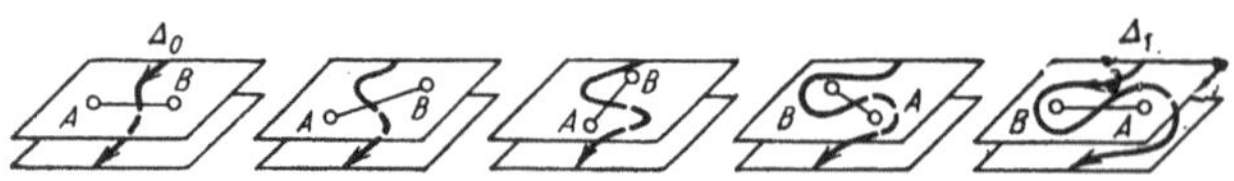

FIG. 11. A monodromy: the path Δ_0 is transformed into Δ_1 as the pair of branch points executes a half revolution.

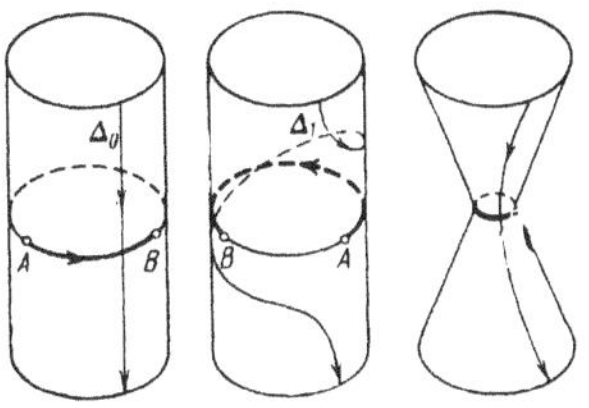

FIG. 12. The twist of a cylinder by monodromy and the constriction of a vanishing cycle as a critical value is circuited or approached.

manifold (the cylinder) such that one of its bases turns by 2π relative to the other (with intermediate parallel planes turning by intermediate angles).

7. VANISHING CYCLES

In our example as the value c approaches the critical value zero, the central parallel ABA of the cylinder $z_1^2 + z_2^2 = c$ (Fig. 12) will constrict to a point (the circle ABA is formed by the real roots of the equation $z_1^2 + z_2^2 = c > 0$). A cycle on a level manifold that constricts to a point as the function value approaches a critical value is termed a *vanishing* cycle at that critical point (in our case, the point $x = 0$).

For a function of two variables with several critical points (assuming that each critical point is nondegenerate), several vanishing cycles will develop on the nonsingular level Riemann surface. To determine them we have to move along the function-value plane from an initial noncritical value to each of the critical values. But if instead of coming in close to the critical values we circuit around them (in little circles, say), then the nonsingular level manifold each time will return to its old position, turning around anew. In place of a single monodromy transformation we will therefore have a whole *monodromy group*.

Consider, for example, the function $f = z_1^3 - 3z_1 + z_2^2$. Apart from the two points $c = \pm 2$, the level manifold $f = c$ will be nonsingular in the plane of the complex variable c. Let us begin, say, with the nonsingular level manifold $f = 0$. This two-sheet covering of the plane, $z_2 = (3z_1 - z_1^3)^{1/2}$, is topologically equivalent to a torus with a side hole (as at a in Fig. 13). As c approaches one of its critical values, say $c = 2$, one pair of branch points will come together (the pair AB) until the torus meridian vanishes (ξ in Fig. 13). If c approaches its other critical value, $c = -2$, the other branch-point pair (BC) will coalesce; in this case the vanishing cycle will be the torus parallel (η in Fig. 13).

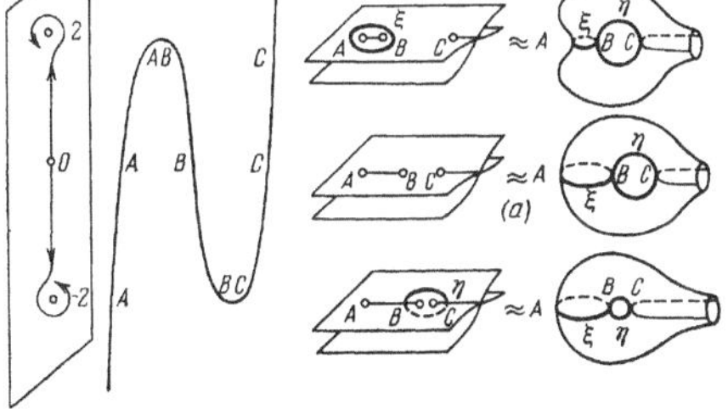

FIG. 13. The vanishing-cycle basis (ξ, η) on a Riemann surface.

We can also readily compute the monodromy group for this example. Any cycle on our level manifold $c = 0$ (a torus with a hole) will be a homologous[2)] linear combination $p\xi + q\eta$ of the meridan ξ and the parallel η, with integer coefficients. As the variable c circuits around the point $c = 2$, the torus will twist, experiencing a monodromy. In the basis (ξ, η) this monodromy operator will have a matrix of the form $\begin{pmatrix} 1 & 1 \\ 0 & 1 \end{pmatrix}$. A circuit of c around the point $c = -2$ will correspond to the monodromy matrix $\begin{pmatrix} 1 & 0 \\ -1 & 1 \end{pmatrix}$. In our example the full monodromy group comprises all possible products of these two matrices and their inverses. This group is the group $SL_2(z)$ of all integer-valued second-order matrices with determinant unity.

In the multidimensional case a level manifold $f(z_1, \ldots, z_n) = c$ will have a real dimension of $2n - 2$ in n-dimensional complex (that is, $2n$-dimensional real) space. The vanishing cycles will have half that dimension: $n - 1$. For instance, at the nondegenerate critical point O of the function $z_1^2 + \ldots + z_n^2$ the real sphere S^{n-1} formed by the real solutions of the equation $z_1^2 + \ldots + z_n^2 = c > 0$ will vanish[8–10] as $c \to 0$.

8. VERSAL DEFORMATIONS

Now let us consider a *degenerate* critical point of a function, say $f = z^4$. Singularities of this type do not occur either in families of generic functions or in general one-parameter families of functions. However, in the two-parameter family

$$F(z, \lambda) = z^4 + \lambda_1 z^2 + \lambda_2 z$$

such a singularity is encountered for zero values of the parameters λ, and any family nearly the same as this one will experience the same type of degeneracy for some nearby value of the parameters. Properly speaking, any family $G(z, \lambda)$ almost identical with the family F may be put into the form

$$G(z, \lambda) \equiv F(h(z, \lambda), \varphi(\lambda)) + c(\lambda),$$

that is, it can be reduced to the family F by a smooth change of variable depending smoothly on the parameters, and by a smooth change of parameters.[11] Such a deformation, to which any other deformation of the function f can be reduced by the substitutions mentioned, is said to be *versal.* The term "versal" comes from the words "universal" and "transversal": the prefix "uni-" is dropped because such a deformation usually is not unique, but versal deformations do prove to be transversal, in the following sense.

Suppose that we have classified in some manner a set of objects (say functions, differential equations, vector fields, matrices, tensors, caustics, wave fronts, ...). We divide the

[2)] Two one-dimensional oriented cycles are said to be *homologous* if the difference between them serves as the oriented boundary of a two-dimensional sheet. This definition is motivated by the fact that closed-form integrals (the circulations of locally irrotational fields) will have equal values along such cycles. Similar definitions apply in the n-dimensional case.

manifold of all possible objects of this kind (say the function space or the matrix space) into parts corresponding to objects of differing classes (such as matrices of differing rank). The bounding hypersurfaces will have codimension one, being determined by a single equation in the space (generally speaking, infinite-dimensional functional space) of all objects of interest to us (Fig. 14).

Generic objects will not lie on the bounding hypersurfaces (generic functions, say, have no degenerate critical points). In the single-parameter families of objects, however, very simple degenerate objects may indeed occur nonremovably. In fact, a single-parameter family will be imaged by a curve in functional space, and this curve may transversally (at a nonzero angle) intersect the bounding hypersurface. In that event, any nearly identical family will contain a similar degeneracy, for any nearby curve will still intersect the bounding hypersurface.

Functions with critical points analogous to $f = z^4$ cannot occur even in single-parameter generic families. In fact, such functions with critical points form in the functional space of all functions a submanifold of codimension *two*, whereas a single-parameter family is imaged in functional space by a curve. A small perturbation can remove this curve from the submanifold of codimension two, just as a curve in 3-space can escape interesecting another curve by means of a suitable wiggle. In typical *two*-parameter families, however, z^4-type functions with singularities are no longer removable, for the two-dimensional surface imaging such a family in functional space will transversally intersect the (codimension 2) manifold of functions with a singularity of the type z^4.

It is this transversality property of a family with respect to a manifold in functional space which ensures that the deformation of a function will be versal—that any small deformation of the function can be reduced to a versal deformation by suitable changes of the coordinates and parameters.

Although physicists have long made use of particularly simple special cases of arguments along these lines, their systematic development represents a mathematical achievement of just the past few years.

It has been found that finite-parameter versal deformations exist for functions of any number of variables having finite-multiplicity critical points, that is, critical points formed through coalescence of a finite number of elementary nondegenerate critical points. For example, the critical points of z^3, a function of one variable, has multiplicity 2, as it is formed by coalescence of two elementary critical points (see Fig. 13). This function has the single-parameter versal deformation $F(z, \lambda) = z^3 + \lambda z$ (the fact that this is the Airy phase integral is no accident). In general, the number of parameters in a versal deformation at a critical point of a function is one smaller than the multiplicity (the number of elementary critical points into which a given compound critical point can be decomposed).

The multiplicity of a critical point is also called the Milnor number, and is commonly designated by letter μ. Examples are:

$$\mu(z^a) = a - 1, \quad \mu(z_1^{a_1} + \ldots + z_n^{a_n}) = (a_1 - 1) \ldots (a_n - 1).$$

9. THE HIERARCHY OF SINGULARITIES

The compound critical points of functions form a hierarchy: simpler points can be produced by decomposing more complicated ones. At the present time this hierarchy has been calculated as far as multiplicty $\mu = 16$; that is, all the critical points that can be formed by coalescence of up to 16 elementary critical points have been described.[11]

The starting point for the hierarchy of compound critical points is a hierarchy of Dynkin diagrams (Fig. 15). The letters here designate the functions

$$A_k = x^{k+1}, \quad D_k = x^2y + y^{k-1}, \quad E_6 = x^3 + y^4,$$
$$E_7 = x^3 + xy^3, \quad E_8 = x^3 + y^5$$

or functions of a greater number of variables obtained by writing out nondegenerate quadratic forms for additional variables (for example,[6] $z_1^4 + z_2^2 + z_3^2$ is of type A_3).

Dynkin diagrams can describe various methods of building compound critical points by combining elementary ones. I shall explain how this can be done by taking a very simple example, the versal deformation of the function A_3:

$$F(x, y, \lambda) = x^4 + y^2 + \lambda_1 x^2 + \lambda_2 x.$$

A value of the parameters [that is, a point in the (λ_1, λ_2) plane] is called a *bifurcation* value if the corresponding function of the variables x, y has nonelementary critical points. In our example the bifurcation values form the semicubical parabola $8\lambda_1^3 + 27\lambda_2^2 = 0$ (Fig. 16). Let us select a generic, nonbifurcation point in the λ plane. The corresponding function will have $\mu = 3$ critical points. We take a noncritical level manifold (topologically equivalent, in our case, to a torus with two holes), and successively allow the function value to approach the three critical values. As a result three different cycles will vanish on the noncritical level manifold. To construct the Dynkin diagram we assign a point to each vanishing cycle and join these points whenever the corresponding cycles intersect (see a review by Husein-Zade[10] for further details). In our example we arrive in this way at the Dynkin diagram for A_3: —·—.

It is the bifurcation values of the versal-deformation parameters for the functions listed above that form the caus-

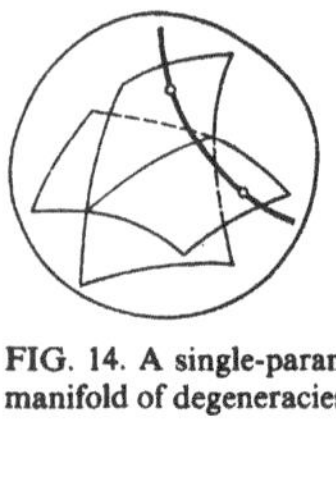

FIG. 14. A single-parameter generic family transversally intersecting a manifold of degeneracies.

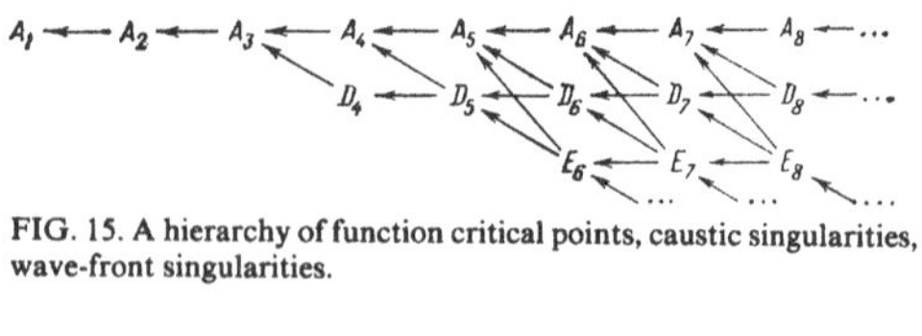

FIG. 15. A hierarchy of function critical points, caustic singularities, and wave-front singularities.

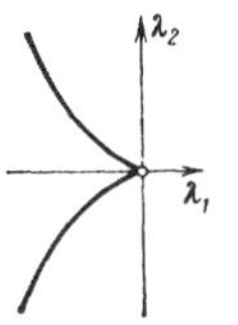

FIG. 16. A semicubical parabola, representing a manifold of cubic, multiple-root polynomials.

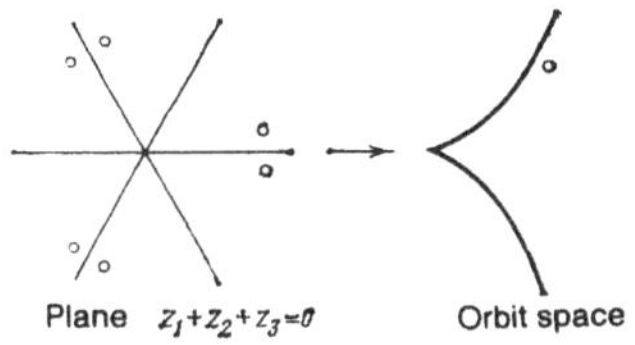

FIG. 17. A Vieta mapping, taking each point into its orbit.

tics of the corresponding Lagrangian singularities. For instance, A_2 is an ordinary point on a caustic; A_3 represents a semicubical parabola for a plane caustic or a cusp edge for a three-dimensional caustic, A_4 is a swallowtail, and D_4 is a pyramid or pucker point. In the complex domain, the last two caustics are equivalent.

To obtain the corresponding wave fronts, one need only expand the space of versal-deformation parameters by an axis of values and note all the critical values of the functions in the family. For example, the critical values of the versal-deformation functions A_3 form a swallowtail [in $(\lambda_1, \lambda_2, F)$ space].

10. REFLECTION GROUPS AND WAVE FRONTS

Wave-front singularities can be derived directly from the Dynkin diagrams by the following construction. To each diagram there corresponds a "kaleidoscope," the term used in mathematics for *a group generated by reflections in Euclidean space*, or a "Coxeter group."[12] In particular, to each vertex in a Dynkin diagram we can assign a vector of length $\sqrt{2}$. If two vertices are joined by an edge, then the angle between the two corresponding vectors is 120°; if no edge is present, the vectors will be orthogonal.

For example, the diagram · — · represents the vectors $(1, -1, 0)$, $(0, 1, -1)$ generating the plane $x + y + z = 0$ in Euclidean 3-space. Similarly, every Dynkin diagram consisting of μ points will determine a "hedgehog" of μ (nonorthogonal) basis vectors in μ-dimensional Euclidean space.

A kaleidoscope is made up of mirrors passing through the origin and orthogonal to the vectors comprising the hedgehog. For instance, the diagram · — · corresponds to a pair of mirrors in the plane, subtending a 60° angle. The group generated by reflections in the mirrors of any of the kaleidoscopes *A*, *D*, *E* is finite. This finiteness condition seriously restricts the position of the mirrors. All mirror positions for which the group is finite have been determined. The positions such that all angles between mirror pairs are 90° or 120° exhaust the mirror positions generating the reflection groups *A*, *D*, *E* and their products.

For each reflection group *A*, *D*, *E* we now construct a wave front. We consider the set of all images of a point in the kaleidoscope mirrors, including multiple reflections. This set is called the *orbit* of the point. The orbit is said to be *regular* if it comprises the maximum possible number of points (equal to the number of elements in the group).

To illustrate, the regular orbit of a point under the reflection group A_2 consists of six points (Fig. 17). If the initial point had been located on a mirror, its orbit would have comprised only three points (a nonregular orbit).

A similar construction can be carried out in the complex domain (by complexifying the initial Euclidean space). Reflections will act on complex points as well, and most complex points have regular orbits.

We now consider the set of all orbits of the group generated by reflections in complex space. This set itself turns out to be a complex space, of the same dimensions.

The effect of the group A_2, for example, may be regarded as a permutation of the three coordinates z_1, z_2, z_3, on the plane $z_1 + z_2 + z_3 = 0$ in three-dimensional (complex) space. Orbits will consist of nonordered triads of complex numbers adding to zero. Such a triad will be uniquely determined by the polynomial $z^3 + \lambda_2 z + \lambda_3$ as its triad of roots. Hence the orbit manifold will be identified with the (λ_2, λ_3) plane of two (complex) variables.

In the manifold of all orbits, nonregular orbits will form a hypersurface. For the example given above, the orbit of the point (z_1, z_2, z_3) will be nonregular if not all the numbers z_k are distinct. Hence the nonregular-orbit manifold will comprise the set of points of the λ plane for which the polynomial with coefficients λ has multiple roots. This set forms a semicubical parabola. We have thereby obtained the A_2 wave front from the corresponding reflection group.

In the general case, an analogous transition can be made from the reflection group to the orbit manifolds and the corresponding wave fronts. We have here a deep generalization of the Vieta theorem and the theory of symmetric functions. A swallowtail, for instance, is obtained from the A_3 reflection group. Since the corresponding algebraic apparatus is quite well developed, one can study the wave-front geometry and rearrangements in detail, and accordingly the geometry of the resultant caustics.

The reflection groups A_k, D_k, E_k themselves can simply be described in terms of phase-function singularities or the asymptotics of oscillating or saddle-point integrals. They represent monodromy groups for the versal deformations (see end of Sec. 8) of phase functions of an odd number of variables (E_8, for example, corresponds to the function $x^3 + y^5 + z^2$).

In the language of Lie-group theory, the reflection groups are obtained in the form of "Weyl groups"; the vanishing cycles correspond to simple roots, the wave fronts describe bifurcations of the Jordan normal forms of matrices, and so on.

The extensive theory that relates caustic and wave-front singularities and integral asymptotics to Lie groups and the groups generated by reflections cannot be surveyed at all adequately in this review.[3,11,13] I would merely mention that the singularities have now been found which correspond to all groups generated by reflections, also including noncrystallographic groups such as the icosahedron symmetry group H_3.

It has been discovered, for example, that the icosahedron controls the singularities for the family of involutes of a plane curve near its inflection point (Fig. 18). That the icosahedron should make an appearance in this situation seems just as surprising as its occurrence in Kepler's mystical law of planetary distances—only here we are talking of a rigorously proved theorem.

The involutes of a plane curve represent the wave fronts of a disturbance propagating in the region bounded by that curve. Consider the graph of the corresponding (multivalue) function of time. To construct such a graph we need merely lift each involute as a whole, along with its analytic continuation, to a height equal to the length of a thread which, if unwound, would form that involute. Figure 19 shows the graph of a multivalued function of time. This surface has two cusp edges (of order 3/2 above the curve itself and order 5/2 above the tangent at the inflection). The surface turns out to be none other than the nonregular-orbit manifold of the icosahedron symmetry group (apart from a smooth change of variables). One can also describe this surface[14] as the ensemble of all tangents to the curve (t, t^3, t^5).

I should emphasize that there seem to be no *a priori* reasons for this relationship between involutes and the icosahedron (between wave fronts and reflection groups). What we have here is a mathematical marvel, just as much so as, say, the connection between problems concerning tangents and those regarding areas. That is why the theory of caustic and wave-front singularities was developed not in the day of Huygens and Newton, who had already begun to look into it, but only in the past decade.

11. BIFURCATION THEORY

Another large arena for applying the concepts of singularity theory is the theory of bifurcations of dynamical systems. Let us begin with the problem of bifurcations in equilibrium positions. Consider the system of differential equations

$$\dot{x} = v(x, \varepsilon),$$

specified by a vector field v in the n-dimensional phase space of points x, the field depending on the parameter ε [or, more generally, on the l-dimensional parameter $\varepsilon = (\varepsilon_1, \ldots, \varepsilon_l)$].

For fixed values of the parameters, the equilibrium positions will be determined by a system of n equations in n unknowns: $v(x,\varepsilon) = 0$. Consider the space of pairs (x,ε), that is, the product of phase space by parameter space. This product has dimensions $n + l$. In the case of a one-parameter system with a one-dimensional phase space, the product will constitute the (x,ε) plane (Fig. 20). On this plane the equation of equilibrium states, $v(x,\varepsilon) = 0$, will specify a curve (or, in general, a submanifold whose dimensionality is equal to the number of parameters). For generic systems this curve will be *smooth*, and will be in "general position" relative to the projection of the parameter on the axis of values. If that is not the case for some particular system, then an arbitrarily small disturbance of the system (that is, of a model of the phenomenon under study) will be capable of qualitatively altering the behavior of the system (Fig. 21).

Accordingly, if at a nongeneral position in some model a bifurcation should occur, like that shown at the left in Fig. 21, one would have to ascertain what special circumstances are responsible. The most common circumstances triggering nongeneral bifurcations are symmetries and a Hamiltonian character for the equations of the problem.[3)] If departures from those special circumstances (symmetries, Hamiltonicity) are indeed insignificant, the problem should be treated in terms of bifurcation theory for systems of that class (symmetric systems, say), with the bifurcations observed being compared against those occurring in generic systems having the same symmetry group.

If on the other hand the degeneracy in the bifurcation process is a chance feature of the model, then the model has to be improved by including some of the small quantities neglected in setting up the equations—and bifurcation theory will suggest which quantities to consider.

The general arguments outlined above, largely due to Henri Poincaré, demonstrate that comprehensive analysis of degeneracies should be accompanied by a study of bifurcations in families have a number of parameters such that a

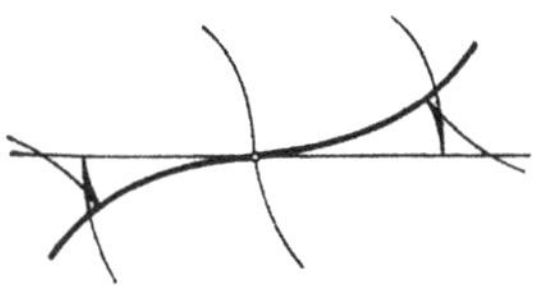

FIG. 18. The family of involutes to a plane curve having an inflection point.

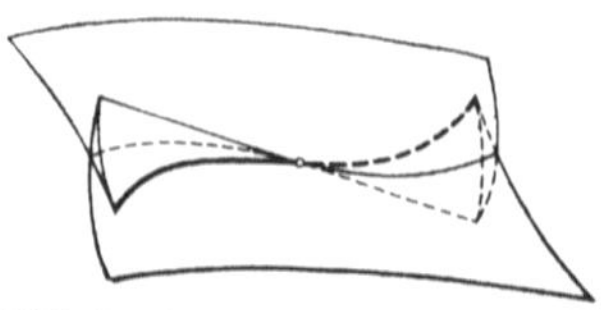

FIG. 19. The nonregular-orbit surface for the icosahedron symmetry group H_3.

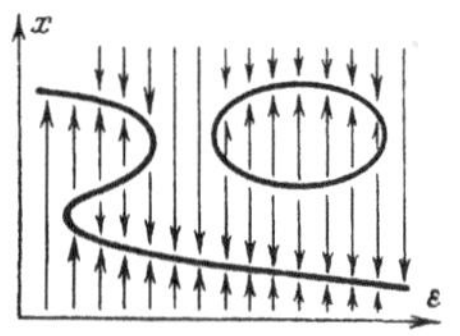

FIG. 20. Bifurcations of equilibrium positions for a generic system.

[3)]An instructive example: for fixed $A \approx 2$ the equation $u_{xxxx} + Au_{xx} + u_x^2 + u = 0$ has whole families of periodic solutions. The reason lies in the symmetry $x \to -x$.

FIG. 21. A small disturbance takes a bifurcation into general position.

given degeneracy is rendered nonremovable by a small disturbance of the family. Thus when studying individual systems it is sufficient to discuss generic cases, neglecting all degeneracies. If there is a single parameter, one should investigate degeneracies of codimension one, that is, those corresponding to a submanifold specified by a single equation in the functional space of all systems. There is no point in studying degeneracies of codimension two except in connection with bifurcations in two-parameter generic families; and so on.

12. EQUILIBRIUM BIFURCATIONS

Although Poincaré's program as described above has long been familiar (in particular, it received wide currency in the schools of A. A. Andronov and L. I. Mandel'shtam), not too much real progress has been made in putting it into practice. Even in the case of just two parameters we are not yet able to cope with all the mathematical difficulties.

The main achievement of singularity theory in this regard is that of sorting out the behavior of equilibrium-position bifurcations (but, unfortunately, not the corresponding phase portraits) in generic systems with an arbitrary finite number of parameters.

As mentioned in Sec. 11, the manifold of equilibrium positions of a generic system forms in (x, ε) space a smooth manifold whose dimensionality is equal to the number l of parameters ε_k (see Fig. 20). Let us consider the projection of that l-dimensional manifold onto the l-dimensional parameter space (along the "x axis," that is, along phase space).

Singularity theory specifies the normal forms to which this projection is brought by a smooth local change of coordinates (depending on the parameters ε) and a smooth change of the parameters ε near each point of (x, ε) space for a generic system.

If, for example, phase space and parameter space have dimension one, as in Fig. 20, then the normal form will become $v(x, \varepsilon) = \pm x^2 + \varepsilon$. As a result, when the single parameter in the generic system varies, only the following bifurcations can occur: coalescence of a stable equilibrium position with an unstable one followed by destruction of both, or creation of paired stable and unstable equilibria. This very simple case had, of course, already been studied by Poincaré in his early work. Modern singularity theory has the advantage of being able to handle analogous problems in higher dimensions.

For instance, if the number l of parameters is arbitrary but, as before, the phase variable x is one-dimensional, then by a theorem of Whitney[1,11] the normal form will be

$$v(x, \varepsilon) = \pm x^{k+1} + \varepsilon_1 x^{k-1} + \ldots + \varepsilon_k, \quad k \leqslant l. \qquad (*)$$

Figure 22 sketches an equilibrium surface $v(x, \varepsilon) = 0$ for the case of two parameters. The projection of this smooth surface on the parameter plane $(\varepsilon_1, \varepsilon_2)$ has a singularity on the fold line (Γ in Fig. 22). The projection of the line Γ onto the parameter plane (the set of critical values of the mapping) is a cusped curve (curve Σ in Fig. 22). This curve in the parameter plane is called the *bifurcation curve* or the *catastrophe line*.

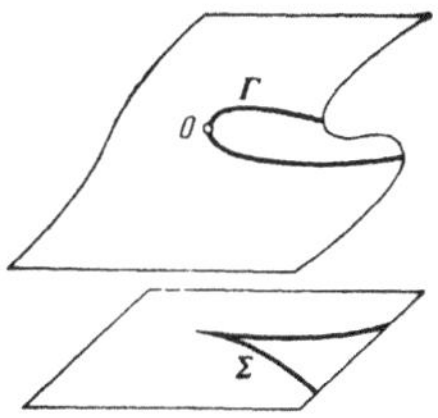

FIG. 22. Map of a Whitney fold.

As a point in the parameter plane approaches the catastrophe line from inside the cusp, two of the three equilibrium positions will coalesce (one stable and one unstable). At that juncture the system will be obliged to jump from the dying equilibrium position, where stability is being lost, to a third and remote equilibrium position. If the model describes, say, an economic system, then the corresponding stable equilibrium position of the economic system should jump to a new state distant from the original one. It was sudden jumps of this kind that prompted Thom[15] to invent the term "catastrophe theory," combining the singularity theory developed by Whitney, the bifurcation theory of Poincaré and Andronov, and their applications (a detailed account will be found in my recent book[16]). As the catastrophe line is approached the stable and unstable equilibria will come together ultimately at infinite speed. That explains why it is so hard to fight against a catastrophe as soon as it seems imminent.

In the neighborhood of the singular point O the singularity of the projected surface (Fig. 22), is called a pleat. This singularity is stable, for any slightly different mapping will have the same type of singularity at some nearby point. Corresponding to this projected pleat in the image plane is the cusp contour. Whitney[1] proved that such folds or pleats are the only types of singularities of projected generic surfaces. Thus the development of a cusp point on the corresponding image curve is a generic phenomenon, not destroyed by a small disturbance of the original projection.

The projection of a generic curve from 3-space onto a plane cannot have cusps, but only self-crossing points. The fact that the image contour does develop a cusp manifests a general principle, according to which "singularities attract singularities"; for the image contour is the projection not of an arbitrary curve but of a curve comprising projected singular points. This principle also accounts for the special character of the singularities in generic caustics and wave fronts (by comparison with the singularities that develop in sets of critical values of generic projections, that is, the singularities in the bifurcation diagrams of the equilibrium positions for general systems of differential equations).

For example, the bifurcation diagrams of three-parameter generic families exhibit only cusp- and swallowtail-type

FIG. 23. Growth of a cycle when an equilibrium position experiences a soft loss of stability.

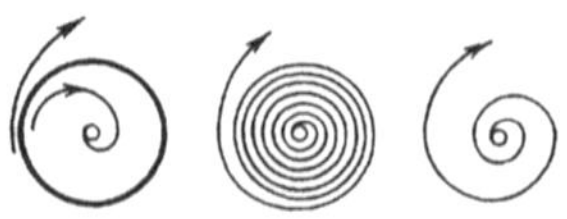

FIG. 24. Constriction of an equilibrium-regime basin in a hard loss of stability.

singularities, whereas generic caustics can also develop pyramid and pucker points (Fig. 5). Equation (*) implies that the singularities in the bifurcation diagrams of, more generally, l-parameter families of differential equations with a one-dimensional phase space will be of the same type as those for the series A wave fronts (Fig. 6).

13. STABILITY LOSS

If phase space is multidimensional, an equilibrium state can lose its stability without the equilibrium itself having to undergo a bifurcation. There is an an alternative possibility: a pair of eigenvalues linearizing an equation at a singular point may make a transition from the stable (left) half-plane, $\mathrm{Re}\lambda < 0$, to the right half-plane.

Andronov had studied the corresponding phase-portrait bifurcation in the plane, publishing the results in the original 1937 edition of his textbook written with Khaĭkin.[17] In 1942 Eberhard Hopf[18] extended part of Andronov's theory to the n-dimensional case, and for that reason in the West this type of bifurcation is usually called the Hopf bifurcation.

In this case the topological normal form becomes $\dot{z} = (i + \varepsilon)z + cz|z|^2$. Depending on the sign of $\mathrm{Re}\, c$, the bifurcation will entail either a "soft" growth of self-oscillations, with an amplitude proportional to the square root of the supercriticality, $\sqrt{\varepsilon}$ (Fig. 23), or a shrinkage to zero of the attraction zone of the equilibrium regime, followed by a "hard" loss of stability and a jump to a different regime (Fig. 24).

This new regime of motion may constitute a new equilibrium position or self-oscillation regime (with strictly periodic oscillations being established), or some more complicated type of motion may arise, described by an attracting set more intricate than a cycle in phase space (that is, by a *strange attractor*).

That complicated "stochastic" motions such as these in determinate dynamical systems could be stable was recognized in the early 1960s by Smale,[19] Sinaĭ,[20] and Anosov.[21] The first efforts to apply these concepts to a description of turbulence date from this same period, or even from 1958-1959, when Andreĭ N. Kolmogorov combined into a single seminar research on the ergodic theory of dynamical systems and on hydrodynamic instability. Yet even today there still do not seem to be any rigorous results concerning the stochasticity of attractor motion for the Navier-Stokes equations. Computer experiments, first performed by Lorenz[22] in 1963, suggest that for at least some attractors the motion will be exponentially unstable.

So far as I am aware, the only definitive results in this field set *upper* limits on the attractor dimension for two-dimensional Navier-Stokes equations in a compact region. Such limits were recently obtained for the first time by Il'yashenko[23] for periodic boundary conditions; M. I. Vishik and A. V. Babin have subsequently extended them to more general attachment conditions. The dimension limits take the form: $\dim \leqslant \mathrm{Re}^p$ (in Vishik and Babin's latest theorems, $p = 4$). It has neither been proved nor refuted that as $\mathrm{Re} \to \infty$ the dimension of *all* attractors for the Navier-Stokes equation will increase without bound. Most specialists feel that no attractor of dimension bounded as $\mathrm{Re} \to \infty$ exists which would attract solutions having almost any initial conditions; but this has not been proved either.

In the case of a soft loss of equilibrium-state stability (such as an interruption of laminar flow), a stable periodic motion (a self-oscillation) will develop. As the parameter continues to vary, the self-oscillatory motion may itself lose stability. The theory of the loss of self-oscillation stability represents another large branch of general bifurcation theory. Some universal singularities have been discovered, not depending on the particular type of system. I cannot describe this theory here in any detail at all (see my book on differential equations[24]), but merely wish to point out the important role of resonances between the self-oscillatory motion occurring at the very outset and small oscillations close to that

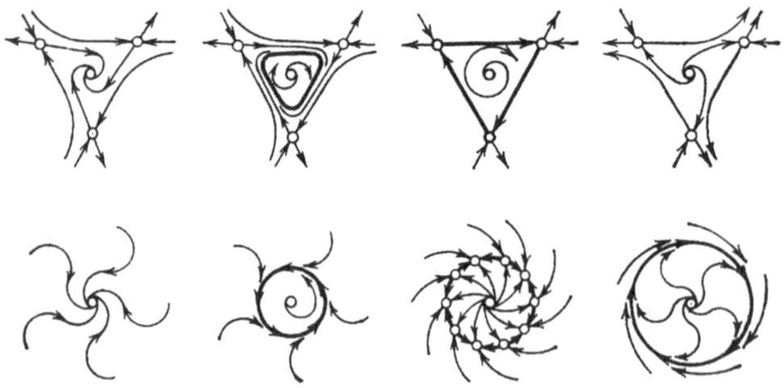

FIG. 25. Typical sequences of phase-oscillation metamorphoses as periodic motions lose their stability near third- and fifth-order resonances.

motion.

It turns out that the chief contribution comes from resonances of the first four orders—the *strong* resonances. Rational numbers with denominator $\geqslant 5$, corresponding to *weak* resonances, may in this context be considered "almost irrational" numbers, whereas the strong resonances will induce qualitatively different "phase oscillations." Figure 25 illustrates the difference between the phase oscillations for the 1:3 and the 1:5 resonances. If self-oscillation stability is lost near a 1:2 resonance, the cycle will double, and usually (although not necessarily) a sequence of universal Feigenbaum doublings[25,26] will be generated.

The path from laminar to turbulent flow may lead either through a chain of soft stability-loss bifurcations [such as the Landau twinnings or bifurcations[27] when n-dimensional tori grow out of $(n-1)$-dimensional ones] or through a hard chain, with the attractor developing far from the laminar flow. And from a physical point of view, the laminar flow need not even lose its stability: it would be enough for its attraction zone to shrink as the Reynolds number increases.

Comprehensive bibliographies of singularity, bifurcation, and catastrophe research will be found in standard texts.[11,16,28–31]

[1]H. Whitney, "Singularities of mappings of Euclidean spaces. I. H. Whitney, Mappings of the plane into the plane," Ann. Math. (2) **62**, 374–470 (1955).

[2]V. I. Arnol'd, "Integrals of rapidly oscillating functions and singularities of Lagrangian-manifold projections," Funkts. Anal. i Prilozh. **6**, No. 3, 61-62 (1972) [Funct. Anal. and Appl. **6**, 222-224 (1973)].

[3]V. I. Arnol'd, "Critical points of smooth functions and their normal forms," Usp. Mat. Nauk No. 185 (**30**, No. 5), 3-65 (1975) [Russ. Math. Surveys **30**, No. 5, 1-75 (1976); *op. cit. infra*,[31] pp. 91–131].

[4]V. I. Arnol'd, "Wave-front evolution and the equivariant Morse lemma," Commun. Pure Appl. Math. **29**, 557-582 (1976).

[5]Ya. B. Zel'dovich, "Fragmentation of a homogeneous medium through gravitation," Astrofizika **6**, 319-335 (1970) [Astrophysics **6**, 164-174 (1973)].

[6]V. I. Arnol'd, "Normal forms of functions near degenerate critical points," Funkts. Anal. i Prilozh. **6**, No. 4, 3-25 (1972) [Funct. Anal. and Appl. **6**, 254-272 (1973)].

[7]V. I. Arnol'd, Mathematical Methods in Classical Mechanics, Nauka, Moscow (1974) [Graduate Texts Math. **60**, Springer (1978)].

[8]J. W. Milnor, Singular Points of Complex Hypersurfaces (Ann. Math. Stud. **61**), Princeton Univ. Press (1968) [Russ. Transl., Mir, Moscow (1971)].

[9]F. Pham, Introduction à l'Étude Topologique des Singularités de Landau, Gauthier-Villars, Paris (1967). [Russ. Transl. Mir, M., 1970].

[10]S. M. Husein-Zade [Guseĭn-Zade], "The monodromy groups of isolated singularities and hypersurfaces," Usp. Mat. Nauk No. 194 (**32**, No. 2), 23-65 (1977) [Russ. Math. Surveys **32**, No. 2, 23-69 (1978)].

[11]V. I. Arnol'd, A. N. Varchenko, and S. M. Husein-Zade, Singularities of Differentiable Mappings [in Russian], Nauka, Moscow (1982).

[12]N. Bourbaki, Groupes et Algebres de Lie, Hermann, Paris (1960-1975) [Russ. Transl., Mir, Moscow (1972)].

[13]V. I. Arnol'd, "The singularities of ray systems," Usp. Mat. Nauk No. 230 (**38**, No. 2), 77-147 (1983) [Russ. Math. surveys **38**, No. 2 (1984, in press)].

[14]O. V. Lyashko, "Classification of critical points of functions on a manifold with a singular edge," Funkts. Anal. i Prilozh. **17**, No. 3, 28 (1983) [Funct. Anal. and Appl. **17**, No. 3 (1984, in press)]; O. V. Lyashko and O. P. Shcherbak (in press).

[15]R. Thom, Stabilité Structurelle et Morphogénèse, Benjamin, New York (1972) [Engl. Transl., Benjamin, Reading, Mass (1975)].

[16]V. I. Arnol'd, Catastrophe Theory, Znanie, Moscow (1981); Moscow Univ. Press (1983) [Engl. Transl. by R. K. Thomas, Springer (1983)].

[17]A. A. Andronov and S. É. Khaĭkin, Theory of Oscillations, United Sci. Tech. Presses, Moscow (1937) [Princeton Univ. Press (1949); 2nd ed., Phys. Math. Press, Moscow (1959) [Addison-Wesley (1966)].

[18]E. Hopf, "Abzweigung einer periodischen Lösung von einer stationären Lösung eines Differential systems," Ber. Math.-Phys. Kl. Sächs. Akad. Wiss. Leipzig **94**, 1 (1942) [Engl. Transl., in: The Hopf Bifurcation[28]].

[19]S. Smale, "Differentiable dynamical systems," Bull. Am. Math. Soc. **73**, 747-817 (1967) [Russ. Transl., Usp. Mat. Nauk No. 151 (**25**, No. 1), 113-185 (1970)].

[20]Ya. G. Sinaĭ, "Dynamical systems with a countably multiple Lebesgue spectrum" [in Russian], Izv. Akad. Nauk SSSR Ser. Mat. **25**, 899-924 (1961).

[21]D. V. Anosov, "Geodesic flows on closed Riemann manifolds of negative curvature" [in Russian], Trudy Mat. Inst. Akad. Nauk SSSR **90** (1967).

[22]E. N. Lorenz, "Deterministic nonperiodic flow," J. Atmos. Sci. **20**, 130-141 (1963).

[23]Yu. S. Il'yashenko, Usp. Mekhan. **5**, No. 1, 31 (1982).

[24]V. I. Arnol'd, Some Supplementary Chapters in the Theory of Ordinary Differential Equations [in Russian], Nauka, Moscow (1978).

[25]M. J. Feigenbaum, "The universal metric properties of nonlinear transformations," J. Stat. Phys. **21**, 669-706 (1979).

[26]M. J. Feigenbaum, "Universal behavior in nonlinear systems," Los Alamos Science **1**, No. 1, 4-27 (1980) [Russ. Transl., Usp. Phys. Nauk **141**, 343 (1983)].

[27]L. D. Landau, Dokl. Akad. Nauk SSSR **44**, 339 (1944).

[28]J. E. Marsden and M. McCracken, The Hopf Bifurcation and its Applications (Appl. Math. Ser. **19**), Springer (1976) [Russ. Transl., Mir, Moscow (1980)].

[29]T. Poston and I. Stewart, Catastrophe Theory and its Applications, Pitman, London (1978) [Russ. Transl., Mir, Moscow (1980)].

[30]R. Gilmore, Catastrophe Theory for Scientists and Engineers, Wiley (1981).

[31]V. I. Arnol'd, Singularity Theory—Selected Papers (London Math. Soc. Lect. Note Ser. **53**), Cambridge Univ. Press (1981) [Engl. Transl., reprinted from Russ. Math. surveys **23-34** (1968-79)].

Translated by R. B. Rodman

SINGULARITIES OF FUNCTIONS, WAVE FRONTS, CAUSTICS AND MULTIDIMENSIONAL INTEGRALS

V. I. ARNOL'D, A. N. VARCHENKO, A. B. GIVENTAL' AND A. G. KHOVANSKII

Moscow State University, Moscow

Abstract

The theory of singularities is a grand-scale generalization of the study of maxima and minima of functions. In recent years a whole arsenal of methods has been developed in this mathematical theory, enabling one to give exhaustive answers to questions of the structure of singularities of wave fronts, caustics, and integrals of rapidly oscillating functions by the method stationary phase, saddle-point integrals with coalescing saddle points.

These methods also apply to the theory of Legendre transformations, to the bifurcation of equilibrium positions and oscillatory regimes of dynamical systems which depend on parameters, to optimization theory, calculus of variations, geometry of Riemannian manifolds, etc.

In this paper we discuss the basic ideas, concepts, and methods of the theory of singularities. The authors have taken measures to separate the conceptual aspects of the paper from the technical details; on the other hand, the text contains concrete results and tables needed for most applications.

From the point of view of mathematics, the success of the theory is based on the connection discovered in recent years between geometrical optics and the theory of asymptotic behavior of integrals and entities seemingly remote from them: simple Lie algebras and the theory of crystallographic groups generated by reflections.

The classification of the simplest singularities of caustics and wave fronts turns out to be related to the classification of compact Lie groups ($A_k = SU_{k+1}$, $D_k = SO_{2k}$, E_6, E_7, E_8), and also to the classification of regular polyhedra in three-dimensional space [i.e., the theory of the discrete subgroups of the group $SU(2)$].

In Section 1 we classify the singularities of systems of rays and fronts in geometrical optics; it is based on the theory of bifurcation of critical points in families of functions that depend on parameters. In Section 2 this information is used for studying short-wave asymptotics in the neighborhood of singularities of caustics. The calculation of the asymptotics of integrals reduces to the stereometry of certain convex polyhedra with integer vertices—the so-called Newton polyhedra of singularities (the vertices of the Newton polyhedra are the powers of monomials appearing in the Taylor series of the singularity). The Newton polyhedra also contain significant information about the topological invariants of manifolds given by algebraic or analytic equations (about the number of handles of Riemann surfaces, the number of solutions of polynomial systems of equations, etc.). Formulas for calculating these invariants are given in Section 3.

Originally publ. in: Mathematical Physics Reviews. Vol. 4 (Soviet Sci. Rev. Sect. C: Math. Phys. Rev., 4), 1-92,
Republished in: Reviews Math. and Math. Phys. 14(2), 1-92,
Reprint by Springer-Verlag Berlin Heidelberg 2018 with kind permission

Section 1 was written mainly by A. B. Givental', Section 2 by A. N. Varchenko, and Section 3 by A. G. Khovanskii.
The authors are grateful to S. P. Novikov, who proposed that they write this survey.

Contents

§1 Singularities of Functions, Caustics, and Wave Fronts

The theory of singularities is concerned with the classification of degeneracies of objects of different types. In this section we shall study the singularities of functions, caustics, wave fronts, and their metamorphoses. A complete classification according to problems of interest would be a limitless task. The guiding thread we shall use through this labyrinth is the *principle of general position*: one should study only those singularities that cannot be eliminated by a slight deformation of the object. This principle is well illustrated by the singularities of level lines on a topographical map which shows peaks, dips and saddles. Upon closer inspection the ridges always turn out to be chains of mountain peaks and saddles.

For further information about the questions discussed in this section and related ones, we refer the reader to (Refs. 14, 12, 9, 57).

1.1 Examples

1.1.1 Wave Fronts and Caustics in the Plane The propagation of disturbances in a medium can be described by means of a system of rays and by using wave fronts. Suppose, for example, that a disturbance propagates towards the interior of an ellipse in the plane with

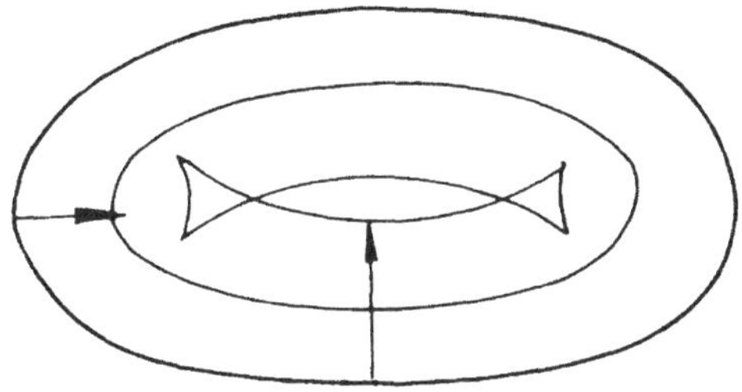

Figure 1

unit velocity. If we mark a segment of length t along the interior normals of the ellipse, we obtain a parallel to the ellipse, an equidistant wave front (Fig. 1). Although the initial front is smooth, after some time the parallel front develops singularities—cusps (the local equation of the front being $y^2 = x^3$ in some curvilinear coordinate system) and self-intersections. These singularities are *stable* (i.e., they cannot be eliminated by a slight change of the initial front). It turns out that all other singularities of the front in a plane (for example, the singularity at the center of a contracting circle) break up into these standard forms when there is a slight shift of the initial front.

A wave front moving in the plane traces out a certain surface in (three-dimensional) space–time. It appears that this surface can be regarded as a wave front (*grand front*) in space–time. The metamorphoses that the moving wave front experiences are identical to the metamorphoses of the sections of the grand front by isochrones (planes with $t = \text{const}$) of space–time. The metamorphosis for the example of the ellipse is shown in Fig. 2. In this case the grand front singularity is called a "swallowtail." The "swallowtail" can be defined as the surface formed in x, y, z space by all the tangent lines to the curve $x = a^2$, $y = a^3$, $z = a^4$—its cuspidal edge. The metamorphosis of Fig. 2 exhausts all generic metamorphoses of wave fronts in the plane.

The envelope of rays normal to the front in the plane is called a *caustic* (see Fig. 3a). The light is concentrated at the points of the caustic. It is easy to show that the cusps on the moving front fill the caustic (Fig. 3b). Thus the caustic is obtained by projecting the cuspidal edge of the grand front from space–time onto ordinary space (Fig. 4). The singularities of the caustic in general position in the plane are exhausted by the cusp points and by the self-intersection points.

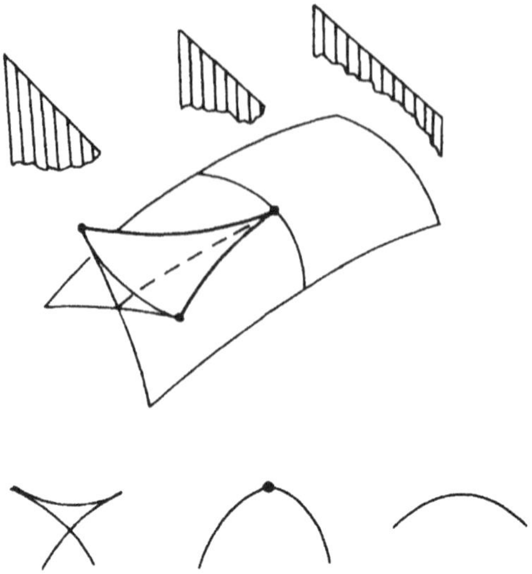

Figure 2

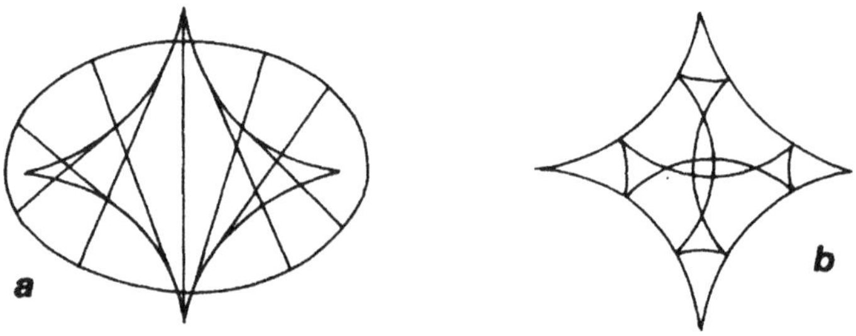

Figure 3

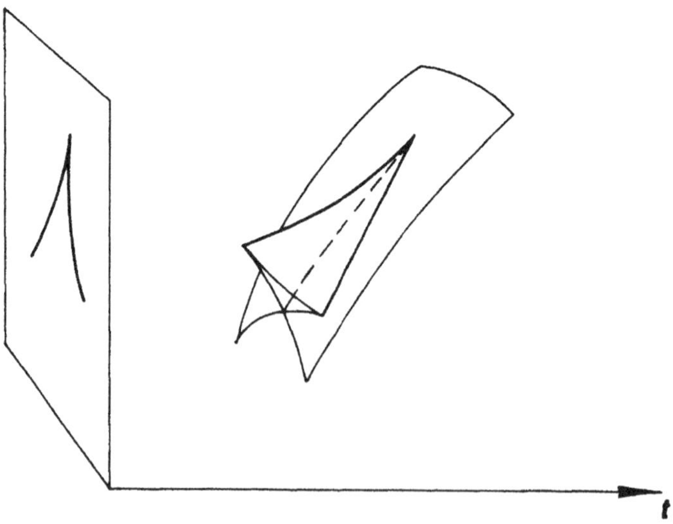

Figure 4

1.1.2 Singularities of a Function of a Single Variable Let $y = f(x)$ be a smooth (i.e., infinitely differentiable) function. A point in the domain of definition of the function is said to be *singular* if the derivative vanishes there. The value of the function at the singular point is called a *critical value*. If the second derivative of the function does not vanish at the singular point, the point is said to be *nondegenerate* or a type-A_1 point. In the neighborhood of a singular point of type A_1 the function can, by a smooth change of the variable, be brought to the form $y = \pm z^2 + \text{const}$. Similarly, if $f'(x_0) = f''(x_0) = 0$, while $f'''(x_0) \neq 0$, then the function can be brought to the form $y = z^3 + \text{const}$ in the neighborhood of the point x_0 (type-A_2 point); if $f'(x_0) = f''(x_0) = f'''(x_0) = 0$, while $f^{IV}(x_0) \neq 0$, the function in the neighborhood of the point x_0 can take the form $y = \pm z^4 + \text{const}$ (type-A_3 point), etc.

A function in general position has only nondegenerate singular points. Degeneracies are eliminated by arbitrarily small changes of the function (here and below a small change refers to the function and all its derivatives). Degenerate singularities appear when we consider families of functions. Thus, in the family $y = x^3 - tx$ of functions* of x with parameter t, for zero value of the parameter t there is a singular point of type A_2 at $x = 0$ (Fig. 5), and any neighboring family has the same singularity for some small value of

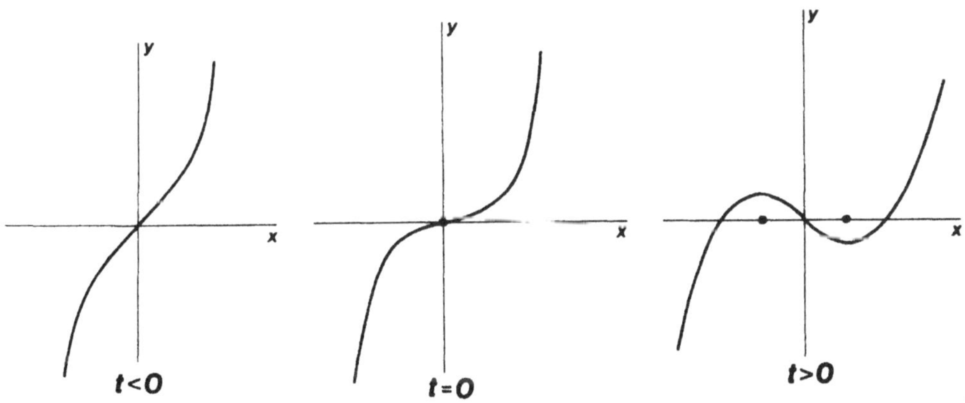

Figure 5

*By a family of functions we mean a function of two arguments, one of which is assumed to be a parameter. The arguments themselves may be points of multidimensional spaces. We shall sometimes call the parameter space the base of the family.

the parameter. In general one-parameter families of functions, A_2 singularities appear only for individual parameter values; in general two-parameter families of functions, A_2 singularities appear for parameter values that form a curve in the parameter plane, while one finds type-A_3 singular points for individual parameter values.

Functions in general position have no singular points with critical values that are *previously assigned* (e.g., zeros): such points are removed by adding a small constant to the function. A one-parameter family of functions in general position, in the neighborhood of a singular type-A_1 point with zero critical value can, by a small change of parameter λ, and a smooth change of variable that is a smooth function of the parameter, be brought to the form $y = \pm z^2 + \lambda$. The general two-parameter family of functions can be brought by these same operations to the form $y = z^3 + \lambda_1 z + \lambda_2$ in the neighborhood of a type A_2 point with critical value zero, and the general three-parameter family, in the neighborhood of a type-A_3 point with critical value zero, can be brought to the form $y = \pm z^4 + \lambda_1 z^2 + \lambda_2 z + \lambda_3$. The families of functions introduced here are said to be *versal deformations* of singular points of types A_1, A_2, A_3, respectively. The set of points in the parameter space of the versal deformation to which there correspond functions with zero critical values, is called the *bifurcation diagram of zeros*. For A_1 the bifurcation diagram of zeros consists of the single point $\lambda = 0$. For A_2, from the condition $y' = 0$ we find: $\lambda_1 = -3z^2$, while from the condition $y = 0$ we get $\lambda_2 = -z^3 - \lambda_1 z = 2z^3$; i.e., the bifurcation diagram of zeros is a semicubic parabola $4\lambda_1^3 + 27\lambda_2^2 = 0$ with cusp at $\lambda_1 = \lambda_2 = 0$ (Fig. 6). The bifurcation diagram of zeros for a singular type-A_3 point is a "swallowtail" (Fig. 7a) in the space of polynomials $z^4 + \lambda_1 z^2 + \lambda_2 z + \lambda_3$, consisting of polynomials with multiple roots. The projection of the singular points of the bifurcation diagram of zeros along the axis of the free term of the versal deformation onto the plane of the other parameters is called the *bifurcation diagram of functions*. For A_3 the bifurcation diagram of functions (Fig. 7b) consists of a semicubic parabola (functions $y = z^4 + \lambda_1 z^2 + \lambda_2 z$, with a singular type-$A_2$ point) and a ray (functions with two singular type-A_1 points with identical critical values). The shaded values of the parameters correspond to functions with three singular points, and the unshaded values correspond to those with one singular point. We see that the

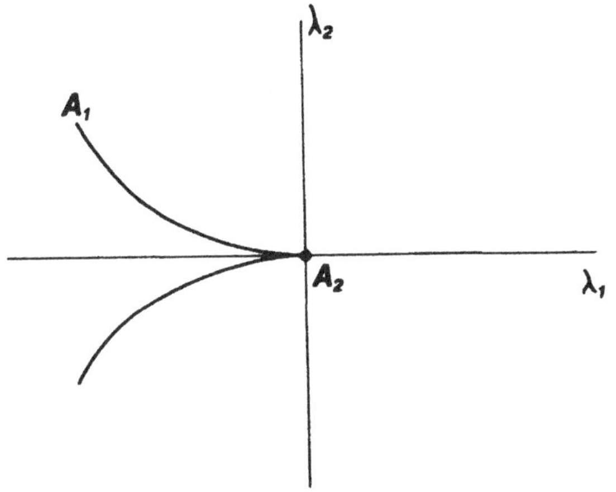

Figure 6

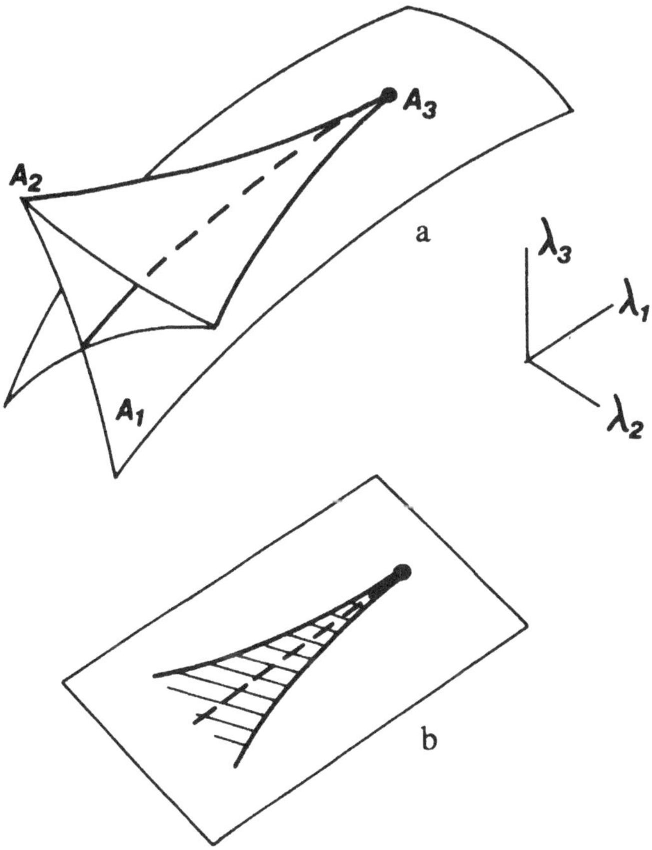

Figure 7

bifurcation diagrams carry considerable information about the breakup of singular points.

1.1.3 Modality, Transversality, and Codimension For greater clarification of the concepts used by us—stability, families, versal deformations, and bifurcation diagrams—we use the following geometric interpretation. We shall picture the objects to be classified (functions, for example) as points of a space (which is actually infinite-dimensional). The classification is a decomposition of the space of all objects (Fig. 8). Objects in general position occupy almost all the space. Degenerate objects occupy the surfaces of separation (the coordinate crossing in Fig. 8) between the regions of objects in general position (the four quadrants). A *stable* object is one for which all the nearby objects are in the same class as it is in (points in quadrants II, III and IV). In Fig. 8 the classes in the first quadrant depend on a continuous parameter (the *modulus*)—the slope angle. The *modality* of an object is the largest number of parameters on

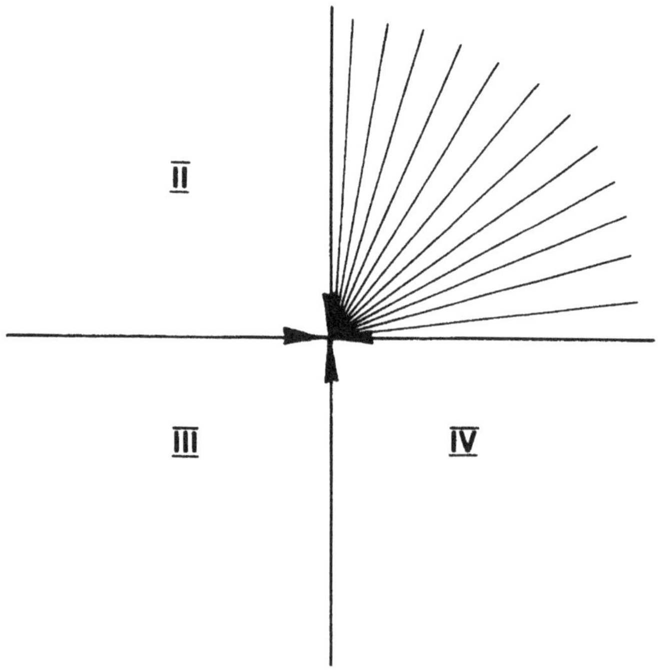

Figure 8

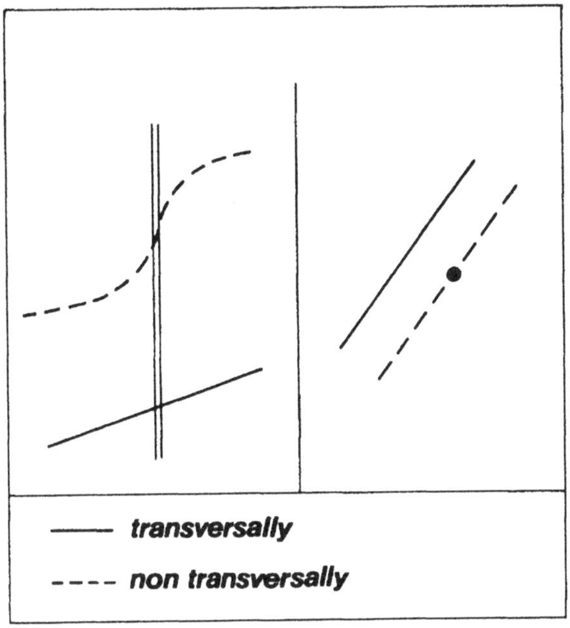

Figure 9

which the classes depend, which intersect a small neighborhood of this object in the space of objects. In Fig. 8 the points of the first quadrant, the positive coordinate semiaxes, and the coordinate origin are unimodal, while the other points are 0-modal. A 0-modal object, i.e., one whose neighborhood includes a finite number of classes, is said to be *simple*. The example of a discrete classification, one in which all objects are simple, gives the classification of the singular points of analytic functions of a single variable: a class of the singular point is determined by the order of the first nonzero term in the Taylor series of the function at that point. A *family* of objects is represented by a curve, surface, etc. in the space of all objects. Families in general position intersect the simple classes *transversally* (i.e., at a nonzero angle, Fig. 9). In particular, in one-parameter families one finds that only those classes cannot be eliminated which are represented as surfaces of *codimension* one (i.e., selected by one equation in the space of objects); in two-parameter families the classes that cannot be eliminated are simple classes of codimension two (two equations, see Fig. 10), etc. In studying finite-parameter families we can ignore sets of infinite codimension in the object space

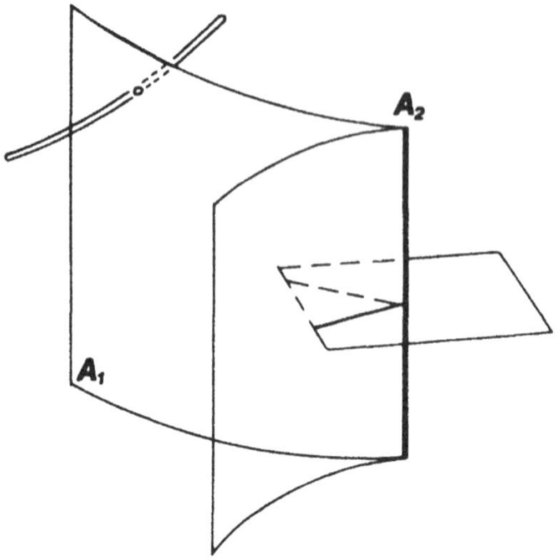

Figure 10

(for example, the set of functions with null Taylor series in the space of smooth functions).

1.2 Critical Points of Functions

1.2.1 Basic Concepts A *singular* or *critical* point of a smooth function is a point in the domain of definition of the function at which its differential vanishes. In considering the behavior of a function in the neighborhood of a critical point, we shall say that two functions are equivalent (more precisely, R-equivalent) if they go over into each other under a suitable, smooth change of variables in the neighborhoods of the critical points.* A critical point of a function is said to be nondegenerate if the quadratic form of the second differential of the function is nondegenerate at that point. A function in the neighborhood of a nondegenerate critical point is equivalent to the quadratic part of its Taylor series (M. Morse), i.e., in a certain coordinate system it has the form

$$\pm x_1^2 \pm \cdots \pm x_n^2 + \text{const}$$

*The term "R-equivalent" comes from the word "right," since only the changes in the variables on the right side of the function symbol in the expression $y = f(x)$ are permitted.

A smooth function in general position has only nondegenerate singular points. Degenerate singular points are eliminated by an arbitrarily small distortion of the function. (Here and below a small change also applies to higher derivatives).

After a small shift (deformation) of the function, a degenerate singular point splits into several nondegenerate points (see Fig. 5). If their number is bounded for any deformation, such a degenerate singular point is said to have *finite multiplicity*. In the neighborhood of a critical point with finite multiplicity a smooth function is equivalent to its Taylor polynomial of sufficiently high order (J. C. Tougeron). If we consider not only real critical points but also complex ones, we find that under deformations of this polynomial the number of singular points into which a degenerate singular point splits is independent of the manner of deformation—so long as all are nondegenerate (just as an n-fold root of a polynomial in one variable splits into n simple roots, which are generally complex). This number is called the *multiplicity* or *Milnor number* of the singularity.

If at a critical point of a function of n variables the quadratic form of the second differential has rank r (i.e., reduces to a sum of r squares), then the function in the neighborhood of this point has, in a certain coordinate system, the form

$$\text{const} \pm x_1^2 \pm \cdots \pm x_r^2 + \varphi(x_{r+1}, \ldots, x_n),$$

where the Taylor series of the function φ at the origin begins with terms of degree three. The number $n - r$ is called the *corank* of the singular point. Functions that may depend on different numbers of variables are said to be *stably equivalent* in the neighborhood of 0 if they become equivalent after adding to them nondegenerate quadratic forms in a suitable number of new variables. Stably equivalent critical points of functions of the same number of variables, with zero second differential, are equivalent.

1.2.2 Simple Singularities and Their Miniversal Deformations The classes of stably equivalent simple critical points of smooth functions with zero critical values form the infinite series $A_\mu^\pm$ ($\mu \geqslant 1$), $D_\mu^\pm$ ($\mu \geqslant 4$) and the classes $E_6^\pm$, E_7, and E_8 that abut each other, as

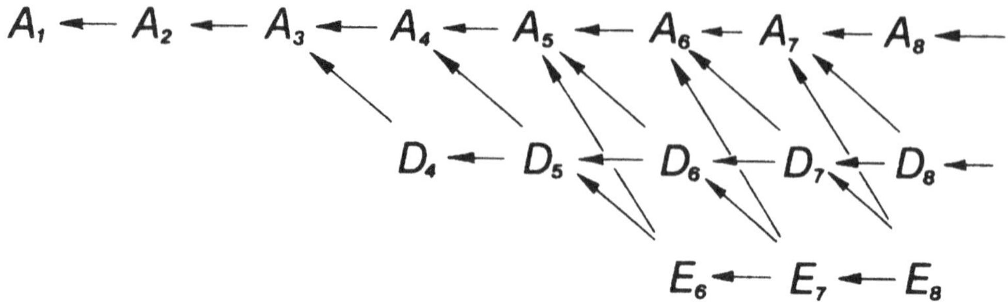

Figure 11

shown in Fig. 11. Below we enumerate examples of the classes of functions with critical point 0:

$$A_\mu^\pm : \pm x^{\mu+1}, \qquad D_\mu^\pm : x^2y \pm y^{\mu-1},$$

$$E_6^\pm : x^3 \pm y^4, \qquad E_7 : x^3 + xy^3, \qquad E_8 : x^3 + y^5.$$

The class A_{2k}^+ coincides with the class A_{2k}^-, and the singular points of the classes A_1^+ and A_1^- are stably equivalent. With these exceptions, the classes enumerated are all different. The subscript gives the multiplicity of the singular point. In families of functions in general position with a number of parameters $l \leqslant 6$, among the singular points with zero critical value one finds only simple singular points of multiplicity $\mu \leqslant l$.

Deformations of a critical point in which it remains unremovable (i.e., every neighboring deformation contains an equivalent critical point) are called *versal* (more precisely, R-versal). Among versal deformations, the *miniversal* deformations have the smallest number of parameters. Infinite-multiplicity critical points which form a set of infinite codimension can be eliminated from the finite-parameter families of functions. The codimension of an R-equivalence class of a finite-multiplicity singular point in the space of all singular points of functions is equal to its multiplicity. A miniversal deformation is represented as a transversal addition to the equivalence class (see Fig. 10), so that the dimension of the base of a miniversal deformation of a singular point is equal to its multiplicity.

For simple singular points of the functions described above one can choose as miniversal the following deformations:

$$A_\mu^\pm : x^{\mu+1} + \lambda_1 x^{\mu-1} + \lambda_2 x^{\mu-2} + \cdots + \lambda_{\mu-1} x + \lambda_\mu,$$

$$D_\mu^\pm : x^2 y \pm y^{\mu-1} + \lambda_1 y^{\mu-2} + \cdots + \lambda_{\mu-1} + \lambda_\mu x,$$

$$E_6^\pm : x^3 \pm y^4 + \lambda_1 y^2 x + \lambda_2 xy + \lambda_3 x + \lambda_4 y^2 + \lambda_5 y + \lambda_6,$$

$$E_7 : x^3 + xy^3 + \lambda_1 y^4 + \lambda_2 y^3 + \lambda_3 y^2 + \lambda_4 y + \lambda_5 + \lambda_6 xy + \lambda_7 x,$$

$$E_8 : x^3 + y^5 + \lambda_1 y^3 x + \lambda_2 y^2 x + \lambda_3 yx + \lambda_4 x + \lambda_5 y^3 + \lambda_6 y^2 + \lambda_7 y + \lambda_8 .$$

For an arbitrary function f with μ-fold singularity at the point 0, a miniversal deformation is constructed as follows. Just like the remainders from the division of integers by some fixed integer, we consider the "remainders from division" of functions by all the partial derivatives of the function f. The remainders form a μ-dimensional space. This means that there exist μ functions $\varphi_1, \ldots, \varphi_\mu$, such that any function defined in the neighborhood of the point 0_x is uniquely representable in the form $\sum c_j \varphi_j + \psi$, where c_j are constants, and ψ is "divisible" by $\partial f/\partial x$: $\psi = \sum h_i \partial f/\partial x_i$ for certain functions $h_1, \ldots, h_n$. For example, for $f = x^3/3$ each function of x has a remainder after division by $\partial f/\partial x = x^2$ of the form $c_1 \cdot 1 + c_2 \cdot x$. A deformation $f + \lambda_1 \varphi_1 + \cdots + \lambda_\mu \varphi_\mu$ is miniversal for the singular point considered. In Fig. 12 we show how one calculates the "remainders" for the function $f = x^3 + xy^3$ with singularity E_7. The monomial $x^m y^l$ corresponds to the point (m, l). The shaded monomials have zero "remainders," a nonzero linear combination of monomials joined by lines is divisible by $\partial f/\partial x$, and the circles mark the basis of "remainders" $\varphi_1, \ldots, \varphi_7$. We are convinced that the Milnor number of singularity E_7 is equal to 7.

An arbitrary deformation of the critical point is obtained from a miniversal deformation by the following construction.

By means of a smooth mapping $t \mapsto \lambda(t)$ of the space T into the base of the family of functions $\mathscr{F}(x, \lambda)$, we can *induce* the new family:

$$G(x, \lambda) = \mathscr{F}[x, \lambda(t)].$$

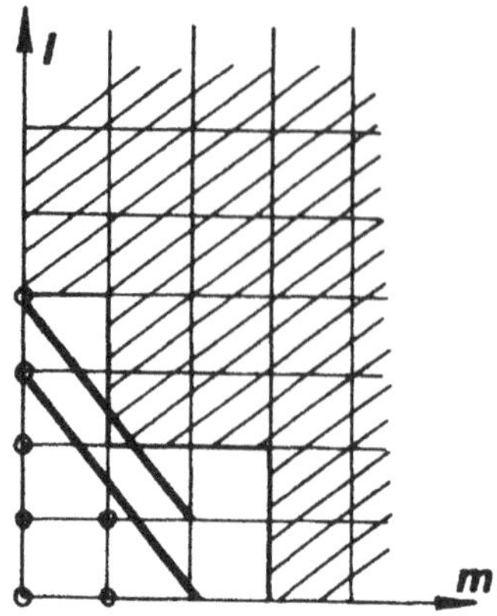

Figure 12

Two families of functions $G(x,\lambda)$ and $\mathscr{F}(x,\lambda)$ are said to be *R-equivalent* if one goes into the other after a change of parameter $\Lambda(\lambda)$ and a change of variables $X_\lambda(x)$, which smoothly depend on the parameter $\lambda: G(x,\lambda) = \mathscr{F}(X_\lambda(x), \Lambda(\lambda))$.

Versality theorem (J. Mather). Any deformation of a finite-multiplicity critical point is equivalent to a deformation induced from a versal deformation. Versal deformations with the same number of parameters are equivalent to each other. Deformations obtained from a versal deformation by adding new parameters on which it is not explicitly dependent are also versal.

1.2.3 Bifurcations and Dynkin Diagrams For the majority of points λ of the base of miniversal deformation of a critical point, the corresponding function f_λ has no singular points with zero critical value. The set of points of the base of miniversal deformation for which the function f_λ has such singular points, is called the *bifurcation diagram of zeros*.

Example: Miniversal deformation of A_μ singularity is the family of polynomials with highest coefficient unity and zero sum of roots; its bifurcation diagram of zeros, which consists of polynomials having multiple roots, is called a generalized swallowtail.

In par. 1 we saw the adjacencies of strata of the bifurcation diagram of zeros carry significant information about the breakups of the initial singular point (see Fig. 7). To enumerate all the strata of bifurcation diagrams of simple singularities, it is convenient to use Dynkin diagrams. The Dynkin diagram of a singularity of multiplicity

μ is a collection of μ points joined by edges, having the following form:

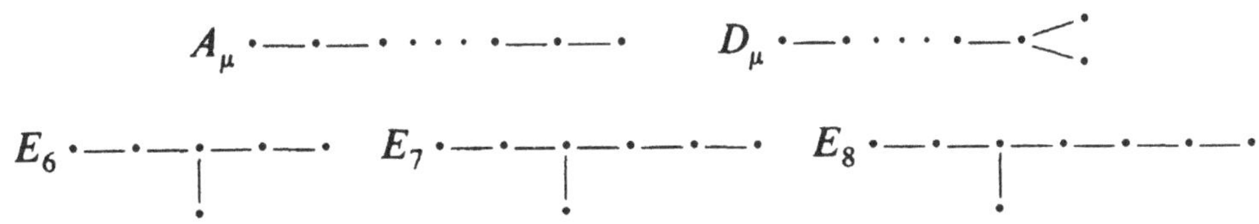

The Dynkin diagrams of classes of simple singularities, whose designations differ only in a $\pm$ sign, are identical. Identification of these classes is actually done if we consider, instead of the classification of singular points of smooth functions, the classification of singular points of complex analytic functions. For example, the functions $x^3 + y^4$ and $x^3 - y^4$, with singular points of types E_6^+ and E_6^-, go over into each other under the substitution $y \to \sqrt{-1}\, y$. All the preceding concepts and results carry over word for word to this case. The classes of stable equivalence of simple singular points of complex-analytic functions with zero critical values form two infinite series A_μ and D_μ and three classes E_μ. The miniversal deformations of complex singularities are the same as above; we need only consider the variables x, y and the parameters λ to be complex. The strata of the complex bifurcation diagram of zeros and the decay of a simple singular point in the complex domain are calculated as follows. The decay due to deformation of a simple critical point into several simple critical points, lying on the *zero* level of the deformed function will be associated with a set of Dynkin diagrams corresponding to the equivalence classes of these critical points. One can show (E. V. Brieskorn) that such a correspondence is a one-to-one correspondence between the possible decays of the initial simple singularity on the zero level surface of the function and the *subgraphs* of its Dynkin diagram, obtained by removing a certain number of vertices and the edges that emerge from them. The complex dimension of a stratum of the bifurcation diagram of zeros, whose points correspond to a given decay, is equal to the number of vertices removed. Inclusion of one subdiagram in another corresponds to the adjacency of the corresponding strata. The decays of a D_4 singularity, the complex dimensions of the strata and their adjacencies are enumerated in Fig. 13.

The bifurcation diagrams of zeros of versal deformations are obtained from the bifurcation diagrams of miniversal deformations by a

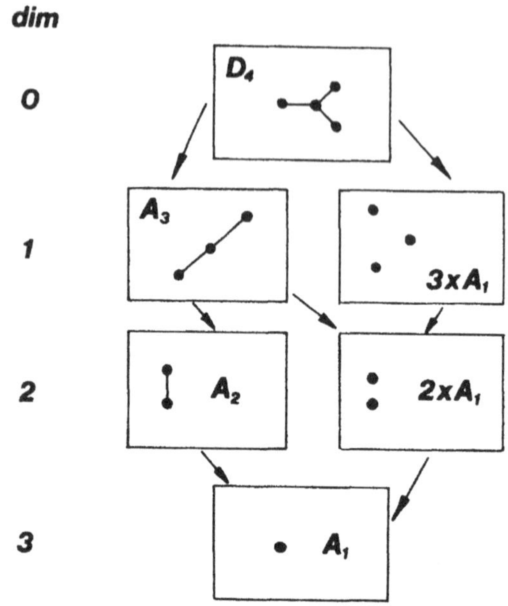

Figure 13

Cartesian product on the space of auxiliary parameters. The decomposition into versal deformations is the same as into miniversal, but the dimensions of the corresponding strata differ in the number of superfluous parameters, so that the codimension of a stratum corresponding to a given decay is independent of the choice of versal deformation and is equal to its codimension in the space of all functions.

1.2.4 Unimodal Critical Points One has now obtained the classification of singularities of functions with a number of moduli $\leqslant 2$ (V. I. Arnol'd). At the same time, one obtains the classification of singular points of multiplicity $\leqslant 16$ and codimension $\leqslant 10$ (for the definition see par. 1.2.5). Below we give the normal forms of unimodal singular points with zero critical value. The parameter a is the modulus.

(a) Parabolic singularities (simple singularities are also called elliptic):

$$P_8 : x^3 + ax^2z \pm xz^2 + y^2z, \qquad a^2 \neq 4, \quad \text{if} \quad +$$

$$X_9 : \pm x^4 + ax^2y^2 \pm y^4, \qquad a^2 \neq 4, \quad \text{if} \quad ++ \quad \text{or} \quad --$$

$$J_{10} : x^3 + ax^2y^2 \pm xy^4, \qquad a^2 \neq 4, \quad \text{if} \quad +$$

Table I

Designation	Normal Form	Restrictions
J_{10+k} $(T_{2,3,6+k})$	$x^3 \pm x^2y^2 + ay^{6+k}$	$a \neq 0, k > 0$
X_{9+k} $(T_{2,4,4+k})$	$\pm x^4 \pm x^2y^2 + ay^{4+k}$	$a \neq 0, k > 0$
$Y_{r,s}$ $(T_{2,r,s})$	$\pm x^2y^2 \pm x^r + ay^s$	$a \neq 0, r, s > 4$
$\tilde{Y}_r$ $(\tilde{T}_{2,r,r})$	$\pm (x^2 + y^2)^2 + ax^r$	$a \neq 0, r > 4$
P_{8+k} $(T_{3,3,3+k})$	$x^3 \pm x^2z + y^2z + az^{3+k}$	$a \neq 0, k > 0$
$R_{l,m}$ $(T_{3,l,m})$	$x(x^2 + yz) \pm y^l \pm az^m$	$a \neq 0, m \geqslant l > 4$
$\tilde{R}_m$ $(\tilde{T}_{3,m,m})$	$x(x^2 \pm y^2 \pm z^2) + ay^m$	$a \neq 0, m > 4$
$T_{p,q,r}$	$axyz \pm x^p \pm y^q \pm z^r$	$a \neq 0, 1/p + 1/q + 1/r < 1$
$\tilde{T}_{p,m}$ $(\tilde{T}_{p,m,m})$	$x(y^2 + z^2) \pm x^p + ay^m$	$a \neq 0, 1/p + 2/m < 1$

(b) Hyperbolic singularities, given in Table I.
(c) Exceptional singularities:

$$K_{12} : x^3 + y^7 + axy^5 \quad K_{13} : x^3 + xy^5 + ay^8 \quad K_{14} : x^8 \pm y^3 + axy^6$$

$$Z_{11} : x^3y + y^5 + axy^4 \quad Z_{12} : x^3y + xy^4 + ay^6 \quad Z_{13} : x^3y \pm y^6 + axy^5$$

$$Q_{10} : x^2z + y^3 \pm z^4 + ayz^3 \qquad Q_{11} : \pm x^2z + y^3 + yz^3 + az^5$$

$$Q_{12} : x^2z + y^3 \pm z^5 + ayz^4$$

$$W_{12} : \pm x^4 + y^5 + ax^2y^3 \qquad W_{13} : \pm x^4 + xy^4 + ay^6$$

$$U_{12} : x(x^2 \pm y^2) \pm z^4 + axyz^2$$

$$S_{11} : x^2z + yz^2 \pm y^4 + ay^3z \qquad S_{12} : x^2z + yz^2 + xy^3 + ay^5$$

In the complex domain the functions in this list, which differ in the $\pm$ sign, are R-equivalent. Representatives of the complex hyperbolic classes can be chosen in the single form:

$$T_{p,q,r} : x^p + y^q + z^r + axyz, \qquad a \neq 0, \qquad \frac{1}{p} + \frac{1}{q} + \frac{1}{r} < 1.$$

Thus the unimodal classes of stable equivalence of singular points of complex-analytic functions form three parabolic one-parameter families P_8, X_9, J_{10}, a three-index series of hyperbolic one-parameter families $T_{p,q,r}$ and 14 exceptional one-parameter families. Some of the adjacencies between unimodal classes are shown in Fig. 14.

1.2.5 R^+-equivalence and Bifurcation Diagrams of Functions So far we have dealt with critical points with fixed (zero) critical value. In

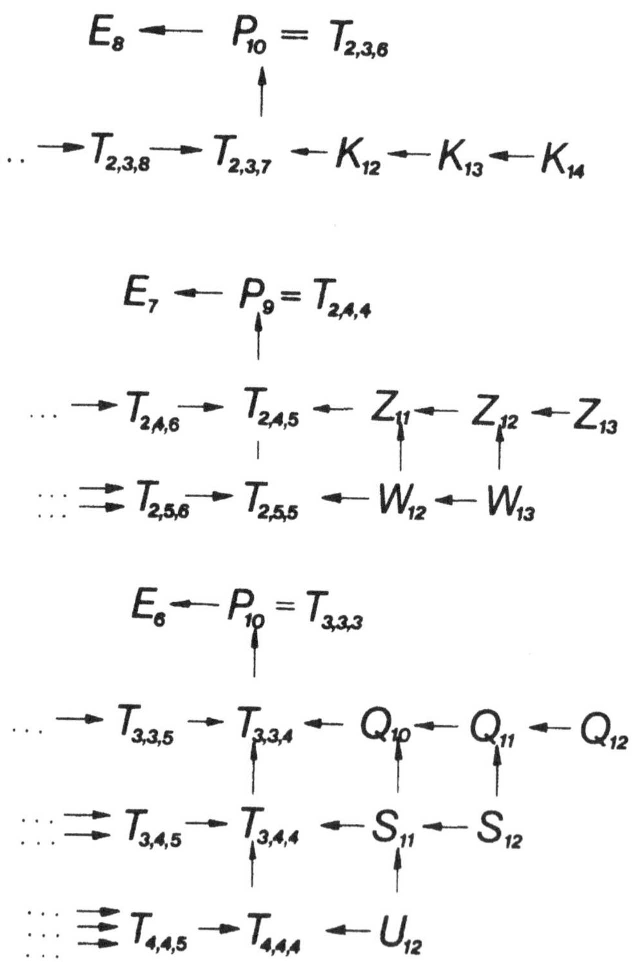

Figure 14

order to remove this restriction we introduce a definition: two functions in the neighborhood of a singular point are said to be R^+-equivalent if they become R-equivalent after adding a suitable constant to one of them. Of course, the R^+-classification of singularities coincides with the R-classification introduced in pars. 1.2.2 and 1.2.4. The whole theory is constructed by analogy with the preceding discussion.

The distinction between the R^+ and R theories begins with the introduction of the concept of versal deformation. An R^+-miniversal deformation is constructed from a set of remainders *that vanish at the singular point*, or, in other words, are obtained from an R-miniversal by dropping the free term. In the basis of R^+-miniversal deformations we distinguish the *bifurcation diagram of functions*, consisting of two strata, called *caustics* and *stratum of intersections*. To the first stratum belong all points λ of the base of R^+-versal deformations for which the function $\mathscr{F}(\cdot,\lambda)$ of the variable x (obtained for a fixed parameter value λ) has degenerate critical points; to the second stratum belong all those points of the basis of R^+-versal deformations for which the function $\mathscr{F}(\cdot,\lambda)$ has multiple critical values. The bifurcation diagram of functions is obtained by projection of the set of singular points of the bifurcation diagram of zeros along the axis of the free term of the R-miniversal deformation (see Fig. 7). The inverse image of a point not lying on the bifurcation diagram of the function is a line parallel to this axis, and intersects the bifurcation diagram of zeros in a finite number of points, equal to the number of (nondegenerate) critical points of the corresponding function. For the A_μ singularity the caustic is the same as the bifurcation diagram of zeros of the $A_{\mu-1}$ singularity, i.e., a generalized swallowtail. This corresponds to a semicubical parabola in Fig. 7. Caustics of classes $D_4^{\pm}$ lying in three-dimensional space, have special names. They are shown in Fig. 15. Characteristic sections of the bifurcation diagrams of functions for A_4 and $D_4^{\pm}$ are shown in Fig. 16.

Two families of functions are said to be *R^+-equivalent* if they become R-equivalent after adding to one of them a constant that depends smoothly on a parameter. As in R-theory, the versality theorem holds: any deformation of a singular point of a function is R^+-equivalent to one induced from an R^+-versal deformation.

Families of functions in general position, depending on $l \leqslant 5$ parameters, have only simple singularities of multiplicity no higher than

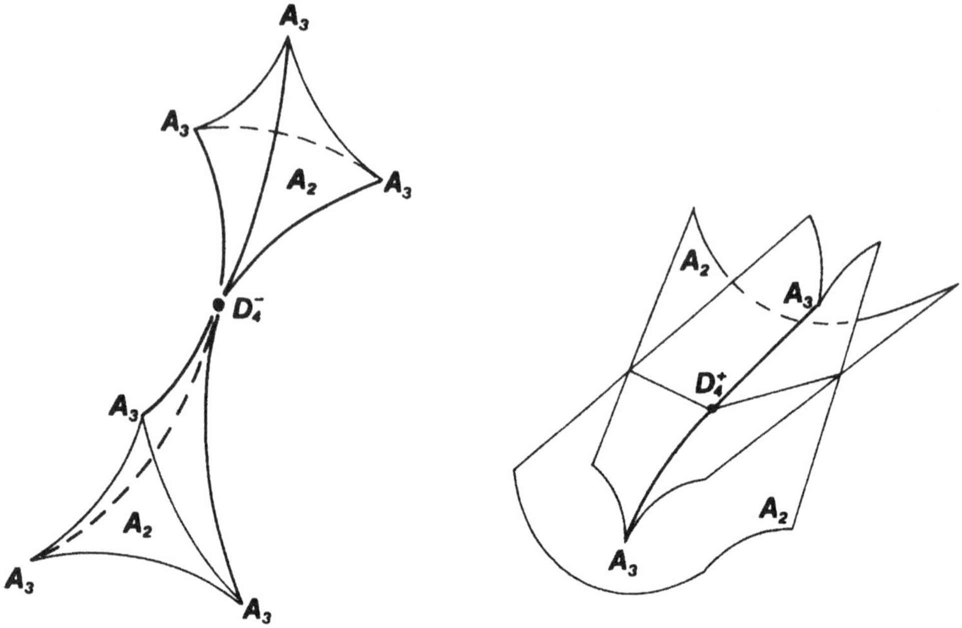

Figure 15

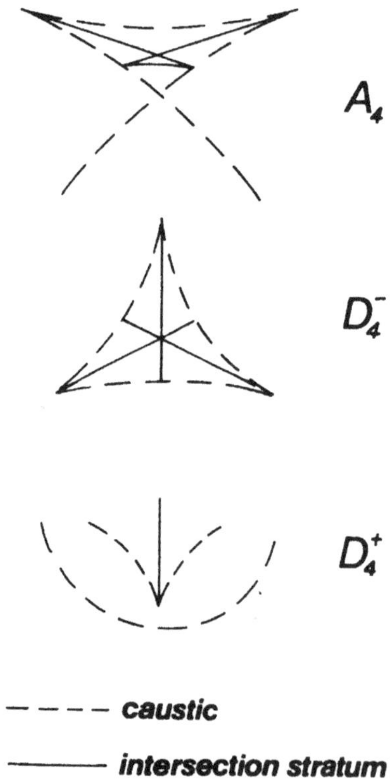

Figure 16

$l+1$. Such families are stable—all neighboring families of functions are R^+-equivalent. For families with a larger number of parameters this is no longer the case. Let us consider as an example the R^+-miniversal deformation of a P_8 singularity (see par. 1.2.4):

$$x^3 + y^3 + z^3 + \lambda_1 xyz + \lambda_2 xy + \lambda_3 yz + \lambda_4 zx + \lambda_5 x + \lambda_6 y + \lambda_7 z.$$

In the family obtained from the miniversal for $\lambda \equiv 0$, the singularity of the function $x^3 + y^3 + z^3$ is removable, but in any neighboring family, e.g., for $\lambda \equiv \text{const} \neq 0$, one finds, instead of it, a singularity of $x^3 + y^3 + z^3 + \lambda_1 xyz$ having the same multiplicity, and differing only in the value of the family modulus λ_1.

We define the *codimension* of a singular point as the codimension in the space of functions with critical point 0 of the set of functions with singular point lying either in the same R^+-equivalence class as the given singularity or in a class differing from the first only in the values of the moduli. One can show (A. M. Gabrielov) that the codimension, multiplicity μ, and modality m of a singular point of a function are connected by the relation $c = \mu - m - 1$. Thus, for singularities of type P_8 the codimension is 6, and these singularities are nonremovable from 6-parameter families of functions.

1.2.6 Singularities of a Hypersurface and V-equivalence of Functions

A *hypersurface* in a space is a set with codimension one; it is given by a single equation $f(x) = 0$. If a point of the hypersurface is not a singular point of the function defining the surface, then by the implicit function theorem, the hypersurface is smooth in the neighborhood of the point. The singular points of the hypersurface with $f = 0$ are those singularities of the function f with zero critical value. But the hypersurface can also be given by a different equation $g(x) \cdot f(x) = 0$, if the function g does not vanish anywhere. We say that the critical points of functions with zero critical values are *V-equivalent* if after a suitable change of variables these functions will differ only by a nonvanishing factor.* The classification of the singular points of hypersurfaces reduces to the V-classification of the critical points of functions.

---

* V-equivalence is an equivalence of the zero level varieties. V is the first letter of the word "variety."

In considering families of hypersurfaces in general position, we again go to the concept of versal deformation. Thus, the general 3-parameter family of plane curves in the neighborhood of a singular point of the curve $y^2 = x^4$ can be brought to the form $y^2 = x^4 + \lambda_1 x^2 + \lambda_2 x + \lambda_3$. The V-miniversal deformation of a finite-multiplicity singular point of the function f has the form $f + \lambda_1\varphi_1 + \cdots + \lambda_k\varphi_k$, where $\varphi_1, \ldots, \varphi_k$ are "remainders from division" of functions defined in the neighborhood of the critical point by the function f and its partial derivatives, i.e., they are obtained by dropping those terms of the R-miniversal deformation which have the same remainders after division by the partial derivatives as functions that are multiples of f.

A function is said to be *quasihomogeneous* if its arguments $x_1, \ldots, x_n$ can be assigned positive (quasi) degree or weights such that all monomials appearing in the Taylor series of this function assume the same quasidegree. The normal forms of simple, parabolic, and exceptional (for $a = 0$) singularities introduced in pars. 1.2.2 and 1.2.4 have this property. For example, the function $f(x, y) = x^2y + y^{\mu-1}$ becomes quasihomogeneous if we set $\deg y = 1/(\mu - 1)$, $\deg x = (\mu - 2)/(2\mu - 2)$. K. Saito has shown that a function in the neighborhood of a singular point is divisible by its partial derivatives if and only if it is R-equivalent to a quasihomogeneous function [for example: $f(x, y) = (\mu - 2)/(2\mu - 2) \cdot x\partial f/\partial x + 1/(\mu - 1) \cdot y\partial f/\partial y$]. Thus, V-versal deformations of quasihomogeneous singularities coincide with R-versal deformations. One can show that in the nonquasihomogeneous case the bifurcation diagram of zeros of an R-versal deformation is obtained from the bifurcation diagram of zeros of a V-versal deformation by taking a Cartesian product with the space of missing parameters.

Two families of functions are said to be *V-equivalent* if they become R-equivalent after multiplication of one of them by a nonvanishing function that depends smoothly on the parameters.

The versality theorem is valid also in V-theory: any deformation of a singular point is V-equivalent to one induced from a V-versal deformation.

In families of hypersurfaces with $l \leqslant 6$ parameters, only simple singular points with multiplicity $\mu \leqslant l$ are unremovable. General families of hypersurfaces with no more than 6 parameters are stable.

The V-classification of singular points of functions differs from the R-classification. For example, in the classes of simple functions there is an identification $A_\mu^+ \sim A_\mu^-$, $D_\mu^+ \sim D_\mu^-$, $E_6^+ \sim E_6^-$. Thus the simple classes of stable V-equivalence of singular points of smooth functions form two infinite series A_μ and D_μ and three classes E_μ. Richer examples are given by the exceptional singularities. For example, in the family K_{12}, for nonzero values of the modulus a the functions $x^3 + y^7 + axy^5$ are V-equivalent to each other (prove!) and are V-nonequivalent to the quasihomogeneous function $x^3 + y^7$.

1.2.7 Boundary Singularities By a submanifold *with a boundary* we mean a half-space selected from the space R^{n+1} with coordinates $x, y_1, \ldots, y_n$ by the inequality $x \geqslant 0$. A point $x = 0$ of the boundary is said to be a *boundary singular* point for a smooth function of the variables x, y, if the level surface of the function that passes through it is not transverse to the boundary (in other words, this point is critical, in the usual sense, for the restriction of the function to the boundary). Two boundary singular points of a function are said to be *equivalent* if the functions in the neighborhood of these points go over into each other under a smooth change of variables that takes the boundary and the half-space $x \geqslant 0$ into themselves. A *simple* boundary singular point with zero critical value is stably equivalent to the singular point 0 of one of the following functions (V. I. Arnol'd):

$$A'_\mu, D'_\mu, E'_\mu : f(x, y_1, y_2) = \pm x + g(y_1, y_2);$$

$$B_\mu : f(x) = \pm x^\mu, \qquad \mu \geqslant 2; \qquad C_\mu : f(x, y) = xy \pm y^\mu, \qquad \mu \geqslant 3;$$

$$F_4 : f(x, y) = \pm x^2 + y^3,$$

where the function g has a simple singular point (of type A_μ, D_μ, or E_μ) at the origin. Miniversal deformations of the simple singular points have the form

$$B_\mu : \pm x^\mu + \lambda_1 x^{\mu-1} + \cdots + \lambda_\mu, \quad C_\mu : xy \pm y^\mu + \lambda_1 y^{\mu-1} + \cdots + \lambda_\mu,$$

$$F_4 : \pm x^2 + y^3 + \lambda_1 xy + \lambda_2 x + \lambda_3 y + \lambda_4, \quad \tilde{A}_\mu, \tilde{D}_\mu, \tilde{E}_\mu : x + \mathscr{F}(y, \lambda),$$

where $\mathscr{F}(y,\lambda)$ is a miniversal deformation of the corresponding singularity of A_μ, D_μ, E_μ.

Functions in general position on a manifold with boundary have only *nondegenerate* singular points (of type A_1 outside the boundary or type A_1' on the boundary). The corresponding *bifurcation diagram of zeros* for the boundary singularity consists of two hypersurfaces: on one of them lie those points λ of the base of miniversal deformation for which the manifold of the zero level of the function $\mathscr{F}(\cdot,\lambda)$ of the miniversal family is singular, on the other lie those points λ for which this manifold is nontransversal to the boundary. Thus, for the miniversal deformation $\mathscr{F}(x,\lambda_1,\lambda_2) = \pm x^2 + \lambda_1 x + \lambda_2$ of a B_2 singularity, the first hypersurface is given by the conditions $x \geqslant 0$, $\partial\mathscr{F}/\partial x = \pm 2x + \lambda_1 = 0$, $\mathscr{F} = \pm x^2 + \lambda_1 x + \lambda_2 = 0$, from which we find $\lambda_2 = \pm\lambda_1^2/4$, $\pm\lambda_1 \leqslant 0$ while the second hypersurface is given by the conditions $x = \mathscr{F} = 0$, from which $\lambda_2 = 0$, i.e., the bifurcation diagram of zeros of B_2 is a semiparabola, tangent to the line. For the B_3 and C_3 singularities, the bifurcation diagrams of zeros are shown in Fig. 17. One can show that the complex bifurcation diagram of zeros looks the same for C_μ as for B_μ, but the first and second hypersurfaces interchange roles. The adjacencies of the simple classes of singular points of functions on a manifold with boundary are shown in Fig. 18.

In the families of functions on manifolds with boundary with $l \leqslant 3$ parameters, the only unremovable singularities are those simple ones with multiplicity $\mu \leqslant l+1$. Nonsimple singular points are unremovable, starting with 4-parameter families of functions. As in par. 1.2.5, we can define R^+-versal deformations of boundary singularities. Pictures of caustics of certain boundary singularities are shown in par. 2.1.5.

The theory of singularities of functions on a manifold with boundary includes the theory of functions on a manifold without a boundary, since the function can also have singularities outside the boundary. On the other hand, the substitution $x = z^2$ converts a function on a manifold with boundary $x = 0$ into a function on a manifold without a boundary, even in the variable z; a change of variables (x, y), which takes the boundary and the half-space $x \geqslant 0$ into themselves, goes over into the change of variables (z, y) taking symmetric points into symmetric points. Therefore, the theory of

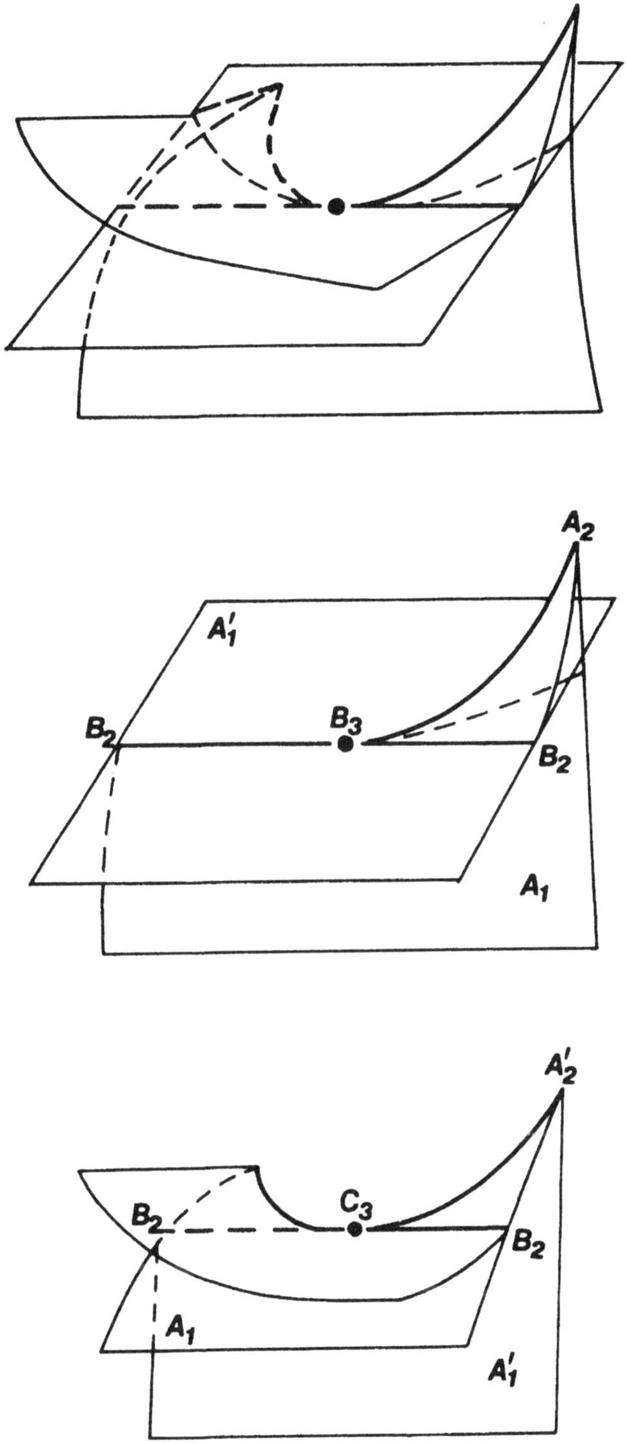

Figure 17

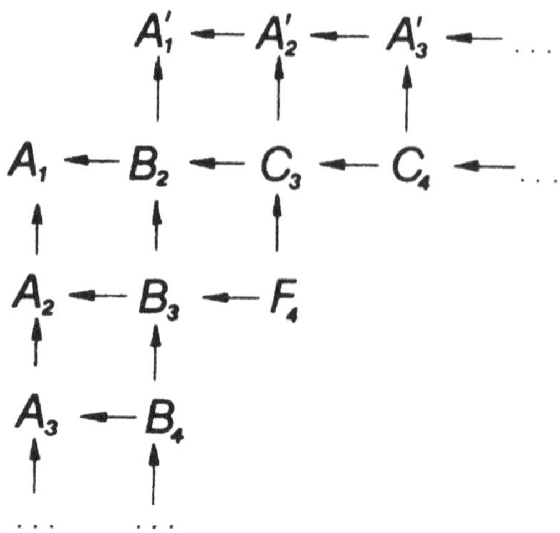

Figure 18

boundary singularities of functions is equivalent to the theory of singularities of even functions (with the corresponding understanding of equivalence).

1.3 Singularities of Wave Fronts

1.3.1 Legendre Mappings A wave front in the plane is a curve with singular points. It turns out that it can be realized as the projection of a *smooth* curve in 3-space onto the plane. Singularities of the front arise where this curve is poorly placed relative to the projection direction. For example, consider the cusp of the wave front $y = \pm x^{3/2}$ (Fig. 19). In the space with coordinates x, y, p, consider the curve $y = \pm x^{3/2}$, $p = dy/dx = \pm \frac{3}{2} x^{1/2}$. It is smooth(!), since it is

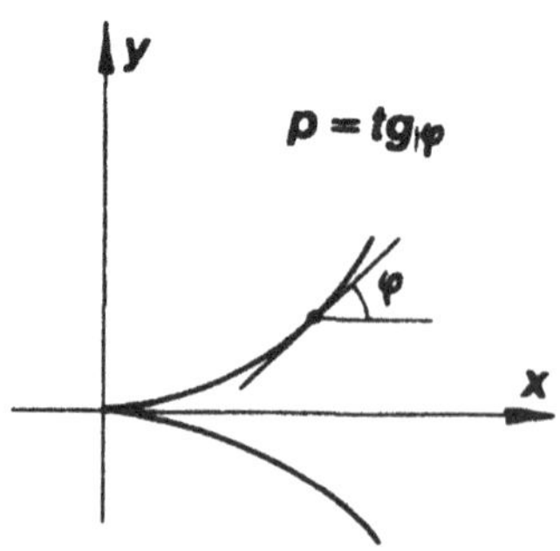

Figure 19

given by the equations $x = 4/9p^2$, $y = 8/27p^3$. The point (x, y, p) of three-dimensional space determines a contact element of the plane—the tangent line with equation $dy = p\,dx$, attached at the point (x, y). Our smooth curve is called a *lift* of the semicubical parabola into the space of contact elements.

We note that the projection of a smooth curve in general position from space onto the plane has as singularities only self-intersections and no cusps. The singularities of wave fronts are singularities of the special class of smooth mappings of n-dimensional manifolds into $(n + 1)$-dimensional ones.

A *contact element* of the space $\mathbb{R}^{n+1}$ is a tangent hyperplane attached at any point. It is given by the equation $dy = p_1 dq_1 + \cdots + p_n dq_n$, where $y, q_1, \ldots, q_n$ is a coordinate system in $\mathbb{R}^{n+1}$, while $p_1, \ldots, p_n$ are the angular coefficients of the hyperplane. The manifold E of all contact elements has dimension $2n + 1$. In the space tangent to E at the point (q, y, p) the equation $dy = p\,dq$ gives a $2n$-dimensional *contact hyperplane*. The field of contact hyperplanes on E actually is independent of the choice of coordinates (q, y) in $\mathbb{R}^{n+1}$.*

The *lift* of a smooth hypersurface in $\mathbb{R}^{n+1}$ consists of all the contact elements tangent to it and has dimension n. Such a hypersurface is locally the graph of a function $y = f(q)$. Its lift is given by the auxiliary equation $p = \partial f/\partial q$, and consequently at each of its points is tangent to the contact hyperplane $dy = p\,dq$. Smooth n-dimensional submanifolds of the space of contact elements, tangent to the field of contact hyperplanes, are called *Legendre manifolds*. It is easy to verify that the fibres of the projection of the space of contact elements in $\mathbb{R}^{n+1}$, consisting of contact elements attached at a given point of the space $\mathbb{R}^{n+1}$, are Legendre fibres. The projection of the space of contact elements into the $(n + 1)$-dimensional space (basis), whose fibres are Legendre fibres, is said to be a *Legendre fibration*.

Suppose that we are given a Legendre submanifold in the space E of a Legendre fibration. Its projection on the base of the fibration is called a *Legendre mapping*.

---

*This field is defined intrinsically as follows: The velocity vector of a moving contact element lies in a hyperplane of the field if the velocity vector of the point of application of the contact element belongs to the contact element.

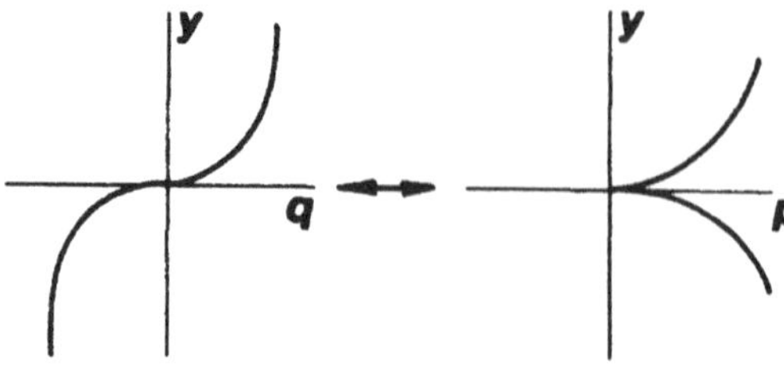

Figure 20

Examples of Legendre mappings: (1) In addition to the space $\mathbb{R}^{n+1}$, we consider the $n + 1$-dimensional space of all hyperplanes in $\mathbb{R}^{n+1}$ (not necessarily passing through the origin). A mapping that associates to a point of a smooth hypersurface in $\mathbb{R}^{n+1}$ the tangent hyperplane at that point is a Legendre mapping. Its image—a hypersurface in the space of all hyperplanes, is called the *Legendre transform* of the original hypersurface and consists of all hyperplanes tangent to it (see Fig. 20).

(2) The mapping that associates to a point of a hypersurface in Euclidean space the end of the exterior normal vector of length t, is a Legendre mapping. This mapping is called a *normal mapping* and its image is called the *equidistant* of the original hypersurface.

The singularities of a Legendre transformation and of the normal mapping of a generic hypersurface are the same as those of a Legendre mapping in general position.

The motion of the wave front in space corresponds to the motion of its lift—a Legendre manifold in the space of contact elements. If the initial front is smooth, then its Legendre manifold remains smooth, even when the front acquires singularities. Thus, *the singularities of wave fronts are the singularities of projections of Legendre submanifolds of the space of contact elements under a Legendre mapping*.

Under a diffeomorphism (smooth mapping) of the space $\mathbb{R}^{n+1}$, the tangent vectors are also transformed, and consequently also the contact elements, so that we may speak of the *lift of a diffeomorphism* of the space $\mathbb{R}^{n+1}$ to a diffeomorphism of the space of contact elements. We shall say that two Legendre mappings are *equivalent* if their projected Legendre manifolds go over into each other under a lift diffeomorphism of the space $\mathbb{R}^{n+1}$. It is obvious that when there is equivalence of Legendre mappings, their fronts also go over into each other.

1.3.2 Generating Families of Hypersurfaces A Legendre manifold, which projects nonsingularly on the base of the Legendre fibration, is given by its front $y = f(q)$ as follows:

$$y = f(q), \qquad p = \partial f/\partial q.$$

It turns out that an arbitrary Legendre mapping can be given analogously by means of a *generating family of hypersurfaces*. It is constructed as follows. Let I be a subset of the set of indices $\{1, \ldots, n\}$, such that the projection $(q, y, p) \mapsto (q_J, y, p_I)$ (where J is the set of residual indices) is nonsingular for the given Legendre submanifold in the neighborhood of the given point (it always exists). The equality

$$d(y - p_I q_I) = -q_I dp_I + p_J dq_J$$

shows that the projection $(q, y, p) \mapsto (-q_J, p_I, y - p_I q_I)$ of the space of contact elements is a Legendre fibration in which our Legendre submanifold has a nonsingular front. Suppose that $y - p_I q_I = f(p_I, q_J)$ is the equation of this front. Then the Legendre manifold is given by the conditions

$$y - p_I q_I = f(p_I, q_J), \qquad q_I = -\partial f/\partial p_I, \qquad p_J = \partial f/\partial q_J.$$

The family $y = \mathscr{F}(p_I, q) = f(p_I, q_J) + p_I q_I$ of hypersurfaces in p_I-space with parameters (q, y) is called a *generating family* of the Legendre mapping whose Legendre manifold is defined by the equations

$$y = \mathscr{F}(p_I, q), \qquad \partial \mathscr{F}/\partial p_I = 0, \qquad p_J = \partial \mathscr{F}/\partial q_J,$$

and whose front consists of those points of the parameter space for which the corresponding hypersurface of the family is singular, i.e., the front is the bifurcation diagram of zeros of the family $y = \mathscr{F}$! In particular, the wave front in general position is the bifurcation diagram of zeros of the family in general position.

Examples and remarks: (1) The family $p^3 + qp - y = 0$ determines the Legendre manifold $q = -3p^2$, $y = -2p^3$ and the front $27y^2 + 4q^3 = 0$.

(2) The front of the family $p_1^4 + q_2 p_1^2 + q_1 p_1 - y = 0$ is the surface of a "swallowtail."

(3) The generating family has a simple physical meaning. Suppose that λ is the point of observation, x the source point, and $\mathscr{F}(x,\lambda)$ the optical path length. Then the wave front at the time t is the set of points λ, for which there exists a point x of the source at which $\mathscr{F}_\lambda = t$ and $\partial\mathscr{F}_\lambda/\partial x = 0$.

THEOREM 1 (L. Hörmander, A. Weinstein) Two Legendre mappings are equivalent if and only if their generating families are V-equivalent.

COROLLARIES (1) The stable singularities of wave fronts are exhausted by the singularities of bifurcation diagrams of zeros of V-versal deformations of functions and their transversal intersections.

(2) Singularities in general position of wave fronts of dimension $n \leqslant 5$ in $n+1$-dimensional space are stable, and are exhausted by the bifurcation diagrams of zeros of V-versal deformations of singularities of functions of types A_μ, D_μ, and E_μ with $\mu \leqslant n+1$.*

(3) The singularities of wave fronts in general position in three-dimensional space are exhausted by cuspidal edges (A_2) and by singularities of the swallowtail type (A_3).*

(4) Wave fronts in general position of dimension $n \geqslant 6$ can be unstable, at least at individual points.

1.3.3 Metamorphoses of Wave Fronts Let us consider a time-dependent wave front. The Cartesian product of spaces which contain the front along the time axis is called *space–time*, whose projection on the time axis is the time function, and the union of all instantaneous fronts is called a *grand front*. The grand front is the front of a certain Legendre mapping in space–time. Suppose that the instantaneous front at the time t is given by the generating family of functions $\mathscr{F}_t(x,\lambda)$ (where λ is a space point). Regarding t as a new parameter, we again determine the formula for the Legendre manifold in the space of contact elements of space–time. Its projection into space–time is the grand front.

---

*It is of course still possible to have transversal intersections of different branches of wave fronts.

We already know that the grand front of a family of fronts in general position, for low dimension, is the bifurcation diagram of zeros of a simple singularity of functions. Thus we must study the generic decompositions of bifurcation diagrams into the level sets of the time function.

Two functions (time functions) in the base of R-versal deformations of the critical point are said to be *equivalent* if they go into each other under a suitable diffeomorphism of the base, preserving the bifurcation diagram of zeros, and under addition to one of them of a suitable constant.

We choose as the miniversal deformations of simple quasihomogeneous singularities the deformations given in par. 1.2.2, so that λ_1 is the coefficient of the "remainder" of largest quasihomogeneous degree. A *special metamorphosis* of the front is one in which the space–time has coordinates $\lambda_1, \ldots, \lambda_\mu, \tau_1, \ldots, \tau_m$, the grand front is

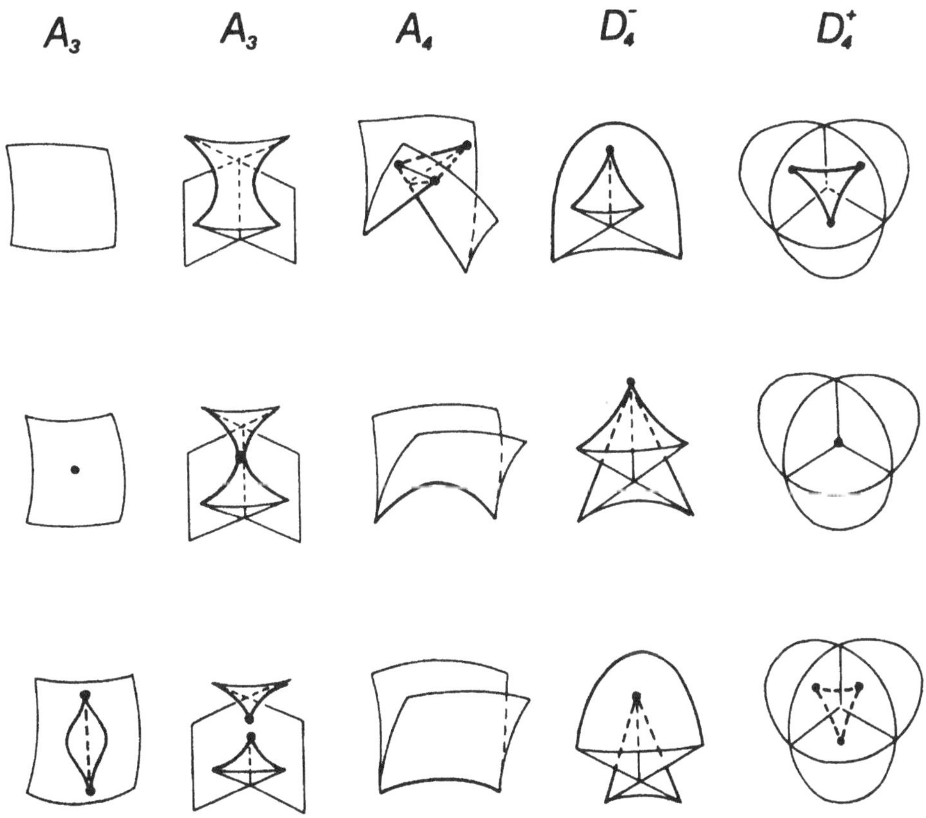

Figure 21

the direct product of the bifurcation diagram of zeros in the space of parameters λ on the parameter space τ, while the time function has the form

$$t = \tau_1 \quad \text{or} \quad t = \pm\lambda_1 \pm \tau_1^2 \pm \cdots \pm \tau_m^2 .$$

THEOREM 2 (V. I. Arnol'd) A generic metamorphosis of wave fronts in spaces of dimension $l \leqslant 5$ is locally equivalent to the special metamorphosis in which $\mu + m = l + 1$.

Under the special metamorphosis with time function $t = \tau_1$, the front is actually not changed. Metamorphosis of the second type for a front in three-dimensional space is shown in Fig. 21.

1.4 Caustics and their Metamorphosis

1.4.1 Geometry of Phase Space A first-order partial differential equation, i.e., a relation $G(x, y, \partial y/\partial x) = 0$ between independent variables x, the function $y(x)$ and its partial derivatives, can be interpreted as a hypersurface $G(q, y, p) = 0$ in the space of contact elements equipped with the field of contact hyperplanes $dy = p\,dq$. The lifts of graphs of solutions of such an equation are the Legendre submanifolds of this hypersurface. An equation in which the required function y does not appear explicitly is called a Hamilton–Jacobi equation. The function G of the coordinates q of configuration space and the momenta p is the Hamiltonian of the mechanical system corresponding to this equation. Just as the mathematical apparatus for our study of wave fronts was the geometry of the space of contact elements, so in the theory of the singularities of caustics we shall deal with the geometry of phase space of Hamiltonian mechanics.

All the concepts of contact geometry have analogs in Hamiltonian mechanics. The role of the field of hyperplanes $dy = p/dq$ is taken by the *action form* $p\,dq$. The Legendre manifolds for projection along the y axis (Fig. 22) go over into *Lagrangian manifolds*—smooth submanifolds with dimension half that of the phase space, on which the action form is locally the differential of a function (in other words, the action integral $\oint p\,dq$ around any small contour is zero). For example,

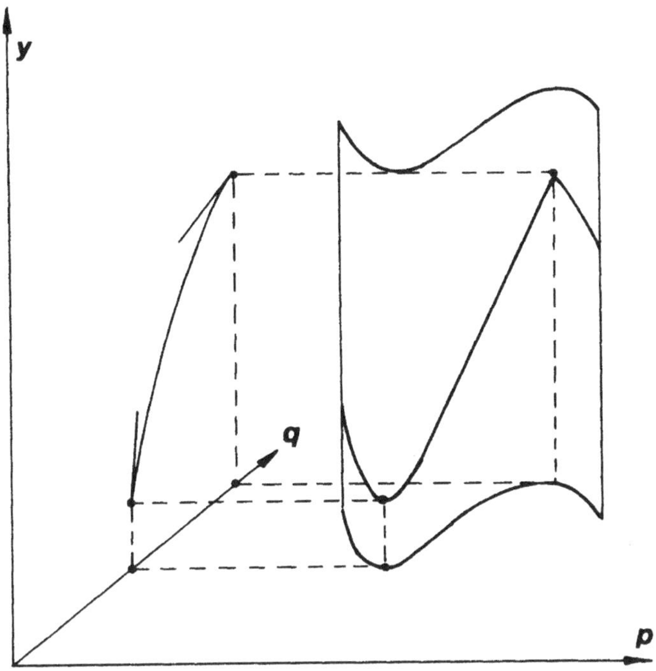

Figure 22

any curve in the phase plane is a Lagrangian curve. The fibration of phase space into Lagrangian fibres is called a *Lagrangian fibration*. An example of a Lagrangian fibration is the projection $(q, p) \mapsto q$ of phase space onto configuration space with fibres $q = \text{const}$.

Suppose that there is a wave front moving in the configuration space. In the extended phase space (q, p, t) this motion corresponds to a "grand" Legendre manifold (t plays the role of y), and in the phase space (its projection) a Lagrangian manifold (Fig. 22). A subset in q-space marked by singularities of the moving front (the two points in Fig. 22) coincides with the critical values of the projection of the Lagrangian submanifold onto the basis of the Lagrangian fibration. We call such a projection a *Lagrangian mapping*, and the set of its critical points, a *caustic*.

1.4.2 Examples of Lagrangian Mappings (1) The mapping of a hypersurface of Euclidean space onto the unit sphere of that space, which associates to a point of the hypersurface the unit normal vector at that point, shifted to the coordinate origin, is called a *Gaussian*

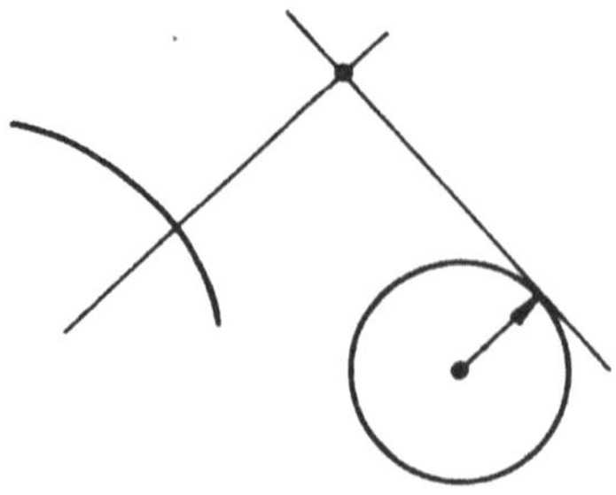

Figure 23

mapping. We represent the Gaussian mapping as the composition of two mappings: the first takes a point of the hypersurface into the point of intersection of the normal with the tangent space to the sphere that is orthogonal to it (Fig. 23), while the second consists of projecting the tangent space of the sphere at its point of application and is a Lagrangian fibration of the phase space of all tangent vectors to the sphere onto the configuration space. One can show that the image of the first mapping is a smooth Lagrangian submanifold of the phase space.

(2) The mapping that associates its terminus to a vector normal to a submanifold of Euclidean configuration space (the vector p, applied at the point q, corresponds to the point $p + q$) is called a *normal mapping*. The set of all normal vectors to the submanifold is a Lagrangian submanifold of the phase space of all vectors, tangent to the configuration space. The critical values of the normal mapping, which are the centers of curvature of the original submanifold, form the caustic of the family of its equidistants (see Fig. 3, in which the submanifold is an ellipse).

(3) Suppose that we are given a function f in Euclidean configuration space. The mapping that associates to a point of the configuration space the value of the gradient of the function at the point is called the *gradient mapping*. The submanifold of phase space consisting of all vectors tangent to the configuration space consisting of the points $(q, p = \partial f/\partial q)$ is a Lagrangian, since the action is the differential of the function f on it: $p\,dq = \partial f/\partial q\,dq = df$. The projection $(q, p) \mapsto p$ of the phase space onto p-space is a Lagrangian fibration.

The Gaussian, normal, and gradient mappings are special classes of mappings of manifolds of the same dimension. One can show that the typical singularities of these three classes coincide with the typical singularities of Lagrangian mappings and, generally speaking, differ from the typical singularities of general mappings of manifolds of the same dimension.

1.4.3 Generating Families of Functions Suppose that we are given a Lagrangian submanifold in the space of a Lagrangian fibration. The function y on this manifold, whose differential is the action $p\,dq$ (in the neighborhood of the particular point) determines the Legendre submanifold of the extended phase space with coordinates (q, y, p) and field of contact hyperplanes $dy = p\,dq$. The generating family of hypersurfaces $y = \mathscr{F}(p_I, q)$ of this Legendre manifold can be regarded as a family of graphs of the family of functions $\mathscr{F}(p_I, q)$. The family $\mathscr{F}$ is called the *generating family of functions* of the given Lagrangian mapping. In terms of its generating family, the Lagrangian manifold is defined by the equations:

$$\partial\mathscr{F}(p_I, q)/\partial p_I = 0, \qquad p_J = \partial\mathscr{F}/\partial q_J$$

(Here I and J are a decomposition of the indices $\{1, \ldots, n\}$, such that the functions (p_I, q_J) form a coordinate system in our Lagrangian manifold at the point considered), while the caustic consists of those points of the configuration space of parameters for which the functions of the family have degenerate singular points, i.e., it is exactly the caustic of the bifurcation diagram of the functions of the family $\mathscr{F}$.

Just as in the case of wave fronts, the physical meaning of the generating family $\mathscr{F}(x, q)$ is that it is the optical path length from the source point x to the observation point q.

We say that two Lagrangian mappings are *Lagrange equivalent* if there exists a diffeomorphism of the phase space that takes the fibres of the Lagrangian fibration into fibres that preserve the action integral over any closed contour and which maps the first Lagrangian submanifold onto the second. Since Lagrange equivalence takes the fibres of a Lagrangian fibration into fibres, it induces a diffeomorphism of the base of the fibration, taking one caustic into another. The converse is not true: from diffeomorphism of caustics it does not follow that the Lagrange mappings are Lagrange equivalent.

Theorem 3 (L. Hörmander, A. Weinstein) Lagrange mappings are Lagrange equivalent if and only if their generating families of functions are R^+-equivalent.

Corollaries (1) The generating families of functions of stable local Lagrangian mappings are exhausted by the R^+-versal deformations of singularities of the functions.

(2) Singularities of caustics in general position in a space of dimension $n \leqslant 5$ are stable and exhausted by the caustics of R^+-versal deformations of singularities of functions such as A_μ, D_μ, and E_μ, with $\mu \leqslant n + 1$.*

(3) Singularities of caustics in general position in three-dimensional space are exhausted by the cuspidal edges (A_3) and by the singular points such as the "swallowtail," "pyramid," and "purse" (A_4, D_4^-, D_4^+, see Fig. 15).*

(4) Caustics in general position in a space of dimension $n \geqslant 6$ may be unstable, at least at individual points.

When the wave front moves, its cuspidal edges slip along the caustic. We obtain a decomposition of the caustic by isochrones. In Fig. 24 we show such a decomposition of the swallowtail (see Fig. 21,

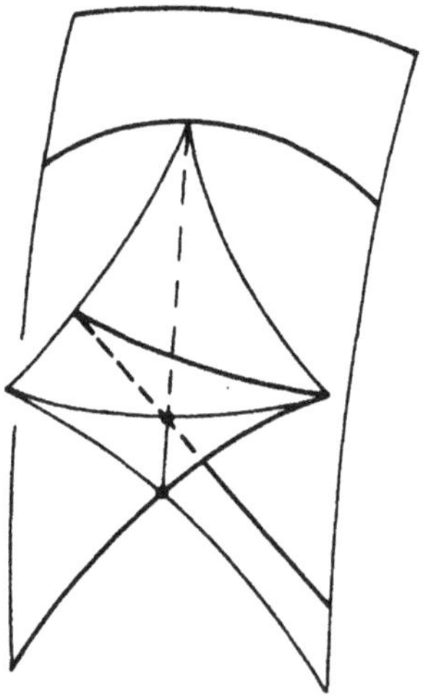

Figure 24

*Transverse crossings of different branches of caustics are also possible.

A_4). This decomposition of the swallowtail into curves differs from the decomposition of the "grand front" in Fig. 2 into instantaneous plane fronts: the cuspidal edges of the moving front do not have self-intersections, and pass twice through each point of self-intersection of the swallowtail, with a time interval of the order of $t^{5/2}$, where t is the distance to the vertex of the swallowtail.

1.4.4 Metamorphoses of Caustics Let us consider a medium of noninteracting and noncolliding particles, moving in configuration space by inertia with a velocity field $v(q)$. After a time t, the particle from point q goes to the point $q + t \cdot v(q)$. We obtain a family of mappings g_t of the configuration space into itself, which is a gradient mapping if the velocity field is potential:

$$v = \partial S/\partial q, \qquad g_t : q \mapsto \partial\left(q^2/2 + t \cdot S(q)\right)/\partial q.$$

Let us assume that the particles of the medium, as before, do not collide, but are moving in a potential force field, which possibly depends on the time. Since during the motion of a Hamiltonian system the action integral over a closed contour is conserved, the mapping through the time t, which is determined by the motion of our system, will be a Lagrangian mapping, provided that the initial velocity field is potential.

If the particles at the initial moment were distributed in space with a positive density, then the density will change as a result of the motion. The faster particles will begin to overtake the slower ones (Fig. 25) and, after the time t, the density ρ will become infinite at the points of the caustic of the mapping for time t (i.e., at the points of

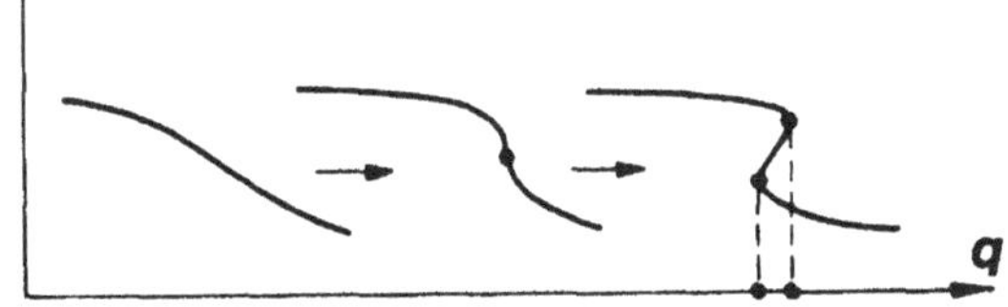

Figure 25

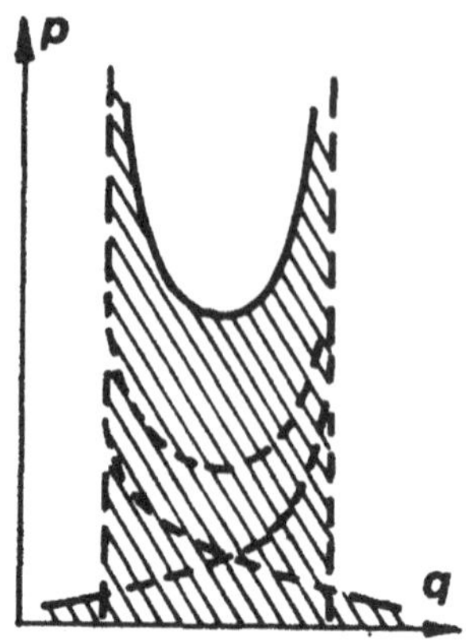

Figure 26

collision of infinitely close particles, Fig. 26). We thus arrive at the problem of describing the metamorphosis of the caustic in time.*

Considering the time as a new parameter of the generating family of functions $\mathscr{F}_t$ of the Lagrangian mapping, we get the generating family of a "grand" Lagrangian mapping of space-time into itself. The problem of describing the general metamorphoses of instantaneous caustics then reduces to the problem of describing the metamorphoses of sections of the "grand caustic" in general position in space–time by the level hypersurfaces of the time function. Two time functions are said to be *equivalent* if they go over into each other under some diffeomorphism of space–time that preserves the "grand caustic" and under a diffeomorphism of the time axis. A list of normal forms of time functions for cases when the grand caustic has singularities A_μ has been calculated by V. I. Arnol'd, and for D_μ by V. M. Zakaljukin. The grand caustic can be described as the set of critical values of the mapping $(x, \lambda, \tau) \mapsto (y = \partial\mathscr{F}/\partial x, \lambda, \tau)$, where in case A_μ

$$\mathscr{F} = \pm x^{\mu+1} + \lambda_1 x^{\mu-1} + \cdots + \lambda_{\mu-2} x^2, \qquad \lambda \in \mathbb{R}^{\mu-2}, \quad \tau \in \mathbb{R}^m,$$

* According to Ya. B. Zeldovich, at an early stage in the existence of the universe, the velocities of the particles formed a potential field, and the appearance of caustics during further motion of the particles is responsible for the present large-scale clumped structure of the universe (extreme nonuniformity in the distribution of the accretions of galaxies).

and in case D_μ

$$\mathcal{F} = x_1^2 x_2 \pm x_2^{\mu-1} + \lambda_1 x_2^{\mu-2} + \cdots + \lambda_{\mu-3} x_2^2, \qquad \lambda \in \mathbb{R}^{\mu-3}, \quad \tau \in \mathbb{R}^m$$

(where the dimension of the space–time is $\mu - 1 + m$).

THEOREM 4 (1) The time function t in general position can be reduced to the form

$$\text{for } A_\mu : t = \tau_1 \quad \text{or} \quad t = \pm\lambda_1 \pm \tau_1^2 \pm \cdots \pm \tau_m^2,$$

$$\text{for } D_\mu : t = \tau_1 \quad \text{or} \quad t = \pm\lambda_1 + y_1 + a\lambda_2 \pm \tau_1^2 \pm \cdots \pm \tau_m^2$$

by a caustics metamorphoses equivalence (for $\mu = 4$, $a\lambda_2$ must be replaced by ay_2, $y_k = \partial\mathcal{F}/\partial x_k$).

(2) In general one-parameter families of caustics in spaces of dimension $n \leqslant 4$, one meets only metamorphoses of types A_μ and D_μ with $\mu - 2 + m = n$.

In Figs. 27, 28 and 29 we show metamorphoses of caustics in general position in the plane and in space.

Returning to the consideration of a collisionless medium, we shall characterize the singularity of the caustic of the transformation after time t by the average density in an ϵ-neighborhood of the point we

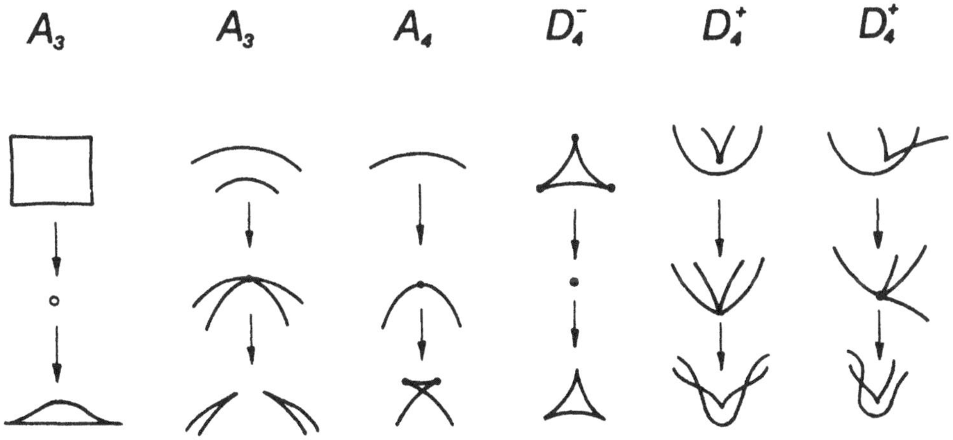

Figure 27

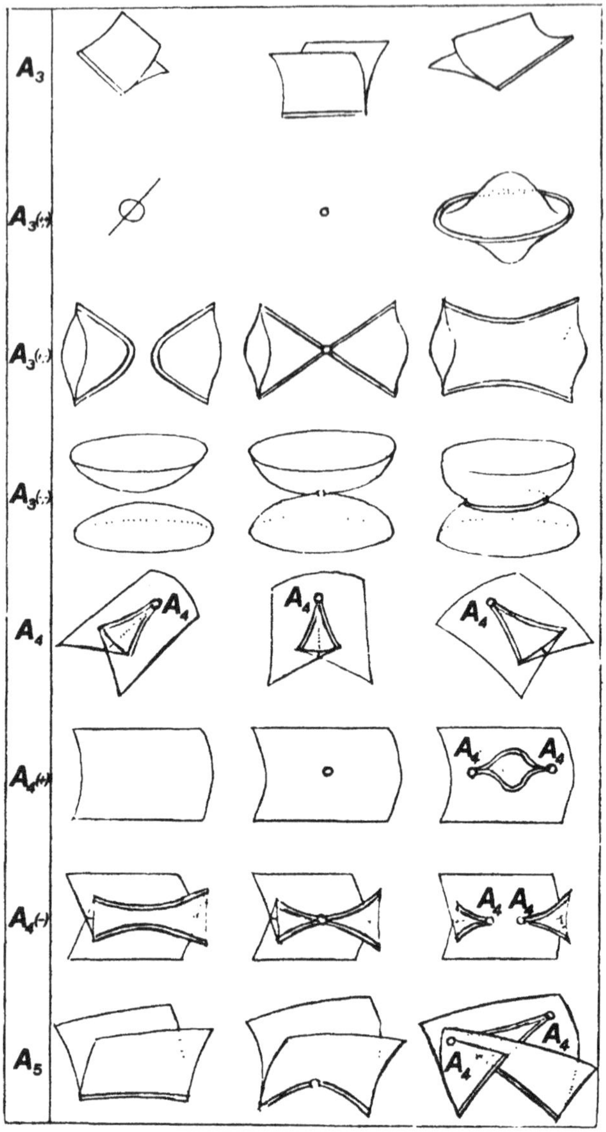

Figure 28

consider (the ratio of the mass in this neighborhood to its volume). At points of the caustic the average density tends to infinity, when the radius of the neighborhood ϵ goes to zero. In Table II we give the order of magnitude of the average density at different points of the caustic. Singularities A_5 and D_5 appear only at individual times.

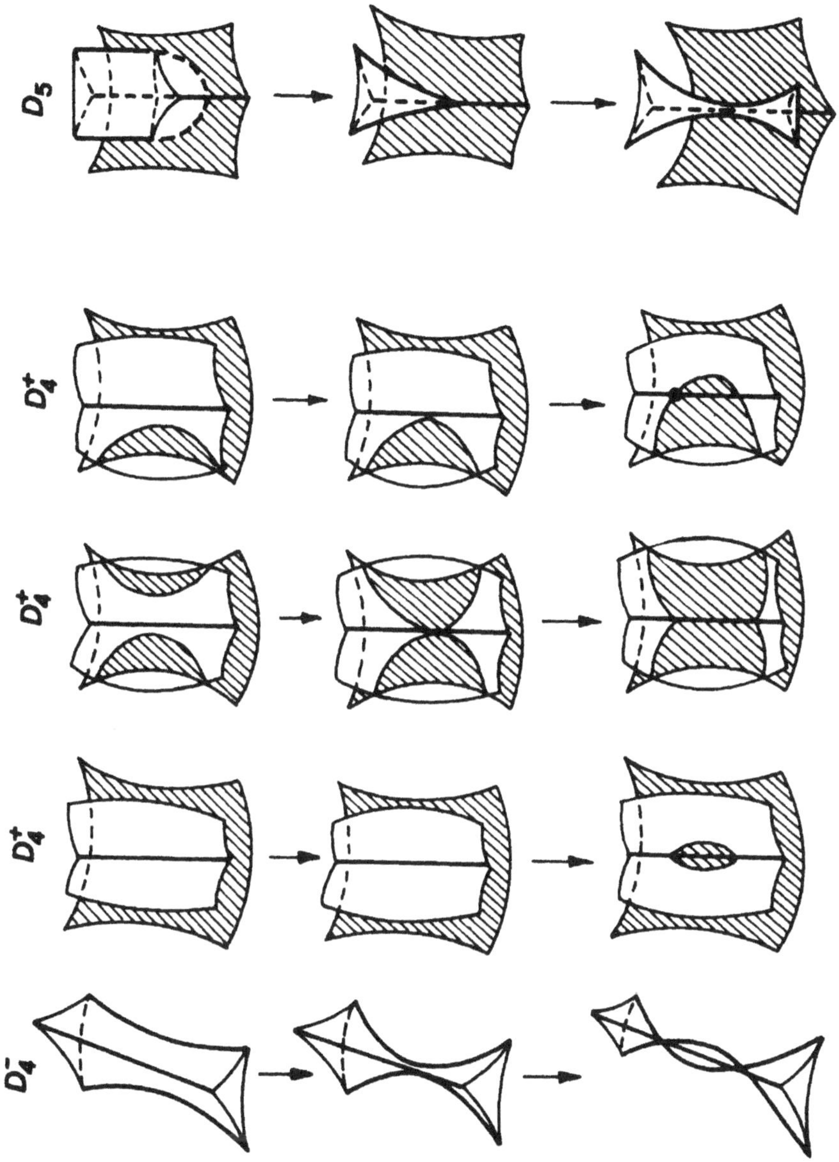

Figure 29

Table II

Caustic	Cusp	Swallowtail	Pyramid Purse	A_5	D_5
$\epsilon^{-1/2}$	$\epsilon^{-2/3}$	$\epsilon^{-3/4}$	ϵ^{-1}	$\epsilon^{-4/5}$	$\epsilon^{-1}\ln\epsilon^{-1}$

§2 Integrals of the Stationary Phase Method

This section is concerned with the study of the asymptotics of *oscillatory integrals*, i.e., integrals of the form

$$I(\tau) = \int_{\mathbf{R}^n} e^{i\tau f(x)}\varphi(x)\,dx_1 \ldots dx_n, \tag{1}$$

for large values of the real parameter τ. Here f and φ are smooth functions which are called, respectively, the *phase* and the *amplitude*. According to the principle of stationary phase, the main contribution to the asymptotic comes from the neighborhood of the critical points of the phase. In this section we discuss the application of the theory of singularities to the study of asymptotics, and present results of calculations of asymptotics in the simplest generic situations, and indicate formulas for calculating asymptotics.

2.1 Examples

2.1.1 Oscillation Integrals and Short-Wave Oscillations Problems in optics, acoustics, quantum mechanics, various branches of mathematics (theory of partial differential equations, probability theory, number theory) require analyzing oscillatory integrals for large values of the parameter.

Example: Consider a surface in space. Assume that each point of the surface radiates a spherical wave of fixed frequency and fixed wave length. Assume that the wavelength is small compared to the size of the surface and to the rate of change of the wave amplitude as we vary the point on the surface. We shall be interested in the overall oscillatory regime at space points outside the surface.

We select a point on the surface and describe the sum of the waves coming from a small neighborhood of this point to a fixed point of the space. This sum is determined by the manner in which the distance from the surface point to the space point varies for small changes of the surface point. If the straight line along which the wave propagates from the surface point to the space point is not perpendicular to the surface, then the arriving waves cancel out each other. If the line along which the wave propagates is perpendicular to the surface, then many waves will arrive at the space point from the selected point of the surface with approximately equal phases. The area of the points on the surface from which waves arrive with approximately equal phases is proportional to the wavelength (in fact, this area is proportional to the area of the set of points at which the modulus of a quadratic expression in the local coordinates is less than the wavelength). Consequently, the contribution of this point of the surface to the total oscillatory regime is proportional to the wavelength. It may turn out that the change in distance to the space point for small variations of the surface point has a higher than second order of smallness (this occurs when the quadratic form for the change in distance is degenerate). In this case the area of the set of points of the neighborhood, from which waves with approximately equal phase arrive, is even greater, and is proportional to the wavelength to a lower power than 1.

The total oscillation at the point y is given by the function

$$e^{2\pi i \omega t} \int_S \frac{e^{2\pi i \|x-y\|/\lambda} \varphi(x)}{\|x-y\|}\, dx$$

where t is the time, ω is the frequency, λ is the wavelength, φ is the amplitude, dx is the element of area of the surface, and S is the surface that radiates the wave. Thus the total oscillation is given by an oscillatory integral, in which the role of the large real parameter is played by a quantity inverse to the wavelength, while the phase is the distance function from the surface point to the fixed point in space. The preceding considerations show that the main contribution to the total oscillation (the oscillatory integral) comes from the neighborhood of critical points of the phase. If all the critical points of the phase are nondegenerate, the contribution to the total vibration from each of them is proportional to the wavelength. If the phase has degenerate critical points, then the contribution from their small neighborhoods is greater, the order of their contribution being proportional to some power of the wavelength smaller than 1. As a rule, the function on the surface, which is equal to the distance to the fixed space point, has only nondegenerate critical points. A space point is said to be *caustic* or *focal* if the function on the surface, equal to the distance to the space point, has a degenerate critical point. The caustic points form a new surface in space, called the *caustic*. At points of the caustic, the total oscillation has a nonstandard large value. If the surface radiates light waves, the caustic is a surface of points with nonstandard brightness. It can be seen on a wall illuminated by rays that are reflected from a concave surface, e.g., the surface of a cup.

Many examples of problems in which the need arises to study asymptotics of oscillating integrals are given in the papers of M. Berry and J. Nye, cited in the bibliography. See, for example, Ref. 46, where the photographs of caustics formed by rays of light refracted in a drop of water are shown.

2.1.2 The Principle of the Stationary Phase States The main contribution to an oscillatory integral comes from the neighborhoods of critical points of the phase; if the phase of the oscillating integral has no critical points on a support of the amplitude, then as the parameter in the integral tends to infinity the integral will tend to zero faster than any power of the parameter.

If we assume that the integral is one-dimensional, we can integrate it by parts:

$$\int e^{i\tau f(x)}\varphi(x)\,dx = -1/(i\tau)\int e^{i\tau f(x)}(\varphi(x)/f'(x))'\,dx.$$

Repeating the integration a sufficient number of times, we obtain our assertion. The multidimensional case is easily reduced to the one-dimensional case.

2.1.3 Fresnel Integral An oscillatory integral whose phase has only nondegenerate critical points is called a *Fresnel integral*.

Example: Consider an oscillatory integral with phase x^2. Figure 30 shows the graph $y = \cos(\tau x^2)\varphi(x)$ of the real part of the integrand.

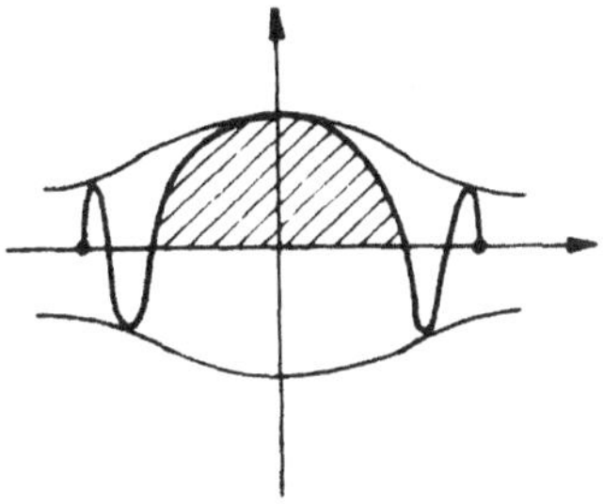

Figure 30

It is clear that for large values of the parameter τ the integral is proportional to the area under the first peak of the graph, i.e., it is proportional to $\varphi(0)\tau^{-1/2}$.

Consider the multiple Fresnel integral

$$\int_{\mathbb{R}^n} e^{i\tau f(x)}\varphi(x)\,dx_1 \ldots dx_n .$$

Assume that the phase of the integral has a nondegenerate critical point at the origin and amplitude zero in the neighborhood of other critical points of the phase.

THEOREM 1 (see Ref. 10) As the parameter τ of the integral tends to $+\infty$, the integral is representable in the form

$$\varphi(0)(2\pi/\tau)^{n/2}\exp(i\tau f(0)$$
$$+ i\pi/4 \operatorname{Sign} f''_{xx}(0))|\det f''_{xx}(0)|^{-1/2} + O(\tau^{-n/2-1}).$$

where f''_{xx} is the matrix of second partial derivatives of the phase, Sign f''_{xx} is the signature* of this matrix.

Example: If $f = x_1^2 + \cdots + x_k^2 - x_{k+1}^2 - \cdots - x_n^2$, then the leading term in the asymptotic expression is

$$\varphi(0)(\pi/\tau)^{n/2}\exp(i\tau f(0) + i\pi(2k-n)/4).$$

*The *signature* of a nondegenerate symmetric matrix is the difference in the numbers of its positive and negative eigenvalues.

2.1.4 Caustics In applications the phases and amplitudes of the oscillatory integrals, as a rule, depend on auxiliary parameters. Let us consider such an integral.

Assume that the phase, considered as a family of functions depending on parameters, is a family of functions in general position (see Section 1). In this case the integral is of Fresnel type for almost all values of the parameters, and at these values is of order $\tau^{-n/2}$.

Parameters for which the phase has a degenerate critical point form a hypersurface in the parameter space. This hypersurface is called a *caustic* (see par. 1.4). For caustic values of the parameters the order of magnitude of the integral is determined by the degenerate critical points of the phase.

2.1.5 Asymptotics of Oscillatory Integrals near a Caustic Assume that for some selected value of the auxiliary parameters the phase of the oscillatory integral has a single critical point, and that the phase, considered as a family of functions depending on the parameters, is a family of functions in general position. In this case the caustic in the neighborhood of the selected parameter is said to be *elementary*.

Examples of elementary caustics, occurring for numbers of parameters equal to 2 and 3, are shown in Figs. 31 and 32. In these figures, we indicate near each part of the caustic the type of degenerate critical points of the phase occurring for these caustic values of the parameters. For example, $A_2 + A_2$ means that the phase has two critical points of type A_2, while the remaining critical points of the phase are nondegenerate. Each degenerate critical point of the phase contributes of order $\tau^{\beta - n/2}$ to the integral. The number β for critical points of types A_μ and D_μ is equal, respectively, to $(\mu - 1)/(2\mu - 2)$ and $(\mu - 2)/(2\mu - 2)$ (see, below, Theorem 7).

THEOREM 2 (see Refs. 12 and 13) For phases depending in a general way on two or three parameters, each elementary caustic is locally diffeomorphic to one of the caustics shown in Figs. 31 and 32.

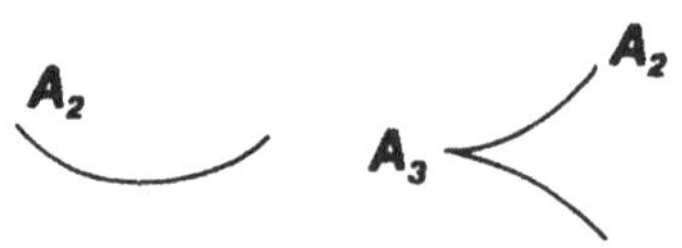

Figure 31

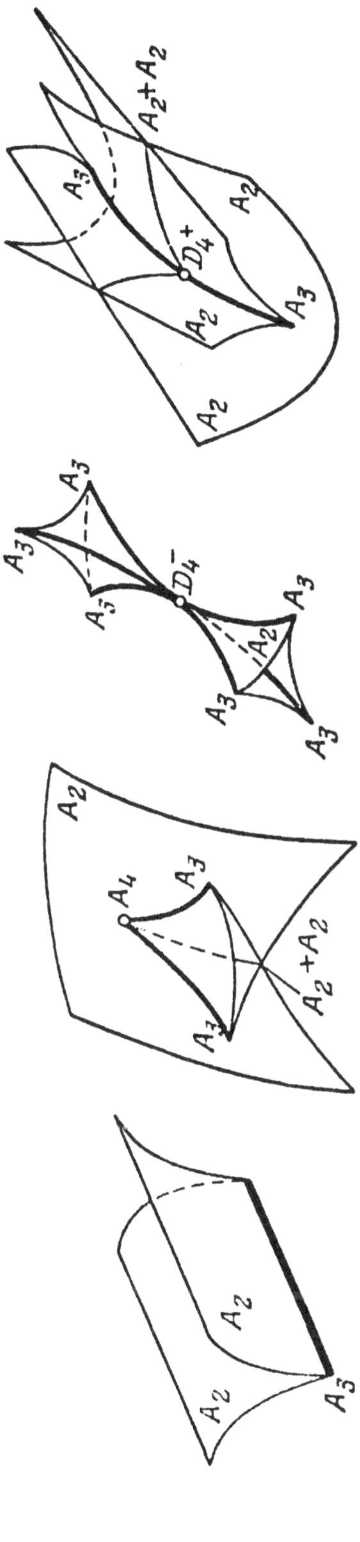

A2+A2
A3
A2
D4+
A3
A2
A2
A3
A3
D4−
A3
A2
A3
A3
A3
A2
A4
A3
A2+A2
A3
A2
A2
A3

Figure 32

Parameters that go over into each other under a local diffeomorphism correspond to integrals of the same order.

For example, if the space is illuminated by a surface, the illumination of a point in general position is of order λ, where λ is the wavelength emitted by the surface. The illumination of the points of some surfaces in space (surfaces of caustics) have order $\lambda^{5/6}$. The illumination of particular curves (cuspidal edges of a caustic) have order $\lambda^{3/4}$. Individual points of the caustic—the vertices of swallowtails, pyramids and purses have illumination of order $\lambda^{7/10}$, $\lambda^{2/3}$, and $\lambda^{2/3}$, respectively.

2.1.6 Oscillatory Integrals over a Half-Space We return to the example in par. 2.1.1. Assume that the radiating surface is opaque to the emitted waves. The waves from the visible parts of the surface arrive at a selected point of the space. Thus the total oscillation at the space point is expressed as the sum of oscillatory integrals, each of which is taken over part of the surface. Thus, in studying short-wave oscillations it is important to know how to calculate the asymptotics of oscillatory integrals over regions with boundaries. Let us examine the case of a smooth boundary.

Consider an oscillatory integral over part of the space $\mathbb{R}^n$ given by the condition that the first coordinate be positive, where the phase and amplitude are assumed to be smooth functions over all space.

In this situation we have the principle of stationary phase: if the phase of the oscillatory integral has no critical points in the integration half-space, and if the restriction of the phase to the boundary of the half-space also has no critical points, then as the parameter τ of the oscillatory integral goes to infinity, the integral tends to zero faster than any power of the parameter.

Assume that the phase of the oscillatory integral over the half-space has no critical points on the boundary of the half-space. Assume that all its critical points inside the half-space are nondegenerate and that the critical points of its restriction to the boundary of the half-space are also nondegenerate. We shall call such an oscillating integral a *Fresnel integral* over a half-space.

THEOREM 3 (see Ref. 10) Assume that the phase of a Fresnel integral over a half-space has no critical points in the half-space of the integration (right up to the boundary). Assume also that the restriction of the phase to the boundary has a single, nondegenerate, critical point at the origin. Then as the parameter τ of the integral tends to $+\infty$, the integral is representable in the form

$$\varphi(0)(2\pi/\tau)^{(n-1)/2}(1/i\tau)\exp(i\tau f(0)+(i\pi/4)\operatorname{Sign}\tilde{f}''_{x'x'}(0))$$

$$\cdot|\det \tilde{f}''_{x'x'}(0)|^{-1/2}+O(\tau^{-(n+1)/2-1}),$$

where $\tilde{f}''_{xx}$ is the matrix of second partial derivatives of the restriction of the phase to the boundary.

Let us assume that the phase and amplitude of the oscillatory integral over the half-space depend on additional parameters. Assume that the phase, considered as a family of functions which depends on the parameters, is a family of functions in general position. In this case, the integral is a Fresnel integral for almost all values of the parameters. The set of those values for which the integral is not of Fresnel type forms a hypersurface in the parameter space, called a *caustic*.

Assume that for a selected value of the parameters the phase has a single critical point on the boundary of the half-space of integration. A caustic in the neighborhood of such a value of the parameters is called *elementary*.

Examples of elementary caustics occurring for the cases of 2 and 3 parameters are shown in Figs. 33 and 34. For a caustic value of the parameters either the phase has a degenerate critical point in the half-space of integration or the restriction of the phase to the boundary has a degenerate critical point. In the figures, near each part of

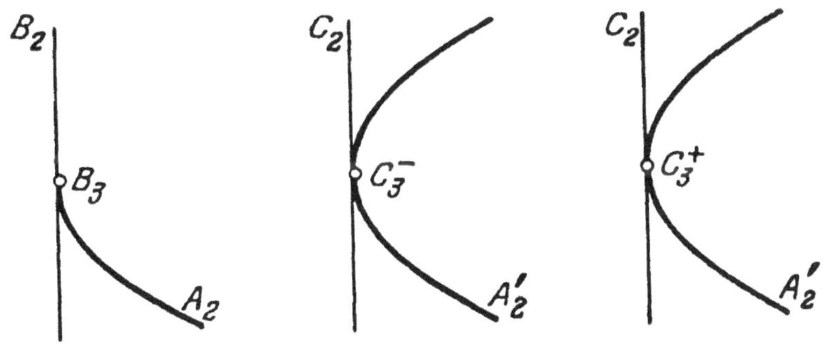

Figure 33

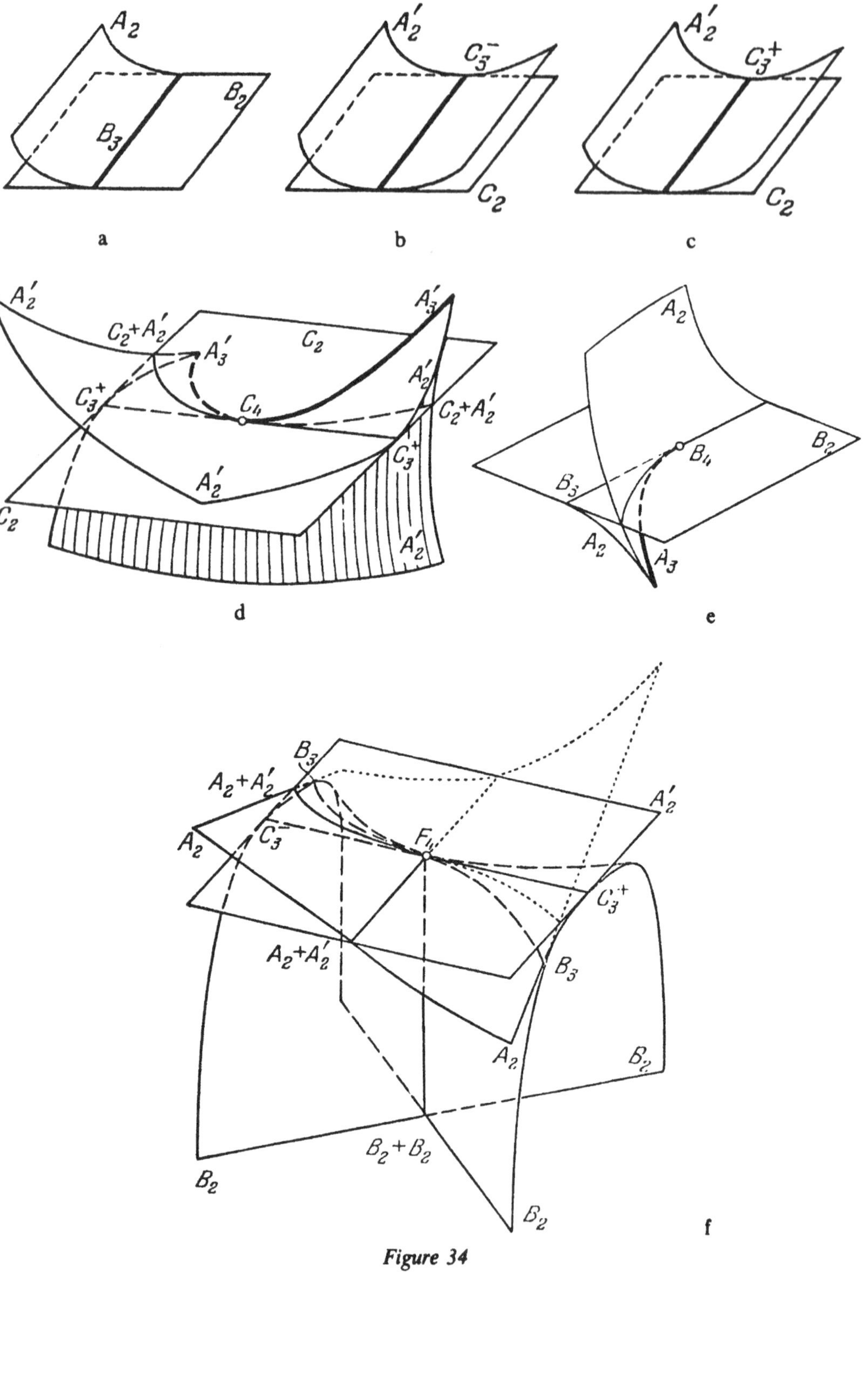
A2
B2
B3
A'2
C3-
C2
A'2
C3+
C2
a
b
c
A'2
C2+A'2
A'3
C2
A'3
A'2
C3+
C4
C2+A'2
C3+
A'2
C2
A'2
d
A2
B4
B2
B3
A2
A3
e
B3
A2+A'2
A'2
A2
C3-
F4
C3+
A2+A'2
B3
A2
B2
B2+B2
B2
B2
f

Figure 34

the caustic we give the notation for these degenerate points. For the normal forms of the critical points designated on the caustics, see Section 1, and also in Refs. 9 and 12. Each degenerate critical point contributes to the integral of order $\tau^{\beta-n/2}$. The number β for critical points of types A'_{μ}, D'_{μ}, B_{μ}, $C_{\mu}^{\pm}$ and F_4 are equal, respectively, to $-1(\mu+1), 1/(\mu-1), (\mu-1)/2\mu, 0, \frac{1}{4}$ (see Ref. 13).

THEOREM 4 (see Refs. 9 and 12) For the phases of oscillatory integrals over a half-space, which depend in general fashion on two or three parameters, each elementary caustic is locally diffeomorphic to one of the caustics shown in Figs. 33 and 34. The parameters that go over into each other under a local diffeomorphism correspond to integrals of equal order of magnitude.

2.1.7 Zones of Light, Shadow, and Penumbra (according to P. K. Mandrykin) Let us assume that the phase and amplitude of the oscillatory integral depend on additional parameters. Consider the parameter space and a caustic located in it. Consider an arbitrary value of the parameters outside the caustic. The phase of the oscillatory integral corresponding to these parameters has only nondegenerate critical points, or has no critical points at all. In the first case, the oscillatory integral is of order $\tau^{-n/2}$, where n is the dimension of the space of integration; in the second case, as $\tau \to +\infty$ the oscillatory integral tends to zero faster than any power of the parameter τ. In this context, we call the region outside the caustic a light region if the values of the parameters for this region correspond to an oscillatory integral whose phase has at least one critical point; we call this region a region of shadow, if the parameters for this region correspond to oscillatory integrals whose phase has no critical points.

Example: In the pictures of caustics corresponding to critical points of types A_2, A_4, and D_4^+, the shadow regions are located below the caustics. The remaining regions outside the caustics, corresponding to critical points of types A_2, A_3, A_4, and $D_4^{\pm}$, are zones of light.

Let us now assume that the oscillatory integral is an integral over a half-space. Consider an arbitrary parameter value outside the caustic. There are three possibilities:

(1) The phase of the integral which corresponds to a selected parameter value has at least one critical point in the half-space of integration. In this case the integral has order $\tau^{-n/2}$.

(2) The phase of the integral which corresponds to the particular parameter value has no critical points in the integration half-space, but the restriction of the phase to the boundary of the half-space has at least one critical point. In this case the integral has order $\tau^{-(n+1)/2}$.

(3) The phase of the oscillatory integral which corresponds to the particular parameter values has no critical points in the integration half-space, and its restriction to the boundary of the half-space also has no critical points. In this case the integral tends to zero faster than any power of τ as the parameter $\tau \to +\infty$.

In accordance with these three possibilities, the regions into which the caustic divides the space of the additional parameters are called *zones of light, penumbra, and shadow*.

Example: The zones of penumbra and shadow in the figures of caustics corresponding to the critical points B_3, $C_3^{\pm}$, C_4, B_4, and F_4 are arranged in the following way. In Fig. 34a the penumbra is above the caustic. In Fig. 34b it is below it. In Fig. 34c the shadow zone is above the caustic; and the penumbra is between the two sheets of the caustic. In Fig. 34d the penumbra is on one side of the plane of the caustic. In Fig. 34e the penumbra is on the upper right hand side of the plane of the caustic. In Fig. 34f the shadow is above the whole caustic; the penumbra is behind the caustic below the plane of the caustic; another penumbra region is on the right hand side between the plane of the caustic and the surface with an edge.

2.2 *Oscillatory Integrals Corresponding to Caustic Values of the parameters*

2.2.1 THEOREM 5 (on asymptotic expansion, see Refs. 8, 13, 7 and 39) Assume that the phase of the oscillatory integral in Eq. (1) is an analytic function in the neighborhood of its critical point x^o. Then the oscillatory integral is expanded in an asymptotic series

$$\exp(i\tau f(x^o)\sum_{\alpha}\sum_{k=0}^{n-1} a_{k,\alpha}(\varphi)\tau^{\alpha}(\ln \tau)^k, \qquad \text{as} \quad \tau \to +\infty, \tag{2}$$

if the amplitude is equal to zero outside a sufficiently small neighborhood of this point. Here the parameter α runs through a finite set of arithmetic progressions, depending only on the phase and consisting of negative rational numbers.

Example (see Ref. 10): For a one-dimensional oscillatory integral with phase $x^{\mu+1}$, the asymptotic series has the form

$$\exp(i\tau f(x^o))\sum_{l=1}^{\infty} a_l\tau^{-l/(\mu+1)}$$

Example (cf. [10]): For a nondegenerate critical point of the phase depending on n variables, the asymptotic series has the form

$$\exp(i\tau f(x^o))\sum_{l=0}^{\infty} a_l\tau^{-n/2-l}$$

In the asymptotic series for an oscillatory integral the phase and the amplitude are not on an equal footing: the phase determines the power of the parameter, while the amplitude determines the coefficients for the powers of the parameter. The dependence on the phase is the more important factor. In the example of par. 2.1.1 for the radiation of waves by a surface, the phase corresponds to the geometry of the surface, while the amplitude corresponds to the intensity of the radiation.

2.2.2 Singularity and Oscillation Indices The fundamental characteristic of the asymptotic series in Eq. (2) is the power of the parameter in the maximal term of the series.

The *oscillation index* of a critical point of the phase is the maximum number α for which some of the coefficients $a_{k,\alpha}(\varphi)$ are not equal to zero for some amplitudes φ located in a sufficiently small neighborhood of the critical point of the phase.

Example: The oscillation index of a critical point of the phase $\pm x_1^{\mu+1} \pm x_2^2 \ldots \pm x_n^2$ is $-1(\mu+1)-(n-1)/2$.

The oscillation index of the critical point x^o of the phase f for the variables $x_1, \ldots, x_n$ is greater by $\frac{1}{2}$ than the oscillation index of the critical point $(x^o,0)$ of the phase $f \pm z^2$ for $(n+1)$ variables. To eliminate the dependence on quadratic terms, we use the notion of a singularity index.

The *singularity index* of a critical point of the phase is the oscillation index plus $n/2$.

The singularity indices of the critical points of the phases f and $f \pm z^2$ are equal. The singularity index of a nondegenerate critical point is zero.

2.2.3 Tables of Singularity Indices (see Refs. 7 and 13) Below we give in tabular form the singularity indices of the critical points classified in Section 1. In the first row we give the designation of the critical point, in the second row we give the singularity index. The meaning of the tables is the following: if, in the neighborhood of a critical point, the phase is brought to the tabulated form by changes of the integration variables, then its singularity index is equal to the singularity index of the tabulated function.

Table III

A_k	D_k	E_6	E_7	E_8
$\frac{k-1}{2k+2}$	$\frac{k-2}{2k-2}$	$\frac{5}{12}$	$\frac{4}{9}$	$\frac{7}{15}$

Simple Singularities

$P_8, X_9, J_{10}, J_{10+k}, X_{9+k}, Y_{r,s}, \tilde{Y}_r, P_{8+k}, P_{l,m}, \tilde{R}_m, T_{p,q,r}, \tilde{T}_{p,m}$
$\frac{1}{2}$

E_{12}	E_{13}	E_{14}, Q_{10}	Z_{12}	Z_{13}, Q_{11}	W_{12}	W_{13}, S_{11}	Q_{12}	S_{12}	U_{12}
$\frac{11}{21}$	$\frac{8}{15}$	$\frac{13}{24}$	$\frac{6}{11}$	$\frac{5}{9}$	$\frac{11}{20}$	$\frac{9}{16}$	$\frac{17}{30}$	$\frac{15}{26}$	$\frac{7}{12}$

Unimodal Singularities

From these tables and the classification of critical points given in Section 1, it follows that (see Ref. 6):

For $l \leqslant 10$ and $k \geqslant 3$, the highest singularity index unremovable in families in general position of phases of k arguments which depend on l parameters has the form

$$\beta_l = \frac{1}{2} - \frac{1}{N}$$

where the number N is given by the Table:

l	0	1	2	3	4	5	6	7	8	9	10, $k=3$	11, $k=3$	10, $k>3$
N	+2	+3	+4	+6	+8	+12	∞	∞	−24	−16	−12	−8	−6

For example, for phases which depend in general manner on one, two, or three parameters, the order of the oscillatory integral for an arbitrary value of the parameters is no higher than $\tau^{\beta - n/2}$, where the numbers β are $1/6$, $1/4$ and $1/3$, respectively.

2.2.4 Formulation of Results and the Newton Polyhedra Let us formulate rules allowing us to calculate the oscillation indices. These rules are formulated in terms of the Newton polyhedra for the Taylor series of the critical points of the phase.

The Newton polyhedron is a convex polyhedron formed by the exponents of monomials that appear in the Taylor series.

If we consider a class of critical points of the phase with a fixed Newton polyhedron, it turns out that almost all critical points of the class have the same oscillation index. We give a formula expressing this common oscillation index in terms of the geometry of the Newton polyhedron. Exceptions are critical points at which the coefficients of the Taylor series satisfy explicitly written algebraic conditions.

It is useful to consider a class of critical points with a fixed Newton polyhedron in studying the discrete characteristics of critical points. As a rule, the characteristic takes a single value for almost all points of the class, this common value being expressed simply in terms of the geometry of the Newton polyhedron (see Section 3).

2.2.5 Definition of the Newton Polyhedron Consider the power series $f = \sum a_k x^k$ in n variables, where $k = (k_1, \ldots, k_n)$, and $x^k = x_1^{k_1} \ldots x_n^{k_n}$. We mark in the space $\mathbb{R}^n$ those points k with integer coordinates for which the coefficient a_k is not zero. All these points lie in the positive octant, i.e., the octant consisting of points with nonnegative coordinates. At each of these points we draw a parallel positive octant. The *Newton polyhedron of the series* is the convex hull of the union of all these octants.

Examples: The Newton polyhedra for the series

$$f = x_1^4 + 2x_1^2x_2^2 + x_2^4 + x_2^5, \qquad g = x_1^4 - 2x_1^2x_2^3 + x_2^6 + x_1^3x_2^3,$$

$$h = x_1^4 + x_1^2x_2^2 + x_2^6$$

are shown in Fig. 35.

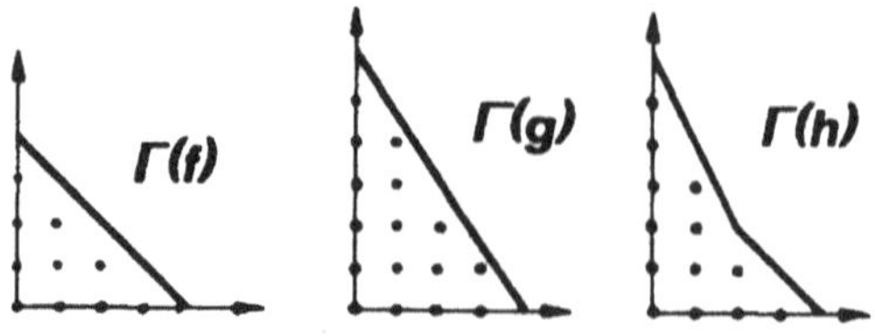

Figure 35

2.2.6 Distance to the Polyhedron and Remoteness of the Polyhedron Consider the bisectrix of the positive octant in $\mathbb{R}^n$, i.e., the line consisting of points with equal coordinates. The bisectrix (diagonal) intersects the boundary of the Newton polyhedron at exactly one point. This point is called the *center of the boundary* of the Newton polyhedron. A coordinate of the center is called the *distance to the Newton polyhedron.* The *remoteness of the Newton polyhedron* is the reciprocal of the distance, taken with a minus sign.

Example: The distances to the Newton polyhedra for the series f, g, and h of the preceding example are, respectively, 2, 12/5, and 2; their remotenesses are, respectively, $-1/2$, $-5/12$, and $-1/2$.

The further the Newton polyhedron is located from the coordinate origin, the greater its remoteness. We say that a Newton polyhedron is *remote* if its remoteness is greater than -1. In other words, a Newton polyhedron is remote if it does not include the point $(1, \ldots, 1)$.

2.2.7 Principal Part of a Power Series; Its Nondegeneracy Let us consider a power series f and its Newton polyhedron.

For any face γ of the Newton polyhedron we call γ-part of the series that part consisting of monomials whose powers are on the face γ; each monomial enters with the same coefficient that it had in the original power series. If the face γ is bounded, then the γ-part is a polynomial. We denote the γ-part of the series by f_γ.

The *principal part* of a power series is the polynomial consisting of all monomials whose powers belong to bounded faces of the Newton polyhedron; here the monomials enter with the same coefficients as they had in the original power series.

Example: For the series f, g, and h, of the preceding example, the principal parts are, respectively, the polynomials

$$x_1^4 + 2x_1^2x_2^2 + x_4^2, x_1^2 - 2x_1^2x_2^3 + x_2^6, x_1^4 + x_1^2x_2^2 + x_2^6.$$

The principal part of a power series is said to be *nondegenerate* if for any closed bounded face γ of the Newton polyhedron of the series, the polynomials $\partial f_\gamma/\partial x_1, \ldots, \partial f_\gamma/\partial x_n$ have no common zeros in $(\mathbb{R}\backslash 0)^n$ (i.e., in the complement of the union of the coordinate hyperplanes).

Example: The principal parts of the series f, h of the preceding example are nondegenerate, while the principal part of the series g of the preceding example is degenerate.

It has been proved that there are few series with degenerate principal parts. Namely, *the set of degenerate principal parts is a proper algebraic subset in the space of all principal parts, corresponding to a given polyhedron.*

2.2.8 The Oscillation Index of a Critical Point of the Phase is Determined by the Remoteness of the Newton Polyhedron of its Taylor Series

THEOREM 6 (see Refs. 7 and 13) Suppose that the phase is an analytic function in the neighborhood of its critical point with zero critical value. Assume that the principal part of the Taylor series of the phase at this critical point is nondegenerate, and that the Newton polyhedron of the series is remote. Then the oscillation index of the critical point of the phase is equal to the remoteness of the Newton polyhedron.

Example: The degenerate critical point of the phase $x_1^{k_1} + x_2^{k_2}$ satisfies the conditions of the theorem. Its oscillation index is equal to $-1/k_1 - 1/k_2$.

Example: The critical points at the coordinate origin of the phases f and h in the example on p. 55 satisfy the conditions of the theorem. Their oscillation indices are, respectively, $-\frac{1}{2}, -\frac{1}{2}$.

According to Theorem 6, the oscillation index of a critical point of the phase can be expressed in terms of the Newton polyhedron of its Taylor series, if the principal part of its Taylor series is nondegenerate, and if the Newton polyhedron of the Taylor series is remote. A coordinate system in which the Taylor series has such properties does not always exist. For example, it does not exist for the critical point at the origin of the function g of the example on p. 55. Nevertheless, for the critical points of the phase depending on two variables one can drop the assumption of the existence of such a coordinate system.

By the *remoteness of a critical point* of the phase with zero critical value we mean the maximum remoteness of the Newton polyhedra of the Taylor series of the phase at the critical point in all systems of local analytic coordinates in the neighborhood of the critical point.

A system of coordinates is said to be *adapted* to a *critical point* of the phase if the remoteness of the Newton polyhedron of the Taylor series of the phase at the critical point in this system of coordinates has the maximum possible value, equal to the remoteness of the critical point.

THEOREM 7 (see Refs. 7 and 13) Suppose that the phase is an analytic function of two variables in the neighborhood of its critical point with zero critical value. Then the oscillation index of the critical point is equal to its remoteness.

The following assertion (see Ref. 7) is useful for finding adapted coordinate systems: A coordinate system is adapted to a critical point of the phase of two variables if at least one of the following conditions is satisfied:

(1) The center of the boundary of the Newton polyhedron of the Taylor series at the critical point in this coordinate system is a vertex of the polyhedron.

(2) The center of the boundary of the Newton polyhedron lies on an unbounded edge of the Newton polyhedron.

(3) The center of the boundary of the Newton polyhedron lies on a bounded edge of the Newton polyhedron and neither the tangent or cotangent of the angle formed by the edge and the coordinate axis in $\mathbb{R}^2$ is equal to an integer.

Example: Consider the functions g, h of the example on p. 55. The coordinate system x_1, x_2 is adapted to the critical point at the origin for these functions [as shown, respectively, by points (3) and (1) of the preceding assertion]. According to Theorem 7, the oscillation indices are $-5/12$ and $-1/2$, respectively.

In par. 2.2.3 we gave tables of singularity indices of the critical points classified in Section 1. The singularity indices of the critical points in the table were calculated from the oscillation indices. The oscillation indices were calculated using Theorems 6 and 7. Theorems 6 and 7 are applicable to the critical points classified in Section 1, since each of the critical points listed there either is a nondegenerate principal part of the Taylor series, or is a function of two variables, written in an adapted coordinate system.

2.2.9 Gelfand–Leray Form In studying oscillatory integrals, the following device, which reduces a multidimensional oscillatory integral to a one-dimensional integral is very useful. Let us consider the oscillatory integral

$$\int_{\mathbf{R}^n} e^{i\tau f(x)}\varphi(x)\,dx_1 \ldots dx_n .$$

We reduce the integral to a multiple integral, in which we first integrate over a hypersurface of constant phase, and then over the remaining variable, the value of the phase. To do this, we introduce new variables in the integral, one of which is the phase.

We shall make two comments. First, the phase can be regarded as a variable only away from its critical points. Therefore, we remove from the space the union of hypersurfaces of critical levels of the phase. The union of these hypersurfaces has zero measure and does not affect the integral. Second, for the integration over the level hypersurfaces we do not need to know each of the other new variables. It is sufficient to know on the level hypersurfaces a form of the $(n-1)$-dimensional density, which after multiplication by the differential of the phase becomes the form of the spatial volume. Such a density form is called a *Gelfand–Leray form* and is denoted by $dx_1 \wedge \cdots \wedge dx_n/df$.

Thus, the oscillatory integral is transformed to the form

$$\int_{-\infty}^{+\infty} e^{i\tau t}\left(\int_{f=t} \varphi\, dx_1 \wedge \cdots \wedge dx_n/df\right)dt.$$

In this representation the oscillatory integral is the Fourier transform of the function given by the inner integral. The function of one variable, defined by the inner integral, is called the *Gelfand–Leray function*.

The Gelfand–Leray function is smooth away from the critical values of the phase. In the neighborhood of a critical value of the phase the Gelfand–Leray function is expanded in an asymptotic series of the form

$$\sum_{\alpha}\sum_{k=0}^{n-1} a_{k,\alpha}(t-t_0)^{\alpha}(\ln(t-t_0))^{k}.$$

Knowing the asymptotic series of the Gelfand–Leray function, we can determine the asymptotic series of the oscillatory integral. Conversely, the asymptotics of the oscillatory integral give information about the asymptotics of the Gelfand–Leray function.

2.2.10 Volume of the Set of Smaller Values Let us assume that the phase has an isolated minimum and that the minimum value of the phase is equal to zero. We also assume that the amplitude in the neighborhood of the point of the minimum is identically equal to one. We denote by J the Gelfand–Leray function, and consider a new function

$$V(t) = \int_0^t J(s)\, ds.$$

It is obvious that, for negative values of the argument, this function is equal to zero, while for small positive values of the argument this function is equal to the volume of the set of points at which the phase takes on values less than the given one. Thus the asymptotics of the function giving the volume of the set of smaller values determines the asymptotics of the oscillatory integral in the case where the phase has an isolated minimum, and the amplitude is equal to a constant in the neighborhood of a point of minimum phase.

We give the order of magnitude of the rates at which the volume of the set of smaller values tends to zero for the simplest isolated minimum points.

Table IV, prepared by V. A. Vasiliev (see Ref. 4), gives the normal forms of germs of smooth functions in the neighborhood of points of minima (to within addition of constants and positive definite quadratic forms in additional variables). The reduction to normal form is accomplished by a smooth change of the independent variables. The column labelled l in the table indicates the number of parameters of the family beyond which the points of a minimum of the particular type become unremovable by small variations of the family. In general families of functions with $l < 16$ parameters, minimum points which are not equivalent to those enumerated in the table do not occur.

For small positive t the leading term in the asymptotics of the

Table IV

Designation	Normal Form	Restrictions	l
A_{2k-1}	x^{2k}	$k \geqslant 1$	$2k-2$
$X_{1,0} = T_{2,4,4}$	$x^4 + ax^2y^2 + y^4$	$a > -2$, $a \neq 2$	7
$X_{1,2r} = T_{2,4,4+2r}$	$x^4 + x^2y^2 + ay^{4+2r}$	$a > 0, r \geqslant 1$	$7+2r$
$Y^1_{2r,2q} = T_{2,4+2r,4+2q}$	$x^{4+2r} + ax^2y^2 + y^{4+2q}$	$a > 0, r, q \geqslant 1$	$7+2r+2q$
$\tilde{Y}^1_{r,r} = \tilde{T}_{2,4+r,4+r}$	$(x^2+y^2)^2 + ay^{4+r}$	$a \neq 0, r \geqslant 1$	$7+2r$
$W_{1,0}$	$x^4 + (a+by)x^2y^3 + y^6$	$a(-1)^q < 0$	12
$W^{\#}_{1,2q}$	$(x^2+y^3) + (a+by)x^2y^{3+q}$	$q \geqslant 1$	$12+2q$

Here a and b are real parameters.

volume $V(t)$ of the set of smaller values has the form

$$\text{const}(\ln t)^k t^{-\beta+n/2},$$

where the number β is given in the table:

A_{2k-1}	$X_{1,0}, X_{1,2r}, Y'_{2r,2q}, \tilde{Y}^1_{r,r}$	$W_{1,0}, W^{\#}_{1,2q}$
$(k-1)/2k$	$1/2$	$7/12$

The number k is equal to 1 for critical points $X_{1,2r}$ and $Y'_{2r,2q}$; for the other critical points $k = 0$.

We formulate a general theorem about the calculation of the rate of fall to zero of the volume of the set of smaller values.

THEOREM 8 (see Refs. 7, 4 and 13) Assume that an analytic function has an isolated minimum, and that the minimum value is equal to zero. Then as $t \to +0$ the function V for the volume of the set of smaller values is expanded in an asymptotic series:

$$\sum_{\alpha} \sum_{k=0}^{n-1} a_{k,\alpha} t^{\alpha} (\ln t)^k.$$

Here the parameter α runs through a finite number of arithmetic progressions that consist of positive rational numbers. If, in addition, we know that the Taylor series of the function at the point of the minimum has a nondegenerate principal part, then the power α of the maximal term of the asymptotic series is equal to minus the remoteness of the Newton polyhedron of the Taylor series.

2.2.11 Area of a Level Surface Consider a function having an isolated minimum. Suppose that the minimum value of the function is zero. In this case the level hypersurface for a small positive value of the function is a compact manifold. Let us calculate the $(n-1)$-dimensional volume of the compact level manifolds. We are interested in the asymptotic of this volume as the level tends to zero.

Example: For an isolated minimum point of the function

$$x_1^{a_1} + \cdots + x_n^{a_n}$$

where $a_1 \leqslant a_2 \leqslant \cdots \leqslant a_n$ are positive even integers, the leading term of the asymptotic of the $(n-1)$-dimensional volume of the manifold at level t, as $t \to +0$, has the form const. t^α, where $\alpha = 1/a_2 + \cdots + 1/a_n$.

In Ref. 13 a general formula is given for calculating asymptotics of the $(n-1)$-dimensional volume of level manifolds as the level tends toward the critical level.

2.2.12 Set of Points with a Small Gradient Another characteristic of a critical point, like the one considered above, is the rate of tendency to zero of the volume of the set of points in which the *length* of the gradient is smaller than a given value, as this value tends to zero.

In the neighborhood of a selected critical point of the function, for each small positive t we consider the volume $V(t)$ of the set of those points of the neighborhood at which the square of the length of the gradient is less than t. We are interested in the asymptotic of the volume as $t \to +0$.

Example: For critical points of types A_k, D_4, and D_k $(k > 4)$ and E_6, E_7 and E_8, the leading term of the asymptotic series of the function V, as $t \to +0$, has the form const. $t^{-\alpha + n/2}(\ln t)^k$, where (k, α) are, respectively,

$$((k-1)/2k, 0), (1/2, 0), (1/2, 1), (7/12, 0), (5/8, 0), (5/8, 0).$$

This proves Table II (p. 42).

To calculate the asymptotics of the volume of the set of points with a small gradient, we can use Theorem 8, which is used for the square of the length of the gradient.

2.3 *Comparison of Oscillatory Integrals for Different Values of the Auxiliary Parameters*

2.3.1 Uniform Estimates In addition to the asymptotics of individual oscillatory integrals, it is frequently useful to know uniform estimates of oscillatory integrals that depend on auxiliary parameters.

Suppose that $f:\mathbb{R}^n \to \mathbb{R}$ is a smooth function. A *deformation* of it is any smooth function $F:\mathbb{R}^n \times \mathbb{R}^l \to \mathbb{R}$, equal to the function f for zero value of the second argument.

We say that at the critical point x^o of the phase f there is a uniform estimate with index α, if for any deformation F of the phase f there is a neighborhood in $\mathbb{R}^n \times \mathbb{R}^l$ of the point $x^o \times 0$, such that for any smooth function φ equal to zero outside this neighborhood and for any positive ϵ, there is a number $C(\epsilon,\varphi)$ for which

$$\left|\int_{\mathbb{R}^n} e^{i\tau F(x,\lambda)}\varphi(x,\lambda)\,dx_1 \ldots dx_n\right| < C(\epsilon,\varphi)\tau^{\alpha+\epsilon}$$

for all positive τ. The lower bound of such numbers α is called the *uniform oscillation index* of the phase at the critical point.

It is obvious that the uniform oscillation index is no smaller than the individual index.

We can assume that the individual oscillation index is equal to the uniform oscillation index, i.e., that the oscillatory integral admits an estimate from above, which is uniform in the additional parameters, by a quantity which is proportional to the magnitude of the integral for the initial value of the additional parameters.

For this assumption to be valid it is necessary that the individual oscillation index be upper semicontinuous for continuous changes of the critical point. Namely, it is necessary that the oscillation index for a complicated critical point be no less than that for a simpler critical point obtained by the decomposition of the more complicated one. Analysis of the table of singularity indices and the known adjacent critical points, as classified in Section 1, shows that such semicontinuity holds for the critical points classified in Section 1.

Theorem 9 For simple (Refs. 5, 40), parabolic (Ref. 50) and hyperbolic critical points of the series $T_{p,q,r}$ (Refs. 2, 3) the uniform oscillation index is equal to the individual index.

It was reported in Ref. 1 that the uniform oscillation index is equal to the individual index for all critical points of the phase for the case of two variables.

Corollary of Theorem 9 For critical points that are unremovable, in families of phases in general position, which depend on no more than seven parameters, the uniform oscillation index is equal to the individual index.

According to Theorem 9, as we move along the caustic corresponding to one of the critical points listed in the theorem, the intensity of the short-wave oscillation at the limit point is no less than the intensity at a nearby prelimit point. Surprisingly, this phenomenon does not occur for all caustics. Namely, there are examples of highly degenerate critical points of the phase, for which the uniform oscillation index is greater than the individual oscillation index (Ref. 7). The codimensions of the critical points of the examples are at most 73, i.e., such points are removable by a small deformation from the families of functions with numbers of parameters smaller than 73.

According to the constructed examples, there exist critical points and their deformations that have the following property. The individual oscillation index of the critical point for a particular value of the deformation parameter is smaller than the individual oscillation index of the critical point for the general value of the deformation parameter, i.e., the magnitude of the oscillatory integral for the particular parameter value is essentially smaller than the magnitude of the oscillatory integral for the general value of the parameter.

It is interesting *whether this phenomenon can be observed physically* as a subset of the caustic that is darker than its surroundings. As already mentioned, such a phenomenon is not observed on caustics in general position in low-dimensional spaces (Theorem 9 and its corollary).

2.3.2 Estimates in the Mean Consider an oscillatory integral that depends on auxiliary parameters:

$$I(\tau,\lambda) = \int_{\mathbb{R}^n} e^{i\tau F(x,\lambda)}\varphi(x,\lambda)\, dx_1 \ldots dx_n .$$

If the value λ of the parameters does not belong to a caustic, the integral is of Fresnel type and has order $\tau^{-n/2}$. If the value of parameters values belongs to a caustic, the order of magnitude of the integral is increased.

As the following result shows, the order of magnitude of the average energy of short-wave oscillation near the caustic is the same as outside the caustic, i.e., the portion of points with large energy of oscillation is insignificant.

THEOREM 10 (see Ref. 40) Denote by $\sum$ the set of critical points of the phase, i.e.,

$$\sum = \{(x,\lambda) \mid \partial f/\partial x_j(x,\lambda) = 0, j = 1, \ldots, n\}.$$

Assume that $\sum$ is a submanifold, i.e., the differentials $d(\partial F/\partial x_j)$, $j = 1, \ldots, n$ are linearly independent at points of the set $\sum$. Assume that the amplitude of the oscillatory integral is equal to zero outside a sufficiently small neighborhood of one of the points of the set $\sum$. Then, as $\tau \to +\infty$ we have the asymptotic expansion

$$\int |I(\tau,\lambda)|^2 \, d\lambda \approx \sum_{l=n}^{\infty} a_l \tau^{-l},$$

where the numerical coefficients depend on the amplitude; in particular, the highest coefficient a_n is proportional to the integral of the square modulus of the amplitude over the critical set $\sum$.

Example: The versal deformation F of a finite multiplicity critical point satisfies the conditions of Theorem 10.

2.3.3 Number of Integer Points in a Region In recent years the use of the theory of singularities has been extended to include a number of new applications to problems of interaction of latticelike and smooth objects. One of the areas in which such interactions occur is the theory of resonances in nonlinear systems. A resonance between frequencies ω_k is an integer relation $\sum n_k \omega_k = n_0$. Such a relation determines a hypersurface in the space of frequencies. A system that depends on parameters is represented as a surface in the space of frequencies (its dimensionality, which depends on the number of parameters, is assumed to be smaller than the number of frequencies).

The question of how close a resonance of intermediate order is to a typical point of the hypersurface is important for the theory of nonlinear oscillations. The answer to it depends on the location of the surface relative to its tangent plane (inflexions and flatness of the surface cause sticking at resonances). For example, estimates of the rate of evolution of action variables of Hamiltonian systems that are close to integrable (N. N. Nekhoroshev) depend on the "steepness" of the unperturbed Hamiltonian function.

The area described—the theory of Diophantine approximations on submanifolds—has not yet been developed sufficiently, but it is already clear that the theory of singularities of wave fronts (i.e., singularities of Legendre transformations) plays a fundamental role in it.

Another area of interaction of smooth and integer-valued structures is the asymptotics of the number of points close to a smooth surface. Suppose that G is a region of volume V in Euclidean space $\mathbb{R}^n$. Then λG is the region G stretched out λ times, $N(\lambda)$ is the number of integer points in λG, $R(\lambda) = \lambda^n V - N(\lambda)$ is the error in the asymptotic formula $N(\lambda) \approx \lambda^n V$ $(\lambda \to \infty)$. Estimating $R(\lambda)$ is a classical problem, which was studied by Gauss but which has not been solved even for a circle (where it is conjectured that R is of order $\lambda^{+1/2}$).

In this area the modern theory of oscillatory integrals leads to new results. In order to understand their significance, we note first that $|R(\lambda)| \leqslant c\lambda^{n-1}$ (this is how the area of the boundary of λG increases) and that the flatnesses of the boundary will increase R. For a sphere with center at 0, $|R(\lambda)| \geqslant c\lambda^{n-2}$ for certain sufficiently large λ (the number of integer points in the layer between spheres of radii λ and $\lambda + d$ is of order $\lambda^{n-1}d$, while the number of spheres containing integer points is of order λd, so that there is a sphere that contains no fewer than $c\lambda^{n-2}$ points).

Using the classification of singularities, Colin de Verdier proved the estimate $|R(\lambda)| \leqslant c\lambda^{n-2+2/n+1}$ for the region bounded by a smooth hypersurface in general position in $\mathbb{R}^n$ for $n \leqslant 7$ (Ref. 50).

The remainder R can be regarded as the sum of a large (order λ^{n-1}) number of terms with mathematical expectation zero (they correspond to cells cut out by the boundary of the region λG). If these terms were independent, their sum, from the laws of statistics, would be a quantity of order $\lambda^{(n-1)/2} \ll \lambda^{n-2}$ $(n > 3)$.

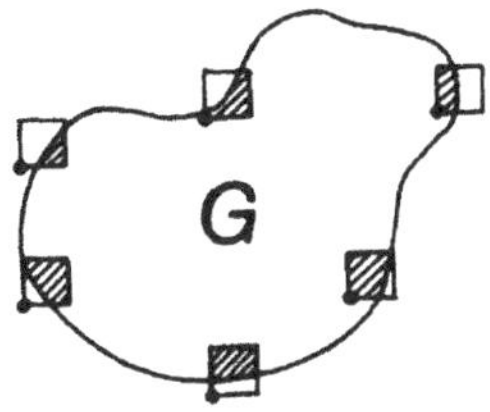

Figure 36

The theory of asymptotics of oscillatory integrals provides a basis for these arguments.

THEOREM 11 (see Refs. 30 and 31) The mean square value of $|R(\lambda)|$ over all lattices obtained from the lattice of integer points by rotations and displacements, admits an estimate from above of order $\lambda^{(n-1)/2}$ for any bounded region in $\mathbb{R}^n$ with a smooth boundary.

§3 The Geometry of Formulas

The number of integer points lying within a polyhedron, the volume of a polyhedron—these and other geometric quantities are encountered in answering various questions of the theory of singularities, in algebra and analysis. In this section we present examples of this kind.

3.1 The Newton Polyhedra

The Newton polyhedron of a polynomial which depends on several variables is the convex hull of the powers of the monomials appearing in the polynomial with nonzero coefficients. The Newton polyhedron generalizes the notion of degree and plays an analogous role. It is well known that the number of complex roots of a system of n equations of identical degree m in n unknowns is the same for nearly all values of the coefficients, and is equal to m^n (Bézout's theorem). Similarly, the number of complex roots of a system of n equations in n unknowns with the same Newton polyhedron is the same for nearly all values of the coefficients and is equal to the volume of the Newton polyhedron, multiplied by $n!$ (Kushnirenko's theorem, see 3.1.1).

The level line of a polynomial in two complex variables is a

Riemann surface. For nearly all polynomials of fixed degree n, the topology of this surface (the number of handles g) is expressed in terms of its degree, and does not depend on the values of the coefficients of the polynomial: $g = (n-1)(n-2)/2$. In the more general case, where instead of polynomials of fixed degree we consider polynomials with fixed Newton polyhedra, all the discrete characteristics of the manifold of zeros of the polynomial (or several polynomials) are expressed in terms of the geometry of the Newton polyhedra. Among these discrete characteristics are the number of solutions of a system of n equations in n unknowns, the Euler characteristic, the arithmetic and geometric genus of complete intersections, and the Hodge number of a mixed Hodge structure on the cohomologies of complete intersections.

The Newton polyhedron is defined not only for polynomials but also for germs of analytic functions. For germs of analytic functions in general position, with given Newton polyhedra, one can calculate the multiplicity of the zero solution of a system of analytic equations, the Milnor number and ζ function of the monodromy operator, the asymptotics of oscillatory integrals (see section 2), the Hodge number of the mixed Hodge structure in vanishing cohomologies, and in the two-dimensional and multidimensional quasihomogeneous cases, one can calculate the modality of the germ of the function.

In the answers one meets quantities characterizing both the sizes of the polyhedra (volume, number of integer points lying inside the polyhedron) as well as their combinatorics (the number of faces of various dimensions, numerical characteristics of their contacts).

In terms of the Newton polyhedron one can construct explicitly the compactification of complete intersections, and the resolution of singularities by means of a suitable toric manifold.

Thus, the Newton polyhedra connect algebraic geometry and the theory of singularities to the geometry of convex polyhedra. This connection is useful in both directions. On the one hand, explicit answers are given to problems of algebra and the theory of singularities in terms of the geometry of polyhedra. We note in this connection that even the volume of the convex hull of a system of points is a very complicated function of their coordinates. Therefore, the formulation of answers in numerical terms is so opaque that without knowing their geometric interpretation no progress is possible. On the other hand, algebraic theorems of general character (the Hodge theorem on the index of an algebraic surface, the Riemann-Roch theorem) give significant information about the geometry of polyhedra. In this way one obtains, for example, a simple proof of the Aleksandrov-Fenchel inequalities in the geometry of convex bodies.

The Newton polyhedra are also met in the theory of numbers (Ref. 17) in real analysis (Ref. 29), in the geometry of exponential sums (Refs. 32, 60), in the theory of differential equations (Refs. 18, 19, 36). In this paragraph we present formulations of some theorems about Newton polyhedra. More details can be found in Refs. 4, 7, 15, 12–13, 15–29, 32–34, 41, 58, 60–61.

3.1.1 The Number of Roots of a System of Equations with a Given Newton Polyhedron According to Bézout's theorem, the number of nonzero roots of the equation $f(z)=0$ is equal to the difference between the highest and the lowest powers of the monomials appearing in the polynomial f. This difference is the volume of the Newton polyhedron (in the present case, the length of a segment). In the following two paragraphs we present the generalizations of this theorem to the case of arbitrary Newton polyhedra.

Let us start with definitions. A *monomial* in n complex variables is a product of the coordinates to integer (possibly negative) powers. Each monomial is associated with its *degree*, an integer vector, lying in n-dimensional real space, whose components are equal to the powers with which the coordinate functions enter in the monomial. A *Laurent polynomial* is a linear combination of monomials. The *support of the Laurent polynomial* is the set of powers of monomials entering in the Laurent polynomial with nonzero coefficients. (The Laurent polynomial is an ordinary polynomial if its support lies in the positive octant.) The *Newton polyhedron* of the Laurent polynomial is the convex hull of its support. It is much more convenient to consider the Laurent polynomials not in C^n but in the $(C\backslash 0)^n$-dimensional complex space, from which all the coordinate planes have been eliminated. With each face Γ of the Newton polyhedron of the Laurent polynomial f we associate a new Laurent polynomial, which is called the *restriction of the polynomial to the face*, denoted by f^Γ and defined as follows: only those monomials appear in f^Γ that have powers lying in the face Γ, with the same coefficients that they have in f.

Now let us consider a system of n Laurent equations $f_1 = \cdots = f_n = 0$ in $(C\backslash 0)^n$ with a common Newton polyhedron. The restricted system, $f_1^\Gamma = \cdots = f_n^\Gamma = 0$ corresponding to each face Γ of the polyhedron. The restricted system actually depends on a smaller number of variables, and in the case of general position is incompatible in $(C\backslash 0)^n$. We say that a system of n equations in n unknowns with a common Newton polyhedron is *regular* if all the restrictions of this system are incompatible in $(C\backslash 0)^n$. The following theorem of Kushnirenko holds: The number of solutions in $(C\backslash 0)^n$, counted with their multiplicities, of a regular system of n equations in n unknowns with a common Newton polyhedron is equal to the volume of the Newton polyhedron multiplied by n!

Example: The Newton polyhedron of the polynomial of degree m in n unknowns is the simplex $0 \leqslant x_1, \ldots, 0 \leqslant x_n$, $\sum x_i \leqslant m$ (we assume that the polynomial contains all monomials of degree $\leqslant m$). The volume of such a simplex is $m^n/n!$. The number of roots of the total system of n equations of degree m in n unknowns, according to Kushnirenko's theorem, is equal to m^n. This answer agrees with Bézout's theorem. If the polynomial does not contain all monomials of degree less than or equal to m, then the Newton polyhedron can be smaller than the simplex, so the number of solutions, calculated from Kushnirenko's theorem, can be smaller than the number m^n calculated from Bézout's theorem. Because of the absence of the monomials, certain infinitely distant points may be roots of the system of equations. Bézout's theorem, which calculates the number of roots of the system in projective space, takes into account these parasitic roots, while Kushnirenko's theorem does not.

Remark: The proof of Kushnirenko's theorem is found in Ref 41. Another proof can be extracted from the recent theorem of Atiyah on symplectic actions of tori (Ref. 62). We associate with each n-dimensional, integer-valued, convex polyhedron a symplectic "Veronese manifold" and a set of n commuting Hamiltonians on it, for each of which the motion is periodic with a period independent of the trajectory. The "moment mapping" corresponding to these Hamiltonians, according to Atiyah's theorem, maps the Veronese manifold on some convex polyhedron. It is not difficult to calculate that this polyhedron coincides with the original polyhedron. The calculation of the number of roots of a typical system of equations with a given

Newton polyhedron reduces to determining the volume of the Veronese manifold, which is easily done with the help of Atiyah's theorem: This volume is proportional to the volume of the polyhedron. The Veronese manifold is constructed from the convex integer polyhedron as follows. Consider the projective "space of monomials" CP^{N-1}, the number of whose homogeneous coordinates is equal to the number N of integer points lying inside and on the faces of Δ. By a *Δ-Veronese manifold* we mean the image of the mapping $(C\backslash 0)^n \to CP^{N-1}$, which associates the point z of $(C\backslash 0)^n$ with the point of projective space whose homogeneous coordinates coincide with the values at the point z of monomials whose degrees lie within Δ and on its faces. Since the Δ-Veronese manifold is algebraic, the numbers of points of its intersection with the planes of complementary dimension are the same for almost all planes. According to the Buffon-Crofton formula this number is proportional to the volume of the manifold. But a plane of codimension n in the space of monomials CP^{N-1} corresponds to a system of n equations with the Newton polyhedron Δ. The number of points of intersection of a plane with the Δ-Veronese manifold is equal to the number of solutions of the system. Therefore, the number of solutions of the system is proportional to the volume of the Veronese manifold. Atiyah's theorem is used to calculate this volume. A real n-dimensional torus acts in the space $(C\backslash 0)^n$: each of the n coordinates can be multiplied by a number with absolute value equal to one. This action of the torus on $(C\backslash 0)^n$ could be carried over to the Δ-Veronese manifold and is symplectic there. Thus, Atiyah's theorem is applicable; it follows from this theorem that the volume of the Δ-Veronese manifold is proportional to the volume of the original Newton polyhedron Δ.

3.1.2 How Does One Find the Number of Solutions of a System of n Equations in n Unknowns with Different Newton Polyhedra? Here is the answer to this question for a system in general position with fixed Newton polyhedra: the number of solutions not lying on the coordinate planes is equal to the mixed volume of the Newton polyhedra, multiplied by $n!$ Below we shall give the definition of mixed volume and describe explicitly the conditions for degeneracy.

The *Minkowski sum* of two subsets of a linear space is the set of sums of all pairs of vectors, in which one vector of the pair lies in one subset and the second vector in the other. The product of a subset and a number can be determined in a similar manner. The Minkowski sum of convex bodies (convex polyhedra, convex polyhedra with vertices at integer points) is a convex body (convex polyhedron, convex polyhedron with vertices at integer points). The following theorem holds:

MINKOWSKI THEOREM The volume of a body which is a linear combination with positive coefficients of fixed convex bodies lying in R^n is a homogeneous polynomial of degree n in the coefficients of the linear combination.

Definition The *mixed volume* $V(\Delta_1, \ldots, \Delta_n)$ *of the convex bodies* $\Delta_1, \ldots, \Delta_n$ in R^n is the coefficient in the polynomial $V(\lambda_1\Delta_1 + \cdots + \lambda_n\Delta_n)$ of $\lambda_1 \times \cdots \times \lambda_n$, divided by $n!$ (here $V(\Delta)$ is the volume of the body Δ).

The mixed volume of n identical bodies is equal to the volume of any one of them. The mixed volume of n bodies is expressed in terms of the usual volumes of their sums in the same way as the product of n numbers is expressed in terms of the n-th powers of their sums. For example, for $n = 2$,

$$ab = \tfrac{1}{2}\left[(a+b)^2 - a^2 - b^2\right]$$

$$V(\Delta_1, \Delta_2) = \tfrac{1}{2}\left[V(\Delta_1 + \Delta_2) - V(\Delta_1) - V(\Delta_2)\right]$$

Similarly, for $n = 3$,

$$V(\Delta_1, \Delta_2, \Delta_3) = \frac{1}{3!}\left[V(\Delta_1 + \Delta_2 + \Delta_3) - \sum_{i<j} V(\Delta_i + \Delta_j) + \sum V(\Delta_i)\right]$$

Example: Suppose that Δ_1 is the rectangle $0 \leqslant x \leqslant a$, $0 \leqslant y \leqslant b$ and Δ_2 is the rectangle $0 \leqslant x \leqslant c$, $0 \leqslant y \leqslant d$. The Minkowski sum $\Delta_1 + \Delta_2$ is the rectangle $0 \leqslant x \leqslant a + c$, $0 \leqslant y \leqslant b + d$. The mixed volume $V(\Delta_1, \Delta_2)$ is equal to $ad + bc$. The number $ad + bc$ is the permanent of the matrix $\left[\begin{smallmatrix} a & c \\ b & d \end{smallmatrix}\right]$ (the definition of the permanent differs from that of the determinant only in that all the terms in the permanent have a plus sign). In the multidimensional case the mixed volume of n parallelepipeds with sides parallel to the coordinate axes is also equal to the permanent of the corresponding matrix.

Let us consider a system of n Laurent equations $f_1 = \cdots = f_n = 0$, with Newton polyhedra $\Delta_1, \ldots, \Delta_n$. Below we define the regularity condition for such systems.

BERNSHTEIN'S THEOREM The number of solutions in $(C\backslash 0)^n$ (which take account of the multiplicity) of a regular system of n equations in n unknowns, is equal to the mixed volume of the Newton polyhedra of the equations of the system, multiplied by $n!$

Example: The number of roots of a general system of polynomial equations, in which the i-th variable enters in the j-th equation with a power no higher than $a_{i,j}$, is equal to the permanent of the matrix $(a_{i,j})$, multiplied by $n!$

Kushnirenko's theorem coincides with Bernshtein's theorem for equations with identical Newton polyhedra. We now proceed to the definition of a *regular system of equations*. We first define the truncations of a system associated with a function ξ. We take an arbitrary linear function ξ on the space R^n in which the Newton polyhedra lie. We denote by f^ξ the restriction of the Laurent polynomial f to that face of its Newton polyhedron on which the linear function ξ takes its maximum value. We associate the restricted system $f_1^\xi = \cdots = f_n^\xi = 0$ with the system of equations $f_1 = \cdots = f_n = 0$ and the linear function ξ. For a nonzero function ξ the restricted system actually depends on a smaller number of variables. Therefore, in the case of general position such a system is inconsistent in $(C\backslash 0)^n$. A given system of equations has only a finite number of restricted systems (if the polyhedra of all the equations coincide, then the truncations correspond to the faces of the common polyhedron). A system of n equations in n unknowns is said to be regular if its restrictions for all nonzero functions ξ are inconsistent in $(C\backslash 0)^n$. It is just such a system to which Bernshtein's theorem is applicable.

We note that if for each nonzero linear function ξ, the maximum in one of the polyhedra is attained at a vertex then the regularity conditions are automatically satisfied. [In this case the truncated system contains an equation which is contained in a monomial set equal to zero; this equation has no solutions in $(C\backslash 0)^n$.] For example, in the case of $n = 2$, the regularity conditions are satisfied automatically, if the two Newton polyhedra on the real plane have no parallel sides.

3.1.3 The Newton Polyhedron of the Germ of an Analytic Function In this paragraph we discuss the calculation of the multiplicity of a root of a system of equations, the Milnor number, and the modality of a function in two variables, in terms of Newton polyhedra.

We begin with definitions. To each power series in n variables we associate its support, the set of powers of monomials appearing in the series with nonzero coefficients. To each point of the support we associate an octant with vertex at that point, and consisting of all points for which each coordinate is not smaller than the corresponding coordinate of the vertex. The *Newton polyhedron* of the series is the convex hull of the union of all octants with vertices on the support of the series. *The Newton diagram of the series* is the union of compact faces of the Newton polyhedron. A Newton polyhedron of the series is said to be *suitable* if it intersects all the coordinate axes.

By the *volume* $V(\Delta)$ *of a suitable polyhedron* we mean the volume of the (nonconvex) region between the origin and the faces of the polyhedron in the positive octant R^n_+. Minkowski's theorem holds for the volumes of suitable polyhedra: the volume of a linear combination, with positive coefficients, of suitable polyhedra (just like the usual volume of polyhedra) is a polynomial in the coefficients of the linear combination. Therefore, we can define *a mixed volume for suitable polyhedra*, which is a verbatim repetition of the usual definition of mixed volume (see par. 3.1.2). For example, in the plane the mixed volume $V(\Delta_1, \Delta_2)$ of two suitable polyhedra Δ_1, Δ_2 is defined by the formula

$$2V(\Delta_1, \Delta_2) = V(\Delta_1 + \Delta_2) - V(\Delta_1) - V(\Delta_2)$$

The following local variant of Bernshtein's theorem (see par. 3.1.2) holds.

THEOREM The multiplicity of the origin as of a regular solution of a system of n analytic equations of n unknowns with suitable Newton polyhedra is equal to the mixed volume of the suitable Newton polyhedra, multiplied by $n!$

Example: The multiplicity of the origin as of a regular solution of the system $x_1^a + x_2^b = x_1^c + x_2^d = 0$ with positive powers a, b, c, d, with $ad \neq bc$, is equal to $\min(ad, bc)$.

We give a precise definition of regularity of the origin. The linear function ξ is said to be negative if all its values in the positive octant are negative (i.e., $\xi = \sum a_k, x_k,\ a_k < 0$). The restriction f^{ξ} of the analytic function f to negative functions ξ is its truncation to that face of the Newton polyhedron at which the function ξ attains its maximum. A null solution of a system of n analytic equations $f_1 = \cdots = f_n = 0$ with suitable Newton polyhedra at the origin is said to be *regular* if the restricted of the system, $f_1^{\xi} = \cdots = f_n^{\xi} = 0$, is inconsistent in $(C\backslash 0)^n$ for all negative functions ξ.

We now give the formula for the Milnor number μ of the germ of an analytic function. First, we shall give several definitions and notations. We say that the germ of an analytic function f with suitable Newton polyhedron Δ is Δ-*nondegenerate* if, for any compact face Γ of the Newton polyhedron, the system $z_1(\partial f^{\Gamma}/\partial z_1) = \cdots = z_n(\partial f^{\Gamma}/\partial z_n) = 0$ is inconsistent in $(C\backslash 0)^n$ (here f^{Γ} is the truncation of f to the face Γ). We give some geometric notations: for a suitable Newton polyhedron $\Delta \subset R^n$, we denote by Δ^I its intersection with the coordinate plane R^I in R^n, by $d(I)$ the dimension of this coordinate plane and by $V(\Delta^I)$ the $d(I)$-dimensional volume of the (nonconvex) region in the positive octant R^I_+ between the origin and the boundary of Δ^I.

THEOREM For a Δ-nondegenerate function f with a suitable Newton polyhedron Δ, the Milnor number is

$$\sum d(I)!\cdot(-1)^{n-d(I)}V(\Delta^I) + (-1)^n$$

where the summation is taken over all intersections Δ^I of the suitable Newton polyhedron with the coordinate planes.

Example: For almost all functions of two variables with a given suitable Newton polyhedron, $\mu = 2S - a - b + 1$, where S is the area under the polyhedron, a and b are coordinates of the points of the polyhedron on the axes (see the figure).

The *modality* $m(f)$ of a function f is the dimension of the space of orbits of the group of diffeomorphisms in the space of convergent power series which has no free and no linear terms in the neighborhood of the orbit of the point f.

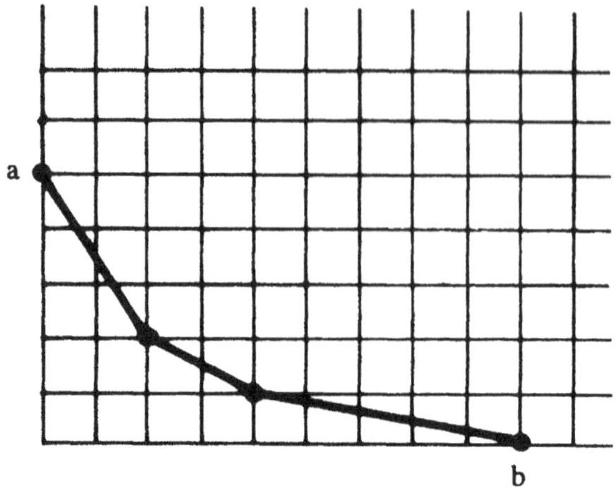

Calculation of the Milnor number.

$$\mu = 2S - a - b + 1$$
$$= 2 \cdot 12\tfrac{1}{2} - 5 - 9 + 1 = 12$$

THEOREM (KUSHNIRENKO) The modality of a Δ-nondegenerate function of two variables with suitable Newton polyhedron is equal to the number of integer points lying on the Newton diagram and below it, for which each coordinate is not less than two.

3.1.4 Complete Intersections Consider in $(C\backslash 0)^n$ a system of k Laurent equations $f_1 = \cdots = f_k = 0$ with Newton polyhedra $\Delta_1, \ldots, \Delta_k$. For Laurent polynomials in general position with given Newton polyhedra, the discrete invariants of the set f solutions are identical and are expressed in terms of the polyhedra. Below we give an explicit description of the conditions for nondegeneracy and describe the Euler characteristic and the homotopy of the complete intersection.

Definition A system $f_1 = \cdots = f_k = 0$ is said to be *nondegenerate for its Newton polyhedra* if for any linear function ξ the following (ξ) condition is satisfied: for any solution z of the truncated system $f_1^\xi = \cdots = f_k^\xi = 0$, lying in $(C\backslash 0)^n$ the differentials df_i^ξ are linearly independent in the tangent space at the point z.

Each of the (ξ) conditions is satisfied for almost all Laurent polynomials with fixed polyhedra (this is the Sard-Bertini theorem).

Although formally the nondegeneracy conditions are described for any linear function ξ, actually the different conditions are finite in number (there exist only a finite number of different truncated systems). Therefore the nondegeneracy conditions are satisfied almost everywhere. Among the nondegeneracy conditions there is the (0) condition (for the null function ξ), according to which the system $f_1 = \cdots = f_k = 0$ determines a nonsingular $(n-k)$-dimensional manifold in $(C\backslash 0)^n$. We note that for $k = n$ the class of regular systems is broader than the class of nondegenerate systems: for regular systems the (0) condition does not have to be satisfied (regular systems can have multiple roots).

THEOREM The Euler characteristic of the nondegenerate complete intersection $f_1 = \cdots = f_k = 0$ in $(C\backslash 0)^n$, $(k \leqslant n)$, with the Newton polyhedra $\Delta_1, \ldots, \Delta_k$ is equal to $(-1)^{n-k} \cdot n! \cdot \sum V(\Delta_1, \ldots, \Delta_k, \Delta_{i_1}, \ldots, \Delta_{i_{n-k}})$, where the sum is taken over all sets $1 = i_1 \leqslant \cdots \leqslant i_{n-k} \leqslant k$.

We note one special case of this theorem.

COROLLARY The Euler characteristic of a hypersurface in $(C\backslash 0)^n$ defined by a nondegenerate equation with fixed Newton polyhedron is equal to the volume of the Newton polyhedron multiplied by $(-1)^{n-1} \cdot n!$

Another consequence of this theorem is Bernshtein's theorem (see par. 3.1.2): for $k = n$, the nondegenerate complete intersections consist of points, and the Euler characteristic is equal to the number of points (strictly speaking, Bernshtein's theorem is slightly stronger than this corollary, since it is applicable to a degenerate regular system).

THEOREM (on homotopy type) Suppose that the Newton polyhedra Δ_i have a total dimension $\dim \Delta_i = n$ and the complete intersection $f_i = \cdots = f_k = 0$ in $(C\backslash 0)^n$ is nondegenerate. Then for $k < n$ the complete intersection is connected and has the homotopy type of a bouquet of $(n-k)$-dimensional spheres. The number of spheres in the bouquet is equal to $n! \sum V(\Delta_1, \ldots, \Delta_k, \Delta_{i_1}, \ldots, \Delta_{i_n}) - (-1)^{n-k}$ (the summation is taken over all sets $1 \leqslant i_1 \leqslant \cdots \leqslant i_{n-k} \leqslant k$).

COROLLARY The cohomology groups of a nondegenerate complete intersection with Newton polyhedra of complete dimension are different from zero only in dimensions $n - k$ and 0, where for $n > k$ the zero-measure cohomology group is one-dimensional.

3.1.5 Genus of Complete Intersections The formulas given below for the genus of complete intersections are generalizations of the following formulas for Abelian and elliptic integrals. Let us consider a Riemann surface $y^2 = P_3(x)$ (the complex phase curve of motion of a point in a field with a cubic potential). This Riemann surface, which is diffeomorphic to the torus, exhibits a single, everywhere holomorphic, differential form dx/y (the differential of the time of motion along the phase curve). In the case of a potential of degree n, the curve $y^2 = P_n(x)$ is diffeomorphic to the sphere with g handles, where g is connected with n either by the formula $n = 2g + 1$, or the formula $n = 2g + 2$ (depending on the parity of n). The basis of holomorphic forms in this case is given by g forms of the type $x^m dx/y$, $0 \leqslant m < g$. The number g is the genus of the curve. The Newton polyhedron of the curve $y^2 = P_n(x)$ is a triangle with vertices $(0,0)$, $(0,2)$, and $(n,0)$. There are exactly g points with integer coordinates strictly inside this triangle. In terms of these points, one can give a basis for the space of holomorphic forms: the point $(1,a)$, which lies inside the triangle, corresponds to the form $x^{a-1}dx/y$. We give below the generalization of this procedure for constructing a basis of holomorphic forms for the multi-dimensional case.

The nondegenerate complete intersections $f_1 = \cdots = f_k = 0$, where the f_i are Laurent polynomials, are smooth algebraic affine manifolds. In the cohomologies of such manifolds there is an additional structure, namely, the mixed Hodge structure. The discrete invariants of such a structure are calculated in terms of the Newton polyhedra. We consider only the calculation of the arithmetic and geometric genus which are invariants of this kind.

We first recall some definitons and general statements. Suppose that Y is a nonsingular (possibly noncompact) algebraic manifold. The set of holomorphic p-forms on Y, which extend holomorphically to any nonsingular algebraic compactification, is automatically closed, and it realizes the zero class of homology of the manifold Y only if it is equal to zero. Forms of this kind form a subspace in the p-dimensional cohomologies of the manifold Y. We denote the dimension of this subspace by $h^{p,0}(Y)$. The *arithmetic genus* of the manifold Y is the alternating sum $\sum(-1)^p h^{p,0}(Y)$ of the numbers $h^{p,0}(Y)$. The *geometric genus* of the manifold Y is the number $h^{n,0}(Y)$, where n is the complex dimension of the manifold Y.

Now we turn to the Newton polyhedron. We shall use the *characteristic $B(\Delta)$ of integer polyhedra*. Here is its definition. Suppose that Δ is a q-dimensional polyhedron with vertices at integer points, lying in R^n and R^q is a q-dimensional subspace, containing Δ. The number $B(\Delta)$ is defined as the number of integer points lying strictly within the polyhedron Δ (in the geometry of the subspace R^q), multiplied by $(-1)^q$.

THEOREM The arithmetic genus of the nondegenerate complete intersection $f_1 = \cdots = f_k = 0$ in $(C\backslash 0)^n$, $(k \leqslant n)$, with the Newton polyhedra $\Delta_1, \ldots, \Delta_k$ is

$$1 - \sum B(\Delta_i) + \sum_{j>i} B(\Delta_i + \Delta_j) - \cdots + (-1)^k B(\Delta_1 + \cdots + \Delta_k)$$

COROLLARY The geometric genus of the nondegenerate complete intersection (for $k < n$) with the polyhedra of full dimensionality is

$$(-1)^{n-k}\Big(-\sum B(\Delta_i) + \sum_{j>i} B(\Delta_i + \Delta_j) - \cdots$$
$$+(-1)^k B(\Delta_1 + \cdots + \Delta_k)\Big)$$

PROOF OF THE COROLLARY According to the theorem on the homotopic type of the complete intersection (see par. 3.1.4), $h^{0,0} = 1$ and $h^{p,0} = 0$ for $0 < p < n - k$. The corollary now follows from the calculation of the arithmetic genus.

We now give a complete discription of the holomorphic forms of highest dimension which could be extended holomorphically to the compactification, for the case of a nondegenerate hypersurface $f = 0$ in $(C\backslash 0)^n$ with the Newton polyhedron Δ of complete dimension. For each integer point, lying strictly within the Newton polyhedron Δ, we denote by w_a the n-form on the hypersurface $f = 0$, defined by the formula

$$w_a = z_1^{a_1} \times \cdots \times z_n^{a_n} \frac{dz_1}{z_1} \wedge \cdots \wedge \frac{dz_n}{z_n} \Big/ df; \qquad a = a_1, \ldots, a_n$$

THEOREM The forms w_a lie in the space of forms extendable holomorphically on the compactification of the manifold $f = 0$; they are linearly independent and generate that space. In particular, the geometric genus of the hypersurface is equal to $|B(\Delta)|$.

Example: Consider a curve in the plane determined by the equation $y^2 = P_n(x)$. The interior monomials for the Newton polyhedron of this curve have the form yx^a, where $1 \leqslant a < n/2$. The forms w_a corresponding to the monomials coincide [on the curve $y^2 = P_n(x)$] with the forms $x^{a-1}dx/2y$. Thus, for the curve $y^2 = P_n(x)$ we get the usual description of all Abelian differentials.

3.1.6 Total Intersections in C^n It is customary to consider systems of equations not in $(C\backslash 0)^n$, but in the usual complex space C^n. Calculations with Newton polyhedra also are carried out in this situation. The answers here are more messy. As an example we consider the calculation of the Euler characteristic of a hypersurface in C^n. Suppose that f is a polynomial in n complex variables with a nonzero free term and Newton polyhedron Δ. We introduce the following notation: Δ^I is the intersection of the Newton polyhedron Δ with the coordinate plane R^I in R^n, $d(I)$ is the dimension of this plane, and $V(\Delta^I)$ is the $d(I)$-dimensional volume of the polyhedron Δ^I.

THEOREM Let f be a nondegenerate polynomial with the Newton polyhedron Δ which has a nonzero free term. Then $f = 0$ is a nonsingular hypersurface in C^n, intersecting transversally all the coordinate planes in C^n. The Euler characteristic of this hypersurface is equal to $\Sigma(-1)^{d(I)-1}d(I)!\,V(\Delta^I)$, where the summation runs over all intersections of the Newton polyhedron Δ^I with the (nonzero) coordinate planes. This theorem follows from the calculation of the Euler characteristic in $(C\backslash 0)^n$ (see par. 3.1.4) and from the additivity of the Euler characteristic. We also note that the formula given for a hypersurface is analogous to the formula for the Milnor number (see par. 3.1.3).

3.2 Index of a Vector Field

The investigation of the topology of real algebraic curves, surfaces, etc., is a necessary stage in the study of singularities in analysis (the Taylor expansion approximates all objects by algebraic ones, and the topology of these objects has a decisive influence over the topology of the analytic objects).

Unfortunately, very little is known about the topology of real algebraic manifolds (even curves) of high degree. Algebraic curves of degree 3 and 4 were studied by Newton and Descartes, but what a plane algebraic curve of degree 8 might look like is not known even today (a curve of sixth degree can't have more than 11 ovals, and in that case only one of them contains other ovals inside it, namely, 1, 5 or 9 others—this was discovered in 1970).

The theory of singularities enables one to give some restrictions on the topology of real algebraic manifolds. For example, from the expression for the Poincaré index of a singular point of the vector field in terms of the local algebra of the singular point there follows the inequality of Petrovskii-Oleinik for the Euler characteristic of algebraic manifolds of fixed degree.

3.2.1 The Index of a Homogeneous Singular Point What values can the index of an isolated singular point of a vector field in R^n, whose components are homogeneous polynomials of degree $m_1, \ldots, m_n$, have? Theorem 1 formulated below gives a complete answer to this question.

We introduce the following notations. $m = m_1, \ldots, m_n$ is a set of natural numbers; $\Delta(m)$ is a parallelepiped in R^n, defined by the inequalities $0 \leqslant y_1 \leqslant m_1 - 1, \ldots, 0 \leqslant y_n \leqslant m_n - 1$; $\mu = m_1 \cdot \ldots \cdot m_n$ is the number of integer points in the parallelepiped $\Delta(m)$; $\prod(m)$ is the number of integer points in the central section $y_1 + \cdots + y_n = \frac{1}{2}(m_1 + \cdots + m_n - n)$ of the parallelepiped $\Delta(m)$.

THEOREM 1 (Refs. 33 and 34) The index ind of an isolated singular point 0 of a vector field in R^n, whose components are homogeneous polynomials of degree $m_1, \ldots, m_n$, satisfies the inequality $|\text{ind}| \leqslant \prod(m)$ and the congruence $\text{ind} \equiv \mu \bmod 2$. There are no other restrictions on the index.

The proof of the estimate of the index in Theorem 1 is obtained from the signature formula for the index (see par. 3.2.5).

We note that the algebraic formula for the number $\prod(m)$ is by no means simple. Just the simple geometric interpretation of this number enables one to present examples of vector fields with all indices not forbidden by Theorem 1 (Ref. 33).

3.2.2 Petrovskii-Oleinik Inequalities Consider a real nonsingular projective hypersurface A of degree $m + 1$, given by a homogeneous polynomial f in n variables.

The Petrovskii-Oleinik inequalities consist of the following:

$$|\chi(A) - 1| \leqslant N_n(m) \qquad \text{if } n \text{ is even;}$$

$$|\chi(B_+) - \chi(B_-)| \leqslant N_n(m) \qquad \text{if } n \text{ and } m \text{ are odd.}$$

Here χ is the Euler characteristic, B_+ and B_- are parts of RP^{n-1} given by the conditions $f \geqslant 0$, $f \leqslant 0$, respectively, while the number $N_n(m)$ is the number of integer points in the central section of the cube with side m [i.e., $N_n(m) = \prod(\underbrace{m, \ldots, m}_{n \text{ times}})$].

It is not difficult to show that both for even and odd n the left sides of the Petrovskii-Oleinik inequalities are equal in modulus to the index of the gradient of the homogeneous polynomial f giving the hypersurface. Thus, the Petrovskii-Oleinik inequalities represent a special case of the inequality $|\text{ind}| \leqslant \prod(m)$ of Theorem 1: this inequality must be applied to the index of the gradient vector field of the homogeneous polynomial of degree $m + 1$.

The Petrovskii-Oleinik inequalities were discovered in 1949 (Ref. 35). The relationship between these inequalities and the estimate of the index of a vector field was established in 1978 (Ref. 34).

3.2.3 Index of a Polynomial Field What values can the total index of singular points of a vector field in the region R^n have, if the region is defined by a polynomial inequality $P_0 < 0$, and if we know the degrees of the components of the field and the degree of the polynomial P_0? In this paragraph we formulate a complete answer to this question.

We denote by ind the sum of the indices of all the singular points of the field V in R^n, and by ind^+ and ind^- the sums of the indices of all the points of V in the regions $P_0 > 0$ and $P_0 < 0$. We shall say that the pair V, P_0 has a degree not exceeding (equal to) m, m_0, where $m = m_1, \ldots, m_n$, if the degree of the i-th component of the field does not exceed (is equal to) m_i, and the degree of the inequality $P_0 < 0$ does not exceed (is equal to) m_0. We shall say that the pair V, P_0 is *nondegenerate* if, first, the real hypersurface $P_0 = 0$ does not pass through singular points of the field V and if, secondly, the real singular points of the field V have finite multiplicity and lie "in a finite part of the space R^n" (the last condition means that the system $x_0 = \bar{P}_1 = \cdots = \bar{P}_n = 0$ has only the zero solution. Here the $\bar{P}_i$ are homogeneous polynomials of degree m_i in $n+1$ variables, equal to P_i when $x_0 \equiv 1$.)

We supplement the geometric definitions of par. 3.2.1: $\prod(m, m_0)$ is the number of integer points of the parallelepiped $\Delta(m)$ satisfying the inequalities

$$\tfrac{1}{2}(m_1 + \cdots + m_n - n - m_0) \leqslant y_1 + \cdots + y_n$$
$$\leqslant \tfrac{1}{2}(m_1 + \cdots + m_n - n + m_0)$$

$O(m, m_0)$ is the number of integer points of the parallelepiped satisfying the inqualities

$$\tfrac{1}{2}(m_1 + \cdots + m_n - n - m_0) \leqslant y_1 + \cdots + y_n$$
$$\leqslant \tfrac{1}{2}(m_1 + \cdots + m_n - n)$$

We note that $O(m, m_0) = \tfrac{1}{2}(\prod(m, m_0) + \prod(m))$, and that $\prod(m) \equiv \prod(m, m_0) \equiv \mu (\operatorname{mod} 2)$.

THEOREM 2 (Ref. 33) For the nondegenerate pair V, P_0 with degrees m, m_0 the numbers $a = \operatorname{ind}$, $b = \operatorname{ind}^+ - \operatorname{ind}^-$ and $c = \operatorname{ind}^+$ satisfy the inequalities $|a| \leqslant \prod(m)$, $|b| \leqslant \prod(m, m_0)$, $|c| \leqslant O(m, m_0)$ and the congruences $a \equiv b \equiv c \equiv \mu \bmod 2$. Conversely, for any number a (number b, number c) satisfying these restrictions, there exists a nondegenerate pair V, P_0 of degrees m, m_0 for which $\operatorname{ind} = a$ ($\operatorname{ind}^+ - \operatorname{ind}^- = b$, $\operatorname{ind}^+ = c$).

We note that Theorem 1 is a special case of Theorem 2 (the case where the field V has homogeneous components and a single singular point at the origin).

We can also evaluate the index for vector fields V with "infinitely remote singular points." The number ind^+ is defined if the region $P_0 > 0$ contains only isolated singular points of the field V. The number ind is defined if all the singular points of the field V are isolated.

THEOREM 3 (Ref. 33) Suppose that for the pair V, P_0 of degrees not exceeding m, m_0, the number ind$^+$ is defined. Then if $m_0 + \cdots + m_n \equiv n (\bmod 2)$ then the modulus of the number ind$^+$ does not exceed $O(m, m_0)$. In this case no other restrictions exist on the number ind$^+$. If $m_0 + \cdots + m_n \not\equiv n (\bmod 2)$ and m_0 is odd, then the modulus of the number ind$^+$ does not exceed $O(m, m_0 + 1)$. In this case there exists a pair V, P_0 with the extremal ind$^+ = \pm O(m, m_0 + 1)$.

COROLLARY Suppose that V is a vector field of degree not exceeding $m = m_1, \ldots, m_n$ with isolated singular points. Then, for $m_1 + \cdots + m_n \equiv n (\bmod 2)$ the estimate $|\text{ind}| \leqslant \prod(m)$ holds, while for $m_1 + \cdots + m_n \not\equiv n (\bmod 2)$ we have the estimate $|\text{ind}| \leqslant O(m, 1)$. Both are exact.

3.2.4 Inverse Jacobian Theorem To study a multiple singular point one uses a small deformation to dissociate it into nonmultiple singular points. The key information about such a dissociation is provided by the theorem on the inverse Jacobian, which is also of intrinsic interest. The inverse Jacobian theorem is applied, for example, in the proof of the signature formula for the index (see par. 3.2.5). We give a formulation of this theorem and derive the classic Euler-Jacobi classification formula from it.

Suppose that a system of n holomorphic equations in n complex unknowns depends on parameters, where for general values of the parameters it has only nonmultiple roots, while for exceptional values of the parameters (which form, in general, a hypersurface in parameter space) certain roots merge and become multiple. At a multiple root the Jacobian of the system is equal to zero, so when the nonmultiple roots of the system coalesce, the Jacobian of the system tends to zero at the coalescing roots. The reciprocal of the Jacobian is defined only at nonmultiple roots, and for coalescing roots tends to infinity. It turns out, however, that when one sums the reciprocals of the Jacobian over all the roots of the system, all the infinities cancel out, and the sum remains finite. We give an exact formulation of this theorem. Suppose that $U \subseteq C^n$ is a domain in C^n, $f: U \to C^n$ is a holomorphic mapping, $\mathscr{J} = \det(\partial f / \partial x)$ is the Jacobian of this mapping, V is a region in the image space that does not intersect the image of the boundary of the region U, and $h: U \to C$ is a holomorphic weight function.

Theorem (about the inverted Jacobian) The function

$$\varphi(a) = \sum_{f(x)=a} h(x)\mathcal{J}^{-1}(x) \tag{1}$$

defined on regular values $a \in V$ of the mapping f, is extendable holomorphically to the whole region V [the summation in (1) is taken over all roots of the system $f(x) = a$ lying in the region U].

One of the corollaries of the inverted Jacobian theorem is the *Euler-Jacobi formula*. We recall this formula and give its derivation from the theorem.

Suppose that for the system

$$P_1 = \cdots = P_n = 0 \tag{2}$$

of n polynomial equations of degrees $m_1, \ldots, m_n$ in n complex unknowns, all the roots are nonmultiple and "lie in the finite part" of the complex space. Suppose that h is an arbitrary polynomial in n variables, whose degree is less than the degree of the Jacobian $\mathcal{J}$ of the system, (i.e., less than $m_1 + \cdots + m_n - n$). In this case the Euler-Jacobi formula

$$\rho = \sum_{P(x)=0} h(x)\mathcal{J}^{-1}(x) \tag{3}$$

holds [where the summation goes over all roots of the system (2)].

The proof is obtained by applying Theorem 1 to a special mapping of the $(n+1)$-dimensional space into itself with a special weight function. The first n components of this mapping are homogeneous polynomials in the $n+1$ variables, which coincide with the polynomials of the system (2) on the hyperplane $x_{n+1} = 1$, the last component is the coordinate function x_{n+1}. The weight function is a homogeneous polynomial coinciding with the polynomial h on the hyperplane $x_{n+1} = 1$.

The inverse image of the point $(0, \epsilon)$ of this mapping consists of points $(\epsilon x, \epsilon)$, where x is a root of the system (2). A calculation shows that

$$\varphi(0,\epsilon) = \rho\epsilon^{p-(\Sigma m_i - n)}$$

where φ is the function defined by formula (1) for the mapping and for the weight function constructed above, ρ is a number defined by formula (3), and p is the degree of the weight function. As ϵ tends to zero, the function φ must remain bounded. If the degree of the weight polynomial is less than the degree of the Jacobian, then this is possible only if $\rho = 0$. Thus the formula of Euler-Jacobi is proved. We note that this formula is used in estimating the index of a polynomial vector field (see par. 3.2.3) and in the Petrovskii-Oleinik proof of their inequalities (Ref. 35).

3.2.5 General Theorems about the Index In this paragraph we give a formulation of the signature formula for the index and prove the estimate of the index of a singular point of a field with homogeneous components from Theorem 1 of par. 3.2.1.

Suppose that $f:(R^n,0)\to(R^n,0)$ is a mapping with finite multiplicity and Q is the local algebra of this mapping. We denote by $\mathscr{J}$ the image of the Jacobian of the mapping f in the local algebra Q.

ASSERTION 1 The image of the Jacobian $\mathscr{J}$ in the local algebra is not equal to zero.

Suppose that l is an arbitrary linear real-valued function on the local algebra Q. We define a bilinear form B_l on the local algebra by the formula $B_l(a,b) = l(a \cdot b)$.

THEOREM 1 [signature formula for the index (Refs. 54 and 55)] For any linear function l that is positive on the image of the Jacobian $\mathscr{J}$, the signature of the bilinear form B_l is equal to the index of the singular point 0 of the mapping f.

COROLLARY If in the μ-dimensional local algebra Q of the mapping f there exists an m-dimensional linear subspace L, the product of any two of whose elements is zero, then the modulus of the index of the point 0 does not exceed $\mu - 2m$.

PROOF OF THE COROLLARY The space L is a zero space for the bilinear form B_l. The modulus of the signature of the quadratic form on a μ-dimensional space having a zero m-dimensional subspace does not exceed $\mu - 2m$.

Let us now prove the estimate of the index of a singular point with homogeneous components $m_1, \ldots, m_n$ given in par. 3.2.1.

As the generators of the local algebra of this singular point we may take monomials whose degrees in the i-th variable are strictly less than m_i for any $i \leqslant n$. Such monomials are in one-to-one correspondence with the integer points of the parallelepiped $\Delta(m)$ (see par. 3.2.1). Consider the R-linear subspace L, which is spanned by the monomials whose degrees are greater than half their possible maximum $[d > \frac{1}{2}(\sum m_i - n)]$. The product of any two monomials of this subspace is equal to zero in the local algebra.

Consequently, $|\text{ind}| \leqslant \mu - 2\dim_R \bar{L}$.

The right side of this inequality is equal to the number of base monomials of average degree, i.e., is equal to $\prod(m)$ (see par. 3.2.1).

The estimate of Theorem 1 is proven. In conclusion, we give two general estimates of the index.

THEOREM 2 The index of a singular point of multiplicity μ in an n-dimensional space does not exceed $\mu^{1-1/n}$.

This theorem is derived from the signature formula using the Teissier inequalities (Ref. 59).

THEOREM 3 (Ref. 34) The modulus of a singular point of a gradient vector field in an even number of variables does not exceed the number $h_1^{n/2,n/2}$.

The number $h_1^{n/2,n/2}$ in the theorem is one of the Hodge-Steenbrink numbers (Ref. 34), which characterizes the complex geometry of the germ of the hypersurface $f = 0$. For the gradient of a homogeneous function f the estimate of the theorem coincides with the estimate of Theorem 1 (see par. 3.2.1).

3.3 *Few-Term Equations*

The topology of objects given by algebraic equations (real algebraic curves, surfaces, singularities, etc.) becomes complicated rapidly as the degree of the equation increases. It has become clear recently that the complexity of the topology depends on the number of monomials they contain, rather than on the degree of the equations: the theorems formulated below estimate the complexity of the topology of geometric objects in terms of how cumbersome their equations are.

3.3.1 Real Few-Term Equations The following theorem of Descartes is well-known. The number of positive roots of a polynomial in one real variable does not exceed the number of changes of sign in the sequence of its coefficients (zero coefficients are dropped from the sequence).

COROLLARY (Descartes' estimate) The number of positive roots of a polynomial is less than the number of terms in it.

A. G. Kushnirenko suggested that polynomials with a small number of terms be called fewnomials. Descartes' estimate shows that irrespective of the power of a few-term polynomial which may be arbitrarily large), it has few positive roots.

THEOREM 1 (Ref. 37) The number of nondegnerate real roots of a system of n fewnomial equations in n unknowns, containing no more than k monomials (independent of the degree of the fewnomials) is estimated from above by a certain function φ_1 of n and k.

THEOREM 2 (Ref. 37) The sum of the Betti numbers of the $(n - m)$-dimensional real manifold, determined by a nondegenerate system of m fewnomial equations, containing no more than k monomials, is estimated from above by some function φ_2 of n and k.

We give the familar estimate for the function $\varphi_1 : \varphi_1(n, k) < (n + 1)^k 2^{[k(k+1)/2]+n}$. This estimate is still far from exact. According to an unproved (but irrefutable) conjecture of A. G. Kushnirenko, an exact estimate of the number of solutions in the positive octant is

$$\prod_{1 \leqslant i \leqslant n} (k_i - 1),$$

where k_i is the number of monomials appearing in the i-th equation.

3.3.2 Complex Few-Term Polynomials The complex roots of the simplest two-term equation $x^N - 1 = 0$, with increasing N, are uniformly distributed in argument. The theorem formulated below shows that the roots of any nondegenerate system of fewnomial equations are also uniformly distributed in argument.

Let us consider a regular system of equations $f_1 = \cdots = f_n = 0$ in $(C\setminus 0)^n$, with fixed Newton polyhedra $\Delta_1, \ldots, \Delta_n$ (see par. 3.1.2). In writing these equations we encounter only k monomials (i.e., the union of the supports of these equations contains only k points). We denote by $N(f, G)$ the number of solutions of this system, whose arguments lie in a given domain G of the torus $T = \{\varphi_1, \ldots, \varphi_n\}$ mod 2π. For the number $\prod(\Delta^*, \partial G)$ defined below, which depends only on the geometry of the Newton polyhedra and on the domain G, the following theorem is valid.

THEOREM (Ref. 60) There exists a function φ of n and k such that for every regular system of n equations with k monomials the relation

$$\left|N(f, G) - \frac{n!}{(2\pi)^n} V(G) V(\Delta_1, \ldots, \Delta_n)\right| \leqslant \varphi(n, k) \prod(\Delta^*, \partial G)$$

is valid. Here $V(G)$ is the volume of the region $G \subset T^n$, and $V(\Delta_1, \ldots, \Delta_n)$ is the mixed volume of the Newton polyhedra of the system.

We give the definition of the number $\prod(\Delta^*, \partial G)$. Suppose that Δ^* is the domain in R^n defined by the inequalities $\{\varphi \in \Delta^* \mid |m\varphi| < \pi/2$ for all integer-valued vectors m, lying in the union of the supports of the Laurent polynomials $f_1, \ldots, f_n\}$. The number $\prod(\Delta^*, \partial G)$ is defined as the minimum number of domains, which are equal to within a parallel displacement of the region Δ^*, necessary to cover the boundary ∂G of the region $G \subset T^n$.

Let us consider some special cases of the theorem: (1) the domain G coincides with the torus T^n. In this case $V(G) = (2\pi)^n$, $\prod(\Delta^*, \partial G) = 0$, and the theorem coincides exactly with Bernshtein's theorem (see par. 3.1.2); (2) the domain G contracts to a point $0 \in T^n$. In this case, $V(G) \to 0$, $\prod(\Delta^*, \partial G) = 1$, and the theorem coincides exactly with Theorem 1 (par. 3.3.1); (3) we simultaneously increase the polyhedra Δ_i (without overlapping them), without increasing the number of monomials k and without changing the domain G. In this case the mixed volume $V(\Delta_i, \ldots, \Delta_n)$ is of the order of the nth power of the size of the polyhedra Δ_i, while the number $\prod(\Delta^*, \partial G)$ is of the order of the $(n-1)$-power of the size. In this case the theorem shows that the roots are distributed uniformly in argument.

3.3.3 Generalizations The idea of few-term polynomials consists in the fact that the intersections of real level lines of "sufficiently simple" functions should be "sufficiently simple." At present, this notion has a basis not only for few-term polynomials, but also for real Liouville functions (Refs. 38 and 61), and for the even wider class of real, transcendent Pfaffian functions (Refs. 37 and 60). The class of Pfaffian functions is sufficiently wide: for example, if $f(x, y)$ is a Pfaffian function, then a solution of the differential equation $y' = f(x, y)$ is a Pfaffian function also. The construction of these functions is based on the fact that the solutions of the first-order equations do not oscillate (in contrast to the solutions of second-order equations such as $y'' = -y$).

The theorem on complex fewnomials has also been generalized (Ref. 60): This generalization allows one to use the theorem in estimating the number of zeros of the linear combinations of exponentials e^{ax}, where a is a real vector.

REFERENCES

1. Karpushkin, V. N., Uniform estimates of oscillating integrals in R^2, *Dokl. Akad. Nauk SSSR*, **254**, N1, 28–31 (1980).
2. Karpushkin, V. N., Uniform estimates of oscillating integrals, *Usp. Mat. Nauk* **36**, N4, 213 (1981).
3. Karpushkin, V. N., Uniform estimates of oscillating integrals with a parabolic or hyperbolic base. *Trud. sem. I. G. Petrovskii*, **9** (1983).
4. Vasil'ev, V. A., Asymptotic behavior of exponential integrals, Newton diagrams and classification of minimum points. *Funkts. Anal.* **11**, N3, 1–11 (1977).
5. Vinogradov, I. M., The method of trigonometric sums in the theory of numbers, Moscow, Nauka, 1971.
6. Arnol'd, V. I., Remarks on the method of stationary phase and Coxeter numbers, *Usp. Mat. Nauk* **28**, N5, 17–44 (1973).
7. Varchenko, A. N., Newton polyhedra and estimates of oscillating integrals. *Funkts. Anal.* **10**, N3, 13–38 (1976).
8. Bernshtein, I. N. and Gelfand, S. I., Meromorphy of the functions P^λ. *Funkts. Anal.* **3**, N1, 84–86 (1969).
9. Arnol'd, V. I., Critical points of functions on manifolds with boundary, simple Lie groups B_k, C_k, F_k, and singularities of evolutes. *Usp. Mat. Nauk* **13**, N5, 91–105 (1978).
10. Fedoryuk, M. V., Saddle point method, Moscow, Nauka 1977.
11. Landau and Lifshitz, Mechanics, Moscow, Nauka, 1965.

12. Arnol'd, Varchenko and Gusein-Zade, Singularities of differentiable mappings, Vol. 1, Moscow, Nauka 1982.
13. Arnol'd, Varchenko and Gusein-Zade, idem, vol. 2, Moscow, Nauka 1983.
14. Arnol'd, V. I., Catastrophe theory, Moscow, Znanie, 1981.
15. Bernshtein, Kushnirenko and Khovanskii, Newton polyhedra, *Usp. Mat. Nauk* **31** N3, 201–202 (1976).
16. Kushnirenko, A. G., The Newton polyhedron and the number of solutions of equations in *n* unknowns, *Usp. Mat. Nauk* **30**, 32, 302–303 (1975).
17. Chebotarev, N. G., The "Newton polyhedron" and its role in the contemporary development of mathematics. Collected works, vol. 3, Moscow-Leningrad, Acad. Press, 1950.
18. Brüno, A. D., Local method of nonlinear analysis of differential equations, Moscow, Nauka 1979.
19. Berezovskaya, F. S., Index of a stationary point of a vector field in the plane, *Funkts. Anal.* **13**, N2, 77 (1979).
20. Kushnirenko, A. G., Newton polyhedra and Milnor numbers, *Funkts. Anal.* **9**, N9, 74–75 (1976).
21. Bernshtein, D. N., Number of roots of a system of equations, *Funkts. Anal.* **9**, N3, 1–4 (1975).
22. Kushnirenko, A. G., Newton polyhedra and Bézout's theorem, *Funkts. Anal.* **10**, N3, 82–83 (1976).
23. Danilov, V. I., Netwon polyhedra and vanishing cohomologies, *Funkts. Anal.* **13**, N2, 32–47 (1979).
24. Khovanskii, A. G., Newton polyhedra and toric manifolds, *Funkts. Anal.* **11**, N4, 56–67 (1977).
25. Khovanskii, A. G., Newton polyhedra and genus of complete intersections, *Funkts. Anal.* **12**, N1, 51–61 (1978).
26. Khovanskii, A. G., Newton polyhedra and the Euler–Jacobi formula, *Usp. Mat. Nauk* **33**, N6, 245–246 (1978).
27. Danilov, V. I., Geometry of toric manifolds, *Usp. Mat. Nauk* **33**, N2, 85–135 (1978).
28. Khovanskii, A. G., Geometry of convex manifolds and algebraic geometry, *Usp. Mat. Nauk* **34**, N4, 160–161 (1978).
29. Khovanskii, A. G., Newton polyhedra and the index of a vector field, *Usp. Mat. Nauk* **36**, N4, 234 (1981).
30. Varchenko, A. N., On the number of integer points in a region, *Usp. Mat. Nauk* **37** (1982).
31. Varchenko, A. N., On the number of integer points in families of homothetic regions, *Funkts. Anal.* **17**, (1983).
32. Kazarnovskii, B. Ya., On the zeros of exponential sums, *Dokl. Akad. Nauk SSSR* **257**, N4, 804–808 (1981).
33. Khovanskii, A. G., Index of a polynomial vector field, *Funkts. Anal.* **13**, N1, 49–58 (1979).
34. Arnol'd, V. I., Index of singular points of a vector field, the Petrovskii–Oleinik inequalities and mixed Hodge structures, *Funkts. Anal.* **12**, N1, 1–14 (1978).
35. Petrovskii and Oleinik, On the topology of real algebraic surfaces, *Izv. Akad. Nauk SSSR. ser. mat.* **13**, 389–402 (1949).
36. Bogdanov, R. I., Local orbital normal forms of vector fields in the plane, *Trud. sem. I. G. Petrovskii*, N5, 51–84 (1975).
37. Khovanskii, A. G., A class of systems of transcendental equations, *Dokl. Akad. Nauk SSSR* **255**, N4, 804–807 (1980).

38. Gelfond, O. A., Khovanskii, A. G., On real Liouville functions, *Funkts. Anal.* **14**, N2, 52–53 (1980).
39. Atiyah, M. F., Resolution of singularities and division of distributions, *Comm. Pure Appl. Math.* **23**, N2, 145–150 (1970).
40. Duistermaat, J., Oscillatory integrals, Lagrange immersions and unfoldings of singularities, *Comm. Pure Appl. Math.* **27**, N2, 201–281 (1974).
41. Kouchnirenko, A. G., Polyèdres de Newton et nombres de Milnor, *Invent. Math.* **32**, 1–31 (1976).
42. Nye and Potter, The use of catastrophe theory to analyse the stability and toppling of icebergs, *Annals of Glagiology* **1**, 49–54 (1980).
43. Nye, Cooley and Thorndike, The structure and evolution of flow fields and other vector fields, *J. Phys. A: Math. Gen.* **11**, N8, 1455–1490 (1978).
44. Nye, J. F., The motion and structure of dislocations in wavefronts, *Proc. Roy. Soc.* (*London*) **A378**, 219–239 (1981).
45. Nye, J. F., Optical caustics from liquid drops under gravity: observations of the parabolic and symbolic umbilics, *Phil. Trans. Roy. Soc.* (London) **292**, 25–44 (1979).
46. Nye, J. F., Optical caustics in the near field from liquid drops, *Proc. Roy. Soc.* (London) **A361**, 21–41 (1978).
47. Berry and Upstill, Catastrophe optics: morphologies of caustics and their diffraction patterns, Progress in Optics, XVIII, North-Holland, 1980.
48. Berry, M. V., Singularities in waves and rays, Les Houches Summer School, 1980; Amsterdam, North-Holland, 1981.
49. Berry, M. V., Waves and Thom's theorem, *Adv. Phys.* **25**, 1–26 (1976).
50. Colin de Verdier, Y., Nombre de points entiers dans une famille homothétique de domains de R^n, *Ann. Scient. Ecole Norm. Super. ser. a* **10**, 559–575 (1977).
51. Nye and Thorndike, Events in evolving three-dimensional vector fields, *J. Phys. A. Math. Gen.* **13**, 1–14 (1980).
52. Randol, B., On the Fourier transform of the indicator function of a planar set, *Trans. AMS* **139**, 271–278 (1969).
53. Randol, B., On the asymptotic behaviour of the Fourier transform of the indicator function of a convex set, *Trans. AMS* **139**, 279–285 (1969).
54. Eisenbud and Levine, The topological degree of a finite C^∞-map germ, *Ann. Math.* **106**, N1, 19–38 (1977).
55. Khimshiashvili, G. N., On the local degree of a smooth mapping, *Repts. Acad. Georgian SSSR*, **85**, N2, 309–311 (1977).
56. Arnol'd, V. I., Mathematical Methods of Classical Mechanics, Springer Verlag, New York–Heidelberg–Berlin (1980).
57. Arnol'd, Shandarin and Zeldovich, The large scale structure of the universe, I. General properites; one- and two-dimensional models, Geophysical and Astrophysical Hydrodynamics (1982).
58. Teissier, B., Variétés toriques et polytopes, *Sem. Bourbaki*, 33e annee, N565, 1980/81.
59. Teissier, B., Appendix to Levine's–Eisenbud's article, *Ann. Math.* **106**, 39 (1977).
60. Hovansky, A., Sur les racines complexes de systèmes d'équations algébraiques ayant un petit nombre de monômes, *C. R. Acad. Sc. Paris* **292**, 937–940 (1981).
61. Khovanskii, A., Théoréma de Bézout pour les fonctions de Liouville, preprint IHES/M/81/45, Septembre, 1981, Bures-sur-Yvette, France.
62. Atiyah, M. F., Convexity and commuting hamiltonians, *Bull. London Math. Soc.* **14**, part 1, N46 (1982).

Remarks on the perturbation theory for problems of Mathieu type

V.I. Arnol'd

CONTENTS

Introduction

For the Mathieu equation

$$\ddot{x} + (\omega^2 + \varepsilon \cos t)x = 0$$

the width of each parametric resonance zone (or forbidden zone, in the terminology of solid state physics) decreases like a power with decreasing depth of modulation ε, and *the exponent is proportional to the number of the zone* (Fig. 1).

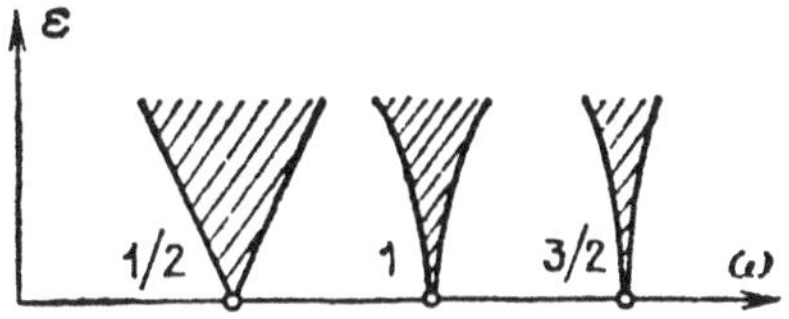

Fig. 1.

It follows, in particular, from general theorems in the present article that the zones of instability for Hill's equation

$$\ddot{x} + (\omega^2 + \varepsilon a(t))x = 0$$

have the same property when the variable coefficient a is *any trigonometric polynomial* (the exponent for the width of the zone decreases in inverse proportion to the degree of the polynomial).

Originally publ. in: Usp. Mat. Nauk 38(4), 189-203, © 1983 Russ. Acad. Sciences
English transl.: Russ. Math. Surveys 38(4), 215-233, © 1983 London Mathematical Society.

For the general Hill's equation (with any periodic coefficient a) the situation is completely different; for a typical a the width of any zone decreases like the *first* power of ε.

In view of the general algebraic nature of the proof, it is applicable also to many other problems in which the perturbation is a trigonometric polynomial. Linear equations or systems with constant coefficients in the leading term and with coefficients in the form of trigonometric polynomials in the lower-order terms thus have special properties, and can be called *equations of Mathieu type.*

For example, consider the equation in the Zel'dovich problem of the existence of a steady-state kinematic magnetic dynamo:

$$H^{\cdot} = \{v, H\} + D\Delta H,$$

where H is an unknown 2π-periodic (in (x, y, z)) magnetic field of divergence zero that is carried by a 2π-periodic velocity field v of an incompressible fluid and diffuses with diffusion coefficient D ($\Delta = -\mathrm{rot\ rot}$ is the Laplacian and $\{\cdot,\cdot\} = \mathrm{rot}\ [\cdot, \cdot]$ denotes the Poisson bracket).

The equation is of Mathieu type if the components of the field v are trigonometric polynomials (for example, for the velocity field

$$\cos y + \sin z, \ \cos z + \sin x, \ \cos x + \sin y$$

considered in [1], which exponentially expands the particles of the fluid). It was in investigating this example that I observed the general properties described above of equations of Mathieu type.

In student work done in 1959 [2] I investigated the family of mappings of the circle onto itself of the form

$$x \mapsto x + a + \varepsilon \cos x$$

and established that the zone of resonance m/n in the (a, ε)-plane (Fig. 2) reaches the point $a = 2\pi m/n$ on the axis $\varepsilon = 0$ by a narrow tongue whose width decreases like ε^n as $\varepsilon \to 0$ (this is stated without proof in [2]).

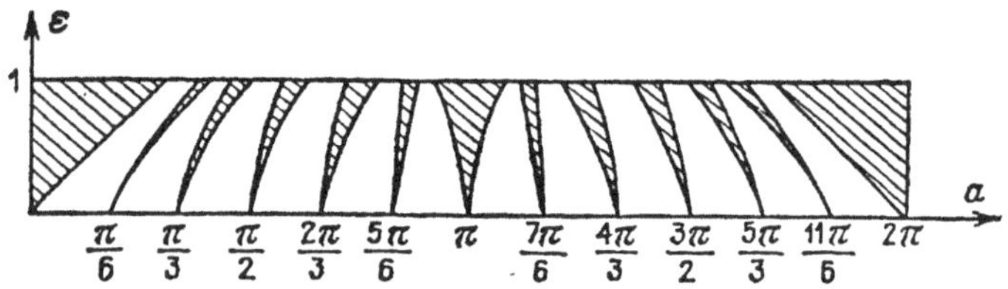

Fig. 2.

The theory of equations of Mathieu type suggests that here also the cosine can be replaced by a trigonometric polynomial of arbitrary degree p (with the corresponding replacement of ε^n by ε^r, where $r = -[-n/p]$).

Thus, the problem of zones of resonance for such special mappings of the circle also turns out to be "a problem of Mathieu type"

I did not try to formulate a single abstract theorem which would yield the above properties of resonance zones at once both for Hill's equations of Mathieu type and for mappings of the circle onto itself, although the algebraic reason for the power decrease in the width of a zone with the amplitude of modulation was obviously the same in the two cases (the conjecture that such a general theorem must exist was communicated to me by Gel'fand as far back as 1959 in a discussion of the results in [2]).

It gives me special pleasure to include a proof of the theorem from the 1959 work, carried out under the direct guidance of Kolmogorov, in the present article, which is dedicated to A.N. Kolmogorov on the occasion of his 80th birthday.

§1. Rayleigh-Schrödinger series for perturbations of a simple eigenvalue

We need some (simple) properties of the expansion of an eigenvalue of a matrix $\Lambda+\varepsilon A$, where Λ is a diagonal matrix, in a series of powers of the small parameter ε. Let λ_i denote the eigenvalues of the unperturbed operator Λ, and V_i the eigenspaces, so that

$$\Lambda = \oplus\Lambda_i, \quad \Lambda_i = \lambda_i E^i\colon V_i \to V_i, \quad \lambda_i \neq \lambda_j.$$

We decompose the matrix of the perturbing operator into the corresponding blocks

$$A_{ij}\colon V_i \to V_j.$$

Definition. The *quiver* of the family $\Lambda+\varepsilon A$ is defined to be the directed graph with vertices corresponding to the V_i (they are denoted by i) and with an edge going from i to j when the operator A_{ij} is non-zero.

Example. For the Mathieu operator

$$\Lambda + \varepsilon A = d^2/dt^2 + \varepsilon \cos t$$

defined in the space of 2π-periodic functions of t the space V_k is spanned by $e^{\pm ikt}$, and A is the operator of multiplication by $\cos t$. Therefore, the quiver has the form

$$\overset{0}{\cdot} \rightleftarrows \overset{1}{\cdot} \rightleftarrows \overset{2}{\cdot} \rightleftarrows \cdots$$

In quivers for operators of Mathieu type each vertex is directly connected with a finite (and even uniformly bounded) number of vertices. This is because a product of trigonometric polynomials is again a trigonometric polynomial.

We return to the general family $\Lambda+\varepsilon A$.

Definition. The *chronological product* of a path $\gamma = (i \to j \to \ldots \to k \to l)$ along arrows of a quiver is defined to be the product of the operators corresponding to the arrows, in the order indicated by the arrows:

$$\Pi_\gamma(A) = (A_{kl}\cdot \ldots \cdot A_{ij})\colon V_i \to V_l.$$

It is easy to prove the following result (by comparing the coefficients of ε, ε^2, ...; see §4).

***Theorem* 1.** *The perturbation of a simple eigenvalue is given by the series*

$$\lambda = \lambda_0 + \varepsilon\alpha_1 + \varepsilon^2\alpha_2 + \ldots,$$

in which the coefficient of the k-th power of the perturbation is equal to the sum

$$\alpha_k = \sum \Pi_\gamma(A)Q_\gamma(\Lambda), \quad |\gamma| = k,$$

over all loops of length k made up of arrows in the quiver and beginning and ending at the point 0.

Here $\Pi_\gamma(A)$ is the chronological product of a loop, and $Q_\gamma(\Lambda)$ is a universal (independent of A) rational function of those eigenvalues of the unperturbed operator that correspond to the vertices in the loop γ.

Remark 1. The explicit form of the rational function Q is not needed for the present, but it is given in §4.

Remark 2. If the quiver is locally finite, then there are finitely many terms in the sum for α_k, so the formal series for λ can be written. Its convergence for operators of Mathieu type is ensured by the ellipticity of the principal term, but we shall regard this series as formal (or we shall assume that the operators are finite-dimensional).

§2. Perturbations of a multiple eigenvalue

If λ_0 is a multiple eigenvalue, then under a perturbation it does not vary smoothly, but the corresponding invariant space V_0 and the restriction of the operator to this space do vary smoothly. The expansions of these objects in power series in ε are given by precisely the same formulae as the perturbations of an eigenvector and an eigenvalue when the latter is simple.

We look for the perturbed invariant space in the form of the graph of a mapping

$$\varepsilon B^1 + \varepsilon^2 B^2 + \ldots, \quad B^k\colon V_0 \to \oplus V_i, \quad i \neq 0.$$

The condition that the graph be invariant is written as the operator equation

$$(\Lambda + \varepsilon A)(E^0 \oplus \varepsilon B^1 + \ldots) = (E^0 \oplus \varepsilon B^1 + \ldots)(\lambda_0 E^0 + \varepsilon\alpha_1 + \ldots),$$

where the $\alpha_k : V_0 \to V_0$ are unknown linear operators, as are the B^k, and E^0 is the identity transformation on V_0.

The sum in the last set of parentheses plays the role of the perturbed eigenvalue. Indeed, this operator, which acts in the space V_0, is similar to the restriction of $\Lambda + \varepsilon A$ to the perturbed invariant space. Therefore, its spectrum coincides with the part of the spectrum of $\Lambda + \varepsilon A$ which arose from λ_0. We shall call it an *eigenoperator* of the family $\Lambda + \varepsilon A$.

***Theorem* 2.** *The coefficient $\alpha_k : V_0 \to V_0$ in the expansion $\lambda_0 E^0 + \varepsilon\alpha_1 + \varepsilon^2\alpha_2 + \dots$ of an eigenoperator of the family $\Lambda + \varepsilon A$ is given by the same formula in Theorem 1 as the coefficient α_k in the expansion of a simple eigenvalue.*

Indeed, this operator equation can be solved in precisely the same way as the scalar equation. For example, equating the coefficients of ε, we find that

$$\Lambda B^1 + AE^0 = E^0\alpha_1 + B^1\lambda_0,$$

from which, representing B^1 in the form $B^1 = \oplus B_i^1$, where $B_i^1 : V_0 \to V_i$ for $i \neq 0$, we find that

$$\alpha_1 = A_{00}, \quad B_i^1 = A_{0i}/(\lambda_0 - \lambda_i).$$

Equating the coefficients of ε^2, we obtain

$$\Lambda B^2 + AB^1 = E^0\alpha_2 + B^1\alpha_1 + B^2\lambda_0.$$

Equating the V_0-components of the left-hand and right-hand sides, we find that

$$\alpha_2 = \sum A_{i0} B_i^1 = \sum A_{i0} A_{0i}/(\lambda_0 - \lambda_i).$$

And equating the V_m-components for $m \neq 0$, we find that

$$(\lambda_0 - \lambda_m) B_m^2 = (AB^1)_m - (B^1\alpha_1)_m,$$

from which

$$B_m^2 = \frac{\sum A_{im} A_{0i}}{(\lambda_0 - \lambda_m)(\lambda_0 - \lambda_i)} - \frac{A_{0m} A_{00}}{(\lambda_0 - \lambda_m)^2}, \qquad i \neq 0,$$

and so on (the general formula for the denominators of all the terms is written out in §4).

§3. Widths of forbidden zones of even Hill's equations of Mathieu type

Let us consider Hill's equation

$$d^2x/dt^2 + (\omega^2 + \varepsilon a(t))x = 0, \quad a(t + 2\pi) \equiv a(t).$$

The zones of parametric resonance (or *forbidden zones*) are defined as the regions in the plane of the parameters (ω, ε) in which the multipliers (the eigenvalues of the monodromy operator for the given equation with periodic coefficients) are greater than 1 in modulus.

***Theorem* 3.** *Assume that the coefficient a is an even trigonometric polynomial of degree p:*

$$a(t) = \sum a_s e^{ist}, \text{ the } s \text{ are integers, } a_s = 0 \text{ for } |s| > p, \text{ and } a_{-s} = a_s.$$

Then the width of the N-th forbidden zone decreases no more slowly than $C\varepsilon^r$ as $\varepsilon \to 0$, where $r = -[-N/p]$.

Proof. According to the general theory of Hill's equation, the multipliers are equal to -1 or 1 on the boundary of a zone, depending on the parity of the number of the zone. Therefore, the boundaries of the zones can be determined with the help of the eigenvalues $\lambda^{\pm}$ of the operator

$$d^2/dt^2 + \varepsilon a(t)$$

on the space of 2π-periodic functions (for even zones) or on the space of 4π-periodic functions which change sign under translation by 2π (for odd zones):

$$\omega^{\pm} = +\sqrt{-\lambda^{\pm}}.$$

For $\varepsilon = 0$ the N-th zone contracts to the point $\omega_N^{\pm} = N/2$, and the corresponding pairs of eigenfunctions have the form

$$e_n, e_{-n}, \text{ where } e_n = e^{int}, \quad n = N/2, \quad N = 0, 1, 2, \ldots$$

(the eigenvalue is simple when $N = 0$).

To investigate the perturbed eigenvalues λ we apply the general formula in Theorems 1 and 2 to the diagonal operator $\Lambda = d^2/dt^2$ and to the perturbation εA (where A is multiplication by a).

Let V_m denote the space spanned by e_m and e_{-m} (this space corresponds to $\lambda_m = -m^2$, where $m = M/2$, and M runs through either all the even or all the odd non-negative integers, depending on the parity of the number N of the zone under investigation).

***Lemma* 1.** *The matrix of the operator $A_{ij} : V_i \to V_j$ in the chosen basis has the form*

$$(*) \qquad (A_{ij}) = \begin{pmatrix} a_{j-i} & a_{j+i} \\ a_{-j-i} & a_{i-j} \end{pmatrix}.$$

Proof. Multiplication by e_s carries e_i into e_{i+s}. Therefore, in the product of the trigonometric polynomial $a = \sum a_s e_s$ by e_i each term falls in an eigenspace different from V_j except for the terms with $s = j - i$ and $s = -j - i$, and similarly for ae_{-i}.

Lemma 1 gives us the next result.

***Lemma* 2.** *If $i+j > p$, then the matrix (A_{ij}) is scalar: $(A_{ij}) = a_{j-i}E$.*

Indeed, the degree of the polynomial a is equal to p, therefore, $a_{i+j} = a_{-i-j} = 0$. Since a is an even polynomial, the diagonal elements of the matrix (*) are all the same, and this proves Lemma 2.

We arrange the quiver of the family $\Lambda + \varepsilon A$ on the m-axis. Its vertices are all the non-negative integer points if the number (N) of the zone being investigated is even, and all the positive half-integer points m if N is odd. The length of each arrow in the quiver does not exceed the degree p of the perturbing polynomial a.

Definition. An arrow $i \to j$ in a quiver is said to be *remote* if the sum of the coordinates of its initial and terminal points is larger than the degree of the perturbing polynomial a:

$$i + j > p.$$

Example. A remote arrow does not contain the point 0.

The following simple lemma contains the main, "topological" point in the proof of Theorem 3.

Lemma 3. *Consider a loop of k arrows, each of length at most p, that begins and ends at the point n. If $kp < 2n$, then all the arrows in the loop are remote.*

Proof. Suppose that there is a non-remote arrow $i \to j$ in the loop, where $i+j \leqslant p$. We replace the vertex j and all the subsequent vertices and arrows of the loop by the vertices and arrows symmetric with respect to the origin. The arrow $i \to j$ itself is replaced by the arrow $i \to -j$. Its length is not greater than p, because the arrow $i \to j$ is non-remote. Accordingly, we have obtained a path from n to $-n$ made up of k arrows, each of length at most p. The length of the whole path is at most kp. But $kp < 2n$, by hypothesis. Hence, there is no non-remote arrow in the loop.

Completion of the proof of Theorem 3. Let us use Theorem 2. To compute the correction to the repeated eigenvalue $\lambda(0) = -n^2$ we form the eigenoperator (second-order matrix)

$$\lambda(\varepsilon) = \lambda(0)E + \varepsilon\alpha_1 + \varepsilon^2\alpha_2 + \dots$$

The matrix α_k can be expressed as a sum over admissible loops consisting of k arrows in our quiver and going from n to n (in the present notation the point 0 in Theorems 1 and 2 is denoted by n). The length of each arrow is at most p, since the perturbation has degree p. By Lemma 3, all the arrows in such a loop are remote when $kp < 2n$. By Lemma 2, a scalar (in our basis) matrix (A_{ij}) corresponds to a remote arrow $i \to j$. Therefore, all the chronological products in the formula for α_k consist of scalar matrices. Hence, the matrix α_k is scalar as long as $k < 2n/p$. But the difference between the two eigenvalues of the matrix $\lambda(\varepsilon)$ does not exceed $C\varepsilon^K$, where K is the number of the first non-scalar matrix among the α_k. We proved above that $K \geqslant r = -[-2n/p]$. Thus, the distance between the two eigenvalues $\lambda^{\pm}(\varepsilon)$ (and hence the width of the zone) decreases no more slowly than ε^r, which is what we needed to prove.

Remark 1. We have not used the fact that the perturbing polynomial is real: the estimate $|\lambda^+ - \lambda^-| \leqslant C\varepsilon^r$, $r = -[-2n/p]$, for the splitting of the eigenvalue $\lambda = -n^2$ under perturbation by a trigonometric polynomial of degree p has been proved for all even polynomials with complex coefficients.

Remark 2. Without the assumption that the perturbing polynomial a is even, our arguments prove only that the matrices α_k with $k < 2n/p$ are diagonal, and not that they are scalar. In reality the diagonal elements are equal even without the assumption of evenness, but to prove this we have to use not only the form of the numerators but also the denominators of the terms in the Rayleigh-Schrödinger series.

§4. Formulae for the denominators of terms in the perturbation theory series

We return to the perturbation theory series and to the notation in §§1 and 2. In order to write out explicitly the rational function Q in Theorems 1 and 2 we consider *properly formed symbols with parentheses*, of the type $((a))b(c(de))f$ ($a($), $)a($, and $(ab))$ are improperly formed symbols).

Definition. The *length* of a symbol is defined to be the number of letters and left parentheses appearing in it.

Example. Symbols of length 2:

$$ab,\ (a).$$

Symbols of length 3 occur in five forms:

$$abc,\ (a)b,\ a(b),\ (ab),\ ((a)).$$

There are 14 forms of symbols of length 4.

Remark. The numbers of forms of symbols of lengths $n = 1, 2, \ldots$ make up the *Catalan sequence* 1, 2, 5, 14, 42, 132, ...; the n-th number in this sequence is also equal to the number of simplicial decompositions of a convex $(n+2)$-gon by non-intersecting diagonals, and is given by the formula(1) $(2n)!/n!(n+1)!$

Definition. A *filling* of a symbol with parentheses is defined to be the result of substituting in it (in place of the abstract letters a, b, ...) non-zero vertices of the quiver of the family $\Lambda+\varepsilon A$ under investigation (the vertex 0 corresponds to the eigenvalue λ_0 being perturbed).

Example. Fillings of the symbols of length 2 with parentheses have the form

$$ij \text{ or } (i),$$

where $i \neq 0 \neq j$; i can coincide with j, but if $i \neq j$, then the fillings ij and ji are different.

(1)The Catalan numbers appear in the solution of very diverse problems, and are not always identified. For example, a general Riemann surface of genus $2k$ can be represented as a $(k+1)$-sheeted covering of the sphere (with $6k$ branch points) in finitely many ways. Griffiths and Harris [3] find this number (there are 1, 2, 5, 14, ... ways for the values of the genus 2, 4, 6, 8, ...), but do not note that they have obtained the Catalan numbers.

Definition. The *loop* of a filling of length n is defined to be the loop

$$0 \to i_1 \to \ldots \to i_n \to 0,$$

where between the end 0's we have placed in succession (from left to right) the vertices of the filling and 0's wherever there are left parentheses in the filling (the right parentheses are ignored in constructing the loop).

Example. The loop of the filling (i) is $0 \to 0 \to i \to 0$.

Definition. The *diagram of a filling* is obtained from the loop of the filling by adding a dashed arrow leading to each 0 which replaced a left parenthesis, from the vertex immediately preceding (to the left of) the right parenthesis corresponding to this left parenthesis.

Example. The diagram of the filling (i) is

Definition. The *denominator of a filling* is defined to be the product of factors of the form $\lambda_0 - \lambda_i$ (once for each solid arrow of the diagram of the filling that leads to the vertex $i \neq 0$) and factors of the form $\lambda_i - \lambda_0$ (once for each dashed arrow of the diagram that leads from the vertex $i \neq 0$).

Example. The denominator of the filling (i) is $-(\lambda_0 - \lambda_i)^2$, according to the diagram in the preceding example.

A filling is said to be *admissible* if its loop consists of arrows of the quiver.

Theorem 4_k. *The k-th correction to the eigenvalue (or eigenoperator) $\lambda = \lambda_0 E^0 + \varepsilon\alpha_1 + \varepsilon^2\alpha_2 + \ldots$ under the perturbation $\Lambda + \varepsilon A$ is given by the formula*

$$\alpha_k = \sum_{|I|=k-1} \frac{\Pi_{\gamma(I)}(A)}{q_I(\Lambda)},$$

where the sum is over all admissible fillings of all the symbols of length $k-1$ with parentheses, and the contribution of each filling is equal to the ratio of the chronological product of its loop to the denominator of this filling.

Example.

$$\alpha_1 = A_{00}, \quad \alpha_2 = \sum_{i\neq 0} \frac{A_{i0}A_{0i}}{\lambda_0 - \lambda_i},$$

$$\alpha_3 = \sum_{i\neq 0\neq j} \frac{{}_{j0}A_{ij}A_{0i}}{(\lambda_0 - \lambda_i)(\lambda_0 - \lambda_j)} - \sum_{i\neq 0} \frac{A_{i0}A_{0i}A_{00}}{(\lambda_0 - \lambda_i)^2}$$

(the sum of the contributions of the diagrams $\to \to \to$ and

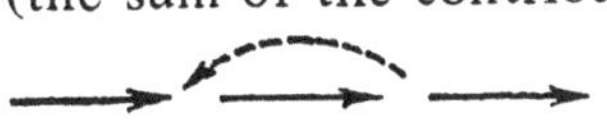

).

Remark 1. Theorems 1 and 2 are obtained from this theorem by summing the contributions of the diagrams with a common loop. Thus, the rational function $Q_\gamma(\Lambda)$ in Theorem 1 is equal to the sum of the reciprocals of the denominators of all the fillings with a given loop γ.

Remark 2. If a loop does not return to the point 0 between its beginning and end, then there are no dashed arrows, and $Q = \prod (\lambda_0 - \lambda_i)^{-1}$ (the product over all internal vertices of the loop).

For example, if $A_{00} = 0$ (which is always easy to achieve by including A_{00} in λ_0), then all the loops of length < 4 in the quiver that go from 0 to 0 do not visit 0 on the way. Consequently, for $A_{00} = 0$

$$\alpha_1 = 0, \quad \alpha_2 = \sum_{i \neq 0} \frac{A_{i0}A_{0i}}{\lambda_0 - \lambda_i}, \qquad \alpha_3 = \sum_{i \neq 0 \neq j} \frac{A_{j0}A_{ij}A_{0i}}{(\lambda_0 - \lambda_i)(\lambda_0 - \lambda_j)}.$$

The formula for α_4 is even more complicated. But the *first non-zero* coefficient in α_k is always equal to the sum of the contributions of the loops that do not visit 0 and, consequently, is given by a simple formula analogous to the ones written out.

Theorem 4 can be proved at the same time as the formula for the corrections to the invariant subspace (Theorem 5). To write this formula we define for each filling I the *path* $\delta(I)$ *of the filling*, which is obtained from the loop of the filling by discarding the last (right-hand) vertex 0 along with the arrow leading to it.

A filling is said to *m-admissible* if its path goes from 0 to m along arrows of the quiver. We look for the invariant space of $\Lambda + \varepsilon A$ obtained by deformation of V_0 in the form of the graph of a mapping

$$B = \varepsilon B^1 + \varepsilon^2 B^2 + \ldots, \quad B^k\colon V_0 \to \oplus V_m, \; m \neq 0.$$

The components of the operator B^k are denoted by

$$B^k_m\colon V_0 \to V_m, \quad m \neq 0.$$

Theorem 5_k. *The operators correcting the invariant space are given by the formula*

$$B^k_m = \sum_{|I|=k} \frac{\Pi_{\delta(I)}(A)}{q_I(\Lambda)},$$

where the sum is over all the m-admissible fillings of length k, and the contribution of each filling is the chronological product of the path of the filling, divided by the deonominator of the filling.

Example 1. The formula written out in § 2 for B^2_m is the sum of the contributions of the fillings *im* and (m).

Example 2. If the eigenvalue λ_0 is simple and ξ_0 is an unperturbed eigenvector, then the perturbed eigenvector has the form

$$\xi_0+\varepsilon\sum\xi_m^1+\varepsilon^2\sum\xi_m^2+\ldots,\quad \xi_m^k=B_m^k\xi_0.$$

Proof of Theorems 4_k *and* 5_k. In the invariance equation

$$(\Lambda+\varepsilon A)(E^0+\varepsilon B^1+\ldots)=(E^0+\varepsilon B^1+\ldots)(\lambda_0E^0+\varepsilon\alpha_1+\ldots)$$

we equate successively the coefficients of ε, ε^2, ... For B^k we obtain the equation

$$B^k\lambda_0-\Lambda B^k=AB^{k-1}-B^{k-1}\alpha_1-\ldots-B^1\alpha_{k-1}-\alpha_k.$$

For this equation to be soluble it is necessary that

$$\alpha_k=\sum A_{m0}B_m^{k-1},\qquad m\neq 0.$$

Therefore, Theorem 4_k follows from Theorem 5_{k-1}. Choosing α_k in this way, we find that

$$(*_k)\quad B_m^k=(\lambda_0-\lambda_m)^{-1}\left(\sum A_{im}B_i^{k-1}-B_m^{k-1}\alpha_1-\ldots-B_m^1\alpha_{k-1}\right),\quad i\neq 0.$$

Computing all the B_m^k successively, we obtain the chronological products and prove Theorems 1 and 2. The explicit form of the denominators can be found by proving that Theorem 5_k follows from Theorems 4_l and 5_l, $0<l<k$.

Indeed, the first term of the expression $(*_k)$ for B_m^k is the sum of the contributions of the m-admissible fillings of length k that are obtained from the i-admissible fillings of length $k-1$, where $i\neq 0$, by adding the symbol m on the right-hand side. Each of the remaining terms, say the one containing $B_m^{k-s}\alpha_s$, is the sum of the products of the contributions from the admissible fillings of length $s-1$ and the m-admissible fillings of length $k-s$, divided by $\lambda_m-\lambda_0$. From two such fillings we can make a filling of length k by appending a parenthesized m-admissible filling of length $k-s$ (from the sum for B_m^{k-s}) on the *right-hand side* of an admissible filling of length $s-1$ (from the sum for α_s). Here the contribution of the resulting filling of length k is exactly equal to the product of the contributions of the component fillings, divided by $\lambda_m-\lambda_0$ (this superfluous factor in the denominator of the filling of length k arises because of the parentheses enclosing the m-admissible filling of length $k-s$). Conversely, each m-admissible filling of length k is obtained—precisely once—under these operations, for there is a unique way to dismantle it from the right-hand end (if a letter is on the right-hand end, then it is removed, and if there is a parenthesis, then the symbol enclosed in it is detached and the corresponding outer parentheses are taken away).

Theorem 5_k thus follows from the preceding Theorems 4_l and 5_l. The assertions 4_1 and 5_1 are obvious, so Theorems 4 and 5 are proved.

Suppose now that $\alpha_1=\alpha_2=\ldots=\alpha_{k-1}=0$. In this case the answer becomes much simpler.

***Theorem* 6.** *The first non-zero correction to the eigenvalue (or eigenoperator) is given by the formula*

$$\alpha_k = \sum_{|\gamma|=k} \frac{\Pi_\gamma(A)}{\prod(\lambda_0-\lambda_i)}$$

(the summation is over the loops of length k in the quiver that start and end at 0 but do not visit 0, and the multiplication in the denominator is over all the internal vertices of the loop).

Proof. In this case the terms with $\alpha_1, \ldots, \alpha_{l-1}$ drop out in the computations according to the formula $(*_l)$ with $l < k$, therefore, $B_i^l = (\lambda_0 - \lambda_i)^{-1}\sum A_{ji}B_j^{l-1}$ $(l = 1, \ldots, k-1)$, which leads to the formula in the theorem.

§5. Conclusion of the investigation of forbidden zones

We now consider the Hill's equation of Mathieu type

$$d^2x/dt^2 + (\omega^2 + \varepsilon a(t))x = 0$$

with an arbitrary trigonometric polynomial of degree p with complex coefficients:

$$a(t) = \sum a_s e^{ist}, \text{ the } s \text{ are integers, and } |s| \leqslant p.$$

In §3 we introduced the eigenvalues $\lambda^\pm$ (which determine the ends of the forbidden zone if the polynomial a is real).

***Theorem* 7.** *The degree of splitting of the eigenvalue $\lambda = -n^2$, where n is a positive integer or a half-integer, admits for small $|\varepsilon|$ the upper estimate*

$$|\lambda^+ - \lambda^-| \leqslant C|\varepsilon|^r, \quad r = -[-2n/p].$$

Proof. Arguing as in the proof of Theorem 3, we get sums over loops of remote arrows as expressions for the coefficients α_k in the expansion of the eigenoperator in a series of powers of ε when $k < 2n/p$. Consequently, all the matrices (A_{ij}) (§3) appearing in the chronological products with such a number k of factors are diagonal and thus commute with one another. This implies that the matrices α_k are also diagonal, but we shall see that they are actually scalar due to special symmetry properties of our expansions.

Definition. Two second-order diagonal matrices are said to be *dual* if they differ only in the order of the diagonal elements.

Suppose that the arrow $i \to j$ is remote, so that the second-order matrix (A_{ij}) is diagonal. The formula given in §3 implies the next result.

***Lemma* 1.** *The diagonal elements (a_{j-i}, a_{i-j}) of the two matrices (A_{ij}) and (A_{ji}) differ only in their order.*

From this it is clear that any polynomial in all such matrices, in which the matrices corresponding to oppositely directed arrows appear symmetrically, is a scalar matrix.

Let us show that in the expression for α_k in Theorem 4 the matrices corresponding to oppositely directed arrows appear symmetrically. It will thereby be proved that the matrices α_k with $k < r$ are scalar, and Theorem 7 obviously follows from this.

Lemma 2. *On the set of all admissible fillings of all symbols of length $k-1$ with parentheses there is an involution such that the contributions to α_k of a filling and its image under the involution are obtained from each other by replacing all the A_{ij} by A_{ji}.*

Theorem 7 follows from the formula in Theorem 4 and Lemma 2, since the matrices α_k are scalar matrices for $k < r$.

Proof of Lemma 2. We break up a filled symbol with parentheses into the fragments between the successive left parentheses. For example, the symbol $a((b)(cd))e(fg)$ is broken up into the fragments a, $b)$, $cd))e$, and $fg)$. Then inside each fragment we write all the vertices in the reverse order, regarding each right parenthesis as associated with the letter closest to it on the left. The fragments a, $b)$, $ed))c$, and $g)f$ are obtained in the example given. We next place left parentheses in the previous places between the fragments. The filling $a((b)(ed))c(g)f$ is obtained in the example.

The constructed operation carries a properly formed symbol with parentheses into the properly formed *dual symbol*, as it does not change the order of the parentheses and does not create pairs of the form ().

A filling dual to an admissible filling is admissible, because the matrices A_{ij} and A_{ji} are either both zero or both non-zero. We compute the contributions of dual fillings.

Lemma 3. *The chronological products of mutually dual fillings appearing in the expansion of α_k for $k < r$ are mutually dual.*

For by Lemma 1, the factors are mutually dual and commute.

Lemma 4. *The denominators of dual fillings coincide.*

Indeed, the denominator of a filling is composed of factors $\lambda_0 - \lambda_i$ with i running through the elements of the filling, and factors $\lambda_i - \lambda_0$ with i running through the elements preceding the right parentheses (where each factor is counted as many times as there are right parentheses immediately after the corresponding i).

Both the set of elements in the filling and the set of elements which preceded right parentheses (taking account of the number of parentheses) are preserved under the operation defined above, as Lemma 4 proves.

Lemmas 3 and 4 imply the equality of the contributions of dual fillings and, hence, Lemma 2 and Theorem 7.

§6. Widths of resonance zones for mappings of the circle

We consider an analytic mapping of the circle onto itself that depends on two parameters a and ε:

$$x \mapsto x + a + \varepsilon f(x), \quad f(x + 2\pi) \equiv f(x).$$

Definition. A point (a, ε) belongs to the *domain of resonance* m/n if some point of the circle is shifted by exactly m revolutions when the mapping is applied n times (that is, if the Poincaré rotation number of the mapping is equal to m/n).

***Theorem* 8.** *Let f be a trigonometric polynomial of degree p. Then the width of the domain of resonance m/n is at most $C\varepsilon^r$, where $-r$ is the integer part of the fraction $-n/p$.*

Moreover, the domain of resonance m/n is bounded by two curves $a = A^{\pm}(\varepsilon)$ such that the terms of degree less than r in the Taylor expansions of $A^{\pm}$ about zero coincide.

Example. Consider the mapping $x \mapsto x + a + \varepsilon \cos x$. In this case the width of the domain of resonance m/n is of order ε^n, as claimed already in [2].

The proof of the theorem is based on an investigation of the r-th variation of the k-th iterate of a rotation. Consider an analytic mapping

$$x \mapsto x + \mu + g(x), \quad g(x + 2\pi) \equiv g(x),$$

that is close to the rotation through the angle $\mu = 2\pi m/n$. Its k-th iterate is close to the rotation through the angle $k\mu$:

$$x \mapsto x + k\mu + {}_k g(x).$$

Let us consider the Taylor expansion at the point $g = 0$ for the operator carrying g into ${}_k g$. We call the term of degree r in this expansion the *r-th variation of the k-th iterate.*

Remark. In order not to have to worry about convergence we can understand the variations in the following formal way. Assume that g is a linear combination of any finite number of terms: $g = \sum \lambda_q h_q$. Then ${}_k g$ is a series in powers of λ, and the r-th variation is the sum of the terms of degree r in λ in this series.

***Lemma* 1.** *The r-th variation of the k-th iterate can be expressed as a polynomial of degree r in terms of g, the shifts of g by the angles $s\mu$ with $0 \leqslant s < k$, and their derivatives of order less than r.*

Proof. This is obvious for $k = 1$. Assume that it is true for the k-th iterate. By definition, for the $(k+1)$st iterate

$$(1) \qquad {}_{k+1}g(x) = {}_k g(x) + g(x + k\mu + {}_k g(x)).$$

Denote the shifts of g by

$$g_k(x) = g(x + k\mu).$$

Then

$$g(x + k\mu + {}_kg(x)) = g_k(x + {}_kg(x)).$$

We expand g_k in the Taylor series $\sum g_k^{(s)}({}_kg)^s/s!$, and substitute for ${}_kg$ the series of variations

$$_kg = {}_kg^1 + {}_kg^2 + \dots$$

By assumption, ${}_kg^i$ is a polynomial of degree i in g, the shifts, and the derivatives of order less than i. The terms with $i \geqslant r$ and the terms with $s \geqslant r$ do not contribute to the r-th variation. Therefore, these transformations of the formula (1) express the r-th variation of the $(k+1)$st iterate as the sum of the r-th variation of the k-th iterate and a sum of products of derivatives of order less than r of the shift of g by the angle $k\mu$ and products of the preceding variations of the k-th iterate. The lemma is proved by induction.

Remark. The first two variations of the k-th iterate can be expressed by simple formulae:

$$_kg^1 = g_0 + g_1 + \dots + g_{k-1},$$
$$_kg^2 = g_1' g_0 + g_2' (g_0 + g_1) + \dots + g_{k-1}' (g_0 + \dots + g_{k-2}).$$

Consider now the n-th iterate of the mapping. Since $\mu = 2\pi m/n$, the n-th iterate of the unperturbed mapping returns all the points of the circle to their places.

On the level of the second term in the perturbation theory the theorem being proved reduces to the following.

Proposition. *Let g be a trigonometric polynomial of degree less than $n/2$, and suppose that ${}_ng^1 = 0$ (that is, the mean value of g is equal to 0). Then ${}_ng^2 = \text{const}$.*

Proof. Consider the difference

$$_ng^2(x \mid \mu) \quad {}_ng^2(x) = g_n' (g_1 + \dots + g_{n-1}) - g_0 (g_1 + \dots + g_{n-1})'.$$

Since $g_n = g_0$ and $g_1 + \dots + g_n = 0$, this difference is equal to zero. Consequently, ${}_ng^2$ is a trigonometric polynomial of degree less than n with period $2\pi/n$, that is, it is a constant. The proposition is proved. It implies our theorem for $r \leqslant 3$.

The proof in the general case uses an analogous argument, formalized by Lemma 2 below.

We consider a one-parameter analytic family of mappings of the circle onto itself:

$$x \mapsto x + \mu + g(\varepsilon, x), \quad g(\varepsilon, x + 2\pi) \equiv g(\varepsilon, x),$$

where $\mu = 2\pi m/n$ and where

$$g = \varepsilon g^1 + \varepsilon^2 g^2 + \dots$$

Lemma 2. *Assume that the n-th iterate of the mapping shifts points by a distance of order ε^r, that is,*

$$_n g(\varepsilon, x) = \varepsilon^r u(x) + o(\varepsilon^r).$$

Then the principal term in the shift is invariant under the rotation through the angle μ:

$$u(x + \mu) \equiv u(x). \tag{2}$$

Proof. Consider the images x_i of a point $x_0 = x$ under the iterates of the mapping. By the condition in the lemma,

$$g(x_0) + g(x_1) + \dots + g(x_{n-1}) = \mu n + \varepsilon^r u(x_0) + o(\varepsilon^r),$$

$$g(x_1) + g(x_2) + \dots + g(x_n) = \mu n + \varepsilon^r u(x_1) + o(\varepsilon^r).$$

Consequently,

$$g(x_n) - g(x_0) = \varepsilon^r [u(x + \mu) - u(x)] + o(\varepsilon^r). \tag{3}$$

On the other hand,

$$x_n = x_0 + 2\pi m + \varepsilon^r u(x_0) + o(\varepsilon^r),$$

and, consequently,

$$g(x_n) - g(x_0) = g'(x_0)\varepsilon^r u(x_0) + o(\varepsilon^r). \tag{4}$$

Since $g' = O(\varepsilon)$, (3) and (4) imply (2).

Proof of the theorem. We consider the following two-parameter family of perturbations of the rotation through the angle $\mu = 2\pi m/n$:

$$x \mapsto x + \mu + H(\alpha, \varepsilon, x), \quad H(\alpha, \varepsilon, x) = \alpha + \varepsilon f(x),$$

where ε and α are small parameters.

Denote the n-th iterate by

$$x \mapsto x + \mu n + {}_nH(\alpha, \varepsilon, x).$$

The resonance condition is the solubility of the equation ${}_nH = 0$ with respect to x. The expansion of ${}_nH$ in a Taylor series in α and ε has the form

$${}_nH = {}_nH^1 + {}_nH^2 + \dots,$$

where each term ${}_nH^i$ is a homogeneous polynomial of degree i in α and ε (the value of the i-th variation of the n-th iterate for $g = \alpha + \varepsilon f$). For example,

$${}_nH^1 = n\alpha + \varepsilon + (f_0 + \dots + f_{n-1}), \text{ where } f_k(x) = f(x + k\alpha).$$

By the implicit function theorem, the equation ${}_nH = 0$ is soluble for $\alpha = \alpha(\varepsilon, x)$:

$$\alpha = \varepsilon v_1(x) + \varepsilon^2 v_2(x) + \dots$$

Let r be the index of the first non-constant coefficient v_i, so that $\alpha(\varepsilon, x) = A(\varepsilon) + \varepsilon^r v(x) + o(\varepsilon^r)$, where A is a polynomial of degree less than r, and v is a non-constant function.

The boundaries of the domain of resonance m/n have asymptotic expressions $\alpha = A(\varepsilon) + \varepsilon^r v_\pm + o(\varepsilon^r)$, where $v_\pm$ are the maximum and minimum of v on the circle. It remains to get a lower estimate of r.

We substitute $A(\varepsilon)$ in place of α in H and denote the resulting function by g:

$$g(\varepsilon, x) = H(A(\varepsilon), \varepsilon, x).$$

After n iterations of the mapping $x \mapsto x + \mu + g(\varepsilon, x)$ we get a rotation through a variable angle of order ε^r. Indeed,

$$\alpha(\varepsilon, x) - A(\varepsilon) = \varepsilon^r v(x) + o(\varepsilon^r),$$

therefore,

$${}_nH(\alpha(\varepsilon, x), \varepsilon, x) - {}_nH(A(\varepsilon), \varepsilon, x) = n\varepsilon^r v(x) + o(\varepsilon^r).$$

But ${}_nH(\alpha(\varepsilon, x), \varepsilon, x) = 0$ by the definition of $\alpha(\varepsilon, x)$. Accordingly,

$${}_ng(\varepsilon, x) = {}_nH(A(\varepsilon), \varepsilon, x) = \varepsilon^r u(x) + o(\varepsilon^r), \quad u = -nv.$$

By Lemma 2, the function u (hence also v) satisfies the periodicity condition $u(x+\mu) \equiv u(x)$.

On the other hand, Lemma 1 gives us the next result.

***Lemma* 3.** *The function u can be expressed as a polynomial of degree at most r in terms of f, shifts of f by angles that are multiples of μ, and their derivatives (of order less than r).*

Proof. If we substitute the series without the free term $\alpha = A(\varepsilon)$ in the Taylor series of ${}_nH(\alpha, \varepsilon, x)$ with respect to α and ε, then the new coefficient of ε^r can be expressed linearly in terms of the old coefficients of terms of degree $\leqslant r$ in α and ε together. By Lemma 1, the old terms of degree i in α and ε can be expressed as polynomials of degree i in terms of shifts of the fucntion $g = \alpha + \varepsilon f$ by angles that are multiples of α and their derivatives with respect to x (for constant α and ε) of bounded order. This proves Lemma 3, because $\partial g/\partial x = \varepsilon \partial f/\partial x$.

Completion of the proof of the theorem. Assume that f is a trigonometric polynomial of degree p and that $rp < n$. Then u is a trigonometric polynomial of degree less than n, by Lemma 3. From Lemma 2 we concluded above that u has period $2\pi m/n$. A trigonometric polynomial of degree less than n with period $2\pi m/n$ is a constant. But u is not a constant (by the choice of r). Hence, $r \geqslant n/p$, and Theorem 8 is proved.

Remark. A diffeomorphism of the circle can have many cycles. Is the number of isolated cycles of a diffeomorphism given by a trigonometric polynomial bounded by a constant depending only on the degree of the polynomial?

This question can be regarded as an analogue of the question in Hilbert's 16th problem on the number of limit cycles of a polynomial differential equation. It has not even been solved for the diffeomorphisms $x \mapsto x + a + b \cos x$.

Conversely, the asymptotic expressions obtained above for the boundaries of domains of existence of cycles probably have analogues in the theory of differential equations with a polynomial right-hand side.

Supplement (30 March 1983)

While the manuscript of the present article was being edited, the author learned of some papers containing other proofs of some of the results: by C. Bloch [5] (Theorems 1, 2, 4–6), D.M. Levy and J.B. Keller [6] (Theorem 3), and H. Hochstadt [7] (Theorem 7).

Bloch labels the terms in the perturbation theory series by diagrams having the form of integer staircases joining the vertices (0, 0) and (n, n) inside the square without dropping below the diagonal.

The number d_n of such diagrams is equal to the corresponding Catalan number ($d_2 = 2$, $d_3 = 5$, ...). Bloch's work thus gives another interpretation of the Catalan numbers. The most natural definition of them is probably as the numbers of words of given lengths in a non-associative monoid with a single generator. It would be interesting to find out whether the non-associative monoid itself is connected with the perturbation theory (or with coverings of Riemann surfaces, where the Catalan numbers are also encountered).

The correspondence between symbols with parentheses and Bloch diagrams is as follows: a left parenthesis means a step upwards, a right parenthesis means a step to the right, and a letter means two steps (first upwards, and then to the right).

Moreover, Manin has communicated to the author that an approach to perturbation theory close to that presented above was worked out for different reasons in a manuscript of M. Wodzycki written in July 1982 and entitled "Variation of the ζ-function of an elliptic pseudodifferential operator".

The author thanks M. Wodzcki, I.M. Gel'fand, A.A. Kiselev, P.A. Kuchment, V.F. Lazutkin, Yu.I. Manin, and S.P. Novikov for useful discussions.

References

[1] V.I. Arnol'd, On the evolution of a magnetic field under the action of transport and diffusion, Collection of papers in memory of V.M. Alekseev, Izdat. Moskov. Gos. Univ., Moscow 1983.

[2] ———, Small denominators. I, Mappings of the circumference onto itself, Izv. Akad. Nauk Ser. Mat. **25** (1961), 21-86. MR **25** # 4113.
= Amer. Math. Soc. Transl. (2) **46** (1965), 213-254.

[3] P. Griffiths and J. Harris, Principles of algebraic geometry, Wiley, New York 1978. MR **80b** # 14001.
Translation: *Printsipy algebraicheskoi geometrii*, Mir, Moscow 1982.

[4] A.A. Kiselev, A diagram technique in the general theory of perturbations and in the adiabatic theory of molecular spectra, in: *Voprosy kvantovoi teorii atomov i molekul* (Questions in the quantum theory of atoms and molecules), No. 1, Izdat. Leningrad. Gos. Univ., Leningrad 1978, 108-143.

[5] C. Bloch, Sur la théorie des perturbations des états liés, Nuclear Physics **6** (1958), 329-347. Nuclear Science Abstracts **12** # 9529.

[6] D.M. Levy and J.B. Keller, Instability intervals of Hill's equation, Comm. Pure Appl. Math. **16** (1963), 469-476. MR **27** # 3875.

[7] H. Hochstadt, Instability intervals of Hill's equation, Comm. Pure Appl. Math. **17** (1964), 251-255. MR **28** # 2310.

Translated by H. McFaden

Moscow State University

Received by the Editors 1 February 1983

VANISHING INFLECTIONS

V. I. Arnol'd

UDC 517.27

1. Formulation of the Result. At a Morsean singular point in general position on a plane curve 6 inflection points merge (two are visible in the figure, the rest are complex). Belows this result of Plücker [1] is carried over to hypersurfaces in $\mathbf{C}^n$.

A point of inflection of such a surface is a point for which the germ of the dual hypersurface is diffeomorphic with the germ of the swallow's tail $\{\lambda: z^{n+1} + \lambda_1 z^{n-1} + \ldots + \lambda_n$ has a multiple root$\}$ at zero.

THEOREM. At a Morsean singular point in general position on a hypersurface in $\mathbf{C}^n$ $(n + 1)!$ points of inflection merge.

2. Origin of $(n + 1)!$. Let $f:(\mathbf{C}^n, 0) \to (\mathbf{C}, 0)$, $df|_0 = 0$. Let $v = \Sigma x_k \partial/\partial x_k + \ldots$ be a vector field with Eulerian linear part at zero. We form the system

$$L_2 = \ldots = L_{n+1} = 0, \text{ where } L_2 = f, \quad L_{i+1} = vL_i \text{ for } i \geqslant 2, \tag{1}$$

by setting equal to zero successive derivatives of f along v.

LEMMA 1. The multiplicity of the zero solution of (1) is not less than $(n + 1)!$ and is equal to $(n + 1)!$ for a pair (f, v) in general position.

Proof. We define M_i by the conditions $M_2 = f$, $M_{i+1} = (i - v)M_i$ for $i \geqslant 2$. The system $M_2 = \ldots = M_{n+1} = 0$ is equivalent with (1). The order of the zero of the function M_i at 0 is not less than i (by Euler's theorem on homogeneous functions). The multiplicity of a zero of a general system of equations of fixed orders is equal to the product of the orders (cf., e.g., [2]).

3. Hesse's Method. To study inflections on a level surface of the function f, we form the bordered Hessian [3]

$$h = \det H, \quad H = \begin{pmatrix} f_{xx} & f_x \\ f_x & 0 \end{pmatrix}.$$

Parabolic points are defined by the equation $h(x) = 0$. The vector $(e, 1)$ belongs to the kernel of the operator H at a parabolic point if and only if the vector e is asymptotic for the level surface.

4. Inflection Conditions. Let f have a Morsean point at 0. Then $\operatorname{Ker} H(0) = \mathbf{C}(0, 1)$. Hence $H(x)$ has a unique eigenvector $\xi(x) = (e(x), 1)$ with eigenvalue $\lambda(x)$, equal to zero at zero; e and λ depend smoothly on x. The linear part of e at zero differs from an Eulerian field by a sign.

Equation of parabolic points: $\lambda(x) = 0$. From $H\xi = \lambda\xi$ it follows that $\lambda = ef$ (derivative of f along e). The system of equations of an inflection consists of the vanishing at 0 of further derivatives of f along e (cf. [4]). This is the system (1) for $v = -e$. The pair (f, v) is not in general position. But it is proved below that the principal homogeneous parts of the functions M_i of Sec. 2 for f in general position are general.

5. Theory of Perturbations. We shall denote the terms of degree i in Taylor series by the left index i. We write the Taylor series of f in the form $x^2/2 + {}_3f + {}_4f + \ldots$ (the quadratic terms are normalized by a linear substitution). We shall understand vectors to be columns so that Euler's theorem assumes the form $\mathbf{x}({}_kf_\mathbf{x}) = k{}_kf$. We write the equation $H\xi = \lambda\xi$ in the form

$$(E + {}_3f_{xx} + \ldots)({}_1e + {}_2e + \ldots) + x + {}_3f_x + \ldots = ({}_1\lambda + \ldots)({}_1e + \ldots),$$
$$(x + {}_3f_x + \ldots)({}_1e + {}_2e + \ldots) = {}_1\lambda + {}_2\lambda + \ldots$$

Moscow State University. Translated from Funktsional'nyi Analiz i Ego Prilozheniya, Vol. 18, No. 2, pp. 51-52, April-June, 1984. Original article submitted December 9, 1983.

Originally publ. in: Funkts. Anal. Prilozh. 18(2), 51-52, © 1984 Russ. Acad. Sciences
English transl.: Funct. Anal. Appl. 18(2), 128-130, © 1984 Consultants Bureau, New York.

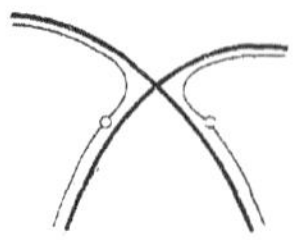

Fig. 1

Comparing homogeneous components we find successively

$$ {}_1\lambda = 0,\ {}_1e = -x,\ {}_2\lambda = -x^2,\quad {}_2e = -{}_3f_x - (-x)\,{}_3f_{xx} = {}_3f_x, $$

and in general

$$ {}_i\lambda = x\,{}_{i-1}e + {}_3f_x\,{}_{i-2}e + \dots + {}_if_x\,{}_1e, \tag{2} $$

$$ {}_ie + {}_3f_{xx}\,{}_{i-1}e + \dots + {}_{i+1}f_{xx}\,{}_1e + {}_{i+1}f_x = {}_2\lambda_{i-2}\,e + \dots + {}_{i-1}\lambda\,{}_1e. \tag{3} $$

6. Investigation of Perturbations. Below, $a \prec E_i$ means that the coefficients of the Taylor series of the component a in x are polynomials in the coefficients of the forms ${}_3f, \dots, {}_if$ (for $i < 3$, are independent of these forms). The notation $a \equiv b \bmod E_i$ means $a - b \prec E_i$.

LEMMA 2_i. ${}_ie \equiv (i-1)\,{}_{i+1}f_x \bmod E_i$.

Proof. 2_1 is obvious. If ${}_je \prec E_{j+1}$ for $j < i$, then from (2), ${}_i\lambda \prec E_i$. Now from (3), ${}_ie \equiv -({}_{i+1}f_{xx}(-x) + {}_{i+1}f_x) \bmod E_i$. But by Euler ${}_{i+1}f_{xx}x = i\,{}_{+1}f_x$. 2_i is proved.

7. Investigation of the Functions M_i. By definition (Secs. 2 and 4), $M_2 = f$, $M_{i+1} = (i + e)M_i$.

LEMMA 3_i. For $j \geqslant i \geqslant 2$, ${}_jM_i \equiv c_{i,j}\,{}_jf \bmod E_{j-1}$, where $c_{i,j} \neq 0$.

Proof. 3_2 is obvious: ${}_jM_2 = {}_jf$, $c_{2,j} = 1$. We derive 3_{k+1} from 3_k. By definition of the functions M,

$$ {}_jM_{k+1} = k\,{}_jM_k + ({}_jM_{kx}\,{}_1e + \dots + {}_kM_{kx}\,{}_{j-k+1}e). $$

Considering that ${}_1e = -x$, we find from Euler's formula that

$$ {}_jM_{k+1} = (k-j)\,{}_jM_k + ({}_{j-1}M_{kx}\,{}_2e + \dots + {}_kM_{kx}\,{}_{j-k+1}e). \tag{4} $$

Here $j \geqslant k+1$, and for $j \leqslant k$, ${}_jM_{k+1} = 0$. It follows from 3_k that

$$ {}_jM_k \equiv c_{k,j}\,{}_jf \bmod E_{j-1},\ ({}_{j-1}M_k, \dots, {}_{k+1}M_k) \prec E_{j-1}. $$

For $k \geqslant 3$ by Lemma 2 we also have $({}_2e, \dots, {}_{j-k+1}e) \prec E_{j-1}$. Hence for $k \geqslant 3$, from (4) and Lemma 3_k it follows that

$$ {}_jM_{k+1} \equiv (k-j)\,c_{k,j}\,{}_jf \bmod E_{j-1},\ \text{т. е.}\quad c_{k+1,j} = (k-j)\,c_{k,j}. $$

For $k = 2$, by Lemma 2, $({}_2e, \dots, {}_{j-2}e) \prec E_{j-1}$ and ${}_{j-1}e \equiv (j-2)\,{}_jf_x \bmod E_{j-1}$. Hence ${}_jM_3 \equiv (2-j)_j f + (j-2)_j f_x x \bmod E_{j-1}$. But ${}_jf_xx = j\,{}_jf$. Thus, ${}_jM_3 \equiv (1-j)(2-j)\,{}_jf \bmod E_{j-1}$. Lemma 3_{k+1} is proved with $c_{k+1,j} = (1-j)(2-j)\dots(k-j)$.

8. Conclusion of the Proof of the Theorem. According to Sec. 2, ${}_jM_i = 0$ for $j < i$. By Lemma 3_i, ${}_iM_i \equiv c_{i,i}\,{}_if \bmod E_{i-1}$, where $c_{i,i} \neq 0$. Hence, changing only terms of degree i in f, we can, without changing ${}_jM_j$ for $j < i$, get M_i with any homogeneous principal part of degree i at zero. Consequently, for f in general position all these principal parts will be in general position. But then the multiplicity of a zero of the system $M_2 = \dots = M_{n+1} = 0$ is equal to $(n+1)!$.

9. Consequences. The number of Morsean singular points in general position with respect to inflections on a surface of degree 4 in $\mathbf{C}^3$ does not exceed 13, because the total number of inflections on a surface of degree d is equal to $2d(d-2)(11d-24) = 320$ [5], and $[320/24] = 13$. It is easy to construct an example with 12 points in the form $a(x) + b(y) + c(z) = 0$.

The maximal number of Morsean singular points on a surface of degree 4 is equal to 16 [6]. But they are not in general position in relation to inflections.

As d increases the number of Morsean points grows no faster than $23d^3/48 + \dots$ [7]. Consequently, asymptotically almost half $(1/2 - 1/44)$ the points of inflection cannot vanish at

Morsean points which are general with respect to inflections, and distance from them is preserved.

LITERATURE CITED

1. J. Plücker, Theorie der algebraischen Kurven, Bonn (1839).
2. V. I. Arnol'd, Usp. Mat. Nauk, 29, No. 2, 17-20 (1974).
3. O. Hesse, Z. Mat., 28, 68-75 (1844).
4. V. I. Arnol'd, Mathematical Methods of Classical Mechanics [in Russian], Nauka, Moscow (1974).
5. A. Clebsch, Z. Mat., 58, 102-110 (1860).
6. E. Kummer, Monatsh. Akad. Berlin, 246-495 (1864).
7. A. N. Varchenko, Dokl. Akad. Nauk SSSR, 270, No. 6, 1294-1297 (1983).

Appendix A to the paper: B.A. Malomed and M.I. Tribelsky "Bifurcations in distributed kinetic systems with aperiodic instability"

In section 3 we have assumed that the stationary solutions (12) to eqs. (1), (3), (4) and (11) to (2) are not formal Fourier series but that they exist indeed. Here we give rigorous proof of the existence of these solutions. Note that the stationary versions of all our equations (1)–(4) are reversible with respect to the change of the sign of the spatial coordinate x. As we have explained in section 2, this property is the principal one. The existence of periodic solutions of sufficiently small amplitude for reversible nonlinear ordinary differential equations is proved in [26, 27]*. "Sufficiently small" means amplitudes which are, generally speaking, small in comparison with $A_{\max}$ (see (20)). Therefore, there remains the problem concerning the existence proof for the whole family of periodic solutions (11), (12), especially because, as we have demonstrated in section 3, only solutions with $A_1^2/(A_1^2)_{\max} \geq 2/3$ are stable in the sense of the temporal evolution. The following passage, kindly written for the present paper by Professor V.I. Arnold, proves that the approximate periodic solutions to reversible equations of the type (11), (12) do correspond to some exact periodic solutions close to them, i.e. this consideration gives the necessary proof.

1) Reversible systems of differential equations possess special properties which are quite distinct from the properties of the general form systems, and in many respects their properties are close to those of the Hamiltonian systems. For instance, periodic solutions to general form systems of differential equations occur as isolated solutions. In Hamiltonian systems, as a rule, periodic solutions form one-parameter families (since periodic solutions, close to a given solution, usually exist on neighbouring isoenergetic surfaces).

The same property is valid for a reversible system of differential equations. The role of reversibility in the periodic motion theory has been already noted by Poincaré and, especially, by Birkhoff [28]. Let us consider the system of differential equations:

$$\frac{\mathrm{d}\boldsymbol{q}}{\mathrm{d}x} = \boldsymbol{g}(\boldsymbol{q}, \boldsymbol{p}),$$
$$\frac{\mathrm{d}\boldsymbol{p}}{\mathrm{d}x} = \boldsymbol{f}(\boldsymbol{q}, \boldsymbol{p}), \tag{A.1}$$

Here $\boldsymbol{q}$, $\boldsymbol{p}$, $\boldsymbol{g}$ and $\boldsymbol{f}$ are n-dimensional vectors. Assume that an involution, i.e. the second order symmetry transformation $\hat{\sigma}$: $(\boldsymbol{q}, \boldsymbol{p}) \to (\boldsymbol{q}, -\boldsymbol{p})$, is defined in the $2n$-dimensional phase space $(\boldsymbol{q}, \boldsymbol{p})$. Let the vector field which determines the motion equations in the phase space changes the sign due to this transformation, i.e. $\boldsymbol{g}$ is an odd function of $\boldsymbol{p}$, and $\boldsymbol{f}$ is an even one: $\boldsymbol{g}(\boldsymbol{q}, -\boldsymbol{p}) = -\boldsymbol{g}(\boldsymbol{q}, \boldsymbol{p})$; $\boldsymbol{f}(\boldsymbol{q}, -\boldsymbol{p}) = \boldsymbol{f}(\boldsymbol{q}, \boldsymbol{p})$. Then the corresponding system of differential equations will be called reversible.

It is clear from the above stated that for reversible systems the involution transformation is equivalent to the transformation $x \to -x$.

Example 1. The Newton equation $\ddot{\xi} = F(\xi, \dot{\xi})$ is reversible if the force F does not depend on the velocities or if it is an even function of the velocity.

Example 2. The equation

$$u^{\mathrm{IV}} + Au'' + Bu + C(u')^2 = 0 \tag{A.2}$$

*This problem was considered also in [32, 33].

is reversible, the involution in the form $u, u', u'', u''' \to u, -u', u'', -u'''$ being generated by the substitution $x \to -x$.

Assume that the reversible system has a periodic solution intersecting the set of the immobile points of the involution. This set is the n-dimensional subspace $\boldsymbol{p} = 0$ in the $2n$-dimensional phase space.

For a general form reversible system, a whole one-parameter family of close periodic solutions (with close, but, generally speaking, different periods) exists in the vicinity of this periodic solution.

Really, let us draw the Poincaré section surface P, say $p_1 = 0$, through the immovable space A ($\boldsymbol{p} = 0$) of our involution. The dimension of P is $2n - 1$. Designate as $\hat{T}$ the Poincaré map $P \to P$. Solutions with initial data on A come back to P and, in a general case, they cut out the submanifold $\hat{T}A$ of the dimension n on P. Two n-dimensional submanifolds A and $\hat{T}A$ which are embedded into a $(2n-1)$-dimensional manifold P must, in a general case, intersect each other in a one-dimensional curve. The points on this curve determine the initial data for the periodic solutions sought for.

Indeed, if $\hat{T}a = b$ is a point on this curve, then, due to the reversibility condition, $\hat{T}^{-1}a = \hat{\sigma}b$, but the point b belongs to the set A of the immovable points of the involution $\hat{\sigma}$. Hence $\hat{\sigma}b = b$, and $\hat{T}^{-1}a = b = \hat{T}a$, i.e. $\hat{T}^2 a = a$.

Note. This consideration warrants the applicability of approximate or asymptotic formulae for the families of periodic solutions to reversible systems (see, e.g., (11) and (12)). If the family is known approximately, and, within the framework of this approximation, A intersects $\hat{T}A$ inside P transversally, then for the exact solution an intersection curve and, consequently, a one-parameter family of periodic solutions also exist (according to the implicit function theorem).

2) Let us now consider a system of differential equations of a general form:

$$\frac{d\boldsymbol{q}}{dx} = \boldsymbol{f}(\boldsymbol{q}), \tag{A.3}$$

where $\boldsymbol{q}$ and $\boldsymbol{f}$ are n-dimensional vectors. If the involution transforming the vector field of the system (A.3) into the field of the opposite sign is generated by the change of the sign of k coordinates: $q_1, q_2, \ldots, q_k$, while the sign of the other l coordinates $(k + l = n)$ is not changed, the consideration analogous to the analysis given above yields a $(k - l + 1)$-parameter family of periodic solutions.

3) A reversible system may have an involution transformation different from the transformation discussed above. For instance, the equation

$$\dot{z} = iz^2$$

is a reversible equation with the involution transformation

$$\hat{\sigma}z = z^*.$$

4) From the examples discussed above it is clear that a reversible system may be non-Hamiltonian. It is important that, in spite of this fact, many properties of Hamiltonian systems can be transferred to reversible systems (for instance, the Birkhoff normal form theory, the bifurcation theory and the Kolmogorov tori theory [29, 30]). In particular, for the equation (A.2) there are two-dimensional Kolmogorov tori in the four-dimensional phase space (on a three-dimensional Poincaré surface they correspond to invariant curves with an irrational winding number; for a given winding number they form a one-parameter family of invariant curves, i.e. an invariant cylinder).

References

[1] Springer Series in Synergetics, vols. 1–12 (Springer, Berlin, 1977–1983).
[2] B.A. Malomed and A.M. Zhabotinsky, in: Synergetics (Proc. Int. Symp. Synergetics and Cooperative Phenomena), pp. 217–226 (Valgus, Tallinn, 1983).
[3] G. Iooss and D.D. Joseph, Elementary Stability and Bifurcation Theory (Springer, New York, 1980).
[4] A.A. Chernov, Usp. Fiz. Nauk (Russian journal "Uspekhi") 100 (1970) 277.
[5] A.S. Mikhailov and I.V. Uporov, JETP 84 (1983) 1481.
[6] B.N. Belintsev, M.A. Livshits and M.V. Volkenstein, Zs. Phys. B 44 (1981) 345; Phys. Lett. A82 (1981) 375.
[7] A.M. Turing, Phil. Trans. Roy. Soc. B 237 (1952) 37.
[8] G.I. Sivashinsky, Physica 4D (1982) 227.
[9] W. Eckhaus, Studies in Non-Linear Stability Theory (Springer, New York, 1965).
[10] S. Kogelman and R.C. DiPrima, Phys. Fluids 13 (1970) 1.
[11] G.I. Sivashinsky, Acta Astronautica 4 (1977) 1177.
[12] G.I. Sivashinsky, SIAM J. Appl. Math. 39 (1980) 67.
[13] S.I. Anisimov, M.I. Tribelsky and Ya.G. Epelbaum, JETP 78 (1980) 1597.
[14] A.A. Vlasov, Many Particles Theory, part 2, chapter 3. (Moscow and Leningrad, 1950).
[15] A.A. Nepomnyashchy, Mekh. Zhidk. i Gaza (Russian journal Fluid and Gas Mechanics) N 3 (1974) 28.
[16] B.A. Malomed, Zhurn. Tekhn. Fiz. (Russian Journal of Technical Physics) 54 (1984) 233.
[17] A.P. Aldushin, B.A. Malomed, and Ya.B. Zeldovich, Doklady AN SSSR 251 (1980) 1102; Combustion and Flame (London) 42 (1981) 3.
[18] A.P. Aldushin and B.A.Malomed, Fizika Goreniya i Vzryva (Russian journal Combustion and Explosion Physics) 17 (1981) 3.
[19] S.I. Anisimov, S.M. Golberg, B.A. Malomed and M.I. Tribelsky, Doklady AN SSSR 262 (1982) 1117.
[20] A.B. Konstantinov, B.A. Malomed and S.M. Golberg, Doklady AN SSSR 270 (1983) 1364.
[21] H. Haken, Synergetics (Springer, Berlin, 1977) chap. 7.
[22] M.I. Tribelsky, Izv. AN SSSR, ser. Fizika (Russian journal USSR Academy of Sciences Proceedings, series "Physics") 46 (1982) 1127.
[23] J.E. Marsden and M. McCracken, The Hopf Bifurcation and Its Applications (Springer, New York, 1976).
[24] B.A. Malomed and I.E. Staroselsky, Khim. Fizika (Russian journal "Chemical Physics") 1 (1983) 985; Phys. Letters 99A (1983) 143.
[25] B.A. Malomed and I.E. Staroselsky, Phys. Letters 99A (1983) 145.
[26] Yu.N. Bibikov, Local Theory of Nonlinear Analytic Ordinary Differential Equations (Springer, Berlin 1979) chaps. 5, 6.
[27] A.D. Brüno, The Local Nonlinear Analysis Method for Differential Equations (Nauka, Moscow, 1979) chap. 3, § 3 (in Russian).
[28] G.D. Birkhoff, Rend. Circ. Mat. Palermo 39 (1915) 1.
[29] J. Moser, SIAM Review 8 (1966) 145.
[30] J. Moser, Math. Ann. 169 (1967) 136.
[31] B.A. Malomed, Physica 8D (1983) 353.
[32] A.N. Sharkovsky, in: Asymptotic Behaviour of Solutions of Differentio-Functional Equations (The Mathematics Institute, Kiev, 1978) pp. 118–142 (in Russian).
[33] R.L. Devaney, Trans. Amer. Mathem. Soc. 218 (1976) 89.

REVERSIBLE SYSTEMS

V.I. ARNOLD

Moscow State University, Moscow, USSR

(In *Nonlinear and Turbulent Processes in Physics*, Gordon & Breach, Harwood Acad. Publishers, New York, 1984, Vol. 3, 1161–1174.[1])

Abstract. An evolutionary system is called reversible if the time reversal accompanied by a suitable phase space reflection preserves the equations of motion. The astonishing similarities between the general pattern of motion in reversible and Hamiltonian systems are discussed; among them there are the existence of Kolmogorov tori and the theory of normal forms, resonances, bifurcations, and stability.

Introduction

The drastic difference between the patterns of motion in Hamiltonian and non-Hamiltonian dynamical systems is well known. For instance, the periodic motions of a generic Hamiltonian system form a one-parameter family (the parameter being the Hamiltonian function's value), while cycles in the phase space of a generic non-Hamiltonian system are, as a rule, isolated.

There exists, however, a class of non-Hamiltonian systems, occurring in many applications, whose behavior is very similar to that of Hamiltonian systems.

Example 1. There are nonisolated families of cycles in the phase space of the equation

$$u'''' + Au'' + Bu + (u')^2 = 0 \qquad (1)$$

(for fixed A and B). We shall see how this nongeneric behavior results from the reversibility (invariance under the reversal of the independent variable) of (1).

The similarity between the Hamiltonian and reversible systems affects many details of their motions. For instance, the behavior of reversible systems at equilibria or in the neighborhoods of periodic motions is described by the same Birkhoff series as that of Hamiltonian systems. The equations of phase oscillations at resonances also coincide, as do the bifurcations' normal forms. Finally, we find the same Kolmogorov invariant tori in the phase spaces of reversible systems as in the Hamiltonian systems' phase spaces.

[1]The Russian version: in *Problems of Nonlinear and Turbulent Processes in Physics*, Notices of the Second International Working Group, Edited by A.S. Davydov and V.M. Chernousenko, Naukova Dumka, Kiev, 1985, Part 2, 15–21. (Editors' note)

Originally publ. in: Nonlinear and Turbulent Processes in Physics (Kiev, 1983). Vol. 3, 1161-1174,

One can see no a priori reason for all these striking similarities, which have been discovered by different authors as a result of calculations that were themselves different in different cases.

(It would be interesting to find analogues to my 1965 conjectures regarding the intersections of Lagrangian manifolds and the periodic solutions of Hamiltonian systems, which were proved recently by C.C. Conley, E. Zehnder, and M. Chaperon.)

Reversible Mappings

A diffeomorphism $\sigma : M \to M$ is called an involution if its square is the identical mapping, $\sigma^2 = 1$.

Example 2. A reflection of Euclidean space $\mathbb{R}^{k+l}$ in the subspace $\mathbb{R}^k$ is an involution, $\sigma(p, q) = (p, -q)$.

Locally (near a fixed point) every involution can be written in such a form by using suitable coordinates (Bochner's theorem).

A diffeomorphism $A : M \to M$ is reversible (with respect to involution σ) if σ transforms A into A^{-1}.

Example 3. A plane reflection in a coordinate line transforms a rotation about the origin into the inverse rotation. For the same reason, a rotation whose angle varies with distance from the origin, $(r, \phi) \mapsto (r, \phi + \alpha(r))$, is reversible.

The reversibility condition $\sigma A \sigma = A^{-1}$ implies $A\sigma A\sigma = 1$. Thus, every reversible diffeomorphism is a product of two involutions.

The converse is also obvious.

Theorem 1 (Birkhoff, 1915). *A product of two involutions is reversible (with respect to each of them).*

Proof. $\sigma^2 = \tau^2 = 1$ implies $\sigma(\sigma\tau)\sigma(\sigma\tau) = 1$ and $\tau(\sigma\tau)\tau(\sigma\tau) = 1$.

Example 4. The Birkhoff billiard mapping transforms a velocity vector at the billiard boundary to the reflected velocity vector at the next reflection point. This mapping is a product of two involutions: the first maps the outgoing velocity vector to the incoming velocity vector at the next reflection point; the second is the reflection with respect to the boundary tangent.

Melrose (1976) extended this Birkhoff construction to obtain an involution of the intersection of any two hypersurfaces of a symplectic manifold; he associated every point of the intersection manifold with the next point of the manifold that belongs to the same phase curve of the Hamiltonian system, defined by the first (the second) hypersurface, to which the initial point belongs.

Both Melrose involutions share a common manifold of fixed points (the "glancing rays" manifold). In other instances, one encounters generic involution pairs.

Involution Pairs

Involution pairs (and hence reversible mappings) appear in several different branches of mathematics, and sometimes quite unexpectedly.

Example 5 (Teixeira, 1982). In the theory of discontinuous ordinary differential equations, the following situation is generic: the surface of discontinuity divides the phase 3-space into two parts; in each part, the system is governed by its own phase-velocity vector field.

The phase curves of the first (second) field are tangent to the discontinuity surface at the points of the first (second) "fold line" (the singular line of projection of the discontinuity surface along the phase curves).

In the general case, the two fold lines intersect at a nonzero angle.

Two involutions of the dividing surface appear in the neighborhood of the intersection point: the first one interchanges the two neighboring points of the intersection of the first vector-field phase curve with the surface; the second, those of the second.

The stability of the equilibrium position at the intersection points depends on the behavior of the iterations of the product of both involutions and, thus, leads to the study of a reversible plane mapping in the vicinity of a fixed point.

Example 6 (Moser and Webster, 1982). Let us consider a real 2-surface in a complex 2-plane space (where the real dimension of the latter is 4). Moser and Webster reduced the search for such a surface biholomorphic invariant at a point where the surface tangent is complex (which is the only difficult case) to the study of a pair of holomorphic involutions of a complex plane at an intersection point of the lines of the involution fixed points. Thus, from the normal form analysis of the reversible mappings, they deduced the normal forms of real surfaces in a complex plane (considered up to biholomorphic changes of plane coordinates).

Example 7. A smooth action of some group on a phase manifold is called a dynamical system whose time is this group.

The reversible mappings can be thought of as actions of an arithmetical progression's symmetry group. Indeed, this group consists of translations and reflections. A reflection is an involution; it reverses the translation.

Reversible Flows

A reversible flow is a dynamical system whose time is the symmetry group of a real line. In other words, a reversible flow is determined by a one-parameter group of phase space diffeomorphisms and by an involution that inverts these diffeomorphisms.

The velocity field of the reversible flow is transformed by this involution into the opposite one. A vector field, changing its sign under the action of the involution (and the differential equation associated with this field) is called reversible (with respect to this involution).

Example 8. The differential equations on the complex z-plane

$$\dot{z} = iz, \qquad \dot{z} = \bar{z}, \qquad \dot{z} = z^2$$

are reversible (with respect to three different involutions). The equation $\dot{z} = \sum a_{k,l} z^k \bar{z}^l$ is reversible with respect to the involution $z \mapsto \bar{z}$ if all the coefficients $a_{k,l}$ have zero real parts.

Example 9. The system of Newton equations is reversible if the applied forces are independent of time and of the velocities or, at least, are even with respect to the velocities:

$$\ddot{x} = F(x, \dot{x}), \qquad F(x, -y) \equiv F(x, y)$$

(the phase space involution changes the signs of all the velocities).

Example 10. An equation of even order is reversible if it is autonomous (does not depend explicitly on the independent variable) and the odd orders derivatives appear only in even combinations (the phase space involution changes the signs of all the derivatives of odd order). For instance, (1) is reversible.

One can see from Examples 8–10 that typical reversible flows are drastically different from generic phase flows in asymmetric systems and are similar to flows in Hamiltonian systems.

For instance, let us consider an equilibrium point of a reversible system in a plane. This point, we suppose, belongs to the line of fixed points of the involution (this is a generic situation). For the linearized system, these points can be generically of either saddle or center type, as is the case with the Hamiltonian systems. The center is conserved even when the nonlinearity of the reversible system is taken into account, since attractors in such a system may not intersect the manifold of fixed points (because the time inversion converts attraction into repulsion).

The example of equation $\dot{z} = z^2$ shows, however, that, in a neighborhood of an equilibrium point, a reversible system may admit no first integrals, nonconstant in this neighborhood, and no integral invariants of positive density.

Periodic Motions

Let us consider a reversible system of differential equations with phase space $\mathbb{R}^{2n}$ and let the manifold of the involution fixed points be $\mathbb{R}^n$. Let us suppose that the system has a periodic solution whose phase curve intersects the manifold of fixed points, $\mathbb{R}^n$, at a nonzero angle. The involution maps this phase curve (the cycle) onto itself with reverse orientation. Thus, there are exactly two involution fixed points on the cycle.

Theorem 2. *For generic reversible systems with the above conditions*: (1) *Symmetric cycles that are close to the given one form a one-parameter family and* (2) *This family is structurally stable, that is, any neighboring reversible system admits a family of symmetric cycles and the families are close to each other.*

Proof. Let Σ^{2n-1} be a σ-invariant section that contains the manifold $\mathbb{R}^n$ of fixed points of the involution σ. Let Σ intersect the given cycle at points a and b transversally (at nonzero angles). Let us consider the Poincaré return mapping $A : \Sigma \to \Sigma$, which transforms a into b and b into a. This mapping is reversible (with respect to σ, acting on Σ). The submanifolds $\mathbb{R}^n$ and $A\mathbb{R}^n$ of the $(2n-1)$-dimensional manifold Σ generically intersect along a curve that contains b. Every point c of this curve belongs to a symmetric cycle. Indeed, $\sigma c = c$, $\sigma(A^{-1}c) = A^{-1}c$, and $A^{-1}c = \sigma A \sigma c$, since A is reversible. Hence, $A\sigma A^{-1}c = c$, $A\sigma(\sigma A \sigma)c = c$, and $A^2 c = c$, that is, c belongs to a cycle.

For a neighboring system, the manifolds $\mathbb{R}^n$ and $A\mathbb{R}^n$ also intersect transversally near b; hence, the family of symmetric cycles still exists.

Remark. For an involution with a fixed point manifold of dimension k embedded in a space of dimension $k + l$, the same reasoning defines a $(k - l + 1)$-parameter family of cycles.

Example 11. For equation (1), $k = l = 2$. Hence, symmetric cycles of (1) are organized into one-parameter families that cannot be destroyed by small perturbations of the system that preserve reversibility.

Neighborhood of a Periodic Motion

Let us consider the case where the phase space dimension is 4 and the involution's fixed point manifold has dimension 2. Equation (1) provides a relevant example.

The dimension of the surface Σ is 3. The involution of this 3-space changes the sign of one coordinate. The reversible Poincaré mapping $A^2 : \Sigma \to \Sigma$ has a line of fixed points that belongs to the 2-surface of fixed points of the involution.

Let us describe the results of applying the Birkhoff–Poincaré theory of normal forms to such reversible mappings. Neglecting for the moment the divergence of the series, we obtain the following picture. The neighborhood of the line of fixed points is (formally) fibered into 2-surfaces invariant under both the Poincaré mapping and the involution. Let us consider one of these surfaces.

We are led to the study of a reversible mapping of a plane in the neighborhood of a fixed point that lies on the invariant curve of the involution. A study of reversible plane mappings near fixed points is of independent interest, as was explained above.

Structure of a Reversible Mapping of a Plane

Depending on the linearization eigenvalues at a fixed point, the mapping may be either hyperbolic (the linearization is a hyperbolic rotation) or elliptic (ordinary rotation). Let us consider the elliptic case.

The mapping is resonant if the rotation angle is commensurable with 2π (equal to $2\pi p/q$; the integer q is called the resonance order). If the mapping is nonresonant, then the formal normal form describes the rotation over a radius-dependent angle. In this case, the fixed point is formally stable (since it is inside the small, formally invariant circles).

For a typical reversible elliptic plane map, there exist genuine (not just formal) invariant Kolmogorov curves in any neighborhood of the fixed point. The mapping's rotation number at such a curve is irrational.

Between these curves, there are "resonant zones" generated by formally invariant curves with rational rotation numbers that are destroyed by the perturbations.

In the resonant zone corresponding to rotation number p/q, the qth iteration of the mapping is close to the mapping of the phase flow of a reversible vector field, which is invariant under the plane rotation through the angle $2\pi p/q$ (this is also true for the more general case of a reversible mapping that is close to the rotation of a plane annulus through an angle that varies with the polar radius).

The reversible plane vector fields, invariant under a rotation through an angle $2\pi p/q$, are interesting objects with nonobvious properties. A reversible field invariant under rotations may, in principle, have complicated singular points and no first integrals. However, in the problem of the resonant zone structure, these cases occur only as exceptions. The typical equation of resonant phase oscillations for a reversible system admits a first integral and displays the same pattern of saddle- and center-type points as the Hamiltonian equation of phase oscillations.

Structure of the Poincaré Mapping

Let us return to the Poincaré mapping of the 3-space. While invariant two-dimensional surfaces exist only formally, the Kolmogorov invariant curves are genuine (since a systematic drift along the line of fixed points is excluded by the reversibility condition). Curves with the same rotation number born at different formally invariant surfaces form a smooth cylindrical 2-surface (we assume a nondegenerate dependence of the formal rotation number on the squared distance from the line of fixed points).

In this situation, the fixed points of the Poincaré mapping have some stability property: escape (and only very slow escape) is possible only along the Kolmogorov surfaces with constant rotation number, and not transversally to them.

There are gaps where the resonant zones are placed between the Kolmogorov surfaces. A Kolmogorov surface may intersect the line of fixed points at the point where the rotation angle of the linearization corresponds to the rotation number of the Kolmogorov curves. In the neighborhood of such a point, the surface looks, generically, like a paraboloid of rotation fibered into its parallels. Indeed, the formal series for the rotation number, dependent on a coordinate z of the fixed point and on the distance r from this point along the formally invariant surface, begins with $\alpha(z) + r^2\beta(z) + \dots$

It may happen that a Kolmogorov surface intersects the line of fixed points twice. In this case, the surface is diffeomorphic to a 2-sphere. This sphere traps the phase points. Thus, we have proved the Lyapunov stability of those fixed points which correspond either to a minimal linearized rotation angle (if β is positive) or to a maximal one (for negative β).

Of course, strong resonances ($q \leqslant 4$) in reversible systems deserve a separate study, similar to that of the theory of strong resonances in Hamiltonian systems.

Applications to Reversible Equations

Let us consider the phase 4-space of the initial reversible equation, say (1). We obtain a one-parameter family of symmetric cycles and a set of invariant Kolmogorov 3-manifolds labeled by the rotation numbers (normally incommensurable with 1) and fibered into Kolmogorov 2-tori. Each torus is filled with orbits of quasi-periodic motion whose ratio of frequencies is constant on every Kolmogorov 3-manifold.

In the resonant zones between these 3-manifolds, we find other Kolmogorov 3-manifolds, also fibered into Kolmogorov 2-tori: they are born from level lines of the first integral of the equation of resonant phase oscillations.

Thus, the general pattern of the motion, at a symmetric cycle, of a reversible system with a phase 4-space and with an invariant 2-surface of the involution is very similar to that of a Hamiltonian system with two degrees of freedom.

However, the existence of first integrals of the Hamiltonian system implies an important difference between these cases. For the Hamiltonian system, the existence of Kolmogorov tori implies (under the usual nondegeneracy conditions) orbital stability of the periodic motion. Meanwhile, in the reversible system, orbits of quasi-periodic motions in the phase space form 3-surfaces that do not isolate the cycle and cannot prevent some orbits from escaping from the cycle along the resonant zone.

It seems reasonable that such an escape, though very slow, is a generic feature of reversible systems, but I have no proof of this conjecture.

Remark. It seems that many properties of reversible systems can be extended to all systems that can be transformed into reversed systems by some diffeomorphism of the phase space (maybe not an involution).

But, at present, the theory of such "weakly reversible" systems seems to be in an embryonic state.

2 Oscillations and Bifurcations in Reversible Systems

V. I. Arnol'd, *Corr. Mem. USSR Acad. Sc. and*
M. B. Sevryuk

2.1. INTRODUCTION

A typical example of a reversible system is the equation

$$\ddddot{x} + A\ddot{x} + Bx + C\dot{x}^2 = 0 \tag{2.1}$$

invariant under the sign change of the independent variable. This equation describes various physical phenomena (see, e.g. the papers by Demekhin [2.19] on a viscous fluid draining off along an inclined plane or by Malomed and Tribel'skii [2.20, 2.28] on the evolution of the fronts of some non-equilibrium first-kind phase transitions, e.g. of the gas combustion or the laser evaporation of a condensed substance). Equation (2.1) with the fixed coefficients A, B, and C has a one-parameter family of cycles. Such families are typical of Hamiltonian systems, while the cycles in the phase space of general non-Hamiltonian systems are, as a rule, isolated.

The similarity between Hamiltonian and reversible systems is not confined to uniting the cycles into families but extends much farther. For example, the evolution of a reversible system near an equilibrium position or a periodic motion is described by the same Birkhoff series as in the case of Hamiltonian systems. For reversible systems an analogue of the Lyapunov theorem is valid that concerns strictly periodic oscillations of a multidimensional nonlinear Hamiltonian system around the equilibrium position in the nonresonant elliptic case. Also coincident are the equations of phase oscillations near resonances and the normal forms of bifurcations. Finally, in the phase space of a reversible system the same invariant Kolmogorov tori can be found as in the case of Hamiltonian systems.

The present paper begins with the definition of reversible systems with discrete and continuous times (Secs. 2.2 and 2.3). Then integrable reversible systems (Sec. 2.4) and their small perturbations (Sec. 2.5) are considered. The results obtained are carried over to a more general class of "weakly reversible" systems (Sec. 2.6). Then we investigate the equilibrium position neighbourhood of reversible (Sec. 2.7) and weakly reversible (Sec. 2.8) systems and study the symmetric, strictly periodic oscillations in reversible systems (Sec. 2.9). A neighbourhood of a closed symmetric trajectory of a reversible system is treated in Sec. 2.12 and appropriate preliminary remarks are exposed in Secs. 2.10 and 2.11. At the beginning

Originally publ. in: Nonlinear Phenomena in Plasma Physics and Hydrodynamics.
Editor: R.Z. Sagdeev, 31-64,
Reprint by Springer-Verlag Berlin Heidelberg 2018 with kind permission

we study all phenomena in the simplest situation of a minimum number of degrees of freedom and then turn to a general multidimensional case as well as to a non-autonomous one (Sec. 2.13).

The paper concludes in a study of the bifurcations of the families of low-amplitude periodic motions of reversible systems oscillating near equilibrium when passing through resonances between small oscillation frequencies. This study reveals unexpected relations of the object at hand with the theory of singularities of differentiable mappings (e.g. in analyzing the resonance 1:1 there emerges the Whitney umbrella) and leads to new results in this general theory (e.g. a rearrangement of the umbrella's cross sections).

The first coauthor is responsible for the statement of the problem, the study of the solutions of reversible systems in the case of small dimensions [2.1] and the formulations of hypotheses on a multidimensional case as well as on the bifurcation normal forms for the families of the low-amplitude periodic motions of reversible systems around the equilibrium position when passing through the resonance 1:1 (Sec. 2.15). The expedience of generalizing the concept of reversibility to that of weak reversibility was also noted in [2.1]. The transfer of the results from [2.1] to the multidimensional case (as well as to the case of an odd-dimensional phase space) and the derivation of proofs were carried out by the second coauthor. The results of Sec. 2.16 (where the passage through further resonances $1{:}N$, $N \geqslant 2$ is studied) and the proofs of the theorems exposed in Sec. 2.15 are also his.

2.2. REVERSIBLE MAPPINGS

Definition. A diffeomorphism $G: M \to M$ of a phase space M onto itself is called an involution if its square is an identical transformation, $G^2 = \mathrm{id}$. The most important is the following example.

Example. The reflection of an Euclidean space $\mathbb{R}^{m+k}$ with respect to a subspace $\mathbb{R}^m$,

$$G: (p, q) \mapsto (p, -q),$$

where $p \in \mathbb{R}^m$, $q \in \mathbb{R}^k$, is an involution.

Locally (i.e. in a neighbourhood of a fixed point) any involution can be written in such a form under an appropriate choice of coordinates (Bochner's theorem).

Definition. A diffeomorphism $A: M \to M$ is called reversible with respect to an involution G if G conjugates A with A^{-1}, i.e. $A^{-1} = GAG$.

Example. A plane rotation is reversible since, by a reflection with respect to a straight line passing through the rotation centre, this rotation is transformed into the inverse one. The same holds for a rotation through a variable (dependent on radius) angle: $(r, \varphi) \to (r, \varphi + \gamma(r))$.

The reversibility condition $A^{-1} = GAG$ that can be rewritten as $AGAG = \mathrm{id}$ shows that AG is an involution and, thus, any reversible diffeomorphism $A = (AG)G$ is a product of two involutions. A converse theorem is also evident:

Theorem (Birkhoff [2.3]). A product of two involutions is reversible with respect to any of them.

Proof. If $\sigma^2 = \tau^2 = \mathrm{id}$, then $\sigma(\sigma\tau)\sigma(\sigma\tau) = \tau(\sigma\tau)\tau(\sigma\tau) = \mathrm{id}$.

Pairs of involutions (and, consequently, reversible mappings) appear in different branches of mathematics (see in [2.1] the discussion of the Birkhoff billiard transformation ([2.4] and [2.5], Ch. 6, §§ 6-9) and of the Melrose billiard transformation [2.6], and pairs of involutions emerging in the studies of discontinuous differential equations* and in the problem of a holomorphic classification of real surfaces in a plane of two complex variables [2.7]).

A reversible mapping can be considered as an action of the symmetry group of arithmetic progression. Indeed, this group consists of translations and reflections, where a reflection acts as an involution reversing the action of a translation.

Definition. Given a certain group H, the action of this group on the phase space by diffeomorphisms is called a dynamic system whose time is H.

From this viewpoint, a reversible mapping is a dynamic system whose time is the symmetry group of arithmetic progression.

2.3. REVERSIBLE FLOWS

Definition. A reversible flow is a dynamic system whose time is the real line symmetry group.

In other words, a reversible flow is specified by a one-parameter group of phase space diffeomorphisms that transforms into the group of inverse diffeomorphisms under a certain phase space involution.

Under this involution the velocity field of a reversible flow transforms into the opposite field.

Definition. A vector field V changing its sign due to an involution G, $G_* V = -V$ (and a differential equation specified by the field), is called reversible (with respect to this involution).

Thus, the velocity field of a reversible flow is reversible too. The converse is also true.

Example 1. The differential equation

$$\dot{z} = \sum_{l,\, s} a_{l,\, s} z^l \bar{z}^s$$

---

* *Teixeira*, private communication (1982).

in the plane of the variable z is reversible with respect to the involution $z \mapsto \bar{z}$ of complex conjugation if and only if all the coefficients $a_{l,s}$ are pure imaginary (Fig. 2.1*a*).

Example 2. The differential equations $\dot{z} = \bar{z}$ and $\dot{z} = z^2$ (Figs. 2.1*b* and 2.1*c*) in the z-plane are also reversible (the corresponding involutions are the reflections with respect to the straight lines $\operatorname{Re} z = \operatorname{Im} z$ and $\operatorname{Re} z = 0$).

Example 3. The system of the Newtonian equations, $\ddot{x} = F(x, \dot{x})$, is reversible if forces are independent of velocities or are even in velocities, $F(x, -y) = F(x, y)$ (a phase space involution changes the signs of all velocities).

Example 4. An even-order equation is reversible if it is autonomous and if odd derivatives enter into it only in even combinations

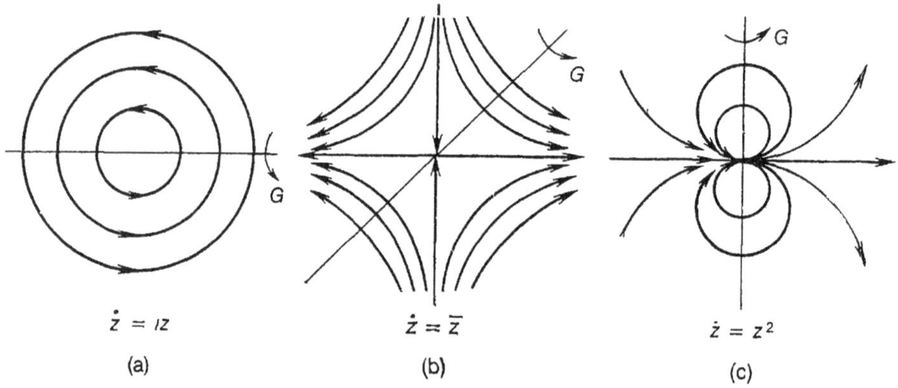

Fig. 2.1

(a phase space involution changes the signs of all odd derivatives). In particular, Eq. (2.1) is reversible.

Note that in all instances of reversible systems and mappings mentioned above the phase space dimension was even and the manifold of fixed points of the involution had a middle dimension.

2.4. INTEGRABLE REVERSIBLE MAPPINGS AND VECTOR FIELDS

As a plane rotates through a variable (dependent on radius) angle around the origin, each circle centered in the origin is invariant both under a mapping itself and an involution reversing it, the restriction of the mapping to any such circle being a conventional rotation.

Generalizing this example to higher dimensions we will give the following definition.

Definition. A reversible mapping (vector field) with the phase space D of an even dimension $2m$ is called integrable if D is foliated into m-dimensional tori invariant under both the mapping itself (the vector field itself) and the involution reversing the mapping (the vector field), the initial mapping (vector field) inducing a quasi-periodic motion on each torus.

By a suitable choice of coordinates each integrable reversible mapping can be written in the form

$$(\varphi, I) \mapsto (\varphi + \omega(I), I)$$

and each integrable reversible differential equation, in the form

$$\dot{\varphi} = \omega(I), \ \dot{I} = 0,$$

in both cases the involution is

$$(\varphi, I) \mapsto (-\varphi, I), \tag{2.2}$$

where $I = (I_1, \ldots, I_m)$ is the torus parameter (an analogue of the action variable in integrable Hamiltonian systems), $\varphi = (\varphi_1, \ldots, \varphi_m)$ is the angular variable on each torus $I = \text{const}$, and $\omega_1(I), \ldots, \omega_m(I)$ are the frequencies.

An integrable reversible mapping (vector field) is called nondegenerate if nondegenerate is the frequency mapping $I \mapsto \omega(I)$, i.e. $\partial\omega/\partial I \neq 0$. Here, after an appropriate renumbering of invariant tori, the initial mapping can be written as

$$(\varphi, I) \mapsto (\varphi + I, I) \tag{2.3}$$

and, respectively, the vector field, in the form

$$\dot{\varphi} = I, \ \dot{I} = 0, \tag{2.3'}$$

with the reversing involution here being that of (2.2). Indeed, the form (2.3), or (2.3′), is used to write the mapping (vector field) in a certain region of the variation of I.

2.5. KOLMOGOROV'S TORI

It happens that in the phase space of a reversible mapping (vector field) that is close to a nondegenerate integrable one invariant (Kolmogorov's) tori exist on which a quasi-periodic motion is again induced (this is fully analogous to the effects arising in the Hamiltonian case).

Theorem. Let be given a nondegenerate integrable and reversible mapping (2.3) or vector field (2.3′) with a reversing involution (2.2). Let us consider a small perturbation of the mapping

$$(\varphi, I) \mapsto (\varphi + I + f(\varphi, I),\ I + g(\varphi, I)), \tag{2.4}$$

and, respectively, that of the vector field

$$\dot{\varphi} = I + f(\varphi, I), \quad \dot{I} = g(\varphi, I), \tag{2.4′}$$

as well as that of the reversing involution

$$(\varphi, I) \mapsto (-\varphi + \tilde{f}(\varphi, I),\ I + \tilde{g}(\varphi, I)), \tag{2.5}$$

such that (2.4), or (2.4′), is reversible (with respect to the involution (2.5)) as before.

Then, if the perturbations f, g, $\tilde{f}$, and $\tilde{g}$ are analytical and sufficiently small, the majority of the invariant tori $I = \text{const}$ of the mapping (2.3) (vector field (2.3′)) under such a perturbation will not decay but only slightly deform. Namely, those tori do not decay which participated in the quasi-periodic motion with "strongly incommensurable" frequencies. On the deformed (Kolmogorov's) tori invariant under both the perturbed mapping (2.4), or the differential equation (2.4′), and the reversing involution (2.5) a quasi-periodic motion is again induced having the same set of frequencies.

For vector fields the theorem was enunciated by Moser in [2.8] and proved by him in [2.9], Secs. 5, 6. It is essential for Moser's proof that the perturbation of the initial integrable and reversible differential equation is not simply small but dependent on the small parameter ε ($\varepsilon = 0$ corresponds to the unperturbed equation). The applications are discussed in [2.8].

The smaller the perturbation the larger the union measure of all Kolmogorov's tori.

For reversible vector fields in a plane (i.e. for a two-dimensional phase space) all those invariant circles do not decay the frequencies of motion along which are sufficiently high and, consequently, the perturbation of a nondegenerate integrable and reversible vector field in the plane remains integrable.

The proof of the theorem is based on the standard (Kolmogorov's) method of constructing an infinite sequence of Newtonian approximations, i.e. changes of variables whose definition domains are more and more contracted to the deformed torus sought for. At each iteration step one has to solve a certain homological equation and it is the verification of its solvability for which one uses the reversibility of the objects considered. An obstruction for the solvability is due to the one-dimensional cohomology of a certain algebraic system (in particular, for discrete time, the cohomology of the group generated by the unperturbed reversible diffeomorphism and by the involution reversing the latter) which thus is trivial.

2.6. WEAK REVERSIBILITY

Many properties of reversible objects are typical of those mappings (vector fields) which transform into the inverse ones (opposite) under the action of some phase space diffeomorphism being not necessarily an involution. Such mappings (vector fields) will be called weakly reversible. Thus, a phase space diffeomorphism A is weakly reversible with respect to another phase space diffeomorphism G if $A^{-1} = GAG^{-1}$ and a vector field V (or the differential equation $\dot{x} = V(x)$) is weakly reversible with respect to some phase space diffeomorphism G, if $G_* V = -V$.

If the given mapping or vector field is reversed by a diffeomorphism G, it is reversed by the diffeomorphism G^{-1} too. If a vector field is weakly reversible with respect to a certain phase space diffeomorphism, then the mapping of the phase flow of this field at any fixed time will be also weakly reversible with respect to the same diffeomorphism (and vice versa).

Kolmogorov's tori prove to exist also in the case when the perturbation of a nondegenerate integrable and reversible mapping (vector field) is assumed to be not reversible but only weakly reversible. In other words, the requirement of the diffeomorphism (2.5) reversing the mapping (2.4) (the vector field (2.4′)) to be involutary is not a necessary one. Moreover, here the restriction of (2.5) to each invariant Kolmogorov's torus of the perturbed mapping (vector field) is an involution. Therefore, if the perturbation is sufficiently small and, consequently, the union measure of invariant tori is positive, then, with analyticity included, this suggests that (2.5) is an involution. In other words, a small, weakly reversible perturbation of a nondegenerate integrable and reversible mapping (vector field) is always a "true" reversible one.

2.7. THE LOCAL THEORY

The theorem of Kolmogorov's tori of reversible mappings (vector fields) has a local analogue, i.e. the presence of invariant tori in a neighbourhood of the equilibrium position.

Consider, e.g. a reversible mapping of a plane in a neighbourhood of a fixed point (the point is also assumed to be fixed with respect to the involution reversing the given mapping).

Depending on the eigenvalues of linearization at the fixed point a general reversible mapping can be hyperbolic (the linearization is a hyperbolic rotation) or elliptic (the linearization is a conventional rotation). The eigenvalues 1 or -1 are of a nongeneral position. Consider the elliptic case in more detail.

Depending on the linearization rotation angle at a fixed point a mapping is called nonresonant (the angle is incommensurable with 2π) or resonant of order q (the angle is $2\pi p/q$).

The formal normal form for a nonresonant mapping is a rotation through an angle varying with radius. Here the fixed point is formally stable since it is surrounded by formally invariant curves.

It appears that a general analytical reversible elliptic mapping of a plane (both in the nonresonant case and in the case of a resonance of order $q \geqslant 5$) also has, in any neighbourhood of a fixed point, "genuine" (and not only formal) invariant Kolmogorov's curves which surround the fixed point. The curves are also invariant under the involution reversing the mapping. The mapping rotation number on such a curve is irrational. As a consequence, we obtain that the fixed point of a general reversible elliptic mapping of a plane is stable.

Remark 1. Here the word "general" means that by no substitution the 3-jet (a cubic Taylor polynomial) of the mapping can be reduced to its linear part (the 1-jet). For any $q \geqslant 3$ there exist examples of reversible elliptic, resonant of order q diffeomorphisms of a plane which do not satisfy this condition and whose fixed point is unstable (analogous examples for Hamiltonian (symplectic) mappings see in [2.11], § 31).

Remark 2. It is well known (see, e.g. [2.10] or [2.11], §§ 32-34) that invariant curves in a neighbourhood of a fixed point (or in an annular domain) exist for any general mapping of the plane that possesses the intersection property, i.e. the image of any curve surrounding the fixed point intersects this curve. For Hamiltonian diffeomorphisms of a plane this property is fulfilled automatically, since the figures bounded by such a curve and its image have the same area. At the same time the examples are known of the reversible diffeomorphisms of a plane which do not have the intersection property in any neighbourhood of the point fixed with respect to the diffeomorphism itself as well as to the involution reversing the latter.

Example (M.B. Sevryuk). Let us denote as ψ a smooth function $\mathbb{R} \to \mathbb{R}$ defined as $\psi(t) = t + c \exp(-t^{-2}) \sin t^{-1}$ for $t \neq 0$ and $\psi(0) = 0$, where

$$0 < c < \left(\max_{t \neq 0} \left| \frac{d}{dt} \exp(-t^{-2}) \sin t^{-1} \right| \right)^{-1}.$$

The function strictly increases on the whole real line and has an inverse smooth function ψ^{-1}.

Let us introduce in a plane a Cartesian coordinate system (x, y) and consider a mapping $(x, y) \mapsto (\psi(x), \psi^{-1}(y))$. The latter is reversible with respect to an involution $(x, y) \mapsto (y, x)$.

If a and b are positive numbers such that $\psi(a) < a$ and $\psi(b) > b$ (a and b can be chosen however small), then the mapping transforms a rectangle with the apices (a, b), $(-a, b)$, $(-a, -b)$, $(a, -b)$ into another rectangle lying strictly inside the initial one.

At the fixed point all the constructed diffeomorphisms of such a kind lose analyticity (and only remain smooth), while the eigenvalues of their linearization are 1 or —1. It is unknown whether there are elliptic (let them be even non-analytical) reversible mappings not possessing the intersection property.

Let us return to a general elliptic mapping of a plane. The fixed zero point of the mapping is surrounded by invariant curves. Be-

Fig. 2.2

tween the curves there are annular resonance zones which are left due to the decay of the formal invariant curves with rational rotation numbers p/q (the resonance p/q zones). A qth mapping iteration in such a zone differs but slightly from the transformation of the phase flow of a reversible vector field invariant under the rotation through the angle $2\pi/q$ and having a nonconstant first integral. As for the Hamiltonian case, these resonance zones contain cycles (fixed points of the qth iteration of the mapping).

More precisely, such zones exist in any neighbourhood of zero. In each of them two cycles (let us call them nodes) necessarily lie on the curve of fixed points of a reversing involution G on either side of zero. Any iteration A^n of the mapping A transforms a node into a point also fixed with respect to A^q, and in a neighbourhood of the latter an involution A^nGA^{-n} acts reversing A^q and leaving this point fixed.

If q is odd, the number of distinct images of the nodes, with the mapping A iterated, is $2q$ (the images of one of the nodes and of the other in going round zero alternate), and if q is even then the above number is q (since for an even q the mapping $A^{q/2}$ transforms the nodes into one another).

However, with A iterated, for an even q placed between any two adjacent (in going round zero) images of the nodes, there exists at least one additional cycle. (Note that for an odd q the resonance zone p/q can also contain the fixed points of A^q that are not the nodal images when A is iterated.) Thus, the total number of the fixed points of the qth iteration of the general mapping A in the resonance zone p/q is no less than $2q$ (for any evenness of q). Overall, the picture of the invariant curves of the iterations of A in a neighbourhood of zero is as shown in Fig. 2.2 (the same as in the Hamiltonian case).

Analogously is arranged the resonance zone p/q in a more general situation of a reversible mapping close to a nondegenerate integrable one in an annular domain.

Consider now a multidimensional case: a reversible mapping of an Euclidean space $\mathbb{R}^{2m}$ $(m \geqslant 2)$ in a neighbourhood of a fixed point lying on an m-dimensional manifold of an involution fixed points. Assume that this mapping is elliptic (i.e. its linearization at a fixed point is a direct product of m rotations) and general (linearization rotation angles have no resonance relation of low order, while the frequencies of motion on the invariant tori of a formal normal form are dependent in a nondegenerate way on the variables numbering the tori).

Then, as in the two-dimensional case, in any neighbourhood of the fixed point, m-dimensional (Kolmogorov's) tori exist that are invariant under both the mapping itself and the involution reversing the latter. The initial mapping on the tori induces a quasi-periodic motion with a discrete time (the proximity to a fixed point plays here, i.e. in the local situation, the role of the smallness of the integrable diffeomorphism perturbation). Resonance zones lie between Kolmogorov's tori.

Let us turn to the case of continuous time, i.e. to reversible flows in a neighbourhood of a singular point fixed under a reversing involution. The isolated singular points of a linear reversible vector field in a plane can only be centres (an example: $\dot{z} = iz$) and saddles (an example: $\dot{z} = \bar{z}$), besides the saddles must be "equilateral" (their eigenvalues are $\pm\lambda$). Moreover, the centre is conserved (transforming neither into a focus nor into a centrefocus) on accounting for the nonlinear terms of the reversible vector field too, since the attractors in a reversible system that intersect the manifold of involution fixed points are impossible (because under time inversion attraction transforms into repulsion). The closed phase curves of such a centre (into which a punctured neighbourhood of a fixed point is foliated) are invariant under involution.

Remark 1. The example $\dot{z} = z^2$ shows that in degenerate cases a reversible system can have no first integral nonconstant in a neighbourhood of a singular point and can conserve no integral invariant with positive density.

Remark 2. An example of an irreversible degenerate Hamiltonian system can be also readily constructed. Let a Hamiltonian H in a plane with the coordinates (x, y) be the product of five linear homogeneous functions, i.e. $H(x, y) = \prod_{i=1}^{5} (a_i x + b_i y)$. If a system with such a Hamiltonian is reversible then the union of the straight lines $a_i x + b_i y = 0$, $i = 1, \ldots, 5$, is invariant under a reversing involution, this being possible only for the exclusive fives $(l_1, \ldots, l_5)$ of straight lines (cross ratios of the fours of the straight lines l_1, l_2, l_3, l_4 and l_1, l_2, l_3, l_5 are independent invariants of the five with respect to the group of linear transformations of a plane and for the "symmetric" fives, invariant with respect to some involution, these two cross ratios are related by a functional dependence).

In a multidimensional case, reversible vector fields in a neighbourhood of a singular point fixed under a reversing involution behave like reversible diffeomorphisms. If the linearization of such a field at a singular point $0 \in \mathbb{R}^{2m}$ $(m \geqslant 2)$ is a direct product of m linear fields with a singular point of the centre type (an elliptic field), then under common nondegeneracy conditions in any neighbourhood of zero there exist m-dimensional (Kolmogorov's) tori invariant under both the phase flow and the involution, and the motion on these tori is quasi-periodic.

Different local versions of the theorem on Kolmogorov's tori for reversible systems of differential equations were analyzed by Bibikov and Pliss (see, e.g. [2.12-2.14] and [2.15], Ch. III, §§ 15-20). More precisely, these papers proved the existence of invariant tori in an arbitrary neighbourhood of a singular point of the analytical system of differential equations, if the latter fulfil some conditions in fact equivalent to reversibility (in some cases these conditions are also satisfied as a result of Hamiltonianity). In [2.13] nonautonomous systems of differential equations (with the right-hand sides being periodic in time) were considered. As to the non-autonomous case, see Sec. 2.13 of the present paper.

2.8. WEAK REVERSIBILITY IN A LOCAL SITUATION

The majority of the results presented in the preceding section can be carried over to the case when in place of reversible objects one considers those being a priori only weakly reversible. Namely, consider an elliptic, weakly reversible general mapping (vector field) in $\mathbb{R}^{2m}$ $(m \geqslant 1)$ in a neighbourhood of a fixed (singular) point, and let the latter be fixed under a reversing diffeomorphism too. In the general case this diffeomorphism turns out to be always an involution so that the weakly reversible mapping (vector field) is actually reversible. Although this result is proved only under some assumptions of a nonresonant and nondegenerate behaviour, examples of elliptic, weakly reversible mappings or vector fields not being reversible are unknown. In the non-elliptic (e.g. hyperbolic) case such examples can be readily constructed even in a two-dimensional situation (moreover, the exemplified mappings and vector fields are general).

Example 1 (M. B. Sevryuk). Let $\alpha(t)$ be an arbitrary real-analytical function preset in a neighbourhood of zero on the real line and satisfying the condition $\alpha(0) \neq 0$. According to the implicit function theorem, the relation $\beta(t)\,\alpha(-t\,\alpha(t)\,\beta(t)) \equiv 1$ defines another real-analytical function $\beta(t)$ in a neighbourhood of zero on the real line. This relation implies that $\alpha(t)\,\beta(-t\,\alpha(t)\,\beta(t)) \equiv 1$.

Consider a mapping $(x, y) \mapsto (x\alpha(xy),\ y\beta(xy))$ in a plane with the Cartesian coordinate system (x, y). The eigenvalues of its linearization are $\alpha(0)$ and $\beta(0) = (\alpha(0))^{-1}$. The mapping is weakly reversible (it is reversed by the diffeomorphism $(x, y) \mapsto (y, -x)$). If the function α is not even, then the mapping is reversed by no zero-conserving involution.

Example 2 (M.B. Sevryuk). Let $f(t)$ be an even and $g(t)$ an odd real-analytical function preset in a neighbourhood of zero on the real line with $f(0) \neq 0$. Consider a vector field with the components $yf(x^2 - y^2) + xg(x^2 - y^2)$ and $xf(x^2 - y^2) + yg(x^2 - y^2)$ in a plane with the Cartesian coordinate system (x, y). The eigenvalues of its linearization are $f(0)$ and $-f(0)$. The vector field is weakly reversible (it is reversed by the same, as in the preceding example, diffeomorphism $(x, y) \mapsto (y, -x)$). If the function g is not identically zero, then this field is reversed by no zero-conserving involution.

2.9. PERIODIC SOLUTIONS

Suppose that a reversible system of differential equations has a periodic solution whose phase curve L is invariant under a reversing involution (such closed phase curves are called symmetric). Then the involution maps L onto itself with an orientation reversed so that on L exactly two points are fixed under the involution.

Conversely, if a closed phase curve (cycle) of a reversible system contains at least one point fixed under a reversing involution then it contains exactly two such points and it is invariant under this involution.

In addition, if some phase curve of a reversible system contains at least two points fixed under a reversing involution, then it is closed, i.e. it is a symmetric cycle (and thus contains exactly two such points). This property is basic in the search for the symmetric closed phase curves of reversible systems: it suffices to find the phase curves intersecting the involution fixed manifold at two points.

Note that any phase curve of a reversible system intersecting the involution fixed point manifold intersects it at a nonzero angle.

Theorem. For reversible systems with a phase space $\mathbb{R}^{2m}$ and a reflection involution with respect to an m-dimensional subspace $\mathbb{R}^m$ the following holds:

(1) Closed symmetric phase curves form a one-parameter family in a neighbourhood of a given general symmetric cycle.

(2) The family is conserved, with a slight deformation, under any small perturbation of the system leaving the latter reversible.

Proof. The given cycle intersects the fixed manifold $\mathbb{R}^m$ at two points, a and b, at a nonzero angle. The phase curves passing through the points of $\mathbb{R}^m$ close to a form an $(m+1)$-dimensional "tube" Φ that generically intersects $\mathbb{R}^m$ at the point b transversally along a curve containing b. Through each point of the latter curve a phase curve passes intersecting $\mathbb{R}^m$ at two points. This curve is, thus, a symmetric cycle.

For adjacent systems, the transversality of the intersection of Φ and $\mathbb{R}^m$ near b is conserved, therefore the cycles will be conserved, too.

This theorem was considered by Devaney [2.16] and for $m = 1$ it was noted by Sharkovskii [2.17, 2.18].

Example. The symmetric cycles of Eq. (2.1) (for which $m = 2$) are included into one-parameter families structurally stable under small perturbations of the equation that do not remove the latter from the reversible class.

To study the solutions of a reversible system in a neighbourhood of a one-parameter family of symmetric cycles, which includes a given cycle, we draw a secant $(2m—1)$-dimensional Poincaré surface Σ, containing a fixed manifold $\mathbb{R}^m$ of the involution G. Let the surface be invariant under G and intersect the cycle at the point a at a nonzero angle. The surface is drawn through one of the points of intersection of the cycle with $\mathbb{R}^m$, e.g. through the point a. The Poincaré return mapping A: $\Sigma \to \Sigma$ transforming a into itself is reversible with respect to the action of G on Σ. For a general cycle the manifolds $\mathbb{R}^m$ and $A\mathbb{R}^m$ of dimension m in the $(2m - 1)$-dimensional space Σ intersect at the point a transversally along a curve passing through a. Each point of the curve is fixed with respect to A and thus belongs to a symmetric cycle. In particular, this again implies the inclusion of the symmetric closed phase curves of a reversible system into one-parameter families.

Here we have the following situation: first, the involution of an odd-dimensional space Σ^{2m-1} that changes the sign of $(m-1)$ coordinates and, second, the reversible Poincaré mapping $A : \Sigma \to \Sigma$ having a whole curve of fixed points that lies in an m-dimensional space $\mathbb{R}^m$ of involution fixed points.

The generalization of the above mentioned results to the case when the dimension of an involution fixed manifold is not necessarily equal to half a phase space dimension (the latter can be odd) is not automatic.

Example. An autonomous equation of odd order $2m+1$ that includes even derivatives only in even combinations is reversible with respect to a phase space involution changing the sign of all even derivatives.

Here the number $(m+1)$ of "odd", i.e. changing their sign under an involution, phase space coordinates (these are matched here by even derivatives) exceeds the number of "even" coordinates, m, by one.

The majority of the above results on Kolmogorov's tori and the local evolution of reversible objects is *mutatis mutandis* conserved only in the case when the number of the "even" coordinates exceeds that of the "odd" ones (i.e. the dimension of the involution fixed manifold equal to the number of "even" coordinates exceeds half the phase space dimension). This is indeed so on the Poincaré surface.

2.10. KOLMOGOROV'S TORI FOR ADDITIONAL "EVEN" COORDINATES

Consider a phase space of a reversible mapping (vector field) with m "even" coordinates and $k < m$ "odd" ones, this space being foliated into k-dimensional tori invariant under both the mapping (phase flow) and the reversing involution. On these tori a quasi-periodic motion proceeds. For a nondegenerate dependence of the frequencies of this motion on the parameters numbering the tori it is possible to choose a phase space coordinates ($\varphi \in T^k$, $I^{(1)} \in \mathbb{R}^k$, $I^{(2)} \in \mathbb{R}^{m-k}$), where the mapping has the following form:

$$(\varphi, I^{(1)}, I^{(2)}) \mapsto (\varphi + I^{(1)}, I^{(1)}, I^{(2)}),$$

and, respectively, the vector field has the form

$$\dot{\varphi} = I^{(1)}, \quad \dot{I}^{(1)} = 0, \quad \dot{I}^{(2)} = 0,$$

while the reversing involution is

$$(\varphi, I^{(1)}, I^{(2)}) \mapsto (-\varphi, I^{(1)}, I^{(2)}).$$

The quasi-periodic motion frequencies on each torus $I^{(1)} = \text{const}$, $I^{(2)} = \text{const}$ only depend on $I^{(1)}$, therefore the tori with the same frequencies are organized into $(m - k)$-parameter families.

It turns out that for a small perturbation of the initial mapping (vector field) leaving it reversible the majority of these families does not decay but is only slightly deformed. Thus, in the phase space of a perturbed mapping (vector field) m-dimensional manifolds exist that are foliated into k-dimensional tori invariant under both the mapping (flow) itself and the involution. On these tori a quasi-periodic motion proceeds with the same (for a fixed m-dimensional manifold) set of frequencies.

Remark. Generally speaking, given additional $(m - k)$ "even" coordinates, weakly reversible perturbations as yet do not ensure the conservation of invariant tori (since they allow a translation along the m-dimensional manifolds foliated into tori with the same frequencies, and given this translation, every torus transforms into another).

2.11. THE LOCAL THEORY FOR ADDITIONAL "EVEN" COORDINATES

This theory is called upon to describe the structure of the Poincaré mapping for reversible differential equations. First, consider the case when the phase space is four-dimensional and the manifold of involution fixed points is two-dimensional, as for Eq. (2.1). We obtain an involution of a three-dimensional space (of the secant Poincaré surface) with one "odd" coordinate and a reversible mapping having a curve of fixed points that lies in the involution fixed plane. The theory of normal forms (of Birkhoff's series), with the reversibility included, shows that at the level of formal series a neighbourhood of the line of fixed points is foliated into two-dimensional, transversal to this line, surfaces invariant under both the mapping and the involution. In the elliptic case (when the linearization of the mapping restriction to each two-dimensional surface is a rotation) of general position on these two-dimensional surfaces Kolmogorov's invariant curves are encountered having an irrational rotation number and surrounding the mapping fixed point that lies on the relevant two-dimensional surface.

It turns out that, although the two-dimensional invariant surfaces exist only formally, Kolmogorov's tori lying on them do exist. Moreover, the curves with the same rotation numbers produced on different formal two-dimensional surfaces are organized together into smooth cylindrical surfaces isolating the line of fixed points from the surrounding space.

At the mapping fixed point where the linearization rotation angle corresponds to the rotation number these surfaces can intersect the line of fixed points. In a neighbourhood of such a point a surface is, generally speaking, diffeomorphic to a paraboloid of revolution (foliated into parallels) since the formal series to express the dependence of the rotation number on the point x on the line of fixed points and the distance r from this point along a formal invariant two-dimensional surface begins with the terms $\alpha(x) + r^2\beta(x) + \ldots$.

It can happen that the surface composed of Kolmogorov's circles will intersect twice the line of fixed points and therefore have the topology of a sphere (Fig. 2.3). Then the sphere locks up the points

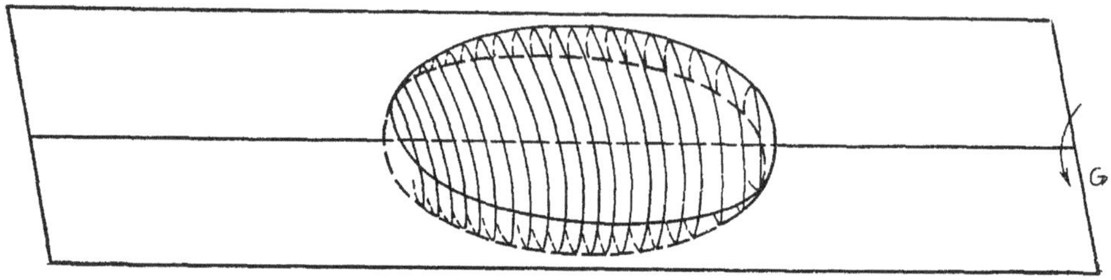

Fig. 2.3

inside it, this enabling us to establish the stability of the fixed points corresponding to either a minimum linearization rotation angle (for positive values of β) or to a maximum one (for negative values of β).

Between the considered surfaces foliated into invariant (under both the mapping and the involution) Kolmogorov's circles with irrational rotation numbers there are left resonance zones. For general mappings, these zones contain fixed points of mapping iterations. The points fixed under the qth mapping iteration are organized into several smooth curves (two of which necessarily lie on an involution fixed surface). The curves can intersect the line of fixed points (where the linearization rotation angle is $2\pi p/q$), where they are tangent to the corresponding formal invariant surface. Some of these curves themselves are surrounded by two-dimensional surfaces foliated into circles invariant under the qth mapping iteration (Fig. 2.4).

For a reversible system in a phase space of an arbitrary even dimension $2m$ (and a reflection involution with respect to an m-dimensional space) we arrive at an involution of a $(2m - 1)$-dimensional space (a secant Poincaré surface) with m "even" and $(m-1)$ "odd" coordinates and at a reversible mapping with the curve of fixed points lying in an involution fixed space $\mathbb{R}^m$. The above results are transferred to the case of $m > 2$ as follows. A neighbourhood of the line of fixed points is formally foliated into transversal to it, $(2m - 2)$-dimensional surfaces which are invariant under both the mapping and the involution. At the pithy level, in the general elliptic case this line is surrounded by m-dimensional surfaces foliated into $(m - 1)$-

dimensional tori which are invariant under both the mapping and the involution and on which a quasi-periodic motion proceeds whose frequencies are the same for a fixed m-dimensional surface.

Now let in an $(m + k)$-dimensional space be preset an involution with m "even" and $k < m$ "odd" coordinates (in the above case $k = m - 1$) and a reversible mapping leaving fixed one of the involution fixed points, say a point a. Generically, the linearization of the

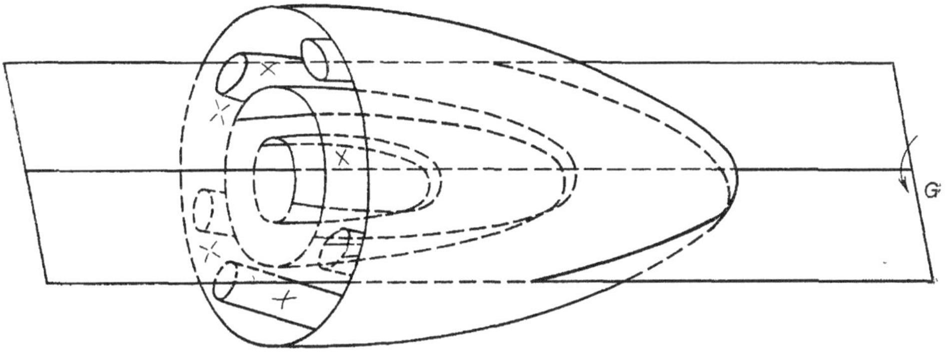

Fig. 2.4

mapping at the point a has r eigenvalues equal to 1 (where $0 \leqslant r \leqslant m - k$) and $(m - k - r)$ eigenvalues equal to -1 (in the above case $r = 1$). It turns out that through the point a there passes an $(m - k)$-dimensional surface invariant under the mapping and lying on the manifold of involution fixed points. The restriction of the mapping to this surface is an involution with r "even" and $(m - k - r)$ "odd" coordinates. Consider the case of $r = m - k$ in more detail. The point a is passed through by an $(m - k)$-dimensional surface of mapping fixed points lying on the manifold of involution fixed points (in the above case, when the reversible mapping was the Poincaré mapping of a reversible equation, the existence of such a line of fixed points was known a priori).

A neighbourhood of this surface is foliated into transversal to it, formal $2k$-dimensional manifolds invariant under both the mapping and the involution. For a general elliptic mapping this surface is surrounded by k-dimensional, invariant under both the mapping and the involution tori on which a quasi-periodic motion proceeds. The tori with the same frequency set are organized into smooth $(m - k)$-parameter families.

For $k = 1$ in the resonance zones between these families one encounters the fixed points of the mapping iterations. The points fixed under the qth mapping iteration are organized into several smooth $(m - 1)$-dimensional surfaces (two of which necessarily lie on the involution fixed m-dimensional surface).

Consider now reversible vector fields. Let in an $(m + k)$-dimensional space be preset an involution with m "even" and $k < m$ "odd" coordinates and a reversible vector field having a singular point on a fixed m-dimensional involution manifold. In a general system this point is passed through by an $(m - k)$-dimensional surface lying on the involution fixed manifold and consisting of field singular points. A neighbourhood of this surface is foliated into transversal to it, formal $2k$-dimensional manifolds that are invariant under both the field and the involution. In a general elliptic system this surface is surrounded by $(m - k)$-parameter families of k-dimensional invariant tori. For $k = 1$ a certain neighbourhood of this surface in a general elliptic system is completely foliated into such families of circles.

2.12. APPLICATION TO REVERSIBLE EQUATIONS

Returning from the geometry of the Poincaré mapping near the line of fixed points to the initial reversible equation we will obtain the following results. To begin with, let the phase space be four-dimensional while the involution be a reflection with respect to a two-dimensional space (as is the case for Eq. (2.1)). We obtain a one-parameter family of symmetric cycles and a set of three-dimensional invariant manifolds numbered by the rotation numbers normally incommensurable with 1 and foliated into two-dimensional Kolmogorov's tori filled with the trajectories of quasi-periodic motions having the fixed frequency ratio on each three-dimensional manifold.

The resonance zones between the manifolds comprise other one-parameter families of symmetric cycles and other three-dimensional manifolds, also foliated into two-dimensional invariant tori. These families emerge from the fixed points of the Poincaré mapping iterations and from the phase oscillations near them.

Thus the general picture of a motion near a symmetric cycle of a reversible system with a four-dimensional phase space and with a two-dimensional manifold of involution fixed points is extremely reminiscent of the picture of the motion of a Hamiltonian system with two degrees of freedom. The only difference is that the Hamiltonian system has the first integral, therefore the existence of Kolmogorov's tori implies (under the usual conditions of nondegeneracy) the stability of the periodic motion.

At the same time, the reversible system has, generally speaking, no first integral. Therefore the existence of quasi-periodic motions (whose phase curves form three-dimensional invariant manifolds foliated into two-dimensional Kolmogorov's tori) does not prevent, in general, some trajectories of the system from escaping the initial periodic motion along a resonance zone.

It seems very likely that such an escape, although very slow, does take place in general reversible systems.

In the case of a $2m$-dimensional phase space and a reflection involution with respect to an m-dimensional subspace we obtain a one-parameter family of symmetric cycles and a set of the $(m + 1)$-dimensional invariant manifolds foliated into m-dimensional Kolmogorov's tori filled with quasi-periodic motion trajectories.

Now let a system with an $(m + k)$-dimensional phase space be given that is reversible under an involution whose fixed manifold is m-dimensional (where not necessarily $k = m$). For $k > m$ the symmetric cycles of such a system of general position are isolated, for $k \leqslant m$ they are organized into the $(m - k + 1)$-parameter families structurally stable under small reversible perturbations (this can be proved by the same reasoning as in the "classical" case of $k = m$). Suppose that $k \leqslant m + 1$, then an $(m - k + 1)$-parameter family of the symmetric cycles of an equation, under the ellipticity of the appropriate Poincaré mapping, is surrounded by the $(m + 1)$-dimensional invariant manifolds foliated into the k-dimensional invariant tori, on which quasi-periodic motions proceed. For $m = 1$, $k = 2$ this implies the stability of a symmetric system's cycle (the Poincaré mapping is a reversible mapping of a plane) in the elliptic case. For two "odd" coordinates ($k = 2$) and an arbitrary m near the general "elliptic" $(m - 1)$-parameter family of the symmetric cycles there exist other stemming from the fixed points of Poincaré mapping iterations, $(m - 1)$-parameter families of symmetric cycles and other $(m + 1)$-dimensional invariant manifolds, also foliated into two-dimensional invariant tori.

2.13. NON-AUTONOMOUS REVERSIBLE SYSTEMS

The above assertions about the invariant tori of reversible differential equations and about their behaviour in a neighbourhood of a singular point can be generalized to the non-autonomous case.

Definition. A differential equation $\dot{x} = V_t(x)$ (generally speaking, non-autonomous) is called weakly reversible if there exists a phase space diffeomorphism G that "reverses" a phase space transformation family corresponding to this equation. In other words, if $\mathscr{F}_{t_1}^{t_2}$ is a phase space transformation in a time period from t_1 to t_2, then for all t_1, t_2 we have

$$G\mathscr{F}_{t_1}^{t_2}G^{-1} = \mathscr{F}_{-t_1}^{-t_2}.$$

This relation is equivalent to the fact that for any fixed moment of time t we will have

$$G_* V_t = - V_{-t}.$$

If the equation is autonomous then we obtain the former definition of a weakly reversible vector field.

If a diffeomorphism G "reverses" the family $\{\mathcal{F}_{t_1}^{t_2}\}$ of phase space transformations, then the diffeomorphism G^{-1} will do the same.

A weakly reversible differential equation is called reversible, if a phase space diffeomorphism, "reversing" it, is an involution.

Of special interest are weakly reversible differential equations whose right-hand sides are periodic in time.

Example. A solution $x = x(t)$ of a reversible, T-periodic in time differential equation with an initial condition $x(0)$ lying on the fixed manifold of a "reversing" involution is T-periodic if and only if $x(T/2)$ is also fixed under this involution (cf. [2.26], [2.27]).

The monodromy operator of a weakly reversible, periodic in time differential equation is also weakly reversible (with respect to the same diffeomorphism). This key remark allows all the results on (weakly) reversible mappings to be immediately carried over to (weakly) reversible, periodic in time differential equations.

In particular, in an extended phase space (whose time is factori-

Table 2.1

	Mappings and phase flows	Non-autonomous differential equations	Flows in a neighbourhood of a periodic solution	
			tori	cycles
Restrictions	$1 \leqslant k \leqslant m$	$1 \leqslant k \leqslant m$	$2 \leqslant k \leqslant m+1$	$1 \leqslant k \leqslant m$
The dimension of Kolmogorov's tori (the last column contains the number of parameters in each cycle family)	k	$k+1$	k	$m-k+1$
The dimension of the manifolds foliated into tori	m	$m+1$	$m+1$	

Remark: The first column relates to reversible mappings and phase flows close to integrable or in a neighbourhood of a fixed point, the second, to the reversible differential equations periodic in time and close to autonomous integrable or in a neighbourhood of a stationary solution, the third, to reversible phase flows in a neighbourhood of a symmetric closed trajectory. The fourth column gives what number of parameters have the families into which symmetric cycles of reversible phase flows are generically organized.

The phase space dimension is $m + k$, the dimension of the manifold of involution fixed points is m.

Kolmogorov's tori for periodic differential equations (the second column) lie in the extended phase space (the time is varying along a circle), in all other cases, in the phase space.

zed by period) of a reversible differential equation close to an autonomous integrable one there exist invariant tori. The same tori also exist near a stationary solution of a reversible differential equation in the nondegenerate elliptic case.

Table 2.1 lists the dimensions of these tori as well as those of the invariant (Kolmogorov's) tori and of the invariant manifolds, into which the tori with the same frequencies (or the same frequency ratios) are organized as dependent on the number of m "even" and k "odd" phase space coordinates in all the preceding situations. The table also shows the number of parameters of the families into which the general symmetric cycles of reversible phase flows are organized and the restrictions, imposed on both the dimension m of an involution fixed manifold and its codimension k in phase space, under which Kolmogorov's tori or non-isolated symmetric cycles exist.

2.14. THE LYAPUNOV-DEVANEY THEOREM

Lyapunov's theorem on one-parameter cycle families near a singular point of a Hamiltonian vector field can be applied to the reversible case as follows ([2.16] and [2.15], Ch. I, §§ 5-6). Consider a reversible system of differential equations in $\mathbb{R}^{2m}$, with a singular point lying on an m-dimensional fixed manifold of the involution G that reverses this system. As in the Hamiltonian case, the number $-\lambda$, together with each eigenvalue λ of a linear reversible system, is also an eigenvalue of this system. This suggests, in particular, that for reversible systems pure imaginary eigenvalues are not an exception (as for the general systems of differential equations) but one of the typical possibilities realized in a stable manner (as for Hamiltonian systems). Let us denote by $\lambda_1, \ldots, \lambda_m, -\lambda_1, \ldots, -\lambda_m$ the eigenvalues of the linearization of vector field prescribing a reversible system at a singular point.

Let λ_1 be nonzero, pure imaginary and none of the ratios λ_j / λ_1 $(j \geqslant 2)$ be integer (the nonresonance condition).

Then the singular point of our system will be passed through by a smooth two-dimensional surface being tangential to the linearization invariant plane that corresponds to the eigenvalues $\pm\lambda_1$. The surface is foliated into symmetric cycles, and as the initial conditions on this surface tend to the singular point, the cycle period tends to $2\pi/|\lambda_1|$.

The studies of resonant Hamiltonian systems are the topics of a major number of papers (see, e.g. [2.23-2.25]). Here we will consider reversible systems near resonances. For simplicity, we will limit ourselves to the case of a four-dimensional phase space (two "even" and two "odd" coordinates). Resonant reversible systems form a set of codimension 1 in the space of all reversible systems. Therefore to study the resonant systems one has to include those into families of reversible systems that depend on a small parameter ε ($\varepsilon = 0$ corresponds to the resonance). The singular point can be treated as independent of ε and placed at zero.

2.15. THE RESONANCE 1:1

Suppose that for $\varepsilon = 0$ our system has a double eigenvalue βi and a double eigenvalue $-\beta i$ ($\beta > 0$). By changing, if necessary, the time scale we can get $\beta = 1$. In the general family for one parameter sign (e.g. when $\varepsilon < 0$) the eigenvalues of the linearization of our system do not lie on the imaginary axis, while for another sign (when $\varepsilon > 0$) these are $\pm \nu_1 i$, $\pm \nu_2 i$, where ν_1 and ν_2 are real and close to 1. The resonance mismatch $|\nu_2 - \nu_1|$ is of the order of $\sqrt{\varepsilon}$. Each symmetric cycle of our system in a neighbourhood of the origin intersects a fixed involution plane at two points. The latter form a certain curve Γ_ε that depends on ε. On the curve Γ_ε, for any ε, an involution σ permutes the points belonging to the same symmetric cycle.

Theorem. When some nondegeneracy conditions imposed on the terms of order $\leqslant 3$ of field Taylor series are fulfilled, then by a suitable choice of a smoothly ε-dependent coordinate system Oxy in a fixed plane of the involution reversing the phase flow and by a change, if necessary, of the ε sign (i.e. by multiplying ε by -1) the equation of the family of curves Γ_ε is reduced to the form

$$(\varepsilon \pm x^2)\, x^2 = y^2, \tag{2.6}$$

where the sign "+" corresponds to the hyperbolic mode and the sign "−", to the elliptic mode (the mode is controlled by the terms of order $\leqslant 3$).

(a) Elliptic mode: $\Gamma_\varepsilon = \{x,\ y : (\varepsilon - x^2)\, x^2 = y^2\}$.

For $\varepsilon \leqslant 0$, Γ_ε is just the origin, i.e. there are no symmetric cycles near the singular point at all. For $\varepsilon > 0$, Γ_ε has the form of a figure-of-eight (Fig. 2.5).

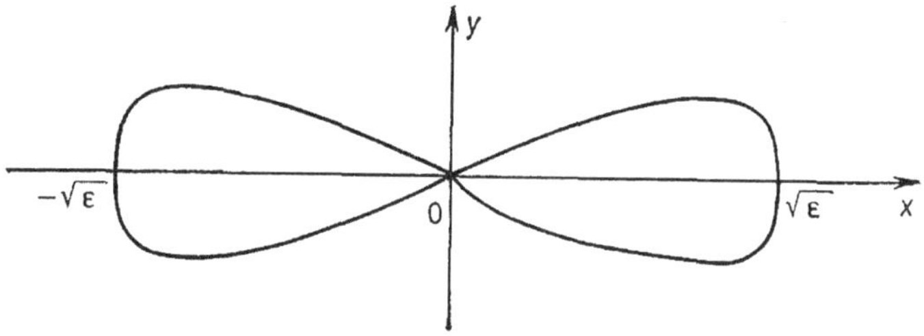

Fig. 2.5

The cycles intersect this figure-of-eight at the points symmetric about the origin.

Thus, when a four-dimensional reversible system passes through the resonance 1:1 in the elliptic mode, on one side of the resonance a foliated into symmetric cycles, two-dimensional surface is split out from the origin, this surface having the topology of a sphere with two identified points (corresponding to the origin). The size of this surface is proportional to the resonance mismatch.

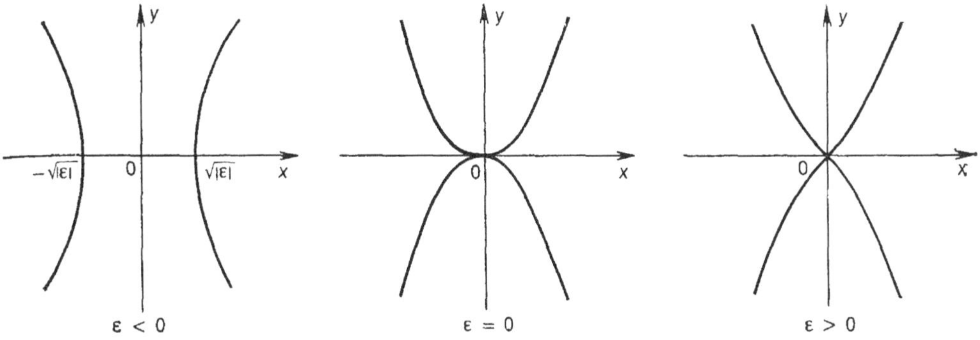

Fig. 2.6

In the paper by Malomed and Tribel'skii [2.20] (see also [2.28]) this surface was approximated for Eq. (2.1) by writing out a few first terms of the series producing the solution.

(b) Hyperbolic mode: $\Gamma_\varepsilon = \{x,\ y : (\varepsilon + x^2)\, x^2 = y^2\}$.

The bifurcation of Γ_ε has the form shown in Fig. 2.6.

In particular, at a resonance instant in the phase space of the system we obtain two tangential at zero, two-dimensional C^1-surfaces foliated into symmetric cycles.

Remark. The union of the curves Γ_ε in a three-dimensional space with coordinates x, y, ε both in the elliptic and in the hyperbolic cases is a surface called the "Whitney umbrella" (or the "Cayley umbrella").

The Whitney umbrella is a surface diffeomorphic to the image of the Whitney mapping $\mathbb{R}^2 \to \mathbb{R}^3$ preset by the following formulas (see [2.2], Ch. 1, Sec. 1.9):

$$v_1 = u_1, \quad v_2 = u_1 u_2, \quad v_3 - u_2^2,$$

where u_1, u_2 are the coordinates in $\mathbb{R}^2$ and v_1, v_2, v_3 are the coordinates in $\mathbb{R}^3$. The equation of the image of this mapping in $\mathbb{R}^3$ is $v_2^2 = v_1^2 v_3$ (the last relation, besides the points of the umbrella, is satisfied by the points of its "stick", $v_1 = v_2 = 0$, $v_3 < 0$, too). The Whitney umbrella is shown in Fig. 2.7.

All the smooth curves on the umbrella passing through its vertex are mutually tangential at this point. The direction of their common tangent (in our coordinates this is the direction of the axis v_1) is thus invariantly related to the umbrella.

In our case an umbrella's neighbourhood in $\mathbb{R}^3$ is foliated into surfaces $\varepsilon = \text{const}$. It is easy to verify that at the umbrella's vertex the derivative of the function ε along the invariant direction is zero.

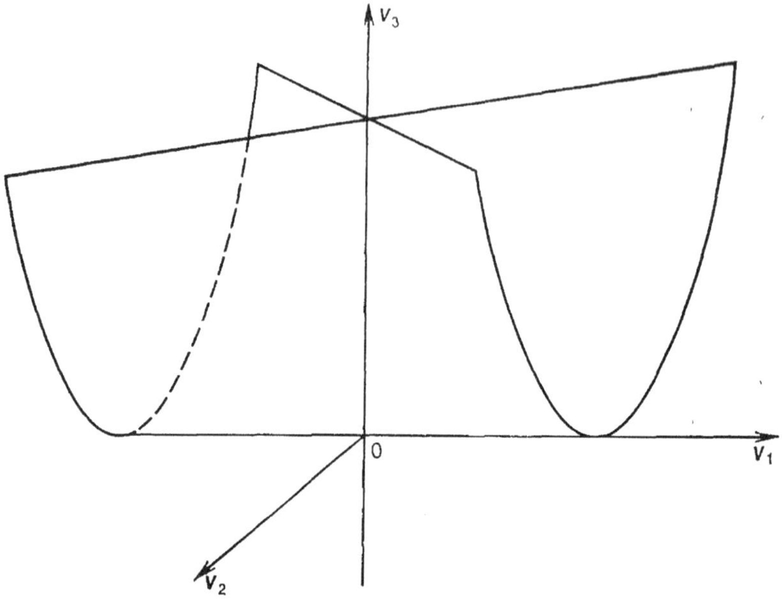

Fig. 2.7

Theorem. In a neighbourhood of the Whitney umbrella in $\mathbb{R}^3$ a function, that is of general position in the class of the functions whose values and whose derivatives along the axis v_1 at zero are both zero, by means of an $\mathbb{R}^3$ diffeomorphism conserving zero and transforming the umbrella into itself, can be reduced to the normal form

$$\pm v_3 \pm v_1^2,$$

where the functions $v_3 + v_1^2$ and $-v_3 - v_1^2$ correspond to the elliptic mode and the functions $v_3 - v_1^2$ and $-v_3 + v_1^2$, to the hyperbolic one.

Remark. Suppose that a quadratic part of the Taylor series of the initial function is $a_1v_2 + a_2v_3 + \sum_{1 \leqslant i \leqslant j \leqslant 3} b_{ij}v_iv_j$. The words "of general position" mean that $a_2 \neq 0$, $a_1^2 \neq 4a_2b_{11}$.

2.16. FURTHER RESONANCES 1:N ($N \geqslant 2$)

After a smooth substitution of the parameter ε and a smoothly ε-dependent change in time scale we can assume that the eigenvalues of our system's linearization at the origin are $\pm i$, $\pm (N + \varepsilon) i$. The symmetric cycles near the origin can be of two types: long-periodic (with a period close to 2π) and short-periodic (with a period close to $2\pi/N$). According to the Lyapunov-Devaney theorem applied to the eigenvalues $\pm(N + \varepsilon)\, i$, for any small value of ε the origin is passed through by a smoothly ε-dependent, two-dimensional surface tangential to the $\pm(N + \varepsilon)i$-invariant plane of the linearization and foliated into short-periodic symmetric cycles. When these cycles contract to a singular point, their periods tend to $2\pi/(N + \varepsilon)$. The surface intersects the involution fixed plane along a smoothly ε-dependent curve $\Gamma_\varepsilon(N)$.

The points of intersection of long-periodic cycles with the involution fixed plane compose a certain ε-dependent curve $\Gamma_\varepsilon(1)$ that bifurcates when ε passes through zero. Our aim is to study these bifurcations.

On the curve $\Gamma_\varepsilon(1)$ an involution σ acts permuting the points which belong to the same long-periodic symmetric cycle. For even N's, the involution is approximately a reflection with respect to the curve $\Gamma_\varepsilon(N)$, and for odd N's, a reflection with respect to the origin. Accordingly, for even N's, the curve $\Gamma_\varepsilon(1)$ is approximately symmetric about $\Gamma_\varepsilon(N)$ while for odd N's, about the origin.

Indeed, suppose that in the fixed plane of the initial involution G a coordinate system is chosen whose abscissa axis is tangential to the $(\pm i)$-eigendirection and ordinate axis is tangential to the $(\pm(N + \varepsilon)\, i)$-eigendirection. Then, if a symmetric long-periodic cycle passes through a point (u, v) in the fixed plane, this cycle will intersect, for the second time, the fixed plane at a point close to the point $(u \cos \pi,\ v \cos N\pi)$, i.e. to that of $(-u,\ (-1)^N v)$.

Analogously, on the curve $\Gamma_\varepsilon(N)$ an involution τ acts that permutes the points belonging to the same short-periodic symmetric cycle. These points are approximately symmetric about the origin.

2.16.1. The Resonance 1:2

Theorem. On fulfilling some nondegeneracy conditions imposed on the quadratic (second-order) terms of the field Taylor series, by a suitable choice of a smoothly ε-dependent coordinate system Oxy in an involution fixed plane, the curve $\Gamma_\varepsilon(N)$ acquires the form $\{x, y : x = 0\}$ (the ordinate axis in Figs. 2.8 and 2.11) and the curve $\Gamma_\varepsilon(1)$ acquires the form

$$\{x,\ \ y : y\,(\varepsilon - y) \pm x^2 = 0\}, \tag{2.7}$$

where the sign "+" corresponds to the hyperbolic mode and the sign "−", to the elliptic mode (the mode is controlled by quadratic terms).

(a) Elliptic mode: $\Gamma_\varepsilon(1) = \{x,\ y : y\,(\varepsilon - y) = x^2\}$.

The bifurcation of $\Gamma_\varepsilon(1)$ is shown in Fig. 2.8. Long-periodic symmetric cycles intersect the involution fixed plane, i.e. that of Fig. 2.8, at the points A and A_1, B and B_1, etc.

The cycle passing through the point C of the intersection between Γ_ε (1) and Γ_ε (N) is not long- but short-periodic. It intersects for the second time the figure plane at a point C_1 on Γ_ε (N) and is resonant (i.e. the rotation angle of the linearization of the restriction of the Poincaré mapping of this cycle to the passing through C, formal two-dimensional invariant surface, in the three-dimensional secant Poincaré surface, is equal to $\pi/2$). Long-periodic cycles form a specific bridge between this resonant, short-periodic, closed trajectory

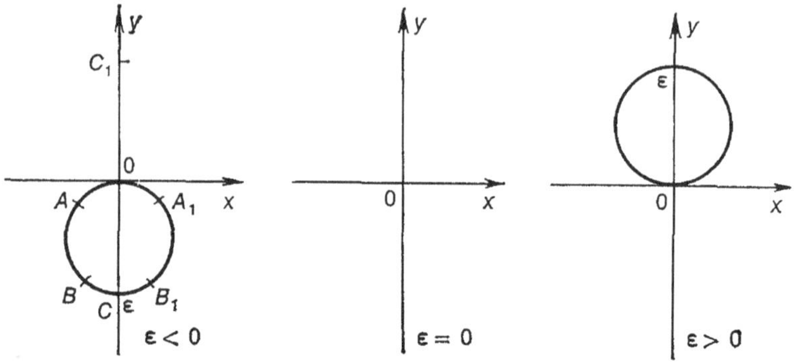

Fig. 2.8

and the singular point (cf. the description of the branching of long-periodic cycles from a resonant short-periodic one in the Hamiltonian case, as shown by Bryuno [2.21]).

The projection of a symmetric cycle, passing through the points A and A_1 placed near the origin, onto a two-dimensional, passing through the origin, plane of general position in the four-dimensional phase space, is nonself-intersecting. The projection of a cycle, passing through the points B and B_1 close to C, goes twice round zero and has an odd number of self-intersection points (Fig. 2.9). The transition proceeds as shown in Fig. 2.10. By Q is denoted the reverse point on the projection at the instant of its reconstruction ("beak").

(b) Hyperbolic mode: $\Gamma_\varepsilon(1) = \{x,\ y : y\,(y - \varepsilon) = x^2\}$.

The bifurcation of Γ_ε (1) is shown in Fig. 2.11.

Long-periodic symmetric cycles intersect the figure plane at the points A and A_1, B and B_1, C and C_1, E and E_1, F and F_1, G and G_1, etc. The cycle passing through the point D of the intersection between Γ_ε (1) and Γ_ε (N) is short-periodic, it intersects for the second time the figure plane at the point D_1 on Γ_ε (N) and is resonant.

The projection of a cycle, passing through the points B and B_1 placed near the origin, onto a zero-containing, two-dimensional plane of general position, is nonself-intersecting. The projection of a cycle, passing through the points A and A_1 placed near D, is a curve like that in Fig. 2.9. As to the projection of the cycles, passing through the points C, C_1 and E, E_1 placed far from the origin, these projections, depending on the two-dimensional direction of the projection, are either nonself-intersecting (and then the projection of a cycle, passing through some intermediate points G and G_1, has a "beak") or curves like a loop in Fig. 2.9 (then the "beak" is inherent in the projection of a cycle passing through some intermediate points F and F_1).

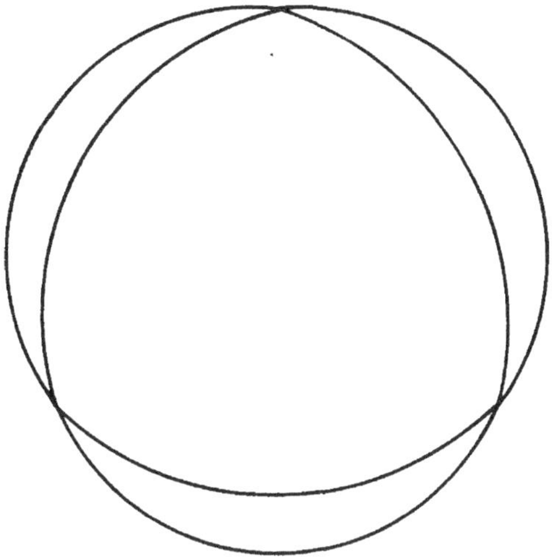

Fig. 2.9

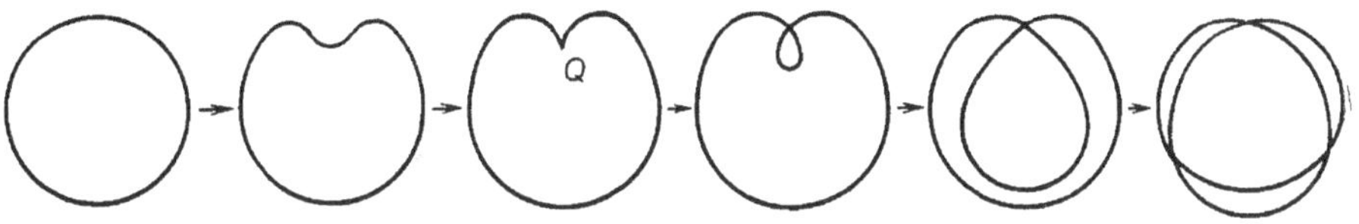

Fig. 2.10

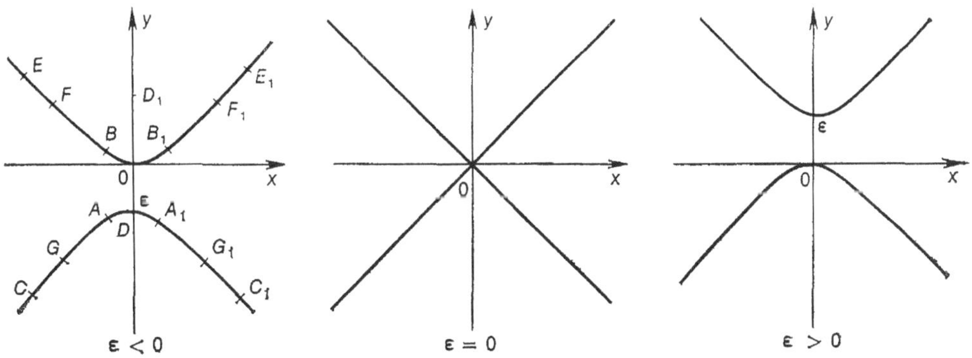

Fig. 2.11

At the resonance instant, in the hyperbolic case the origin of the four-dimensional phase space is passed through by two two-dimensional invariant surfaces having a conical singularity and foliated into long-periodic cycles.

Remark 1. In the elliptic and hyperbolic cases, when a long-periodic cycle approaches a resonant short-periodic one, its projection shown in Fig. 2.9 becomes more and more like a two-pass circle.

Remark 2. The union of the curves $\Gamma_\varepsilon(1)$ in a three-dimensional space with the coordinates x, y, ε in the elliptic and hyperbolic modes is a cone. The union of the curves $\Gamma_\varepsilon(N)$ is a smooth two-dimensional surface (in "normal" coordinates, the plane $x = 0$) whose tangential plane at zero is conjugated with the plane $\varepsilon = 0$ with respect to the cone (more exactly, a straight line, conjugated with each of these planes with respect to the quadratic part of the Taylor series of the equation prescribing the cone, lies in the other

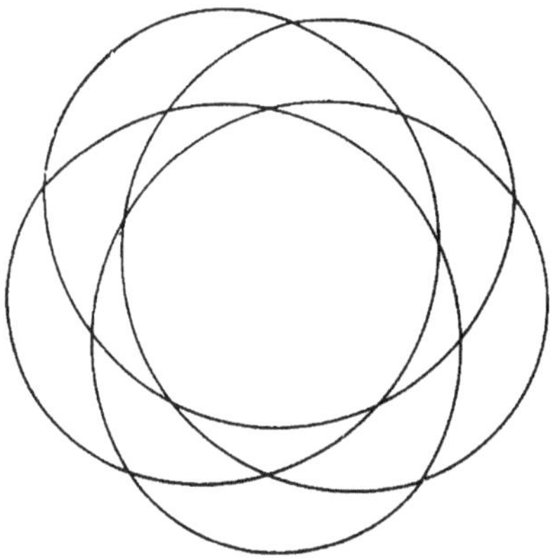

Fig. 2.12

plane). The union of the curves $\Gamma_\varepsilon(N)$ intersects the cone along two curves intersecting at the cone vertex at a nonzero angle (for both modes).

Theorem. Suppose that in $\mathbb{R}^3$ there are given: a cone whose vertex is at zero (according to the Morse lemma we, without any loss of generality, can presume that its equation is $v_1^2 + v_2^2 - v_3^2 = 0$, where v_1, v_2, v_3 are the coordinates in $\mathbb{R}^3$), a smooth, two-dimensional, passing through zero general surface Ξ that intersects the cone along two curves; and a smooth general function ε whose zero level surface passes through zero. Tangential planes to $\varepsilon^{-1}(0)$ and to Ξ at zero are conjugated with respect to the quadratic form $v_1^2 + v_2^2 - v_3^2$. Then, by the $\mathbb{R}^3$ diffeomorphism that conserves zero and maps the cone onto itself, one can reduce the function ε to one of two normal forms: $\varepsilon = v_2$ (hyperbolic mode) or $\varepsilon = v_3$ (elliptic mode) and, simultaneously, the surface Ξ to the normal form $\{v_1, v_2, v_3 : v_1 = 0\}$.

Now let us turn to the resonances $1{:}N$, where $N \geqslant 3$. Figures 2.13-2.18 show the relevant bifurcations of $\Gamma_\varepsilon(1)$. The curve $\Gamma_\varepsilon(N)$ is indicated by the dashed line. A symmetric cycle, passing through the points A and A_1 of the intersection between $\Gamma_\varepsilon(1)$ and $\Gamma_\varepsilon(N)$, is short-periodic. It is resonant and the projections of long-periodic cycles, that branch off from it, onto a zero-containing, two-dimensional plane of general position have the form shown in Fig. 2.12 (overall N turns). The closer these cycles to a resonant short-periodic one the more their projections are similar to an N-pass circle. The "beak" phenomenon is analogous to that for the resonance 1:2.

When the resonant short-periodic cycle is connected with the origin by a "bridge" made of long-periodic symmetric closed trajectories, that branch off from it, then the part of the curve Γ_ε (1) that is an intersection between the bridge and the involution fixed plane is shown in the subsequent figures by the heavy line.

2.16.2. The Resonance 1:3

Theorem. On fulfilling some nondegeneracy conditions imposed on the terms of order $\leqslant 3$ of the field Taylor series, by a suitable choice of a smoothly ε-dependent coordinate system Oxy in the involution fixed plane and by changing, if necessary, the sign of ε the equation of the curves Γ_ε (N) reduces to the form

$$x = y^2 h(y, \varepsilon) + Cy, \tag{2.8}$$

where $C \in \mathbb{R}$, h is a smooth function, and the equation of the curves Γ_ε (1) reduces to the form

$$y(\varepsilon + y^2 + \alpha xy) + x^3 = 0, \tag{2.9}$$

where $\alpha \in \mathbb{R}$, $4\alpha^3 + 27 \neq 0$.

The constant α is controlled by the terms of order $\leqslant 3$.

The inequality $4\alpha^3 + 27 > 0$ corresponds to the elliptic mode, the inequality $4\alpha^3 + 27 < 0$, to the hyperbolic one.

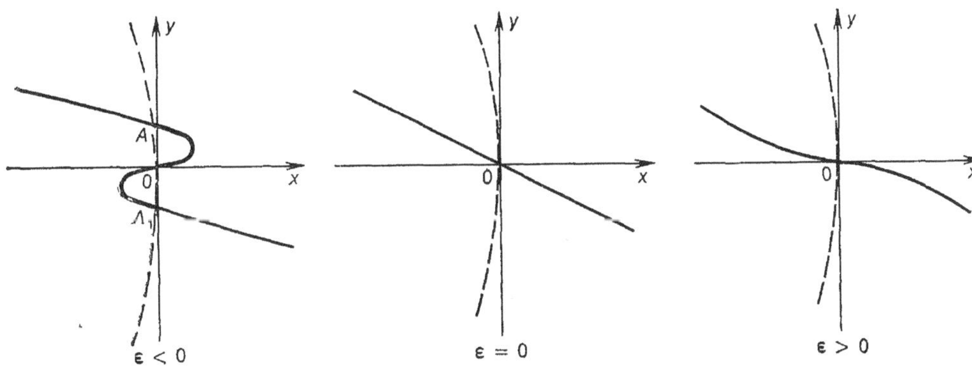

Fig. 2.13

The bifurcation of Γ_ε (1) for the elliptic and hyperbolic modes is shown in Figs. 2.13 and 2.14, respectively. The curves Γ_ε (N) in these figures are shown for the case of $C = 0$. The distances from the origin to A and A_1 for $\varepsilon < 0$ have the order of $\sqrt{|\varepsilon|}$. For $\varepsilon \neq 0$, Fig. 2.14 shows Γ_0 (1) by dots. For $\varepsilon > 0$ the distances ρ_1 and ρ_2 have the order of $\sqrt{\varepsilon}$.

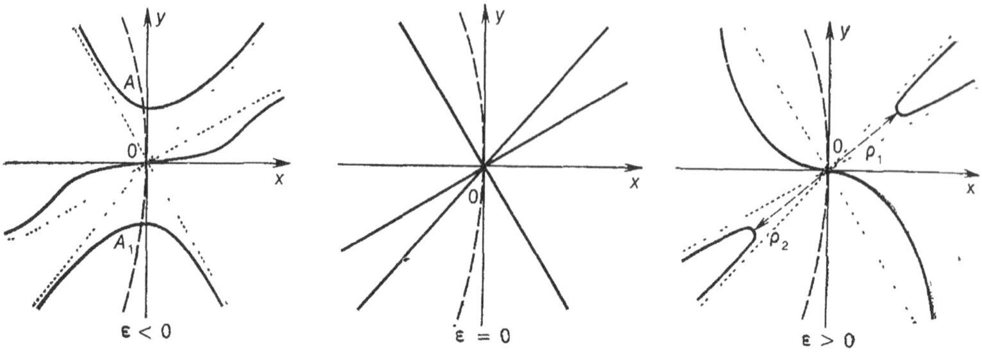

Fig. 2.14

Remark. At the resonance instant the singular point is passed through either by one (elliptic mode) or by three (hyperbolic mode) two-dimensional surfaces with a conical singularity at zero, which are foliated into long-periodic symmetric cycles. An intersection of each of these surfaces with the involution fixed plane is a smooth curve passing through zero.

2.16.3. Higher Resonances $1:N$ $(N \geqslant 4)$

Theorem. On fulfilling some nondegeneracy conditions imposed on the terms of order $\leqslant N$ of the field Taylor series, by multiplying the parameter ε by a suitable nonzero constant and by a suitable choice of a smoothly ε-dependent coordinate system Oxy in the involution fixed plane the equation of the curves Γ_ε (N) acquires the form (2.8), where $C = 0$, and the equation of the curves Γ_ε (1) acquires the form

$$y\,(\varepsilon \pm y^2 - x^2) + x^N = 0, \tag{2.10}$$

where the sign "+" corresponds to the hyperbolic mode and the sign "—", to the elliptic one (the mode is controlled by the terms of order $\leqslant 3$).

The bifurcation of Γ_ε (1) in the elliptic mode is shown in Fig. 2.15 (even N) and in Fig. 2.16 (odd N) and in the hyperbolic mode, in Fig. 2.17 (even N) and Fig. 2.18 (odd N). In the figures corresponding to $\varepsilon > 0$, dots denote the following: in the elliptic case it is the circle $x^2 + y^2 = \varepsilon$, in the hyperbolic case it is the hyperbola $x^2 - y^2 = \varepsilon$. The distances from the origin to A and A_1 for $\varepsilon > 0$ in the elliptic case and for $\varepsilon < 0$ in the hyperbolic case have the order of $\sqrt{|\varepsilon|}$. The distances δ_1 and δ_2 for $\varepsilon > 0$ have the values of the order of $\varepsilon^{(N-1)/4}$.

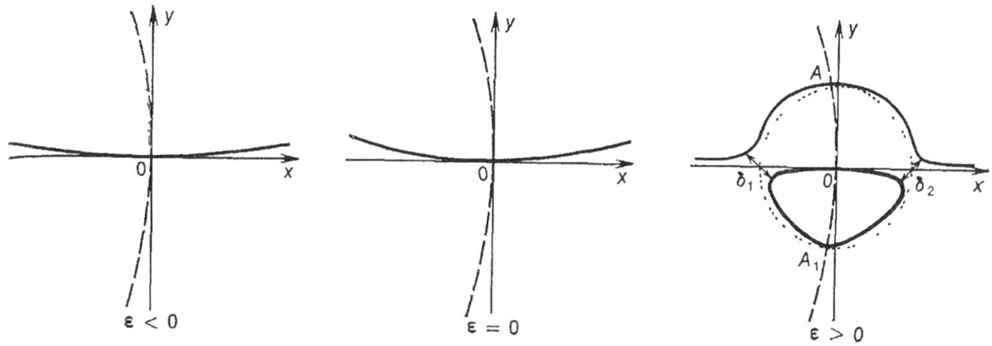

Fig. 2.15

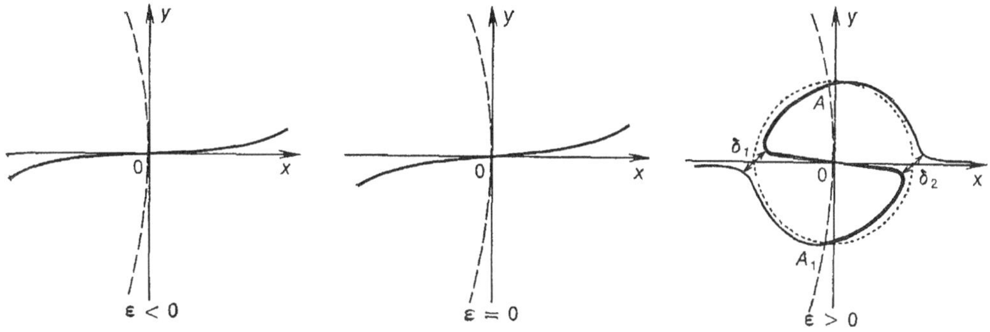

Fig. 2.16

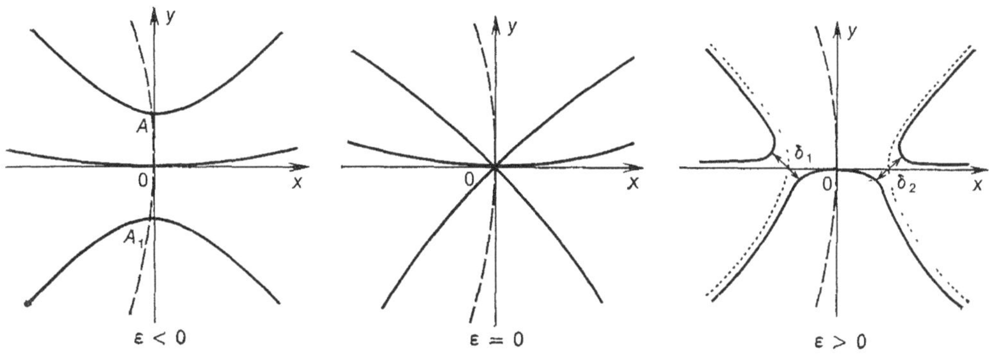

Fig. 2.17

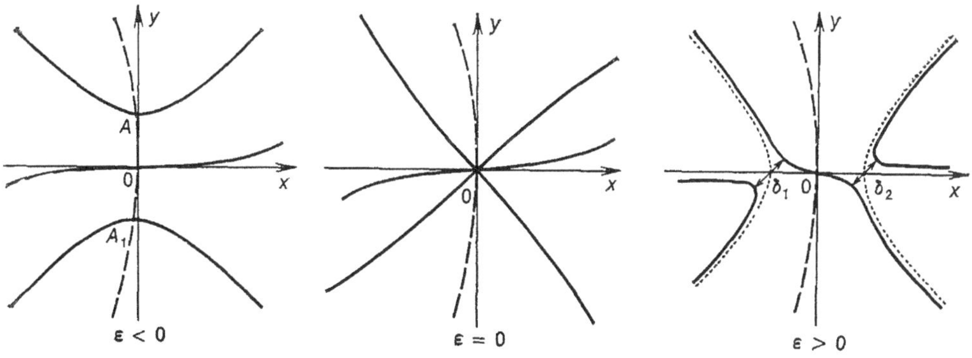

Fig. 2.18

We see that at the instant of a resonance $1{:}N$, where $N \geqslant 4$, the singular point is passed through by a two-dimensional surface foliated into long-periodic symmetric cycles (it is possible to show that its smoothness class at zero is C^{N-3} rather than C^{N-2}), tangential to a $(\pm\, i)$-invariant plane of the vector field linearization at the singular point. Besides, in the hyperbolic mode the singular point is passed through by two additional two-dimensional surfaces foliated into long-periodic cycles and having a conical singularity at this point. An intersection of each of these surfaces with the involution fixed plane is a smooth curve passing through zero (for odd N) or an "angle" composed of two transversal rays with the initial point at zero (for even N).

2.16.4. Some Additional Remarks Concerning Further Resonances

In the case of the resonances $1{:}N$, $N \geqslant 3$, the union of the curves Γ_ε (1) in a three-dimensional space with the coordinates x, y, ε is a surface having at zero the singularity A_{N-1} ([2.2], Ch. 2). Under a suitable choice of the coordinate system v_1, v_2, v_3 the equation of this surface will acquire the form $v_1 v_2 + v_3^N = 0$. This surface is invariantly related to the planes $v_1 = 0$ and $v_2 = 0$ (in a tangential space to $\mathbb{R}^3$, at zero). The zero level surface of the ε-function is tangential to one of these planes.

Theorem. Suppose that in $\mathbb{R}^3$ there are preset a surface $\{v_1, v_2, v_3 : v_1 v_2 + v_3^N = 0\}$ and a general function ε such that $\varepsilon(0) = 0$ and $\varepsilon^{-1}(0)$ is tangential to the plane $v_1 = 0$ at zero. Then by an $\mathbb{R}^3$ diffeomorphism conserving zero and transforming the surface into itself the function ε can be reduced to the normal form

$$\pm\,(v_1 + v_2^2 - \alpha v_2 v_3) \text{ for } N = 3,$$

$$\pm\,(v_1 + v_2^2 + \alpha v_3^2) \text{ for } N \geqslant 4,\ \alpha \in \mathbb{R}.$$

The condition of function generality (nondegeneracy) implies that $4\alpha^3 + 27 \neq 0$ for $N = 3$ and $\alpha \neq 0$ for $N \geqslant 4$.

For $N \geqslant 4$ the function considered can be reduced to the normal form

$$v_1 + v_2^2 \pm v_3^2,$$

where the sign "+" corresponds to the elliptic mode and the sign "−", to the hyperbolic one, if along with the action of the zero-conserving $\mathbb{R}^3$ diffeomorphisms transforming a surface into itself one allows a multiplication of the function by a nonzero constant.

Here the singularity of the pair (a surface of type A_{N-1}, a function) is a real analogue of the critical point of a function on a manifold with a singular boundary of type $\check{A}_3$ for $N = 3$ and of type $I_3(N)$, for $N \geqslant 4$, according to the classification by Lyashko [2.22].

The behaviour of reversible systems at the resonance instant is extremely similar to that of resonant Hamiltonian systems. For example, for a Hamiltonian system (with many degrees of freedom), in the case of the resonance 1:3 between two pure imaginary linearization eigenvalues, depending on Taylor series terms of order $\leqslant 4$ of the Hamiltonian, Roels [2.23] found either one two-dimensional, foliated into long-periodic cycles surface passing through a singular point or three such surfaces. For the resonance $1 : N$, $N \geqslant 4$, also in a multidimensional situation, he found [2.24] a two-dimensional surface, passing through a singular point, foliated into long-periodic cycles and being tangential to the corresponding invariant plane of the linearization. The case of a resonant Hamiltonian system with two degrees of freedom was considered in detail by Henrard [2.25] where an extended bibliography was presented too. Henrard has demonstrated that for the resonance 1:2 in a nondegenerate system near a singular point either there is no family of long-periodic cycles at all or there are two such families, while for the resonance $1:N$, $N \geqslant 3$, the number of the families of long-periodic cycles is either 1 or 3. In [2.25] the subharmonic resonances ($p:q$, where p and q are relatively prime and $q > p > 1$) have been also investigated for which Lyapunov's theorem describes both short- and long-periodic cycles, yet here the existence of the families of "very long-periodic" cycles is possible.

References

2.1. V. I. Arnol'd, in: *Nonlinear and Turbulent Processes*, Acad. Publ., New York, 1984, p. 1161.
2.2. V. I. Arnol'd, A. N. Varchenko, and S. M. Gusein-Zade, in: *Monographs in Mathematics*, Vol. 82, Birkhäuser, Basel, Stuttgart, 1985.
2.3. G. D. Birkhoff, *Rendiconti di Palermo*, 39 (1915).
2.4. G. D. Birkhoff, *Acta Math.*, 48 (1927).
2.5. G. D. Birkhoff, *Dynamical Systems*, AMS Publication, New York, 1927; revised edition, 1966.
2.6. R. B. Melrose, *Invent. Math.*, 37, 165 (1976).
2.7. J. K. Moser and S. M. Webster, *Acta Math.*, 150, 3-4, 255 (1983).
2.8. J. Moser, *SIAM Review*, 8, 2, 145 (1966).
2.9. J. Moser, *Math. Ann.*, 169, 136 (1967).
2.10. J. Moser, *Nachr. Akad. Wiss.*, Göttingen, Math.-Phys. Kl., II *a*, 1 (1962).
2.11. C. L. Siegel and J. K. Moser, *Lectures on Celestial Mechanics*, Springer-Verlag, 1971.
2.12. Yu. N. Bibikov, and V. A. Pliss, *Differ. Uravneniya*, 3, 11, 1864 (1967).
2.13. Yu. N. Bibikov, *Dokl. Akad. Nauk SSSR*, 185, 1, 9 (1969).
2.14. Yu. N. Bibikov, *Differ. Uravneniya*, 7, 8, 1347 (1971).
2.15. Yu. N. Bibikov, in: *Lect. Notes in Math.*, Springer-Verlag, 702 (1979).
2.16. R. L. Devaney, *Trans. Am. Math. Soc.*, 218, 89 (1976).
2.17. A. N. Sharkovskii, in: *The Asymptotic Behaviour of the Solutions of Differential Functional Equations*, Kiev, 1978, p. 118 (in Russian).

2.18. A. N. Sharkovskii, in: *Proc. Eighth Intern. Conf. Nonlinear Oscill.*, Prague, 1073 (1978).
2.19. E. A. Demekhin, *Izv. Akad. Nauk SSSR, Ser. Mekh. Zhidk. i Gaza*, 5, 36 (1983).
2.20. B. A. Malomed and M. I. Tribel'skii, *Dokl. Akad. Nauk SSSR*, 275, 6, 1363 (1984).
2.21. A. D. Bryuno, *Mat. Sbornik*, **83**, 2, 273 (1970).
2.22. O. V. Lyashko, *Funkts. Analiz*, **17**, 3, 28 (1983).
2.23. J. Roels, *J. Diff. Eqs.*, **10**, 3, 431 (1971).
2.24. J. Roels, *J. Diff. Eqs.*, 9, 2, 300 (1971).
2.25. J. Henrard, *J. Diff. Eqs.*, **14**, 3, 431 (1973).
2.26. V. A. Sarychev and V. V. Sazonov, Preprint No. 105, IPM (1976).
2.27. V. A. Zlatoustov and T. A. Shipovskikh, Preprint No. 124, IPM (1984).
2.28. B. A. Malomed and M. I. Tribel'skii, *Physica* D14, 1, 67 (1984).

THE PERIOD MAP AND POISSON STRUCTURES

V.I. ARNOLD

(Joint sessions of the Petrovskiĭ seminar and the Moscow Mathematical Society; meeting on January 19, 1985, section 2.[1])

The *swallowtail* in the three-dimensional space consists of the polynomials $x^4 + ax^2 + bx + c$ that have multiple roots. A Poisson structure is said to be *admissible* (with respect to the swallowtail) if the line of self-intersection of the swallowtail lies in one symplectic leaf.

Theorem 1. *A generic holomorphic admissible Poisson structure can be reduced to the form* $\{a, b\} = \{b, c\} = 0$, $\{a, c\} = 1$ *by a diffeomorphism that is biholomorphic at the origin and preserves the swallowtail.*

Lemma. *A generic function constant on the line of self-intersection of the swallowtail can be reduced, in a neighborhood of the origin, to the form* $b+\mathrm{const}$ *by a local diffeomorphism that preserves the swallowtail (by a holomorphic diffeomorphism, a* C^∞ *diffeomorphism, or a formal one, depending on the class of the function).*

Here "generic" means that the level surface is not tangent to the swallowtail at the origin: $(\partial f/\partial b)(0) \neq 0$.

Consider a local family of nondegenerate holomorphic bivector fields (or 2-forms) in the planes $b = \mathrm{const}$.

Theorem 2. *A family of fields (respectively, forms) can be locally reduced to the form* $\partial_a \wedge \partial_c$ *(respectively,* $\mathrm{d}a \wedge \mathrm{d}c$*) by a diffeomorphism biholomorphic at the origin and preserving the swallowtail.*

Theorem 1, which follows from the Lemma and Theorem 2, provides an axiomatic description of the Poisson structures induced by a general period map according to Varchenko and Givental [1] in the base of the versal unfolding of the singularity A_3: these are exactly all the admissible structures.

References

[1] A.N. Varchenko, A.B. Givental. Period map and intersection form. *Funktsional. Anal. i Prilozhen.*, 16 (1982), no. 2, 7–20. English translation: *Funct. Anal. Appl.*, 16 (1982), no. 2, 83–93.

[1]Translation of Uspekhi Mat. Nauk, **40** (1985), no. 5, p. 236.

Originally publ. in: Usp. Mat. Nauk 40(5), 236,
Translated by Boris Khesin and Andrey Khesin
Springer-Verlag Berlin Heidelberg 2018 with kind permission

FRONTIER PROBLEMS

V.I. ARNOLD

(Published in the collection "Essays on the Development of Mathematics in the USSR. Theoretical Mathematics. Applied Aspects of Mathematics". I.Z. Shtokalo et al. (Eds.), Naukova Dumka, Kiev, 1983, 421–426.)

1. Singularity and bifurcation theory

The singularity and bifurcation theories have been among the most rapidly developing areas of analysis during the last decade. These theories are at the intersection of several mathematical domains: differential and algebraic geometry, topology, theory of complex spaces, theory of differential equations, etc. Their applications embrace a broad scope of problems of mathematical physics: geometric and wave optics, theory of caustics and wave fronts, theory of short-wave asymptotics, the stationary phase method, elasticity theory, problems of control theory and differential games. They are used in the studies of the stability loss of self-oscillations in mechanical and other systems.

The essence of these theories is in the decomposition of the functional space of objects under consideration (say, functions or the right-hand sides of some equations) into the regions of "generic objects" (for instance, functions with nondegenerate critical points or vector fields with nondegenerate singular points) and the hypersurface of degenerate objects (which in the real situation separates the connected components of the region of generic objects). The hypersurface of degenerate objects is itself an infinite dimensional functional space of codimension 1, i.e., it is determined by one equation in the original functional space. On this surface, there are generic points and the remainder of codimension 2 in the original functional space, and so on. Thus the original functional space turns out to be stratified (it is divided into parts of codimension 0, 1, 2, ...).

In applications, generic points are usually most interesting. Indeed, if a point of our functional space is a model of a certain real system, then the model parameters are known only approximately, that is, up to a "small perturbation" of the point. But by such a small perturbation we can shift this point to a generic position by moving it off from the hypersurface of degenerate objects.

Degenerate objects appear in a way unremovable by a small perturbation in the case where we are interested not in an individual object, but in a family of objects depending on parameters. For instance, a one-parameter family is a curve in our functional space. Such a curve, generally speaking, intersects the hypersurface of degenerate objects in its non-singular points and transversally (at a non-zero angle). In this case, every nearby curve also intersects this hypersurface. Therefore, every one-parameter family close to the one under consideration will contain a degenerate object of the given type (though, generally speaking, the degeneration will occur

Originally publ. in: Essays on the Development of Mathematics in the USSR. Editors: I.Z. Shtokalo et al., 421-426,
Translated by Boris Khesin and Andrey Khesin
Springer-Verlag Berlin Heidelberg 2018 with kind permission

not at the same parameter value as in the initial family but for a close value of the parameter). Similarly, in a two-parameter family one can encounter degenerate objects constituting in the functional space a set of codimension 2 (and one cannot get rid of these objects by a small perturbation of the family), and so on.

Thus, while studying various degenerations, the main object of investigation is not an individual object, but a family with parameters whose number is equal to the codimension of the infinite dimensional surface corresponding to the degeneration in question in the functional space. Furthermore, among all families containing a given object or, as they are called, among all deformations (unfoldings) of the object considered, there are special deformations transversal to the infinite dimensional surface described above, the so-called versal deformations. A versal deformation, in a sense, is the "richest" deformation of the object in question; arbitrary deformations can be obtained from the versal ones.

In the functional space of the objects under study, to a versal deformation there corresponds a finite dimensional surface whose dimension is equal to the codimension of the infinite dimensional surface formed by the objects with degenerations of the considered type. The infinite dimensional hypersurface of all degenerations cuts out on this finite dimensional surface (or on the space of parameters of the versal deformation, also called the deformation base) a finite dimensional hypersurface. This hypersurface in the base of the versal deformation is called a bifurcation diagram.

The objective of the singularity and bifurcation theories is the study of stratifications of various functional spaces, appearing in analysis, according to the degeneration types, as well as the study of the corresponding versal deformations and bifurcation diagrams.

Although the described above general program of the theories of singularities and bifurcations goes back to the works by H. Poincaré, and its specific realization started back in the 1930s in topological works by H. Whitney and in the works of A.A. Andronov's school on bifurcation theory, the vigorous development of these theories started only at the end of the 1960s after the works by Thom, Malgrange, Mather, and others. Here, we give a survey of some most important results obtained in this domain by Soviet mathematicians.

General questions of singularity theory. The scheme described above, stratifications of the functional space, versal deformations, bifurcation diagrams, is an optimistic one: the existence of all these objects in specific functional spaces appearing in analysis is far from evident, and there are known examples where they do not exist or possess rather pathological properties. Therefore there arises a question on foundations of the theory which would provide precise definitions of all the objects involved and would prove their properties. One of possible versions of such a foundation for the functional space of differentiable mappings of one finite dimensional manifold into another was proposed by R. Thom in the 1960s: here one examines the topological (rather than differentiable) classification of smooth mappings.

A foundation of the aforementioned theory, however, encountered substantial difficulties: one had to prove the existence of the topological stratification, that is, the algebraicity of the partition of the functional space into mappings of different topological classes of finite codimension and the remainder of infinite codimension.

Within this remainder, the topological type of a mapping can vary continuously as examples show. In a series of papers A.N. Varchenko overcame this difficulty and constructed the required algebraic stratification.

The general singularity theory also includes V.P. Palamodov's proof of the well known formula for the local degree of a holomorphic mapping at a singular point, as well as the formula discovered by G.N. Khimshiashvili (as well as by H. Levine and D. Eisenbud) for the local degree of a smooth mapping, which expresses the index of a singular point of a vector field as the signature of an appropriate quadratic form on the local ring of the singularity.

Theory of critical points of functions. The simplest (nondegenerate) critical points can be reduced to a quadratic normal form. The classification of degenerations has been pushed to critical points of multiplicity 16, which can be obtained by merging no more than 16 nondegenerate critical points. It turned out here that the modality of a degenerate point is particularly important. (Modality is the number of parameters in the normal forms to which one can reduce functions of the class under consideration. More precisely, it is the dimension of the space of orbits for the action of the infinite dimensional group of diffeomorphisms on the space of functions with a critical point at the origin and a critical value equal to 0.)

The simplest degenerate critical points have modality 0. The singularities of modality zero are also said to be simple. Under deformations, simple critical points can turn into critical points of only finite number of types (up to changes of independent variables). All simple critical points have been classified. Their classification turned out to be related (though rather mysteriously) to the classifications of groups generated by reflections, to the classifications of simple Lie algebras and their Weyl groups, as well as to braid theory and the classification of the symmetry groups of regular polyhedra[1] in the conventional Euclidean space. For instance, the list of normal forms of simple functions of three complex variables is as follows:

$$A_k = x^{k+1} + y^2 + z^2, \quad D_k = x^2y + y^{k-1} + z^2,$$
$$E_6 = x^3 + y^4 + z^2, \quad E_7 = x^3 + xy^3 + z^2, \quad E_8 = x^3 + y^5 + z^2.$$

The intersection form on the space of two-dimensional homology of a non-singular level set of such a function is given by the Coxeter graph of the corresponding group generated by reflections. The bifurcation diagram for the level sets is the algebraic variety of irregular orbits of this group on the complexification of the Euclidean space. The fundamental group of the complement to the bifurcation diagram is the Artin–Brieskorn braid group of the type A_k, D_k, or E_k, respectively. The higher homotopy groups of the complement are all trivial. The monodromy group of the family is the natural representation of the braid group in the Weyl group.

The connection with regular polyhedra is as follows. The normal form of each simple singularity is a syzygy relating invariants of one of the discrete subgroups of the group $SU(2)$. The non-simple singularities form a subvariety of codimension 6 in the functional space. The singularities with modality 1 and 2 have also been classified. Their classification turns out to be related to the theory of automorphic forms and to certain discrete subgroups of the group $SU(1,1)$ of motions of the

[1]The "polygons" in the original text was a typo. (Editors' remark)

Lobachevsky plane. The aforementioned classification allows one to study all the singularities which appear in generic families of functions depending on no more than 10 parameters.

The theory described above has been developed in a series of works by G.N. Tyurina, V.I. Arnold, A.M. Gabrièlov, S.M. Guseĭn-Zade, I.V. Dolgachev, A.N. Varchenko, V.S. Kulikov, O.V. Lyashko, and others. There are also related studies by D.B. Fuchs, E.A. Gorin, V.Ya. Lin, and others on braid theory, which are important for other areas of functional analysis as well.

Applications of the theory of critical points. Here we mention just a few applications (out of numerous that exist) of the theory of critical points of functions. In geometric optics, in the theory of short-wave asymptotics, and in other domains of mathematical physics, there appear surfaces with distinctive singular points, so-called caustics and wave fronts. The singularities of generic caustics and wave fronts in low-dimensional spaces can be identified with the bifurcation diagrams of simple singularities in the space of functions. This enables one to construct the theory of singularities of gradient (Lagrangian) mappings and of singularities of Legendre transformations (Legendrian mappings). The stratifications and bifurcation diagrams in the corresponding functional spaces have been studied up to codimension 10, and this leads to a complete description of singularities of caustics and wave fronts in spaces of dimension not exceeding 10.

The same method allows one to examine metamorphoses of caustics and wave fronts in generic families, for instance, metamorphoses of singularities of a moving front. The theory of Lagrangian and Legendrian singularities is being applied to the study of singularities of the maxima of functions depending on parameters and to the study of convex hulls of smooth manifolds. Here one has also obtained classification theorems and normal forms encompassing such a domain in the corresponding functional spaces that its complement has large codimension (in the problem on maximum, the codimension is 17).

Of particular importance is the study of degenerate critical points of functions in relation to the analysis of asymptotics of oscillating integrals which can be calculated by means of the stationary phase method:

$$I(h, \lambda) = \int e^{iS(x,\lambda)/h} \varphi(x, \lambda)\, dx_1 \cdots dx_n, \qquad h \to 0.$$

Here the region in the functional space that has been completely examined extends up to the stratum of codimension 10, so that there are known all the singularities for generic families of phase functions S depending on no more than 10 parameters.

The results described above were obtained in the works by V.I. Arnold, V.M. Zakalyukin, A.N. Varchenko, V.A. Vassiliev, and V.D. Sedykh.

Newton's polyhedra. The study of critical points of functions has led to the discovery of a large number of new explicit formulas. These formulas allow one to compute various discrete (in particular, topological) invariants of analytic objects by means of discrete data, so-called Newton's polyhedra.

A polynomial in one variable has two discrete characteristics: the degree, characterizing its behavior at infinity, and the order, characterizing its behavior at the origin. In the case of a polynomial in several variables, the degree and order are replaced by a more detailed characteristic, namely, the convex integral polyhedron spanned by the exponents of the monomials (with non-zero coefficients) constituting the polynomial. In the same way, a critical point of a function can be associated with the convex polyhedron formed by the exponents of the terms of the Taylor series for the function at the critical point.

It turns out that all "reasonable" discrete invariants of analytic objects can be expressed by simple formulas in terms of geometric characteristics of the corresponding integral Newton polyhedra. More precisely, consider the functional space formed by complex analytic objects with a fixed Newton polyhedron. Then all the discrete invariants of the object will be constant on this entire functional space except for a bifurcation hypersurface of real codimension 2 in the space and hence not dividing the space. Therefore, these invariants can be expressed via the polyhedron itself for almost all objects with the given polyhedron.

Such formulas have been obtained for a large number of invariants including the multiplicity of a singular point of a function or a vector field, the Betti numbers of hypersurfaces and complete intersections, the dimensions of the spaces of holomorphic forms and other Hodge numbers of algebraic varieties and singularities, the exponents of the asymptotics of integrals and the eigenvalues of the monodromy, etc. These results were obtained in the 1970s by A.G. Kushnirenko, A.G. Khovanskiĭ, A.N. Varchenko, V.I. Danilov, and others. Those formulas lead to an interesting and unexpected interrelation of analysis, on the one hand, and the geometry of integral polyhedra (that is, integral linear programming), on the other hand. This interrelation yields new results in both domains. To a large degree, the success has been achieved due to the use of modern tools of algebraic geometry: theory of resolution of singularities and theory of toric varieties (Mumford and others).

Bifurcation theory. Application of general ideas of singularity theory in the theory of differential equations has led to substantial progress in the study of functional spaces of differential equations, e.g., the spaces of vector fields. In particular, one has explored the structure of the bifurcation diagrams near the strata of codimension 1 and near some strata of codimension 2. This has enabled one to make great progress in the study of phenomena related to the stability loss of self-oscillations. There have been completely examined the generic case for a planar system with two zero roots, as well as the bifurcations of normal forms for the loss of stability of a self-oscillation near resonances of orders different from 4.

There is substantial progress in the problem on bifurcations of the complex phase portrait of a differential equation: in the so-called Poincaré case there have been found versal deformations, and in a number of cases outside of the Poincaré domain there have been figured out topological reasons for the divergence of the Poincaré series near resonances. These phenomena turned out to be connected with the bifurcations of elliptic curves on complex surfaces. In the study of the functional space of complex surfaces containing an elliptic curve, one has constructed bifurcation theory similar to the corresponding theory for differential equations. Simultaneously, there has been established algebraic insolubility of a number of classical problems in the theory of differential equations (for instance, of the Lyapunov stability problem for a singular point), that is, there has been proven non-algebraicity of the corresponding bifurcation diagrams in the functional space of all vector fields.

Also closely related to the bifurcation theory, there is the question on versal deformations of the Jordan normal forms of matrices. This is the question on a normal form to which one can reduce a family of matrices, smoothly depending on a parameter, by means of a linear transformation.

The reduction of the matrix of an operator to the Jordan normal form is an unstable operation since the entries of the normal form and the reducing transformation depend on the initial operator in a discontinuous way. Versal deformations of Jordan matrices are the simplest normal forms for which there is no instability in the reduction. Such forms have been found for the complex and real general linear groups and for the complex and real symplectic groups, which is important for applications in the problems of mechanics. The study of the simplest bifurcation diagrams in this situation brings us back to simple critical points of functions (see above).

One can find more details on the scope of questions mentioned here in V.I. Arnold's book "Geometrical Methods in the Theory of Ordinary Differential Equations" (the Russian original under the title "Supplementary Chapters to the Theory of Ordinary Differential Equations" of 1978).

2. Real algebraic geometry

Lately one observed a substantial progress in the classical real algebraic geometry by means of transcendental, functional analytical methods. These achievements are due to the use of a number of transcendental methods, in the first place, the Atiyah–Singer formula for the involution signature, on the one hand, and the Kodaira theory of deformations of Kähler manifolds, on the other hand. First results among many in this direction were obtained by V.I. Arnold in his proof of D.A. Gudkov's conjecture on the arrangement of ovals of planar algebraic curves. In the subsequent series of works V.A. Rokhlin, as well as V.M. Kharlamov, D.A. Gudkov, G.A. Utkin, A.D. Krakhnov, V.I. Zvonilov, O.Ya. Viro, and others obtained an extensive new information on the topology of real algebraic varieties. In particular, V.M. Kharlamov gave a complete answer to the question of the 16th Hilbert problem on possible topology of surfaces of degree 4 in the three-dimensional projective space.

Real algebraic geometry is not only a domain of applications of functional analytical methods (success in the study of algebraic objects is based on the fact that an extension of the functional space by omitting the restriction of algebraicity gives a larger freedom for deformations and homotopies). It is also important by itself for a number of areas of functional analysis, especially in the studies of the geometry of bifurcation diagrams in the functional space. In particular, the works by I.G. Petrovskiĭ on real algebraic geometry, which were generalized and developed by the aforementioned authors, are closely related to the theory of singularities of the fundamental solutions of partial differential equations (namely, to the question on wave diffusion and on lacunas).

3. Quasiclassical asymptotics

The stationary phase method has been used in physics to construct various short-wave asymptotics for quite a while. However, a consistent mathematical theory of related phenomena was invented only during the last 10–15 years; its creation required a substantial development of the functional analytical and especially geometrical side of the question. First studies in this direction were carried out by V.P. Maslov before 1965. The results obtained are at the border of functional analysis per se and mathematical physics. They were described in detail in a series of works (for instance, "Perturbation Theory and Asymptotic Methods" (1965) and "Operator Methods" (1973) by V.P. Maslov, his surveys in the collection "Principal Results of Science and Technology, Current Problems in Mathematics, Volume 8" (1977), "Quasiclassical Approximation for the Equations of Quantum Mechanics" (1976) by V.P. Maslov and M.V. Fedoryuk). Among new concepts that appeared in these studies, the most fundamental ones are the so-called canonical operator and Maslov index. The Maslov index, generalizing the Morse index, is closely related to symplectic geometry and enters all the quantization conditions. It takes into account the quarter-wave discontinuities at the tangency of a ray with the caustic. This index can be defined as the intersection index of a curve on a Lagrangian submanifold in a symplectic space with the hypersurface of singularities of the projection of this submanifold of the phase space to the configuration one. The canonical operator theory was developed and generalized by Hörmander in his theory of "Fourier integral operators". The technique developed by Maslov and Hörmander is widely applied to investigations of wave fronts, propagation of singularities and discontinuities for hyperbolic equations of mathematical physics.

We also mention the studies on the foundation of quasiclassical asymptotics of stationary problems. V.F. Lazutkin showed that for the Laplace operator on a manifold which is ε-close to a manifold with a completely integrable geodesic flow, the quasiclassical approximation properly describes most of the eigenvalues. More precisely, it describes all eigenvalues with the exception of a part whose asymptotical proportion at high frequencies is of order $\sqrt{\varepsilon}$. These results were obtained by means of fusion of the achievements of the theory of short-wave asymptotics and of the ergodic theory of classical dynamical systems.

4. Ergodic theory

In ergodic theory, which intensively progressed in our country by 1967, there was recently obtained a number of fundamental results as well. The theory of approximations of automorphisms by periodic ones developed by A.M. Stepin and others gave the opportunity to solve long-standing questions on classical dynamical systems. The obtained multiplicative ergodic theorem has found broad applications, in particular, in the theory of characteristic Lyapunov exponents and in numerical experiments with dynamical systems (V.I. Oseledets, V.M. Millionshchikov). The studies of D.V. Anosov and others on realization of abstract dynamical systems by diffeomorphisms shed new light on a large scope of problems of the theory of classical dynamical systems. The series of papers by M.I. Brin and Ya.B. Pesin on the theory of partially hyperbolic systems is an important step forward in our understanding of the ergodic properties of dynamical systems afforded by differential equations or mappings. In this respect we would like to mention the formula expressing the entropy in terms of the characteristic Lyapunov exponents (Ya.B. Pesin, V.M. Millionshchikov) and Pesin's theorem on determining the ergodic, Bernoulli, and K-components in a dynamical system. The works by G.A. Margulis on irreducible discrete subgroups of semisimple Lie groups are a rather convincing demonstration of the power of ergodic theory methods. In these works, one employs the multiplicative ergodic theorem to prove the arithmetic nature of the irreducible discrete subgroups with a finite volume of the quotient space and to give a classification of finite dimensional representations of such subgroups. In the proof of the finiteness of the quotient groups of discrete subgroups of the type described above, one examines the invariant (under those subgroups) algebras of measurable sets.

PREFACE TO THE BOOK "SELECTED WORKS BY HERMANN WEYL"

V.I. ARNOLD, A.N. PARSHIN

(Published in the book: Hermann Weyl. Selected Works. Mathematics. Theoretical Physics. Editor: V.I. Arnold. Series "Classics of Science", Nauka, Moscow, 1984, pp. 7–8.)

Hermann Weyl (1885–1955) entered the scientific scene in the very beginning of the 20th century. He is one of those few great scientists who managed to leave an individual stamp in almost all areas of mathematics. Being a deserving successor of his teacher David Hilbert and an outstanding continuer of the traditions of the German mathematical school, Weyl was also a remarkable physicist who greatly contributed to the development of relativity theory and quantum mechanics.

If one also adds to this his deep philosophical interpretation of the fundamentals of science, his numerous general interests, a very original and artistic style of his books and papers, then undoubtedly the personality of Weyl can be regarded as an important phenomenon of the whole German culture of the 20th century.

The present edition is a collection of selected works by H. Weyl which within the limits of the possible reflects the above mentioned sides of his studies. This edition includes famous works by Weyl on representations of Lie groups, differential equations, number theory, and foundations of mathematics as well as his less-known papers, which are still of interest even today. The latter are exemplified, in particular, by his article "Elektron und Gravitation" ("Electron and gravitation") which was ahead of his time and whose real value was manifested only recently in relation to the appearance of gauge theories in particle physics. Finally, popular surveys and biographies, only a small part of which could be included into this book, give some idea about Weyl's mastery as a popularizer and historian of science.

The publication of these works along with the original scientific treatises allows one to better show the influence of Weyl himself, as well as of his predecessors and contemporaries, on the development of science.

A great volume of Weyl's heritage made our choices far from being indisputable. However, we decided to refrain from publication of papers whose Russian translation can be found elsewhere. An exception was made for the memoir "Theorie der Darstellung kontinuierlicher halb-einfacher Gruppen durch lineare Transformationen" ("Representation theory of continuous semi-simple groups through linear transformations") which is apparently the best known work of H. Weyl and the existing translation of which contains large omissions.

At the end of the book there is an article on Weyl's life and work by the well-known French mathematicians C. Chevalley and A. Weil, as well as comments to the papers in this volume and the full bibliography which includes all known to us Russian translations of Weyl's works.

We are deeply indebted to I.S. Alekseev, Yu.A. Danilov, S.S. Demidov, H. Koch, R. Langlands, and O. Neiman for their help in preparation of the present edition.

Originally publ. in: Hermann Weyl. Selected Works. Mathematics. Theoretical Physics.
Editor: V.I. Arnold. Series "Classics of Science", 7-8,
Translated by Boris Khesin and Andrey Khesin
Springer-Verlag Berlin Heidelberg 2018 with kind permission

COMMENTS TO THE PAPER "DAS ASYMPTOTISCHE VERTEILUNGSGESETZ DER EIGENSCHWINGUNGEN EINES BELIEBIG GESTALTETEN ELASTISCHEN KÖRPERS" BY HERMANN WEYL

V.I. ARNOLD, A.N. PARSHIN

(Published in the book: Hermann Weyl. Selected Works. Mathematics. Theoretical Physics. Editor: V.I. Arnold. Series "Classics of Science", Nauka, Moscow, 1984, pp. 446–449.

Weyl's paper was published in Rend. Circ. Mat. Palermo, 1915, vol. 39, pp. 1–50. The English translation of its title: "The asymptotic law of the distribution of the eigenoscillations of an elastic body of arbitrary shape".)

The Weyl formula for the asymptotics of the eigenvalues of the Laplace operator in a domain was one of the sources of quantum mechanics (see *W. Pauli*, Treatises on Quantum Theory, Vol. 1. Series "Classics of Science", Nauka, Moscow, 1975, p. 101;[1] *C.L. Fefferman*, Bull. Amer. Math. Soc., 1983, vol. 9, no. 2, pp. 129–206). On the other hand, the same formula laid out a foundation to a peculiar domain of mathematical analysis, the so-called spectral geometry, where one examines those geometrical characteristics of a domain or Riemannian manifold that can be reconstructed from the spectrum of the Laplace operator (see above).[2] According to a figurative expression of M. Kac, the question can be formulated as follows: "Can one hear the shape of a drum?" [IV, 13].[3] The Weyl formula shows that one can "hear" the volume. On the other hand, J. Milnor constructed two non-isometric 16-dimensional flat tori for which the eigenvalues of the Laplace operator (on functions, as well as on differential forms) coincide (Proc. Nat. Acad. Sci. USA, 1964, vol. 51, no. 4, p. 542). Therefore, it is not always possible to reconstruct the Riemannian metric from the spectrum. For closed surfaces of constant negative curvature, the Laplace spectrum and the lengths of geodesics determine each other (*H. Huber*, Math. Ann., 1959, Bd. 138, Nr. 1, S. 1–26). In this relation I.M. Gelfand posed a question on whether the spectrum uniquely determines the moduli of a Riemannian surface (Proc. Internat. Congr. Mathematicians (Stockholm, 1962), Inst. Mittag-Leffler, Djursholm, 1963, pp. 74–85) and showed that any isospectral deformation of the surface is trivial. H.P. McKean proved that the conjugacy class of the corresponding Fuchsian group in $SL(2,\mathbb{R})$ is determined by the spectrum of the Laplace–Beltrami operator up to a finite number of possibilities (Comm. Pure Appl. Math., 1972, vol. 25,

[1]The German original: *W. Pauli*, Quantentheorie. In: Handbuch der Physik, 1926, Bd. 23, S. 1–278; H. Geiger und K. Scheel (Hrsg.), Springer, Berlin. The reference in question is in § 14 of this survey. (Editors' note)

[2]Here "see above" refers to the previous comment (by W. Müller) in the book to the same paper by H. Weyl. (Editors' note)

[3]Here "[IV, 13]" refers to the bibliography at the end of the book, namely, to the paper by *M. Kac*, Amer. Math. Monthly, 1966, vol. 73, no. 4, part II, pp. 1–23. (Editors' note)

Originally publ. in: Hermann Weyl. Selected Works. Mathematics. Theoretical Physics.
Editor: V.I. Arnold. Series "Classics of Science", 446-449,
(Weyl's paper was first published in: Rend. Circ. Mat. Palermo 39, 1-50 (1915))
Translated by Boris Khesin and Andrey Khesin
Springer-Verlag Berlin Heidelberg 2018 with kind permission

no. 3, pp. 225–246), but then M.-F. Vignéras found an example of two non-conjugate groups with the same spectrum (C. R. Acad. Sci. Paris Sér. A-B, 1978, vol. 287, no. 2, pp. A47–A49). Subtler spectral characteristics of a surface will probably allow one to reconstruct the Riemannian surface (see the discussion in *A.B. Venkov* (op. cit., §§ 6, 11)).[4]

The distribution of the eigenvalues for the membrane of a compact domain in a Euclidean space can be rather irregular. However, the smoothened (averaged) characteristics of the asymptotic behavior of the eigenvalues with large numbers exhibit a relatively regular pattern. The smoothened asymptotics are related to interesting geometric characteristics of the domain boundary. The most standard way of smoothening is a consideration of the trace of the heat operator (see above).[5] We will denote this trace by Z_- for the Dirichlet boundary conditions and by Z_+ for the Neumann boundary conditions. The Minakshisundaram–Pleijel expansion gives the following asymptotic equality (pointed out by R.T. Seeley in a more general situation) for a domain $\subseteq \mathbb{R}^{d+1}$:

$$Z_{\mp}(t) = (4\pi t)^{-d/2} \sum c_k^{\mp} t^{k/2}, \qquad k \geqslant -1.$$

According to the Weyl formula, the first coefficient c_{-1} is equal to the domain volume. The next two coefficients were computed by McKean and Singer (*H.P. McKean, I.M. Singer*, J. Differential Geometry, 1967, vol. 1, no. 1, pp. 43–69), who used the theory of invariants of the orthogonal group in the form described by Weyl in [I, 13]:[6]

$$c_0^{\mp} = \mp\tfrac{1}{4}\sqrt{4\pi} \cdot \text{(the area of the domain boundary)},$$
$$c_1^{\mp} = -\tfrac{1}{6} \cdot \text{(the integral of the mean curvature of the boundary)}.$$

The volume, the area, and the integral of the mean curvature are three coefficients of the polynomial in ε expressing the volume of the ε-neighborhood of a domain in the Euclidean space. These and other results suggest a conjecture that the subsequent coefficients up to c_d can also be expressed, similarly to the coefficients in the formula for the volume of the ε-neighborhood of a domain in the Euclidean space, via integrals of symmetric functions of the principal curvatures over the boundary. (This conjecture of V.I. Arnold (1965) was imprecisely quoted in the paper by McKean and Singer.)

Apparently this conjecture is still neither proven nor disproven. S.A. Molchanov (Uspekhi Mat. Nauk, 1975, vol. 30, no. 1, pp. 3–59) computed one more coefficient for a domain in the three-dimensional space. For instance, for the Dirichlet boundary condition

$$c_2^- = \int k_1 k_2 \, dS/(96\pi) - \int (k_1 + k_2)^2 \, dS/(2^8\pi),$$

---

[4]Here "*A.B. Venkov* (op. cit., §§ 6, 11)" refers to the paper by *A.B. Venkov*, Uspekhi Mat. Nauk, 1979, vol. 34, no. 3, pp. 69–135, cited in W. Müller's comment to the same paper by H. Weyl. (Editors' note)

[5]Here "see above" refers to W. Müller's comment to the same paper by H. Weyl. (Editors' note)

[6]Here "[I, 13]" refers to the bibliography at the end of the book, namely, to the monograph by *H. Weyl*, The Classical Groups. Their Invariants and Representations. Princeton University Press, Princeton, NJ, 1939. (Editors' note)

where k_i are the principal curvatures of the boundary. The first term, according to the Gauss–Bonnet formula, is proportional to the Euler characteristic, while the second one does not have a topological nature. It shows that the Minakshisundaram–Pleijel coefficients in any case differ from the coefficients in the formula for the volume of the ε-neighborhood (not only by factors independent of the domain).

For a planar domain, the coefficients have been computed up to c_5 (*K. Stewartson, R.T. Waechter*, Math. Proc. Cambridge Phil. Soc., 1971, vol. 69, no. 2, pp. 353–363).

On a compact Riemannian d-dimensional manifold without boundary, the Minakshisundaram–Pleijel expansion has the form

$$Z \sim (4\pi t)^{-d/2} \sum_{k=0}^{\infty} c_k t^k \qquad (t \to +0),$$

where the coefficients c_k are expressed as integrals over the manifold of certain functions of the metric and its derivatives. Similar expansions also exist for operators acting in sections of a vector bundle, for instance, on differential forms. H. Donnelly (Invent. Math., 1975, vol. 29, no. 3, pp. 239–243) discovered a connection of these expansions with the Weyl formula for the tube volume. Namely, the coefficient at ε^{m+2k} in the representation of the volume of the ε-neighborhood of a submanifold of codimension m in the Euclidean space as a polynomial in ε turned out to be a linear combination of the coefficients c_k corresponding to the spectra of the Laplace operators on differential forms of all degrees. Furthermore, already the function of the curvature given by the integrand of the coefficient of the Weyl formula for the tube volume can be linearly expressed via the integrands for c_k.

V.S. Buslaev (Dokl. Akad. Nauk SSSR, 1971, vol. 197, no. 5, pp. 999–1002) pointed out that the Minakshisundaram–Pleijel coefficients for the boundary of a convex domain in the Euclidean space appear while one studies the scattering asymptotics. Namely, for the scattering phase he obtained an expansion, which for the Dirichlet boundary condition has the form

$$\ln \det S(k) \sim \sum b_j \bar{c}_j k^{d-j}, \qquad j \geqslant -1,$$

where S is the "scattering matrix" and the numbers b_j do not depend on the domain (see also *A. Jensen, T. Kato*, Comm. Partial Differential Equations, 1978, vol. 3, no. 12, pp. 1165–1195; *A. Majda, J. Ralston*, Duke Math. J., 1978, vol. 45, no. 1, pp. 183–196; 1978, vol. 45, no. 3, pp. 513–536; 1979, vol. 46, no. 4, pp. 725–731). The first term is proportional to the volume domain and can be regarded as an analogue of the Weyl formula for the domain exterior.

These results were generalized to the case of scattering by a non-convex obstacle satisfying the "non-trapping" condition, that is, reflecting all incident rays (*V. Petkov, G. Popov*, Ann. Inst. Fourier (Grenoble), 1982, vol. 32, no. 3, pp. 111–149), to matrix elliptic operators (*V.Ya. Ivriĭ, M.A. Shubin*, Dokl. Akad. Nauk SSSR, 1982, vol. 263, no. 2, pp. 283–284), and to the case of a metric perturbation with compact support, instead of an obstacle (*A. Majda, J. Ralston*, Duke Math. J., 1978, vol. 45, no. 3, pp. 513–536).

R.B. Melrose has discovered another generalization of the Weyl and Minakshisundaram–Pleijel formulas to the case of the domain exterior (*R.B. Melrose*, Comm. Pure Appl. Math., 1980, vol. 33, no. 4, pp. 461–499). Melrose considers the forward scattering amplitude of a plane wave. For short waves the first adjustment to the amplitude of the incident wave is proportional to the area of the shadow of the convex obstacle on a hyperplane orthogonal to the incident ray (*A. Majda, M.E. Taylor*, Comm. Pure Appl. Math., 1977, vol. 30, no. 5, pp. 639–669). The corresponding adjustment to the trace value (i.e., to the amplitude integral over all the directions of the incident ray) is proportional to the volume of the obstacle, since the shadow area averaged over all the projection directions is proportional to the volume of the convex body (see the comment to H. Weyl's paper "On the volume of tubes").

The next terms of the expansion of the forward amplitudes (which is an asymptotic expansion over fractional powers of the wavelength with the denominator 3 and possibly with logarithms) can be expressed via the boundary curvature. The first of these coefficients is obtained by integration along the terminator (i.e., the shadow boundary on the obstacle surface) of the cubic root of the curvature of the obstacle surface in the direction of the incident ray.

Returning to Weyl's studies, we note that in [II, 19][7] he posed the question on the next term in the asymptotic behavior of the number $N(\lambda)$ of eigenvalues $\leqslant \lambda$ (the Weyl problem; the prehistory of Weyl's works in this area of analysis is described in [II, 16, S. 442][8] and [IV, 13][9]). D.G. Vassiliev's recent theorem (Funktsional. Anal. i Prilozhen., 1983, vol. 17, no. 4, pp. 79–81) is the most general result in the solution to the Weyl problem. He considered elliptic differential operators A of order $2m$ on a compact manifold X with boundary Γ and eigenvalue problems with boundary conditions of quite general type. With an operator A one can associate a Hamiltonian system on X (the geodesic flow, if A is the Laplace operator), on the trajectories of which one imposes two conditions: the condition of simple reflection on the boundary Γ, which provides the uniqueness of the reflected trajectory, and the condition of zero measure for the set of so-called absolutely periodic points (in the analytic case the latter condition follows from a simple criterion, namely, non-negativity of a certain Poisson bracket on the boundary). Then

$$N(\lambda) = a\lambda^{n/2m} + b\lambda^{(n-1)/2m} + o\left(\lambda^{(n-1)/2m}\right),$$

where a is expressed in terms of the integral of the principal symbol of the operator A, while b is the integral over $T^*\Gamma$ of the sum of the distribution function for the eigenvalues and the scattering phase of the one-dimensional problem on the half-line for the equation with constant coefficients, obtained from the original boundary problem (at each point of the boundary Γ). This result includes the aforementioned

[7]Here "[II, 19]" refers to the bibliography at the end of the book, namely, to the paper by *H. Weyl*, Über die Randwertaufgabe der Strahlungstheorie und asymptotische Spektralgesetze. Journal für die reine und angewandte Mathematik, 1913, Nr. 143, S. 177–202. (Editors' note)

[8]Here "[II, 16]" refers to the bibliography at the end of the book, namely, to the paper by *H. Weyl*, Das asymptotische Verteilungsgesetz der Eigenwerte linearer partieller Differentialgleichungen (mit einer Anwendung auf die Theorie der Hohlraumstrahlung). Math. Ann., 1912, Bd. 71, Nr. 4, S. 441–479. (Editors' note)

[9]See footnote[3].

theorems of Duistermaat and Guillemin, Seeley, and Ivriĭ[10] (see also the work by *V.M. Babich, B.M. Levitan*, Dokl. Akad. Nauk SSSR, 1976, vol. 230, no. 5, pp. 1017–1020, for domains with geodesically concave boundary).

The question on the asymptotics of the number $N(\lambda)$ on open Riemannian manifolds (of finite volume) is very far from its solution even in the two-dimensional case. The main problem here is the existence of an infinite discrete spectrum of the Laplace operator (the Roelke problem, see *A.B. Venkov*, op. cit., § 11).[11] From the available general theorems on the spectrum of the Laplace operator in a non-compact fundamental domain D of a Fuchsian group (*A. Selberg*, Proc. Internat. Congr. Mathematicians (Stockholm, 1962), Inst. Mittag-Leffler, Djursholm, 1963, pp. 177–189; *L.D. Faddeev*, Trudy Moskov. Mat. Obshch., 1967, vol. 17, pp. 323–350; see also *S. Lang*, $\mathrm{SL}_2(\mathbb{R})$. Addison-Wesley Publishing Co., Reading, Mass., 1975; *P.D. Lax, R.S. Phillips*, Scattering Theory for Automorphic Functions. Annals of Mathematics Studies, No. 87, Princeton Univ. Press, Princeton, NJ, 1976) one cannot obtain the required information on the discrete spectrum, except for the case of arithmetic groups. Recently A.B. Venkov proved (see op. cit.;[12] *A.B. Venkov*, Proc. Internat. Congr. Mathematicians, Vol. 1, 2 (Warsaw, 1983), PWN, Warsaw, 1984, pp. 909–919) infiniteness of the discrete spectrum and the asymptotic formula

$$N(\lambda) = (4\pi)^{-1}\mathrm{vol}(D)\lambda - (4\pi)^{-1}m\sqrt{\lambda}\ln\lambda + O\left(\sqrt{\lambda}\right)$$

for discrete groups Γ_M on the Lobachevsky plane, generated by even products of reflections with respect to the edges of a certain geodesic polygon M with angles of the form π/k where $k > 1$ is an integer or ∞ (m being the number of zero angles in M). In this scope of questions, the fundamental role belongs to the analogy between prime numbers and conjugacy classes of elements in Fuchsian groups, which manifests itself in the definition of the Selberg ζ-function and the studies of its relation to the Laplace operator and its spectrum (with the help of the Selberg trace formula). On similar problems for discrete groups in the three-dimensional hyperbolic space, see *J. Elstrodt, F. Grunewald, J. Mennicke*, Uspekhi Mat. Nauk, 1983, vol. 38, no. 1, pp. 119–147.[13]

We also mention yet another broad domain of research related to estimates of the first eigenvalues of the Laplace operator on Riemannian manifolds via such invariants as the diameter or isoperimetric constant (*S. Gallot*, Séminaire Bourbaki, Vol. 1980/81, Lecture Notes in Math., Vol. 901, Springer, Berlin, 1981, pp. 132–148; Geometry of the Laplace Operator (Proc. Symposium Univ. Hawaii, 1979). Proceedings of Symposia in Pure Mathematics, Vol. 36, Amer. Math. Soc., Providence, RI, 1980; *P. Li, Sh.-T. Yau*, Comm. Math. Phys., 1983, vol. 88, no. 3, pp. 309–318; *G. Pólya*, Proc. London Math. Soc. (3), 1961, vol. 11, no. 1, pp. 419–433).

---

[10]Here V.I. Arnold and A.N. Parshin refer to the papers by *J.J. Duistermaat, V.W. Guillemin*, Invent. Math., 1975, vol. 29, no. 1, pp. 39–79; *R. Seeley*, Adv. in Math., 1978, vol. 29, no. 2, pp. 244–269; *V.Ya. Ivriĭ*, Funktsional. Anal. i Prilozhen., 1980, vol. 14, no. 2, pp. 25–34, cited in W. Müller's comment to the same paper by H. Weyl. (Editors' note)

[11]See footnote[4].

[12]See footnote[4].

[13]This is a translation of a preprint. For corrections see Uspekhi Mat. Nauk, 1983, vol. 38, no. 6, pp. 177–178. (Editors' note)

COMMENTS TO THE PAPER "ON THE VOLUME OF TUBES" BY HERMANN WEYL

V.I. ARNOLD

(Published in the book: Hermann Weyl. Selected Works. Mathematics. Theoretical Physics. Editor: V.I. Arnold. Series "Classics of Science", Nauka, Moscow, 1984, pp. 479–482.
Weyl's paper was published in Amer. J. Math., 1939, vol. 61, no. 2, pp. 461–472.)

The paper by H. Weyl on the volume of tubes had an influence on several branches of mathematics. The multidimensional Gauss–Bonnet formula, essentially contained in this paper,[1] started vigorous development of the differential geometric aspects of the theory of characteristic classes. On the base of that paper, one obtained fundamental results in the theory of "multidimensional variations" due to A.G. Vitushkin and in integral geometry. Complex analogues of Weyl's theorems on the volumes of tubes immediately lead to the Chern classes and have lately been efficiently used in the studies of singularities of analytic varieties (P.A. Griffiths).

A. The relation between the theory of the tube volume and the Gauss–Bonnet formula is as follows. The coefficient at the highest degree of the radius in the formula for the tube volume is a topological invariant. Its computation leads to the Gauss–Bonnet formula, which expresses the integral of the Gauss curvature of a Riemannian compact manifold in terms of its Euler characteristic. Therefore the formula relating these two intrinsic, independent of the embedding, characteristics of a Riemannian manifold was obtained just from the analysis of an embedding into the Euclidean space. This is an instructive example of how leaving the framework of an axiomatic theory and using structures outside of the given problem can facilitate its solution.

The sequence of computations is as follows:

1) For a hypersurface in the Euclidean space the Gauss–Bonnet formula can be easily proven (H. Hopf). Indeed, the integral of the total (Gauss) curvature is equal to the product of the unit sphere volume by the degree of the Gauss map of the hypersurface into this sphere, as the integral of the Jacobian of the Gauss map. On the other hand, the degree of the Gauss map is equal to one half of the Euler

[1]This paper "already contains in principle the Gauss–Bonnet formula for submanifolds of the Euclidean space" [IV, 4]. A detailed derivation of the multidimensional Gauss–Bonnet formula, based on H. Weyl's work on the volume of tubes, can be found in the article by *C.B. Allendoerfer, A. Weil,* The Gauss–Bonnet theorem for Riemannian polyhedra. Trans. Amer. Math. Soc., 1943, vol. 53, no. 1, pp. 101–129.

From the Editors: "[IV, 4]" here refers to the bibliography at the end of the book, namely, to the obituary by *C. Chevalley, A. Weil,* Hermann Weyl (1885–1955). L'Enseignement Mathématique, 1957, tome 3, fasc. 3, pp. 157–187. The statement quoted by V.I. Arnold is on p. 186.

Originally publ. in: Hermann Weyl. Selected Works. Mathematics. Theoretical Physics.
Editor: V.I. Arnold. Series "Classics of Science", 479-482,
(Weyl's paper was first published in: Amer. J. Math. 61(2), 461-472 (1939))
Translated by Boris Khesin and Andrey Khesin
Springer-Verlag Berlin Heidelberg 2018 with kind permission

characteristic of the hypersurface. The latter follows immediately from the Morse theory. [Restrict a generic linear function to the hypersurface. We call the points of its maximum and minimum on the unit sphere the north and south poles. The degree of the Gauss map is equal to the number of preimages of the north (or south) pole counted taking the orientations into account. The contribution of a preimage of the south pole is equal to $(-1)^i$, where i is the Morse index of the critical point of the function on the hypersurface. For each preimage of the north pole, the contribution is the same if the hypersurface dimension is even and is opposite in sign otherwise. The sum of the indices of all critical points is equal to the Euler characteristic of the hypersurface (according to the Euler–Poincaré formula). Therefore the degree of the Gauss map is equal to one half of the Euler characteristic for an even-dimensional hypersurface, as well as for an odd-dimensional one (in the latter case, the Euler characteristic vanishes).]

2) The formula for the tube volume allows one to extend the Gauss–Bonnet formula from hypersurfaces to submanifolds of any codimension in the Euclidean space. Below it will be convenient to assume that the codimension is odd (if needed this can be achieved by multiplying the ambient space by a line).

The boundary of a tubular neighborhood of sufficiently small radius is a smooth hypersurface. The coefficients at the highest powers of the tube radius in the expansion of the volume of a tubular neighborhood of the initial manifold and of the hypersurface differ only by a factor of 2 (here one uses the assumption that the codimension is odd). According to the Weyl formula this coefficient for the initial submanifold, up to a constant factor, is equal to the integral of the Gauss curvature of the submanifold. Therefore the integral of the Gauss curvature of a submanifold of the Euclidean space is proportional to the Euler characteristic of the boundary of a tubular neighborhood and hence to the Euler characteristic of the submanifold itself: $\chi(\text{of the boundary}) = 2\chi(\text{of the submanifold})$, since the codimension m of the submanifold is odd and consequently $\chi(S^{m-1}) = 2$.

3) In order to prove now the Gauss–Bonnet formula for any compact Riemannian manifold, it suffices to isometrically embed the submanifold into the Euclidean space of sufficiently high dimension. The latter is always possible due to the Nash theorem. One can get by without a reference to the Nash theorem: for this it is sufficient to verify that the integral of the Gauss curvature is, as physicists say, a topological charge, that is, it does not change under small variations of the metric. The latter is achieved by applying Stokes' formula to the appropriately expressed derivative of the integral with respect to the parameter.

B. The relation of the theory of tube volumes with calculus of variations and integral geometry is manifested already by the classical Buffon–Crofton theorem where the coefficients of the Weyl formula appear. In the simplest case of a convex body the theorem is as follows. The coefficient of dimension $(\text{length})^p$ in the formula for the neighborhood volume is equal to the volume of the orthogonal projection of the body to a p-dimensional space, averaged over the Grassmann manifold of all the p-dimensional subspaces of the vector space (up to a factor depending on the dimensions only).

There are two generalizations of this theorem to the non-convex case. The first one was constructed by A.G. Vitushkin in his theory of multidimensional variations.[2] In defining a variation of dimension p, the volume of the projection is averaged taking into account multiplicities but discarding orientations. Vitushkin ingeniously applied the characteristics of submanifolds obtained this way to the 13th Hilbert problem to prove impossibility of representation of functions by means of superpositions. The relation of variations with the coefficients of the formula for the tube volume consists in estimating the variations from above via the integrals of the absolute values of the curvatures (see *A.M. Leontovich, M.S. Melnikov*, On the boundedness of variations of a manifold. Trudy Moskov. Mat. Obshch., 1965, vol. 14, pp. 306–337). Lately, the variations of manifolds related to critical points of smooth functions and mappings are being used in singularity theory (see *E.E. Landis*, Estimates of the variations of lesser value sets of a smooth function in a neighborhood of a minimum point. Vestnik Moskov. Univ. Ser. I Mat. Mekh., 1980, no. 3, pp. 24–28; *V.I. Arnold*, Some open problems in singularity theory. In: Theory of Cubature Formulas and Applications of Functional Analysis to Problems in Mathematical Physics, Trudy Sem. S.L. Soboleva, No. 1, Akad. Nauk SSSR, Sibirsk. Otdel., Inst. Mat., Novosibirsk, 1976, pp. 5–15).

In integral geometry instead of the integral of the intersection multiplicity, one averages the integral of the Euler characteristic of the preimage of a point under the projection. In other words, one considers the integral of the Euler characteristic of the intersection of a manifold with all affine planes of a fixed dimension in the Euclidean space (over the natural measure on the manifold of all such planes).

This integral turns out to be proportional to the coefficient in the Weyl formula which has the same dimension (now regardless of the convexity). The coefficient of proportionality, as always, depends on the dimensions only.

C. The relation of the theory of tube volumes with the Chern classes consists in expressing the coefficients of the polynomial equal to the volume of a tubular neighborhood of a complex analytic manifold in terms of the Chern form of the Kähler connection of the submanifold and an appropriate power of the Kähler symplectic structure. More precisely, denote by $\tau_r(M)$ the union of the balls of small radius r in normal planes of a manifold M lying in the complex space $\mathbb{C}^N$ equipped with the standard Hermitian metric (if r is not small, one needs to take points of a tubular neighborhood with multiplicities). For a complex manifold of dimension m and codimension n, the formula for the tube volume is as follows:

$$\mathrm{Vol}\,\tau_r(M) = \sum_{k=0}^{m} c_{k,m,n} r^{2(n+k)} \mu_k(M),$$

where

$$\mu_k(M) = \int_M c_k(\Omega_M) \wedge \varphi^{m-k}.$$

Here φ is the standard Kähler 2-form, Ω_M is the curvature form of the Kähler connection, and c_k are the Chern forms defined by the expansion

$$\det\left(\lambda E + \frac{i}{2\pi}\Omega\right) = \sum_{k=0}^{m} \lambda^{m-k} c_k(\Omega).$$

[2] *A.G. Vitushkin*, On Multidimensional Variations. Gostekhteorizdat (Gosudarstv. Izdat. Tekhn.-Teor. Lit.), Moscow, 1955.

A similar formula also holds in the projective space. For a compact non-singular (and hence algebraic) submanifold of the complex projective space, all the coefficients are of topological nature and can be expressed via the Chern classes $c_k(M) \in H^{2k}(M)$ and the class of the hyperplane section $\varphi \in H^2(M)$. In particular, the volume of a complex algebraic manifold is of topological nature (according to the known Wirtinger theorem, this volume is equal to the degree of the submanifold up to a factor depending on the dimensions, i.e., on the choice of the length unit in the projective space). Thus in this case the formula for the tube volume gives a generalization of the Wirtinger theorem.

In the study of affine varieties and their singularities, the formulas for the volumes of tubes are usually applied to the part of the variety belonging to a ball of large radius (if one examines non-compact varieties) or small radius (if one examines singularities) in $\mathbb{C}^N$. A detailed exposition of the related formulas asymptotical in the radius can be found in the paper by *P.A. Griffiths*, Complex differential and integral geometry and curvature integrals associated to singularities of complex analytic varieties. Duke Math. J., 1978, vol. 45, no. 3, pp. 427–512,[3] and in Griffiths' talk at the Helsinki congress: *P.A. Griffiths*, Some problems in complex analytic geometry with growth conditions. In: Proc. Internat. Congr. Mathematicians (Helsinki, 1978), Acad. Sci. Fennica, Helsinki, 1980, pp. 645–651.

D. The main result of Weyl's paper on the volume of tubes is the independence of the coefficients of the volume expansion on the embedding of a submanifold into the Euclidean space and explicit expressions for these coefficients via the intrinsic geometry of the submanifold, namely, via the components of the curvature tensor. These expressions will apparently become more transparent if one writes them in a coordinate-free and index-free way.

The Riemannian curvature tensor can be regarded as a symmetric linear operator from the second exterior power of the tangent space into itself. This enables one to define exterior powers of this operator, which will be acting from the even exterior powers of the tangent space into themselves. The traces of these operators are just the integrands in the Weyl formulas for the coefficients of the expansion of the tube volumes, i.e., symmetric functions of the principle external curvatures.

---

[3]The first chapter of which is entitled "Hermann Weyl's formula for the volume of tubes and the Gauss–Bonnet theorem".

ON THE PAPERS ON CLASSICAL MECHANICS

A.N. Kolmogorov

My papers on classical mechanics appeared under the influence of von Neumann's papers on the spectral theory of dynamical systems (see reference [1] in paper No. 53) and, particularly, under the influence of the Bogolyubov-Krylov paper of 1937 (see reference [12] in paper No. 53).

I became extremely interested in the question of what ergodic sets (in the sense of Bogolyubov-Krylov) can exist in the dynamical systems of classical mechanics and which of the types of these sets can be of positive measure (at present this question still remains open). To accumulate specific information we organized a seminar on the study of individual examples. My ideas concerning this topic and closely related problems aroused wide response among young mathematicians in Moscow.

This is discussed in greater detail in Arnol'd's commentary.

CLASSICAL MECHANICS

(V.I. Arnol'd)

Kolmogorov's classical papers Nos. 52 and 53 produced a very strong effect on the subsequent development of the theory of dynamical systems, and at present there are dozens of books developing or presenting the material of these papers. In this brief commentary it is impossible to embrace all applications of these results and we confine ourselves to some improvements introduced into the theory after 1954.

1. The basic problem of dynamics

The statement of the problem of the motion of systems that are close to the systems of classical mechanics, including the problem of orbit evolution in the three-body problem, dates back to Newton [1]. Laplace [2] stated explicitly the theorem on stability of the semimajor axes of Keplerian ellipses, which is a forerunner of Kolmogorov's theorem on preservation of tori, but proved

Originally publ. in: A.N. Kolmogorov. Selected Works. Mathematics and Mechanics, 433-444,

English transl.: Selected Works of A.N. Kolmogorov. Vol. I. Mathematics and Mechanics.
Editor: V.M. Tikhomirov, 503-518, © 1991 Kluwer Academic Publishers, Dordrecht Boston London.

it only in terms of approximate perturbation theory. On analyzing numerous attempts to justify and improve Laplace's argument, Poincaré [3] stated the problem in its modern form (to study the motion of a system whose Hamiltonian $W(p) + \theta S(q,p,\theta)$ is periodic in q) and called it the basic problem of dynamics (see [3], Chapter 1, §13). In the papers under consideration Kolmogorov solves this problem for the majority of initial conditions in the generic case ($\det \partial^2 W/\partial p^2 \neq 0$).

2. Rigid body and planetoid

Among the numerous problems of mechanics to which Kolmogorov's theorem on preservation of periodic motions can be applied we consider the following two classical problems:

(a) the problem of motion of an asymmetric heavy rigid body about a fixed point;

(b) the restricted circular three-body problem on the plane motion of a planetoid of negligibly small mass under the attraction of a heavy body S (Sun) and a body J (Jupiter) of small mass rotating around it.

In problem (a) Kolmogorov's theorem implies that the square of the total angular momentum changes little during infinite time, provided that at the initial instant the body rotates sufficiently fast (the kinetic energy is large as compared to the potential energy) (see [4]).

As to the planetoid, its instantaneous Keplerian elliptic orbit only rotates slowly in its plane under the influence of the "planet" J if at the initial instant the orbit does not intersect the circular orbit of the "planet" J and the ratio of the masses of the bodies J and S is sufficiently small.

It also follows directly from Kolmogorov's theorem that the majority of magnetic surfaces are stable relative to a sufficiently small variation of the magnetic field in toroidal systems of TOKAMAK type for non-zero "shear" (the derivative of the frequency ratio with respect to the torus number).

3. Limiting degeneration

Kolmogorov's theorem is not applicable to non-linear oscillations of a Hamiltonian system near equilibrium points or periodic motions (owing to degeneration, called limit degeneration by Born [5] when the tori contract to a point). However, a slight modification of this theorem leads to invariant tori even in the case of limiting degeneration. Therefore the Kolmogorov tori provide solutions, for example, to the Birkhoff problem of stability of general elliptic points of area-preserving mappings of a plane onto itself [6].

The most complete result concerning this problem is due to Moser [7]: the fixed point 0 of an area-preserving smooth mapping determined in polar coordinates by a formula

$$(r, \phi) \mapsto (r, \phi + \alpha + \beta r^2) + O(r^4)$$

is stable if $\alpha \neq 2\pi k/3,\ 2\pi k/4;\ \beta \neq 0$.

It follows from Kolmogorov's theorem, modified for the case of limiting degeneration, that an oscillating point always remains near an equilibrium state that is stable in the linear approximation, or near a periodic motion for the majority of initial conditions close to this motion, provided that the oscillatory Hamiltonian system under consideration satisfies some non-degeneracy conditions ("in general", these are satisfied).

In the case when the dimension of the phase space can be reduced to three, the two-dimensional tori whose existence is guaranteed by the generalized Kolmogorov theorem divide the phase space and therefore provide genuine Lyapunov stability of equilibrium states or periodic motions. Among the classical problems solved in this way we mention the following two:

(a) a proof of the stability of Lagrange periodic solutions of the restricted planar circular three-body problem [8, 9];

(b) a proof of the stability of an inverted friction-free non-linear pendulum under sufficiently fast vertical oscillation of the point of suspension.

4. Self-degeneration

There are many situations in which the unperturbed motion is described by a smaller number of frequencies than the number of the degrees of freedom (self-degeneration in the sense of Born; examples of this kind are Kepler's problem and the problem of charged particle motion in a strong magnetic field).

The extension of Kolmogorov's theory to this case [10] made it possible to prove, in particular, the perpetual adiabatic invariance of the action in non-linear systems with one degree of freedom for a slowly varying Hamilton function [11] as well as the perpetual motion of charged particles within axisymmetric magnetic traps [12].

A combination of these results with the study of oscillation near an equilibrium position (see §3) leads to a proof of the perpetual smallness of variations of the semimajor axes of planet orbits for the majority of initial conditions, such that the initial Keplerian ellipses differ slightly from non-intersecting circles lying in one plane and traversed in the same direction, provided that the masses of the planets are sufficiently small as compared to that of the central body [12].

The exceptional initial conditions leading to substantial changes in the semimajor axes of Keplerian ellipses presumably form a set of positive measure that is everywhere dense in the phase space.

5. Diffusion

If the invariant tori in Kolmogorov's theorem do not divide the phase space (reduced by taking into account the first integrals), then it is possible that after a long time a phase curve issuing from a slot between the tori goes far away from the invariant torus of the unperturbed system containing the initial point of this curve. In this case the values of the action variables (the integrals of the unperturbed system) diverge strongly from their initial values.

Examples of such divergence in a non-degenerate system were constructed in [13]. The average divergence rate in these examples is exponentially small (of the order of $e^{-1/\sqrt{\theta}}$ where θ is the magnitude of the disturbance). This exponentially slow divergence cannot be detected by any approximation of perturbation theory, where power series in θ are used and time intervals are of the order of negative powers of θ; such time intervals are insufficient for the action to go far away from its initial value. However, since the divergence develops in an unpredictable direction (strongly depending on the initial conditions) and has the character of a random walk between the Kolmogorov tori, physicists called this phenomenon diffusion [14].

Nekhoroshev [15–17] established an exponential upper estimate for the average rate of evolution of the action: the variation of the action remains small (of the order of θ^a, $0 < a < 1$) for an exponentially long time (of the order of $\exp(1/\theta^b)$).

The Nekhoroshev theorem explains why the evolution of the action variables cannot be revealed in any approximation of perturbation theory; in particular, in the theorems stated by Laplace and his successors.

Nekhoroshev estimated the parameters a and b in terms of some geometric characteristics of the curvature of the Hamiltonian level surfaces, the so-called steepness characteristics. At present these characteristics have been calculated for generic systems with at most three degrees of freedom [18].

6. Measure of the exceptional set

Kolmogorov's theorem asserts that (in a bounded region of the phase space) the measure of the complement of the invariant tori which are not destroyed under perturbation is small together with θ. A more thorough analysis makes it possible to obtain the upper bound $C\sqrt{\theta}$ for this measure (Lazutkin [19], Svanidze [20], Pöschel [21]). For an analytic system with two degrees of freedom and self-degeneration (one of the frequencies is proportional to θ), Neustadt established the exponential upper bound $e^{-C/\theta}$ for the measure of the complement of the invariant tori.

Estimates for the measure of the exceptional set play an important role in the application of Kolmogorov tori to the construction of quasiclassical asymptotic expansions for the majority of the eigenvalues of the Laplace operator in regions close to integrable billiards (for example, ellipsoids); see [22].

It is clear that the estimate $C\sqrt{\theta}$ is best possible, since $\sqrt{\theta}$ appears in all asymptotic expansions related to random degeneration (in the terminology of Born [5]); Poincaré [3] was already aware of the fact that in the unperturbed problem the destruction of a resonance torus results in a domain of influence of width of the order of $\sqrt{\theta}$ with respect to the action variable. This can easily be verified by averaging the disturbance over a torus of dimension smaller by 1 and integrating the resulting system with one degree of freedom (describing so-called phase oscillation in the resonance). We see that the amplitude of the phase oscillation is of the order of $\sqrt{\theta}$ and the period is of the order of $1/\sqrt{\theta}$.

However, in the indicated domain some phase curves again lie on tori of full dimension (so-called islands in the slots between tori close to unperturbed ones). The question whether the initial conditions corresponding to motions that are not conditionally periodic have positive measure remains open in the generic case (the answer is likely to be affirmative). The existence of such initial conditions was proved by Alekseev [23].

7. Smoothness

The theorem on the preservation of tori was proved by Kolmogorov for analytic systems and he did not assume that this result could be extended to at least infinitely differentiable systems. However, soon after Kolmogorov's method was first described in detail [24], Moser combined Kolmogorov's approach with the Nash smoothing method [25] and managed to prove the theorem on the preservation of tori for a system of finite smoothness [26, 27]. Initially, Moser used derivatives of the Hamilton function of a very high order (334 for $s = 2$), but later [28, 29] the order was reduced to $r > 2s$ where s is the number of degrees of freedom.

As was indicated by Kolmogorov, in the analytic case the dependence of the tori on the frequency of motion over them proved to be, in a sense, Borel monogenic (differentiable on a set that is a Cantor continuum extending to the real domain) [24]. For the smooth case Pöschel [21] proved that the dependence of the tori on the frequency is smooth in Whitney's sense on the Cantor set, provided that the Hamiltonian disturbance is r-smooth and $r > 3s - 1$.

On the other hand, Takens [30] has constructed an example of a 1-smooth mapping of an annulus that is area-preserving and close to rotation through a variable angle but has no invariant curves (4-smoothness would be sufficient for the existence of such curves). In none of the cases is it known exactly how many derivatives are needed for the existence of Kolmogorov tori.

Mather [31, 32] has interpolated the family of invariant curves of a mapping of an annulus to a one-parameter family in which all rotation numbers from the maximum possible value to the minimum possible value are realized (and not only the numbers belonging to the Cantor set as in Kolmogorov's theorem). Some of his "invariant curves" are in fact homeomorphic to the Cantor set and not to the circle. However all these curves are found by means of a unified variational principle introduced by Percival [33] for the approximate numerical calculation of Kolmogorov tori.

8. Non-Hamiltonian systems

The method used by Kolmogorov to construct his tori simultaneously proves many versions of the implicit function theorem in all areas of calculus where the solution of a linearized equation results in loss of smoothness (see [26, 34, 35]).

Two theorems of this kind were proved as early as the forties, namely, Siegel's theorem (asserting that a holomorphic vector field is equivalent to the mapping given by its linear part at a singular point with normally incommensurable eigenvalues [36–38]) and Cartan's theorem [39].

The simplest non-Hamiltonian analogue of Kolmogorov's theorem is found in the theory of differential equations on a two-dimsensional torus (or the theory of diffeomorphic mappings of a circle onto itself). The most complete results in this area were established by Herman [40], who proved that for almost every rotation number an analytic diffeomorphism of a circle is analytically conjugate to a rotation.

In the multidimensional case it has only been proved that the majority of diffeomorphisms of tori that are close to translations are rectifiable [24].

Kolmogorov's method was successfully applied to the problem of finding invariant tori in non-Hamiltonian systems (Bogolyubov [41], Moser [42]). These results are particularly important in the theory of instability of self-excited oscillation since the bifurcation occurring in the generation of a torus from a cycle (and also of tori of higher dimension from tori of lower dimension) is one of the typical phenomena characteristic of the "stochastization" of motion of a dynamical system (for example, the passage from a laminar flow to turbulence in the theory of hydrodynamic stability). Applications of Kolmogorov's method to the theory of bifurcation of tori are discussed in the paper by Chenciner [43].

9. Numerical trials

After Kolmogorov established his theorem, an extensive series of numerical trials was conducted to verify the theorem. The trials were primarily aimed at studying the maximum magnitude of disturbances not destroying the Kolmogorov tori, at finding the measure of the complementary set of the Kolmogorov tori, and at analyzing the ergodic properties of the motion over this set [14, 44].

All the calculations confirm that: (1) the majority of tori are preserved under small disturbances; (2) the measure of the complementary set increases, beginning with the moment when the domains of influence of neighbouring resonances overlap (until this moment the measure of the complementary set is so small that the system can hardly be distinguished from an integrable system); (3) the motion in the complementary sets of the tori is stochastic in nature; namely, the neighbouring trajectories diverge exponentially.

Greene and Percival [45] have numerically examined the boundary of analyticity of Kolmogorov tori (for complex values of the angular variables) and have shown that this boundary is related to ergodic properties of the system in the complex domain and to the asymptotics of nearby periodic motions.

According to the experimental data, the magnitude of disturbances under which the Kolmogorov tori are preserved is not so small as predicted by rigorous proofs of estimates from above for this magnitude. Good agreement with experimental data is shown by the resonance overlap criterion [14, 44].

10. Exponential instability

Systems with stable transitivity and mixing on the energy level surfaces which Kolmogorov discusses at the end of §3 of the lecture at the Amsterdam Congress (paper No. 53) actually exist. Sinai and Anosov proved that geodesic flows on compact manifolds of negative curvature (along each two-dimensional direction) possess these properties (and also certain other stochastic properties, including the Ornstein-Weiss metric isomorphism to the "Bernoulli system" induced by coin tossing) [46–48]. Moreover, these properties are preserved under small perturbations not only in the class of Hamiltonian systems but also in the class of general dynamical systems. [1] The system also preserves its topological structure under perturbations, namely, in spite of the fact that the long-period trajectories form an everywhere dense set and that almost every trajectory is everywhere dense, a small perturbation of the vector field determining the system leaves it homeomorphic to itself (the Smale conjecture, proved by Anosov, [49]).

The discovery of such systems made the ergodic theory of classical dynamical systems extremely important not only for Hamiltonian mechanics but also for the study of phenomena in systems with energy dissipation, which also turned out to be capable of exhibiting stable stochastic behaviour. Incidentally, hydrodynamical systems are of this kind (for them, however, the situation is more complex due to an infinite-dimensional phase space). As early as 1958–59 Kolmogorov combined the ideas of the ergodic theory of classical dynamical

[1] With smooth invariant measure. A system with exponential instability need not have a smooth invariant measure, while, in general, there are a lot of ergodic nonsmooth measures. In the case of discrete time the "Gibbs measure of maximum entropy" is singled out from these measures. For a review of the question of invariant measures see [61].

systems and the theory of hydrodynamical instability in his seminar "Selected problems of calculus" [50].

The programme of this seminar displayed at the notice-board of the Mechanics-Mathematics Department of Moscow University announced the study of "models, albeit idealized, of hydrodynamic instability" from the standpoint of ergodic theory. Here Landau's model was meant primarily. This described the appearance of turbulence as consecutive bifurcations of a laminar flow resulting in periodic and conditionally periodic motions over tori whose dimensions increase together with the Reynolds number.

In Kolmogorov's opinion, the tori with conditionally periodic motions were preferred in the model to other dynamical systems because systems with mixing (such as geodesic flows on manifolds of negative curvature) were unknown to Landau (see §3 of the lecture in Amsterdam). When discussing Landau's model, Kolmogorov stressed that the steady-state motion appearing after stability loss can be generated "far from" the bifurcating "laminar" motion before it loses stability (so that the passage to complex "turbulent" motion can proceed in a jump-like manner as a result of "rigid" rather than "soft" stability loss). This opens up the possibility of appearance of motions unrelated to tori. At that time Kolmogorov also predicted the existence of attracting sets ("attractors" in modern terminology) which are not manifolds but are pathologies at the set-theoretic level (the results of Lorenz' numerical experiments [51] appeared only several years later and, unfortunately, were not known to mathematicians for a long time).

However, along with finite-dimensional attractors, Kolmogorov suggested a model with an attractor in the form of an infinite-dimensional torus resulting from infinitely many bifurcations in the spirit of Landau's model, in the sense that in the experiment a spectrum with an infinite frequency basis is interpreted as a continuous spectrum and the motion is seen as being mixing and turbulent.

Smale's daring conjecture (1961) on the structural stability of dissipative systems with exponential divergence of trajectories was completely unexpected and its proof [49] in 1962 was solid evidence in favour of a model with finite-dimensional attracting set. A large number of papers on these problems have been published since that time (Yudovich, Ladyzhenskaya, Mallet-Paret, Foias, etc.). The finite-dimensionality of an attractor for any Reynolds numbers is now proved in two-dimensional hydrodynamics (first Il'yashenko estimated the dimension of an attractor from above in terms of the fourth degree of the Reynolds number for the Kolmogorov model (see [60]) in hydrodynamics on a two-dimensional torus and then Vishik and Babin estimated it in terms of the fourth degree of the Reynolds number for hydrodynamics on a two-dimensional compact manifold with boundary). As to the three-dimensional case, the question whether the attractor remains finite-dimensional for any Reynolds number is still open.

11. Exchange, capture, and oscillation

The problem of ultimate motions in the three-body problem discussed at the end of §4 of paper No. 53 was later thoroughly studied by Kolmogorov's students, Sitnikov and Alekseev.

Sitnikov [52] was the first to construct oscillatory motions, that is, to find initial conditions such that from time to time one body goes arbitrarily far away from the other two bodies, which remain at a finite distance from one another, and from time to time returns to them. In his example the oscillation occurs both as $t \to -\infty$ and as $t \to +\infty$.

Using his theory of quasi-random dynamical systems, Alekseev [23, 53] found all types of motions whose existence had remained doubtful after the papers of Shmidt and Sitnikov (and was denied by Chazi); namely, he showed that a body coming from afar can be captured by a two-body system so that a three-body system is formed (although the probability of this event is zero) or can start to oscillate, moving from time to time arbitrarily far away from the other two bodies. As $t \to +\infty$ oscillation can also appear in a system which remains bounded as $t \to -\infty$. The only problems remaining unsolved are those on the measure of the set of initial conditions leading to oscillation (which is likely to be positive).

12. Spectral properties

Kolmogorov's conjecture (§5 of paper No. 53) on the stability of a continuous (more precisely, countably-multiple Lebesgue) spectrum was proved by Sinai and Anosov [46, 47]. Thus far the conjecture that a discrete spectrum with a finite number of independent frequencies (not exceeding the phase space dimension) and a countably-multiple Lebesgue spectrum is typical has not been refuted for analytic systems. However, in the infinitely-differentiable case Anosov and Katok [54] constructed diffeomorphisms with a smooth invariant measure (for example, on a circle) which have: (a) a discrete spectrum with any (finite or infinite) prescribed number of basic frequencies; (b) a simple continuous singular spectrum in the absence of mixing; and (c) a combination of the former and latter. These examples are, however, exceptional since the diffeomorphisms constructed belong to a nowhere dense set in the function space of all diffeomorphisms endowed with the topology of C^∞ or C^n.

The assertion of paper No. 51 that an analytic flow on a torus can have a continuous spectrum is proved in [55]. In the case of continuous velocity the spectrum can be mixed [56]. A flow on a torus has no mixing in the rigorous sense and, consequently, no Lebesgue spectrum [57].

The conjecture in paper No. 51 that the divergence of the sum of squares of the Fourier coefficients of a formal solution to an equation with small divisors implies that there is no measurable solution and the spectrum is continuous proved false; Anosov [58] has constructed analytic systems with a measurable solution having neither a square summable solution nor a simply summable solution (the measurable solution is positive and therefore generalizations of the integral do not help here).

On the other hand, conditions of lacunarity type for Fourier series have been found which are sufficient for the non-existence of measurable solutions. It follows from these conditions that, for example, for any right-hand side of an equation with small divisors which is not a trigonometric polynomial there exists an irrational rotation number such that the equation has no measurable solutions [59].

References

1. I. Newton, *Philosophicae naturalis principia mathematica*, London, 1686.
2. P.S. de Laplace, *Traité de mécanique céleste*, Vol. 1, Paris, 1799.
3. H. Poincaré, *Les méthodes nouvelles de la mécanique céleste*, Vol. 1, Paris, 1892.
4. V.I. Arnol'd, 'Proof of a theorem of A.N. Kolmogorov on preservation of conditionally periodic motions under small change in the Hamilton function', *Uspekhi Mat. Nauk* **18**:5 (1963), 13–40 (in Russian).

5. M. Born, *Vorlesungen über Atommechanik*, Springer, Berlin, 1925.
6. V.I. Arnol'd, 'On stability of equilibrium positions of Hamiltonian systems of ordinary differential equations in the general elliptic case', *Dokl. Akad. Nauk SSSR* **137**:2 (1961), 255–257 (in Russian).
7. J. Moser, 'New results on the stability of periodic motions', *Proc. Internat. Congress Math.*, 1962, Uppsala, 1963.
8. A.M. Leontovich, 'On stability of Lagrange periodic solutions in the restricted 3-body problem', *Dokl. Akad. Nauk SSSR* **143**:3 (1962), 525–528 (in Russian).
9. A.P. Markeev, 'On stability of a triangle of points of libration in the circular restricted 3-body problem', *Prikl. Mat. Mekh.* **33**:1 (1969), 112–116 (in Russian).
10. V.I. Arnol'd, 'On generation of a conditionally periodic motion from a family of periodic motions', *Dokl. Akad. Nauk SSSR* **138**:1 (1961), 13–15 (in Russian).
11. V.I. Arnol'd, 'On the behaviour of the adiabatic invariant under slow periodic change in the Hamilton function', *Dokl. Akad. Nauk SSSR* **142**:4 (1962), 758–761 (in Russian).
12. V.I. Arnol'd, 'Small divisors and stability problems in classical and celestial mechanics', *Uspekhi Mat. Nauk* **18**:6 (1963), 92–192 (in Russian).
13. V.I. Arnol'd, 'On instability of dynamical systems with many degrees of freedom', *Dokl. Akad. Nauk SSSR* **156**:1 (1964), 9–12 (in Russian).
14. G.M. Zaslavskii and B.V. Chirikov, 'Stochastic instability of non-linear oscillations', *Uspekhi Fiz. Nauk* **105**:1 (1971), 3–40 (in Russian).
15. N.N. Nekhoroshev, 'On the behaviour of Hamiltonian systems close to integrable ones', *Funktsional. Anal. i Prilozhen.* **5**:4 (1971), 82–83 (in Russian).
16. N.N. Nekhoroshev, 'Exponential estimate of the stability time of Hamiltonian systems close to integrable ones I', *Uspekhi Mat. Nauk* **32**:6 (1977), 5–66 (in Russian).
17. N.N. Nekhoroshev, 'Exponential estimate of the stability time of Hamiltonian systems close to integrable ones II', *Trudy Sem. I.G. Petrovskogo* **5** (1979), 5–50 (in Russian).
18. E.E. Landis, 'Tangential singularities', *Funktsional. Anal. i Prilozhen.* **15**:2 (1981), 36–49 (in Russian).

19. V.F. Lazutkin, 'Existence of caustics for billiard problems in a convex region', *Izv. Akad. Nauk SSSR Ser. Mat.* **37**:1 (1973), 188–223 (in Russian).
20. N.V Svanidze, 'Small perturbations of integrable dynamical systems with integral invariant', *Trudy Mat. Inst. Akad. Nauk SSSR* **147** (1980), 124–146 (in Russian).
21. J. Pöschel, *Integrability of Hamiltonian systems on Cantor sets*, Zürich, 1981.
22. V.F. Lazutkin, *Convex billiard and eigenfunctions of the Laplace operator*, Leningrad Univ. Press, Leningrad, 1981 (in Russian).
23. V.M. Alekseev, 'Quasi-random dynamical systems I–III', *Mat. Sb.* **76**:1 (1968), 72–134; **77**:4 (1968), 545–601; **78**:1 (1969), 227–228 (in Russian).
24. V.I. Arnol'd, 'Small divisors I', *Izv. Akad. Nauk SSSR Ser. Mat.* **25**:1 (1961), 21–86 (in Russian).
25. J. Nash, 'The imbedding problem for Riemannian manifolds', *Ann. of Math.* **63** (1956), 20–63.
26. J. Moser, 'A new technique for the construction of solutions of non-linear differential equations', *Proc. Nat. Acad. Sci. USA* **47** (1961), 1824–1831.
27. J. Moser, 'On invariant curves of area preserving mappings of an annulus', *Nachr. Akad. Wiss. Göttingen. Math.-Phys. Kl.*, (1962), 1–20.
28. H. Rüssmann, 'Kleine Nenner I. Über invariante Kurven differenzierbaren Abbildungen eines Kreisringes', *Nachr. Akad. Wiss. Göttingen. Math.-Phys. Kl.* No. 5, (1970),67–103.
29. J. Pöschel, *Über invariante Tori in differenzierbaren Hamiltonischen Systemen*, Bonn, 1980.
30. F.A. Takens, 'A C^1 counterexample to Moser's twist theorem', *Indag. Math.* **33**:4 (1971), 379–386.
31. J. Mather, *Existence of quasi-periodic orbits for twist homeomorphisms of the annulus*, Preprint, Princeton Univ., 1981.
32. J. Mather, 'Area preserving twist homeomorphism of the annulus', *Comment. Math. Helv.* **54** (1979), 397–404.
33. I.C. Percival, 'Variational principles for invariant tori and Cantori', In: *Symposium on non-linear dynamics and beam-beam interactions: Amer. Inst. Phys. Conf. Proc.* **37** (1980), 302–310.
34. V.I. Arnol'd, 'Singularities of smooth maps', *Uspekhi Mat. Nauk* **23**:1 (1968), 3–44.

35. E. Zehnder, 'Generalized implicit function theorems with applications to some small divisors problem. I', *Comm. Pure Appl. Math.* **28**:1 (1975), 91–140.
36. C.L. Siegel, 'Iteration of analytic functions', *Ann. of Math.* **43** (1942), 607–612.
37. C.L. Siegel, 'Über die Normalform analytischer Differentialgleichungen in der Nähe einer Gleichgewichtslösung', *Nachr. Akad. Wiss. Göttingen. Math.-Phys. Kl.* (1952), 21–30.
38. V.I. Arnol'd, *Supplementary chapters in the theory of ordinary differential equations*, Nauka, Moscow, 1978, pp. 194–199 (in Russian).
39. H.J. Cartan, 'Sur les matrices holomorphes de n variables complexes', *J. Math. Pures Appl.* **19** (1940), 1–26.
40. M.R. Herman, 'Sur la conjugaison differentiable des difféomorphismes du cercle à des rotations', *Publ. Math. IHES* **49** (1979), 5–234.
41. N.N. Bogolyubov, 'On quasiperiodic solutions of problems of non-linear mechanics', In: Proc. 1st Summer Math. School, Kiev, Naukova Dumka **1** (1964), 11–101 (in Russian).
42. J. Moser, 'A rapidly convergent iteration method and non-linear partial differential equations I, II', *Ann. Scuola Norm. Sup. Pisa* **20** (1966), 265–315, 499–535.
43. A. Chenciner, 'Courbes invariantes non normalement hyperboliques au voisinage d'une bifurcation de Hopf dégénérée de difféomorphismes de R^2', *C.R. Acad. Sci. Paris* **292**:10 (1981), 507–510.
44. G. Casati, B.V. Chirikov, J. Ford and F.M. Izrailev, *Stochastic behaviour in classical and quantum Hamiltonian systems*, Lecture Notes in Phys. **93**, 1979.
45. J.M. Greene and I.C. Percival, 'Hamiltonian maps in the complex plane', *Physica* **3D**:3 (1981), 530–548.
46. Ya.G. Sinai, 'Chemical dynamical systems with countable Lebesgue spectrum', *Izv. Akad. Nauk SSSR Ser. Mat.* **30**:1 (1966), 15–68 (in Russian).
47. D.V. Anosov, 'Geodesic flows on closed Riemannian manifolds of negative curvature', *Trudy Mat. Inst. Akad. Nauk SSSR* **90** (1967),
48. D. Ornstein, *Ergodic theory, randomness and dynamical systems*, Univ. Press, Newhaven, 1974.

49. D.V. Anosov, 'Structural stability of geodesic flows on compact Riemannian manifolds of negative curvature', *Dokl. Akad. Nauk SSSR* **145**:4 (1962), 707–709.

50. V.I. Arnol'd and L.D. Meshalkin, 'Kolmogorov's seminar on selected problems in analysis (1958–1959)', *Uspekhi Mat. Nauk* **15**:4 (1960), 247–250.

51. E. Lorenz, 'Deterministic non-periodic flow', *J. Atmos. Sci.* **20** (1963), 130–141.

52. K.A. Sitnikov, 'Existence of oscillating motions in the 3-body problem', *Dokl. Akad. Nauk SSSR* **133**:2 (1960), 303–306.

53. V.M. Alekseev, *Sur l'allure finale du mouvement dans le problème de trois corps* In: Act. Congrès Intern. Math. Nice, 1970, Paris, 1971.

54. D.V. Anosov and A.B. Kotak, 'New examples in smooth ergodic theory. Ergodic diffeomorphisms', *Trudy Moskov. Mat. Obshch.* **23** (1970), 3–36.

55. M.D. Shklover, 'On classical dynamical systems on a torus with continuous spectrum', *Izv. Vuzov. Mat.* **10** (1967), 113–124.

56. A.B. Krygin, 'Example of a continuous flow on a torus with mixed spectrum', *Mat. Zametki* **15**:2 (1974), 235–240.

57. A.V. Kochergin, 'On the absence of mixing for special flows over a twisted circle and for flows on the two-dimensional torus', *Dokl. Akad. Nauk SSSR* **205**:3 (1972), 515–518.

58. D.V. Anosov, 'On an additive functional homological equation related to ergodic twist of a circle', *Izv. Akad. Nauk SSSR Ser. Mat.* **37** (1973), 1259–1274.

59. A.Ya. Gordon, 'A sufficient condition for unsolvability of the additive homological equation related to an ergodic twist of a circle', *Funktsional. Anal. i Prilozhen.* **9**:4 (1975), 71–72.

60. L.D. Meshalkin and Ya.G. Sinai, 'Study of the stability of the stationary solution of a system of equations of planar motion of an incompressible viscous fluid', *Prikl. Mat. Mekh.* **6** (1961), 1140–1143.

61. Ya.G. Sinai, 'Gibbs measures in ergodicity theory', *Uspekhi Mat. Nauk* **27**:4 (1971), 22–64.

SUPERPOSITIONS

A.N. Kolmogorov

Two basic papers of this series (Nos. 55 and 56) stemmed from the student seminar which I led in 1955. Among the participants of that seminar I remember the second year students V.I. Arnol'd and A.A. Kirilov and the third year student S.A. Smolyak. The seminar was devoted to the theory of approximate representation of functions of several variables including problems of approximate nomography. In my introductory lecture I formulated Hilbert's thirteenth problem as a long-term plan which almost surely would not be fulfilled. It is interesting to note that in connection with nomography problems, nomographable functions of two variables, that is, functions of the form $\chi(\phi(x) + \psi(y))$, attracted a certain amount of attention from the very beginning. Functions of this kind proved to be an essential element in the final representation of functions of several variables by functions of one variable and the operation of addition. However, Hilbert's problem was first solved on the basis of completely different ideas, using technique developed by A.S. Kronrod. In this way I proved the central theorem of these papers, which stated that any continuous function of $n \geq 4$ variables can be represented as a superposition of functions of three variables.

By improving the methods of those papers Arnol'd gave the final solution to Hilbert's problem in the form of a theorem asserting that any continuous function of $n \geq 3$ variables can be represented as a superposition of functions of two variables.

Originally publ. in: A.N. Kolmogorov. Selected Works. Mathematics and Mechanics, 444-451,

English transl.: Selected Works of A.N. Kolmogorov. Vol. I. Mathematics and Mechanics.
Editor: V.M. Tikhomirov, 518-527, © 1991 Kluwer Academic Publishers, Dordrecht Boston London.

Only some months later, using completely different construction techniques, did I manage to prove the theorem of paper No. 56, which strengthens Arnol'd's theorem, according to which any continuous function of an arbitrary number of variables can be represented as a superposition of functions of one variable and addition.

SUPERPOSITIONS

(V.I. Arnol'd)

In Kolmogorov's opinion, the proof of the fact that every continuous function of n variables can be represented as a superposition

$$f(x_1,\dots,x_n)=\sum_{q=1}^{2n+1}\chi_q\Big[\sum_{p=1}^{n}\phi^{pq}(x_p)\Big] \tag{1}$$

of continuous functions of one variable and the operation of addition was his most technically difficult achievement. In this commentary basic improvements of the representation (1) are discussed, along with a very brief description of some other aspects of the problem of representability of functions by means of superpositions (cf. [1–3]).

In Kolmogorov's representation (1) the inner functions ϕ^{pq} are fixed and only the outer functions χ_q depend on the represented function f.

1. Representability theorems

Kolmogorov's paper No. 56 was preceded by paper No. 55, in which he proved that any continuous function of four variables can be represented as a superposition of continuous functions of three variables. The method developed in paper No. 55 reduces the representability of functions of three variables as superpositions of functions of two variables to a representability problem for functions defined on universal trees of three-dimensional space. The latter problem was solved by Arnol'd in [34], and thus continuous functions of three variables were for the first time expressed as superpositions of continuous functions of two variables. After that the representation (1) reducing all continuous functions of a finite number of variables to superpositions of continuous functions of one variable and addition was soon obtained by Kolmogorov. In this connection he noted that the constructions used in his paper No. 56 were found by analyzing those employed in No. 55 and [34] and discarding some details unnecessary for the derivation of the final result.

Thorough proofs of Kolmogorov's theorem and the lemmas in his paper of 1957 were published by Tikhomirov [4], Lorenz [2], Sprecher [5], Hedberg [43], and others. Lorenz [2] noted that the outer functions χ_q can be replaced by a single function χ. Sprecher [5] reduced all the inner functions to translations and extensions of a single function ψ with the property that there exist $\epsilon > 0$ and $\lambda > 0$ such that any continuous function of n variables can be represented as

$$f(x_1, \ldots, x_n) = \sum_{q=1}^{2n+1} \chi[\lambda^p \psi(x_p + \epsilon q) + q]. \tag{2}$$

Fridman [6] proved that the inner functions ϕ^{pq} in (1) can be chosen so that they satisfy a "Lipschitz" condition. Sprecher [7] extended this result to the representation (2) (the function ψ can be chosen to satisfy a Lipschitz condition). It follows from Kolmogorov's representation (1) and Bari's representation [8] of any continuous function of one variable as a sum of three superpositions of absolutely continuous functions $\sum f_k \circ g_k$ that all continuous functions of any number of variables can be represented by means of superpositions of absolutely continuous functions of one variable and the operation of addition.

Doss [11] developed the Kolmogorov representation and used Kahane's results [42, 44] to obtain a set of continuous functions f_k and g_k of one variable such that any function continuous on a square can be expanded in a series

$$\sum a_k \exp(i f_k(x) g_k(y)),$$

where $\sum |a_k| < \infty$. He showed that if the inner sum in (1) is replaced by a product, then it is possible to take "almost any" (in the sense of Baire category) functions ϕ, while as χ one can take a function independent of q and with an absolutely convergent Fourier transform.

Ostrand [39] extended the representation (1) to functions defined on an n-dimensional compactum, while Doss [12] extended (1) to functions defined on an unbounded set (in R^n).

2. Non-representability theorems

Bassalygo proved that for any *three* functions ψ_k continuous on a square there exists a continuous function f which cannot be represented as $\sum \chi_k \circ \psi_k$ for any continuous χ_k [9].

A similar expansion into a sum of *five* summands (1) is possible by virtue of Kolmogorov's theorem. Doss [10] showed that it is insufficient to take *four* summands instead of five to represent an arbitrary continuous function of two variables as a sum of the form (1) when the (fixed) inner functions ϕ^{pq} are monotone.

Vitushkin and Khenkin have proved that

(1) for any functions ϕ_k continuous on a square and any continuously differentiable functions ψ_k there exists a function analytic on the square that cannot be represented as a sum of products $\sum \phi_k \cdot (g_k \circ \psi_k)$ for any choice of continuous outer functions g_k of one variable; moreover,

(2) the set of such sums (with all possible continuous g_k) is closed and nowhere dense (in the uniform metric);

(3) there exists a polynomial $(x_1 + \nu x_2)^\mu$ not belonging to this set.

These results can be extended to functions defined on a cube of arbitrary dimension (see [13–15]).

Fridman proved that sets of functions of two variables representable in the form $\sum g_k \circ \psi_k$ with fixed continuously differentiable inner functions ψ_k are nowhere dense in L^2, as are sets of functions of three variables representable as $\sum g_k \circ \psi_k$ where the functions $\psi_k : R^3 \to R^2$ are twice continuously differentiable and the functions $g_k : R^2 \to R$ are continuous [35, 36].

3. Hilbert's thirteenth problem

The exact formulation of Hilbert's problem to which Kolmogorov referred in his papers reads (see [1]): "to show that the equation of the seventh degree

$$f^7 + xf^3 + yf^2 + zf + 1 = 0 \tag{3}$$

cannot be solved by means of any continuous functions depending only on two variables".

A great number of papers are devoted to the representability of functions as superpositions of functions depending on a smaller number of variables and satisfying certain additional conditions (algebraicity, analyticity, smoothness). Hilbert was, of course, aware of the fact that superpositions of discontinuous functions represent all functions of a larger number of variables.

Hilbert also knew about the existence of *analytic* functions of three variables that cannot be represented by any finite superpositions of *analytic* functions of two variables (he indicated this in [1]).

Ostrovski [16] proved that the function

$$h(x,y) = \sum x^k/y^k$$

cannot be represented as a finite superposition of inifinitely differentiable functions of one variable and algebraic functions of an arbitrary number of variables.

Vitushkin [17] proved that there exist p times continuously differentiable functions of n variables that cannot be expressed in terms of finite superpositions of $q \geq 1$ times continuously differentiable functions of $k < n$ variables, if $q/k > p/n$.

Kolmogorov [3] put forward the conjecture that there exist analytic functions of three variables that cannot be expressed as superpositions of continuously differentiable functions of two variables and that there are analytic functions of two variables that cannot be represented by superpositions of continuously differentiable functions of one variable and addition.

The qeustion of representability of a root of equation (3) in the form of a superposition of analytic or algebraic functions remains open.

4. The resolvent problem

In the statement of his problem, Hilbert proceeded from a recult of Tschirnhausen [18], according to which a root of an algebraic equation of degree $n > 5$, that is, a function $f(x_1, \dots, x_n)$ determined by an equation

$$f^n + x_1 f^{n-1} + \dots + x_n = 0, \tag{4}$$

can be expressed as a superposition of algebraic functions of $n-4$ variables. Hilbert assumed that the number $n-4$ cannot be reduced for $n = 6, 7, 8$; he also proved that in order to solve an equation of degree $n = 9$ it suffices to have functions of $n-5$ variables [19]. Wiman [20] extended the latter result to $n > 9$, while Chebotarev [21] reduced the number of variables involved in the representation of functions to $n-6$ for $n \geq 21$ and to $n-7$ for $n \geq 121$.

Chebotarev was the first to attempt to find topological obstructions to the representability of algebraic functions as superpositions of algebraic functions, but his proofs were not convincing (see [22, 24]). Using topological notions (related to the behaviour of a many-valued algebraic function on and near a branching manifold) it is now possible to prove that algebraic functions cannot be represented by complete superpositions of integral algebraic functions (*completeness* means that the represented function must involve *all* the branches of the many-valued functions and not only one of them as, for example, in the formulas expressing solutions to equations of the 3rd and 4th degree).

Certain topological obstructions to the representation by a complete superposition of algebraic functions were constructed in this way (cohomology classes in the complement of the branching manifold which are invariant under the Tschirhausen transformation (see [25–29])). Lin [29] established the following, most complete, result: in any neighbourhood of the origin, for any $n \geq 3$, the root $f(x_1, \ldots, x_n)$ of equation (4) is not a complete superposition of entire algebroid functions of fewer than $n-1$ variables and single-valued holomorphic functions of an arbitrary number of variables.

Thus, from the standpoint of *complete* superpositions of *entire* algebraic functions, even fourth-degree equations cannot be solved without using functions of three variables.

The proof of Lin's theorem can be regarded as a modernization and a kind of rehabilitation of Chebotarev's considerations [23] and Morozov's attempt to correct them [24].

The use of non-entire functions reduces approximately by one the number of variables in the representation. For instance, the root of the equation $f^5 + xf + y = 0$ can be represented by superpositions of algebraic functions of one variable and arithmetic operations including division, and in any neighbourhood of zero it cannot be represented by a superposition of analytic functions of an arbitrary number of variables and algebroid functions of one variable (Khovanskii [45]).

It would be interesting to check whether Deligne's theorems on the behaviour of mixed Hodge structures under algebraic mappings give new obstructions to the representation of a function by a superposition of algebraic functions.

5. Approximations

Returning to Kolmogorov's papers on superpositions, we also mention the problems stated by him in relation to the closedness of the classes of functions representable by superpositions in the uniform metric. The theorem asserting that the function xy, which can be approximated by functions $(x+\epsilon)(y+\epsilon)$

representable in the form $\chi[\phi(x) + \psi(x)]$ cannot be represented in this form on the square $0 \le x, y \le 1$ was quoted in the lecture at the Third USSR Congress of Mathematicians and was proved in [30].

Kreines and Vainshtein [31] proved that the uniform limit of a sequence of functions representable in this form can be represented in the same form, if the limit is monotone in each variable. Mukhin showed [32] that the class of all functions nomographable by single-valued functions is non-closed and nowhere dense. Fridman [36, 37] discussed the closedness of certain linear superpositions on I^2. The best approximation of a continuous function on a rectangle by sums $\phi(x) + \psi(y)$ is always attainable (cf. [32]).

The question whether functions on plane curves are representable in this form was posed by Kolmogorov for trees, and the problem was solved in [34]; the same question for closed curves immediately leads to "small divisors" (see [35]).

The representability of functions on a curve $x_k = u_k(t)$ in the form

$$f(t) = \sum_{k=1}^{n} A_k \phi_k(u_k(t))$$

(with fixed coefficients A_k) is important in the theory of singularities of differentiable mappings. A typical example is the expansion of functions on the semicubical parabola:

$$f(t) = \phi(t^2) + \psi(t^2 + t^3).$$

If a smooth function f admits of such a representation as a formal series, then there also exists a smooth representation (of class C^∞) (Dufour). However, if f is holomorphic, then as a rule, a holomorphic representation need not exist (Voronin [47]).

References

1. *Hilbert's problems*, Nauka, Moscow, 1969, pp.163–170 (in Russian).
2. G.G. Lorenz, 'Metric entropy, width, and superpositions of functions', *Amer. Math. Monthly* **69** (1962), 469–485.
3. D.A. Sprecher, 'A survey of solved and unsolved problems on superpositions of functions', *J. Approx. Theory* **6**:2 (1972), 123–134.
4. V.M. Tikhomirov, 'A.N. Kolmogorov's work on ϵ-entropy of function classes and superpositions of functions', *Uspekhi Mat. Nauk* **18**:5 (1963), 55-92 (in Russian).

5. D.A. Sprecher, 'On the structure of continuous functions of several variables', *Trans. Amer. Math. Soc.* **115** (1965), 340–355.
6. B.L. Fridman, 'Increasing smoothness of functions in Kolmogorov's theorem on superpositions', *Dokl. Akad. Nauk SSSR* **177** (1967), 1019–1022 (in Russian).
7. D.A. Sprecher, 'An improvement in the superposition theorem of Kolmogorov', *J. Math Anal. Appl.* **38** (1972), 208–213.
8. N.C. Bari, 'Mémoire sur la représentation finie des fonctions continues', *Math. Ann.* **103** (1930), 185–248, 598–653.
9. L.O. Bassalygo, 'On the representation of continuous functions of two variables by continuous functions of one variable', *Vestn. MGU Ser. Mat.-Mekh.* **21** (1966), 58–63 (in Russian).
10. R. Doss, 'On the representation of continuous functions of two variables by means of addition and continuous functions of one variable', *Colloq. Math.* **10** (1963), 219–259.
11. R. Doss, 'Representations of continuous functions of several variables', *Amer. J. Math.* **98**:2 (1976), 375–383.
12. R. Doss, 'A superposition theorem for unbounded continuous functions', *Trans. Amer. Math. Soc.* **233** (1977), 197–203.
13. A.G. Vitushkin, 'Proof of the existence of analytic functions of several variables that are not representable as linear superpositions of continuously differentiable functions of a smaller number of variables', *Dokl. Akad. Nauk SSSR* **156**:6 (1964), 1258–1261 (in Russian).
14. G.M. Khenkin, 'On linear superpositions of continuously differentiable functions', *Dokl. Akad. Nauk SSSR* **157**:2 (1964), 283–290 (in Russian).
15. A.G. Vitushkin and G.M. Khenkin, 'Linear superpositions of functions', *Uspekhi Mat. Nauk* **22**:1 (1967), 77–124 (in Russian).
16. A. Ostrovski, 'Über Dirichletsche Reihen und algebraische Differentialgleichungen', *Math. Z.* **8**:3–4 (1920), 241–298.
17. A.G. Vitushkin, *On multidimensional variations*, Gostekhteorizdat, Moscow, 1955 (in Russian).
18. W. Tschirnhausen, 'Methodus auferendi omnes terminos intermedios ex data equatione', *Acta Eruditorum*, 1683.
19. D. Hilbert, 'Über die Gleichung neunten Grades', *Math. Ann.* **97** (1927), 243–250.

20. A. Wiman, 'Über die Anwendung der Tschirnhausen Transformation auf die Reduktion algebraischer Gleichungen', *Nova Acta R. Soc. Sci. Uppsaliensis* vol. extraordin. editum, 1927, pp. 3–8.

21. N.G. Chebotarev, 'On the resolvent problem', *Uchen. Zap. Kazan. Univ.* **114**:2 (1954), 189–193 (in Russian).

22. N.G. Chebotarev, 'The resolvent problem and critical manifolds', *Izv. Akad. Nauk SSSR Ser. Mat.* **7** (1943), 123–146 (in Russian).

23. N.G. Chebotarev, 'The resolvent problem', In: *Collected Works*, Vol. 1, USSR Acad. Sci. Press, Moscow, 1949, pp. 327–340 (in Russian).

24. V.V. Morozov, 'On some questions in the resolvent problem', *Uchen Zap. Kazan. Univ.* **114**:2 (1954), 173–187 (in Russian).

25. V.I. Arnol'd, 'On cohomology classes of algebraic functions invariant under a Tschirnhausen transformation', *Funktsional. Anal. i Prilozhen.* **4**:1 (1970), 84–85 (in Russian).

26. V.I. Arnol'd, 'Topological invariants of algebraic functions II', *Funktsional. Anal. i Prilozhen.* **4**:2 (1970), 1–9 (in Russian).

27. V.Ya. Lin, 'On superpositions of algebraic functions', *Funktsional. Anal. i Prilozhen.* **6**:3 (1972), 77–78 (in Russian).

28. V.Ya. Lin, 'Algebraic functions with universal invariant manifold', *Funktsional. Anal. i Prilozhen.* **6**:1 (1972), 81–82 (in Russian).

29. V.Ya. Lin, 'Superpositions of algebraic functions', *Funktsional. Anal. i Prilozhen.* **10**:1 (1976), 37–45 (in Russian).

30. V.I. Arnol'd, 'On the representation of continuous functions of two variables in the form $\chi[\phi(x)+\psi(y)]$', *Uspekhi Mat. Nauk* **12**:2 (1957), 119–121 (in Russian).

31. I.A. Vainshtein and M.A. Kreines, 'Sequences of functions of the form $f(X(x)+Y(y))$', *Uspekhi Mat. Nauk* **15**:4 (1960), 123–128 (in Russian).

32. O.N. Mukhin, 'Certain properties of classes of surfaces nomographable by single-valued functions', *Vestn. MGU Ser. Mat.-Mekh.* **17**:1 (1962), 24–29 (in Russian).

33. Yu.P. Ofman, 'On best approximation of functions of two variables by functions of the form $\phi(x)+\psi(y)$', *Izv. Akad. Nauk SSSR Ser. Mat.* **25**:2 (1961), 239–252 (in Russian).

34. V.I. Arnol'd, 'On representation of continuous functions of three variables by superpositions of functions of two variables', *Mat. Sb.* **48** (1959), 3–74;

98 (1962), 392 (in Russian).

35. V.I. Arnol'd, 'Small divisors I. On maps of the circle onto itself', *Izv. Akad. Nauk SSSR Ser. Mat.* **25**:1 (1961), 21–86 (in Russian).
36. B.L. Fridman, 'Nowhere dense spaces of linear superpositions of functions of several variables', *Izv. Akad. Nauk SSSR Ser. Mat.* **36**:4 (1972), 814–846 (in Russian).
37. B.L. Fridman, 'Estimate of the dimension of the null space of linear superpositions', *Mat. Sb.* **82**:1 (1970), 111–126 (in Russian).
38. G.G. Lorenz, *Approximation of functions*, Holt, Rinehart and Winston, New York, 1966.
39. P.A. Ostrand, 'Dimension of metric spaces and Hilbert's problem 13', *Bull. Amer. Math. Soc.* **71** (1965), 619–622.
40. J. Sternfeld, 'Dimension theory and superposition of continuous functions', *Israel J. Math.* **20** (1975), 300–310.
41. R. Kaufman, 'Linear superposition of smooth functions', *Proc. Amer. Math. Soc.* **46** (1974), 360–362.
42. J.P. Kahane, 'Sur le réarrangements de fonctions de la classe *A*', *Studia Math.* **31** (1968), 287–293.
43. T. Hedberg, Appendix IIB, In: H.S. Shapiro, *Lecture Notes in Math.* **187** (1971).
44. J.P. Kahane, 'Sur le théorème de superposition de Kolmogorov', *J. Approx. Theory* **13** (1975), 229–234.
45. A.G. Khovanskii, 'On representability of algebroid functions by superpositions of algebraic functions and algebroid functions of one variable', *Funktsional. Anal. i Prilozhen.* **4**:2 (1970), 74–79 (in Russian).
46. D.E. Marshall and A.G. O'Farrell, *Approximation by a sum of two algebras: the lightning bolt principle*, Preprint 1–23, Maynooth College, Co. Kildare (Ireland), 1982.
47. S.M. Voronin, 'Analytic classification of pairs of involutions and its application', *Funktsional. Anal. i Prilozhen.* **16**:2 (1982), 7–15 (in Russian).

Acknowledgements

Springer-Verlag thanks the publishers of V.I. Arnold's papers for having granted permission to reprint them in these Collected Works.

The numbers behind the original sources correspond to the numbers given in the Table of Contents.

We have made every effort to identify the copyright owners for all articles included in this book in order to obtain reprint permission. Some of our requests however remained unanswered. The following list includes all sources, and owners where known.

Reprinted from Crafoord Prize in Mathematics	3
Reprinted from Ergodic Theory and Dynamical Systems	8
Reprinted from Funct. Anal. Appl.	1, 5, 6, 17, 26
Reprinted from Geophys. Astrophys. Fluid Dynamics	10
Reprinted from J. Soviet Math.	11, 16, 22
Reprinted from Nonlinear and Turbulent Processes in Physics	28
Reprinted from Nonlinear Phenomena in Plasma Physics and Hydrodynamics	29
Reprinted from Physica D	27
Reprinted from Proc. Indian Acad. Sci. (Math. Sci.)	2
Reprinted from Proc. of the Internat. Congress Math.	20
Reprinted from Progr. Math.	14
Reprinted from Reviews Math. and Math. Phys.	24
Reprinted from Russ. Math. Surv.	19, 25
Reprinted from Selecta Math. Sov.	12
Reprinted from Selected Works of A.N. Kolmogorov	35, 36
Reprinted from Soviet Phys. Usp.	23
Reprinted from Usp. Mat. Nauk	18

Kolibri 360 GmbH
Ferdinand-Jühlke-Straße 7
99095 Erfurt, Deutschland
produktsicherheit@kolibri360.de